INTERNATIONAL UNION OF PURE AND APPLIED CHEMISTRY

ANALYTICAL CHEMISTRY DIVISION
COMMISSION ON SOLUBILITY DATA

SOLUBILITY DATA SERIES

Volume 26

SULFITES, SELENITES AND TELLURITES

SOLUBILITY DATA SERIES

Volume 1	H. L. Clever, *Helium and Neon*
Volume 2	H. L. Clever, *Krypton, Xenon and Radon*
Volume 3	M. Salomon, *Silver Azide, Cyanide, Cyanamides, Cyanate, Selenocyanate and Thiocyanate*
Volume 4	H. L. Clever, *Argon*
Volume 5/6	C. L. Young, *Hydrogen and Deuterium*
Volume 7	R. Battino, *Oxygen and Ozone*
Volume 8	C. L. Young, *Oxides of Nitrogen*
Volume 9	W. Hayduk, *Ethane*
Volume 10	R. Battino, *Nitrogen and Air*
Volume 11	B. Scrosati and C. A. Vincent, *Alkali Metal, Alkaline Earth Metal and Ammonium Halides. Amide Solvents*
Volume 12	C. L. Young, *Sulfur Dioxide, Chlorine, Fluorine and Chlorine Oxides*
Volume 13	S. Siekierski, T. Mioduski and M. Salomon, *Scandium, Yttrium, Lanthanum and Lanthanide Nitrates*
Volume 14	H. Miyamoto, M. Salomon and H. L. Clever, *Alkaline Earth Metal Halates*
Volume 15	A. F. M. Barton, *Alcohols with Water*
Volume 16/17	E. Tomlinson and A. Regosz, *Antibiotics: I. β-Lactam Antibiotics*
Volume 18	O. Popovych, *Tetraphenylborates*
Volume 19	C. L. Young, *Cumulative Index: Volumes 1–18*
Volume 20	A. L. Horvath and F. W. Getzen, *Halogenated Benzenes, Toluenes and Phenols with Water*
Volume 21	C. L. Young and P. G. T. Fogg, *Ammonia, Amines, Phosphine, Arsine, Stibine, Silane, Germane and Stannane in Organic Solvents*
Volume 22	T. Mioduski and M. Salomon, *Scandium, Yttrium, Lanthanum and Lanthanide Halides in Nonaqueous Solvents*
Volume 23	T. P. Dirkse, *Copper, Silver, Gold, and Zinc, Cadmium, Mercury Oxides and Hydroxides*
Volume 24	W. Hayduk, *Propane, Butane and 2-Methylpropane*
Volume 25	C. Hirayama, Z. Galus and C. Guminski, *Metals in Mercury*
Volume 26	M. R. Masson, H. D. Lutz and B. Engelen, *Sulfites, Selenites and Tellurites*

Selected Volumes in Preparation

E. Tomlinson, *Antibiotics: II. Peptide Antibiotics*

H. L. Clever and C. L. Young, *Methane*

H. Miyamoto, *Copper and Silver Halates*

J. W. Lorimer, *Beryllium, Strontium, Barium and Radium Sulfates*

H. L. Clever and C. L. Young, *Carbon Dioxide*

SOLUBILITY DATA SERIES

Editor-in-Chief
A. S. KERTES

Volume 26

SULFITES, SELENITES AND TELLURITES

Volume Editor

MARY R. MASSON
University of Aberdeen
Scotland, UK

Co-Editors

H. D. LUTZ
University of Siegen
Federal Republic of Germany

B. ENGELEN
University of Siegen
Federal Republic of Germany

Contributors

M. R. MASSON, H. D. LUTZ
and
B. ENGELEN

PERGAMON PRESS

OXFORD · NEW YORK · BEIJING · FRANKFURT
SÃO PAULO · SYDNEY · TOKYO · TORONTO

U.K. — Pergamon Press, Headington Hill Hall, Oxford OX3 0BW, England

U.S.A. — Pergamon Press, Maxwell House, Fairview Park, Elmsford, New York 10523, U.S.A.

PEOPLE'S REPUBLIC OF CHINA — Pergamon Press, Qianmen Hotel, Beijing, People's Republic of China

FEDERAL REPUBLIC OF GERMANY — Pergamon Press, Hammerweg 6, D-6242 Kronberg, Federal Republic of Germany

BRAZIL — Pergamon Editora, Rua Eça de Queiros, 346, CEP 04011, São Paulo, Brazil

AUSTRALIA — Pergamon Press Australia, P.O. Box 544, Potts Point, N.S.W. 2011, Australia

JAPAN — Pergamon Press, 8th Floor, Matsuoka Central Building, 1-7-1 Nishishinjuku, Shinjuku-ku, Tokyo 160, Japan

CANADA — Pergamon Press Canada, Suite 104, 150 Consumers Road, Willowdale, Ontario M2J 1P9, Canada

First edition 1986

Library of Congress has catalogued this serial title as follows:

Solubility data series. – Vol. 1– – Oxford; New York: Pergamon, c 1979–
v.; 28 cm.
Separately cataloged and classified in LC before no. 18.
ISSN 0191–5622 = Solubility data series.
1. Solubility–Tables–Collected works.
QD543.S6629 541.3'42'05–dc19 85–641351
AACR 2 MARC-S

British Library Cataloguing in Publication Data

Sulfites, selenites and tellurites.—(Solubility data series; v. 26)
1. Sulphites—Solubility 2. Selenites—Solubility 3. Tellurites—Solubility
I. Masson, Mary R. II. Lutz, H. D.
III. Engelen, B. IV. Series
546'.72 QD181.S1

ISBN 0-08-032517-3

Printed in Great Britain by A. Wheaton & Co. Ltd., Exeter

CONTENTS

INTERNATIONAL UNION OF PURE AND APPLIED CHEMISTRY
IUPAC Secretariat: Bank Court Chambers, 2–3 Pound Way,
Cowley Centre, Oxford OX4 3YF, UK

FOREWORD

If the knowledge is
undigested or simply wrong,
more is not better

How to communicate and disseminate numerical data effectively in chemical science and technology has been a problem of serious and growing concern to IUPAC, the International Union of Pure and Applied Chemistry, for the last two decades. The steadily expanding volume of numerical information, the formulation of new interdisciplinary areas in which chemistry is a partner, and the links between these and existing traditional subdisciplines in chemistry, along with an increasing number of users, have been considered as urgent aspects of the information problem in general, and of the numerical data problem in particular.

Among the several numerical data projects initiated and operated by various IUPAC commissions, the *Solubility Data Project* is probably one of the most ambitious ones. It is concerned with preparing a comprehensive critical compilation of data on solubilities in all physical systems, of gases, liquids and solids. Both the basic and applied branches of almost all scientific disciplines require a knowledge of solubilities as a function of solvent, temperature and pressure. Solubility data are basic to the fundamental understanding of processes relevant to agronomy, biology, chemistry, geology and oceanography, medicine and pharmacology, and metallurgy and materials science. Knowledge of solubility is very frequently of great importance to such diverse practical applications as drug dosage and drug solubility in biological fluids, anesthesiology, corrosion by dissolution of metals, properties of glasses, ceramics, concretes and coatings, phase relations in the formation of minerals and alloys, the deposits of minerals and radioactive fission products from ocean waters, the composition of ground waters, and the requirements of oxygen and other gases in life support systems.

The widespread relevance of solubility data to many branches and disciplines of science, medicine, technology and engineering, and the difficulty of recovering solubility data from the literature, lead to the proliferation of published data in an ever increasing number of scientific and technical primary sources. The sheer volume of data has overcome the capacity of the classical secondary and tertiary services to respond effectively.

While the proportion of secondary services of the review article type is generally increasing due to the rapid growth of all forms of primary literature, the review articles become more limited in scope, more specialized. The disturbing phenomenon is that in some disciplines, certainly in chemistry, authors are reluctant to treat even those limited-in-scope reviews exhaustively. There is a trend to preselect the literature, sometimes under the pretext of reducing it to manageable size. The crucial problem with such preselection - as far as numerical data are concerned - is that there is no indication as to whether the material was excluded by design or by a less than thorough literature search. We are equally concerned that most current secondary sources, critical in character as they may be, give scant attention to numerical data.

On the other hand, tertiary sources - handbooks, reference books and other tabulated and graphical compilations - as they exist today are comprehensive but, as a rule, uncritical. They usually attempt to cover whole disciplines, and thus obviously are superficial in treatment. Since they command a wide market, we believe that their service to the advancement of science is at least questionable. Additionally, the change which is taking place in the generation of new and diversified numerical data, and the rate at which this is done, is not reflected in an increased third-level service. The emergence of new tertiary literature sources does not parallel the shift that has occurred in the primary literature.

With the status of current secondary and tertiary services being as briefly stated above, the innovative approach of the *Solubility Data Project* is that its compilation and critical evaluation work involve consolidation and reprocessing services when both activities are based on intellectual and scholarly reworking of information from primary sources. It comprises compact compilation, rationalization and simplification, and the fitting of isolated numerical data into a critically evaluated general framework.

The *Solubility Data Project* has developed a mechanism which involves a number of innovations in exploiting the literature fully, and which contains new elements of a more imaginative approach for transfer of reliable information from primary to secondary/tertiary sources. *The fundamental trend of the Solubility Data Project is toward integration of secondary and tertiary services with the objective of producing in-depth critical analysis and evaluation which are characteristic to secondary services, in a scope as broad as conventional tertiary services.*

Fundamental to the philosophy of the project is the recognition that the basic element of strength is the active participation of career scientists in it. Consolidating primary data, producing a truly critically-evaluated set of numerical data, and synthesizing data in a meaningful relationship are demands considered worthy of the efforts of top scientists. Career scientists, who themselves contribute to science by their involvement in active scientific research, are the backbone of the project. The scholarly work is commissioned to recognized authorities, involving a process of careful selection in the best tradition of IUPAC. This selection in turn is the key to the quality of the output. These top experts are expected to view their specific topics dispassionately, paying equal attention to their own contributions and to those of their peers. They digest literature data into a coherent story by weeding out what is wrong from what is believed to be right. To fulfill this task, the evaluator must cover *all* relevant open literature. No reference is excluded by design and every effort is made to detect every bit of relevant primary source. Poor quality or wrong data are mentioned and explicitly disqualified as such. In fact, it is only when the reliable data are presented alongside the unreliable data that proper justice can be done. The user is bound to have incomparably more confidence in a succinct evaluative commentary and a comprehensive review with a complete bibliography to both good and poor data.

It is the standard practice that the treatment of any given solute-solvent system consists of two essential parts: I. Critical Evaluation and Recommended Values, and II. Compiled Data Sheets.

The Critical Evaluation part gives the following information:

(i) a verbal text of evaluation which discusses the numerical solubility information appearing in the primary sources located in the literature. The evaluation text concerns primarily the quality of data after consideration of the purity of the materials and their characterization, the experimental method employed and the uncertainties in control of physical parameters, the reproducibility of the data, the agreement of the worker's results on accepted test systems with standard values, and finally, the fitting of data, with suitable statistical tests, to mathematical functions;

(ii) a set of recommended numerical data. Whenever possible, the set of recommended data includes weighted average and standard deviations, and a set of smoothing equations derived from the experimental data endorsed by the evaluator;

(iii) a graphical plot of recommended data.

The Compilation part consists of data sheets of the best experimental data in the primary literature. Generally speaking, such independent data sheets are given only to the best and endorsed data covering the known range of experimental parameters. Data sheets based on primary sources where the data are of a lower precision are given only when no better data are available. Experimental data with a precision poorer than considered acceptable are reproduced in the form of data sheets when they are the only known data for a particular system. Such data are considered to be still suitable for some applications, and their presence in the compilation should alert researchers to areas that need more work.

The typical data sheet carries the following information:

(i) components - definition of the system - their names, formulas and Chemical Abstracts registry numbers;
(ii) reference to the primary source where the numerical information is reported. In cases when the primary source is a less common periodical or a report document, published though of limited availability, abstract references are also given;
(iii) experimental variables;
(iv) identification of the compiler;
(v) experimental values as they appear in the primary source. Whenever available, the data may be given both in tabular and graphical form. If auxiliary information is available, the experimental data are converted also to SI units by the compiler.

Under the general heading of Auxiliary Information, the essential experimental details are summarized:

(vi) experimental method used for the generation of data;
(vii) type of apparatus and procedure employed;
(viii) source and purity of materials;
(ix) estimated error;
(x) references relevant to the generation of experimental data as cited in the primary source.

This new approach to numerical data presentation, formulated at the initiation of the project and perfected as experience has accumulated, has been strongly influenced by the diversity of background of those whom we are supposed to serve. We thus deemed it right to preface the evaluation/compilation sheets in each volume with a detailed discussion of the principles of the accurate determination of relevant solubility data and related thermodynamic information.

Finally, the role of education is more than corollary to the efforts we are seeking. The scientific standards advocated here are necessary to strengthen science and technology, and should be regarded as a major effort in the training and formation of the next generation of scientists and engineers. Specifically, we believe that there is going to be an impact of our project on scientific-communication practices. The quality of consolidation adopted by this program offers down-to-earth guidelines, concrete examples which are bound to make primary publication services more responsive than ever before to the needs of users. The self-regulatory message to scientists of the early 1970s to refrain from unnecessary publication has not achieved much. A good fraction of the literature is still cluttered with poor-quality articles. The Weinberg report (in 'Reader in Science Information', ed. J. Sherrod and A. Hodina, Microcard Editions Books, Indian Head, Inc., 1973, p. 292) states that 'admonition to authors to restrain themselves from premature, unnecessary publication can have little effect unless the climate of the entire technical and scholarly community encourages restraint...' We think that projects of this kind translate the climate into operational terms by exerting pressure on authors to avoid submitting low-grade material. The type of our output, we hope, will encourage attention to quality as authors will increasingly realize that their work will not be suited for permanent retrievability unless it meets the standards adopted in this project. It should help to dispel confusion in the minds of many authors of what represents a permanently useful bit of information of an archival value, and what does not.

If we succeed in that aim, even partially, we have then done our share in protecting the scientific community from unwanted and irrelevant, wrong numerical information.

A. S. Kertes

PREFACE

SCOPE OF THE VOLUME

This volume deals with the solubilities of the sulfites, selenites and tellurites of the alkali metals and ammonium, the alkaline-earth metals, and manganese, iron, cobalt, nickel, copper, silver, zinc, cadmium, mercury and lead. Solubility data for binary systems and for all types of multicomponent systems are included. In all systems, one of the solvent components is water; therefore in the few cases where solubilities in other solvents are given, the systems are treated as aqueous multicomponent systems.

NOMENCLATURE

According to the fully systematic IUPAC nomenclature (1), sulfite is sulfate(IV) [or strictly trioxosulfate(IV)], pyrosulfite is disulfate(IV)(2-) or disulfite, selenite is selenate(IV), tellurite is tellurate(IV), etc. The systematic names have not been used in the text, because the trivial names are well-established ones, and are less likely to be misunderstood. Some of these trivial names are IUPAC "accepted names" (1), but others, such as tellurite, are not.

GENERAL COMMENTS

The literature has been covered up to 1984, and as far as we are aware the entire literature has been covered. However, the editors will be grateful if any omissions are brought to their attention, so that they may be included in any updates to the volume.

The solubilities of hydrated compounds are always given in terms of the unhydrated components. When solubility data in the literature were expressed in units of mass % (% w/w in the older literature) or related quantities, or when amount-of-substance concentrations were reported together with the densities of the solutions, conversions to molality (units - mol kg^{-1}) were made by the compilers/evaluators, with use of IUPAC-recommended atomic masses. In many of the compilation sheets, the molalities have been calculated by computer, and expressed in a uniform format with three figures after the decimal point. This sometimes has resulted in the appearance of too many significant figures. We ask the reader to forgive this time-saving device. In the evaluations, the correct number of significant figures is always given.

Phase diagrams are presented for as many as possible of the ternary systems, and also for some quaternary ones. Many of these were plotted by computer from the original data, irrespective of whether or not figures were provided in the original papers. All the computer plots were produced to a uniform format; this means that the diagrams may be superimposed for comparison purposes. For the sake of clarity, the axes are not labelled with numbers; in all cases the scales indicated by the tick marks run from 0 to 100 on each axis, and the units are mass %. For some systems, the phase diagrams given in the original papers are reproduced with the permission of the relevant copyright holders.

For "soluble systems", most of the solubility data are expressed in units of mass %, and conversions were made as stated above. For "sparingly soluble systems", analyses are usually reported as amount-of-substance concentrations, and the actual solubilities are expressed as solubility products, or in terms of other forms of equilibrium constants.

The solubility products quoted are usually analytical "concentration solubility products". Otherwise, definitions like "activity solubility product" or "solubility product based on the activities" or "concentration solubility product, corrected for hydrolysis effects" are given. Only rarely have "thermodynamic" constants been obtained, and usually conditions were insufficiently well defined to allow calculation of thermodynamic constants from the corresponding concentration constants by making activity corrections. The nomenclature used for solubility product constants and other equilibrium constants is that of the IUPAC "Orange Book" (2); otherwise, a definition is provided.

ERRORS

Errors have been expressed in various ways, depending on the information provided in the original paper. Whenever possible, an estimated value of the standard deviation, s, of the analyses is given, as recommended in the "Orange Book" (3).

EXPERIMENTAL METHODS

In the most common method for determination of solubility, aqueous solutions or pure water are saturated with one or more solid component(s) under isothermal conditions, and after equilibrium is reached, the compositions of the solutions and of the solid phases are determined analytically. This "isothermal method" is also referred to as the "saturation method" in this volume.

A procedure referred to very frequently is the "remainders" or "wet residues" method of Schreinemakers, which was originally described in 1893 (4). This indirect method for determination of compositions of solid phases involves the analysis of solid plus adhering mother liquor, followed by plotting on the phase diagram, and extrapolation of the line joining the solution-composition point to the wet-solid-composition point. When several such lines are drawn, they should intersect at the point corresponding to the composition of the salt. A fuller description is given by Findlay (5).

All the anions discussed in this volume are anions of weak acids. Thus, for H_2SO_3, $pK_{a1} = 1.8$, $pK_{a2} = 6.8$; for H_2SeO_3, $pK_{a1} = 2.6$, $pK_{a2} = 8.3$; for H_2TeO_3, $pK_{a1} = 6.1$, $pK_{a2} = 9.6$; etc. Therefore, any solution of a salt will undergo hydrolysis to some extent. Generally, this is neglected with the "soluble systems", but with the "sparingly soluble systems", hydrolysis can have a considerable influence on the equilibria, and ought to be taken into account. Where authors have not considered the effect of hydrolysis, or where the calculations have been made in a manner that appears to be incorrect, the compiler has attempted to remedy matters.

The methods used to deal with the quite complicated equilibria are those of Ringbom (6), involving the concepts of "conditional constants" and "side-reaction coefficients".

The conditional stability constant β'_n for a mononuclear metal-ligand complex is, in general form:

$$\beta'_n = \frac{[ML_n]}{[M'][L']^n} \qquad (1)$$

where [M'] and [L'] are the "conditional concentrations" of metal and ligand. That is, [M'] is the total concentration of metal ion that has not reacted with the main ligand L, including any bound as hydroxo-complexes; and [L'] is the concentration of ligand L not bound to M, whether L is protonated or non-protonated, or in the form of complexes with other metals. The ratios $[M']/[L] = \alpha_M$ and $[L']/[L] = \alpha_L$ can be calculated from the stability constants of all the complexes formed in side-reactions; α_M and α_L are called "side-reaction coefficients", or just α-coefficients.

In this volume, in nearly all cases, the side-reactions of interest are the reactions of the anions (ligands) with protons. For these reactions

$$\alpha_{L(H)} = 1 + [H^+]K_1 + [H^+]^2K_1K_2 + \ldots \qquad (2)$$

where the K's are the protonation (i.e. association) constants. When written in terms of the dissociation constants for a weak dibasic acid, equation (2) becomes

$$\alpha_{L(H)} = 1 + \frac{[H^+]}{K_{a1}} + \frac{[H^+]^2}{K_{a1}K_{a2}} \qquad (3)$$

If we now return to the general case, substituting for [M'] and [L'] in equation (1) leads to

$$\beta'_n = \frac{[ML_n]}{[M]\,\alpha_M \cdot [L]^n\alpha_L{}^n} = \frac{\beta_n}{\alpha_M\,\alpha_L{}^n} \qquad (4)$$

or

$$\log \beta'_n = \log \beta_n - \log \alpha_M - n \log \alpha_L \qquad (5)$$

Although this brief introduction has been given for a generalized stability constant, the Ringbom treatment is equally applicable to solubility equilibria, with the definition of "conditional solubility products". A more detailed discussion of this topic has been given by Inczédy (7).

The hydrolysis phenomenon has frequently been utilized as a method for increasing the solubility of a sparingly soluble salt (by dissolution in acid) to a value at which the concentrations become more readily determinable. That is, experimental determinations are made of the conditional solubility product at various pH values, then the solubility product in pure water is calculated from equation (4) or (5). However, good estimates are required for the acid-dissociation constants if a good value for the pure-aqueous solubility product is to be obtained.

No corrections to allow for hydrolysis were done for the systems involving sulfite and a bivalent metal. Here, usually the total amount of dissolved sulfite is given.

PROCEDURE FOR EVALUATIONS

In the alkali-metal and ammonium systems, the procedure was as follows. The data for a system were collected, and analysed by use of the statistical package MINITAB (8). Regression analysis was done for various polynomial functions, and the best was selected by consideration of the values of *s*, the standard deviation of the dependent variable about the regression line, and the *t*-ratios for each coefficient (which show whether the value of the coefficient is statistically significant, or whether it should be better set to zero). Also, histograms and plots of the standardized residuals were examined, because for a correctly selected function these should have an approximately Gaussian distribution. When the best function had been selected, any points which had standard residuals greater than +2 or less than −2 were rejected, then the regression analysis was repeated. Usually, more points then had to be rejected; the procedure was repeated until all the remaining points had standard residuals in the range −2 to +2. Sometimes, several points from a single paper had to be rejected, and it was then decided that the whole set must be unreliable, so it was removed. After rejection of points was completed, a further check was made that the best function had been selected. The final smoothing equation then derived was used to calculate recommended or tentative solubility values for an appropriate series of temperatures, and computer-drawn diagrams showing the data and the calculated regression line(s) were then prepared.

For systems involving sulfite and bivalent metals, a somewhat different procedure was used. After collecting all data for a system, the "best" papers according to consistency of data, determination method, etc. were selected. From the results of these papers, tentative or recommended solubility data were derived. If possible, then a linear multiple regression analysis was done for the function

$$\log X = A + B/T + C \log T \qquad (6)$$

with $1/T$ and $\log T$ as independent variables. For these functions, values for the coefficients A, B, and C, and the correlation coefficients are given, and graphs were drawn.

REFERENCES

1. *Nomenclature of Inorganic Chemistry*, IUPAC Definitive Rules 1970, Butterworths, London, 1971.

2. Irving, H.M.N.H.; Freiser, H.; West. T.S. *Compendium of Analytical Nomenclature* IUPAC Publication, Pergamon, Oxford, 1978. p. 54.

3. *Idem. ibid.* p. 9.

4. Schreinemakers, F.A.H. *Z. Phys. Chem.* 1893, 11, 75.

5. Findlay, A. *The Phase Rule and its Applications 5th Ed.* Longmans, Green and Co., London, 1923. p. 246. (Also Findlay A.; Campbell, A.N.; Smith, N.O. *9th Ed.* Dover, New York, 1951.)

6. Ringbom, A. *Complexation in Analytical Chemistry* Interscience, New York, 1963.

7. Inczédy, J. *Analytical Applications of Complex Equilibria* Horwood, Chichester, 1976.

8. *MINITAB RELEASE 82/1*, copyright Penn State University.
Ryan, T.A. Jr.; Joiner, B.L.; Ryan, B.F. *Minitab Student Handbook* Duxbury Press, N. Scituate, Mass., 1976.

Acknowledgements: the editors gratefully acknowledge the advice and comments of members of the IUPAC Commission on Solubility Data, with special thanks to Dr. Mark Salomon, Prof. J.W. Lorimer, Prof. A.S. Kertes, and Dr. C. Vincent. Thanks are also due to Dr. R.A. Chalmers and Prof. Yu. Zolotov for help in obtaining copies of papers not available in the U.K., Mrs. N. James for photocopying, Dr. K.L. Loening of Chemical Abstracts Services for supplying C.A. Registry Numbers, Mr. I. Noble for assistance with on-line literature searches, Mrs. Erika Olm for preliminary typing of some compilations/evaluations, Mrs. Gisela Potyka for preparing drawings, the West Seaton computer staff for operating the Calcomp plotter, and Mrs. S. Alexander for her most efficient typing of the camera-ready copy.

The editors are also pleased to thank the following copyright-holders who have given us permission to reproduce Figures, subject to acknowledgement of copyright: Academia, Prague (Fig. on p. 376), The Canadian Journal of Chemistry (Figs. on p. 69), The National Association of Corrosion Engineers (Fig. on p. 224), H. Sano, Osaka (Fig. on p. 175), SNTL, Prague (Fig. on p. 175), The Technical Association of the Pulp and Paper Industry (Figs. on pp. 162 and 167) and Verlag Chemie GmbH (Figs. on pp. 254, 263, 276 and 297).

INTRODUCTION: THE SOLUBILITY OF SOLIDS IN LIQUIDS

Nature of the Project

The Solubility Data Project (SDP) has as its aim a comprehensive search of the literature for solubilities of gases, liquids, and solids in liquids or solids. Data of suitable precision are compiled on data sheets in a uniform format. The data for each system are evaluated, and where data from different sources agree sufficiently, recommended values are proposed. The evaluation sheets, recommended values, and compiled data sheets are published on consecutive pages.

This series of volumes includes solubilities of solids of all types in liquids of all types.

Definitions

A *mixture* (1,2) describes a gaseous, liquid, or solid phase containing more than one substance, when the substances are all treated in the same way.

A *solution* (1,2) describes a liquid or solid phase containing more than one substance, when for convenience one of the substances, which is called the *solvent* and may itself be a mixture, is treated differently than the other substances, which are called *solutes*. If the sum of the mole fractions of the solutes is small compared to unity, the solution is called a *dilute solution*.

The *solubility* of a substance B is the relative proportion of B (or a substance related chemically to B) in a mixture which is saturated with respect to solid B at a specified temperature and pressure. *Saturated* implies the existence of equilibrium with respect to the processes of dissolution and precipitation; the equilibrium may be stable or metastable. The solubility of a metastable substance is usually greater than that of the corresponding stable substance. (Strictly speaking, it is the activity of the metastable substance that is greater.) Care must be taken to distinguish true metastability from supersaturation, where equilibrium does not exist.

Either point of view, mixture or solution, may be taken in describing solubility. The two points of view find their expression in the quantities used as measures of solubility and in the reference states used for definition of activities and activity coefficients.

The qualifying phrase "substance related chemically to B" requires comment. The composition of the saturated mixture (or solution) can be described in terms of any suitable set of thermodynamic components. Thus, the solubility of a salt hydrate in water is usually given as the relative proportion of anhydrous salt in solution, rather than the relative proportions of hydrated salt and water.

Quantities Used as Measures of Solubility

1. *Mole fraction* of substance B, x_B:

$$x_B = n_B / \sum_{i=1}^{c} n_i \qquad (1)$$

where n_i is the amount of substance of substance i, and c is the number of distinct substances present (often the number of thermodynamic components in the system). *Mole per cent* of B is 100 x_B.

2. *Mass fraction* of substance B, w_B:

$$w_B = m'_B / \sum_{i=1}^{c} m'_i \qquad (2)$$

where m'_i is the mass of substance i. *Mass per cent* of B is 100 w_B. The equivalent terms weight fraction and weight per cent are not used.

3. *Solute mole (mass) fraction* of solute B (3,4):

$$x_{S,B} = n_B / \sum_{i=1}^{c'} n_i = x_B / \sum_{i=1}^{c'} x_i \qquad (3)$$

where the summation is over the solutes only. For the solvent A, $x_{S,A} = x_A$. These quantities are called *Jänecke mole (mass) fractions* in many papers.

4. *Molality* of solute B (1,2) in a solvent A:

$$m_B = n_B/n_A\ M_A \qquad \text{SI base units: mol kg}^{-1} \qquad (4)$$

where M_A is the molar mass of the solvent.

5. *Concentration* of solute B (1,2) in a solution of volume V:

$$c_B = [B] = n_B/V \qquad \text{SI base units: mol m}^{-3} \qquad (5)$$

The terms molarity and molar are not used.

Mole and mass fractions are appropriate to either the mixture or the solution points of view. The other quantities are appropriate to the solution point of view only. In addition of these quantities, the following are useful in conversions between concentrations and other quantities.

6. *Density*: $\rho = m/V$ SI base units: kg m^{-3} (6)

7. *Relative density*: d; the ratio of the density of a mixture to the density of a reference substance under conditions which must be specified for both (1). The symbol $d^t_{t'}$ will be used for the density of a mixture at t°C, 1 atm divided by the density of water at t'°C, 1 atm.

Other quantities will be defined in the prefaces to individual volumes or on specific data sheets.

Thermodynamics of Solubility

The principal aims of the Solubility Data Project are the tabulation and evaluation of: (a) solubilities as defined above; (b) the nature of the saturating solid phase. Thermodynamic analysis of solubility phenomena has two aims: (a) to provide a rational basis for the construction of functions to represent solubility data; (b) to enable thermodynamic quantities to be extracted from solubility data. Both these aims are difficult to achieve in many cases because of a lack of experimental or theoretical information concerning activity coefficients. Where thermodynamic quantities can be found, they are not evaluated critically, since this task would involve critical evaluation of a large body of data that is not directly relevant to solubility. The following discussion is an outline of the principal thermodynamic relations encountered in discussions of solubility. For more extensive discussions and references, see books on thermodynamics, e.g., (5-10).

Activity Coefficients (1)

(a) *Mixtures*. The activity coefficient f_B of a substance B is given by

$$RT\ \ell n(f_B x_B) = \mu_B - \mu_B^* \qquad (7)$$

where μ_B is the chemical potential, and μ_B^* is the chemical potential of pure B at the same temperature and pressure. For any substance B in the mixture,

$$\lim_{x_B \to 1} f_B = 1 \qquad (8)$$

(b) *Solutions*.

(i) *Solute substance, B*. The molal activity coefficient γ_B is given by

$$RT\ \ell n(\gamma_B m_B) = \mu_B - (\mu_B - RT\ \ell n\ m_B)^\infty \qquad (9)$$

where the superscript $^\infty$ indicates an infinitely dilute solution. For any solute B,

$$\gamma_B^\infty = 1 \qquad (10)$$

Activity coefficients y_B connected with concentration c_B, and $f_{x,B}$ (called the *rational activity coefficient*) connected with mole fraction x_B are defined in analogous ways. The relations among them are (1,9):

$$\gamma_B = x_A f_{x,B} = V_A^*(1 - \sum_s c_s)y_B \qquad (11)$$

or

$$f_{x,B} = (1 + M_A \sum_s m_s)\gamma_B = V_A^* y_B/V_m \quad (12)$$

or

$$y_B = (V_A + M_A \sum_s m_s V_s)\gamma_B/V_A^* = V_m f_{x,B}/V_A^* \quad (13)$$

where the summations are over all solutes, V_A^* is the molar volume of the pure solvent, V_i is the partial molar volume of substance i, and V_m is the molar volume of the solution.

For an electrolyte solute $B \equiv C_{\nu+}A_{\nu-}$, the molal activity is replaced by (9)

$$\gamma_B m_B = \gamma_\pm{}^\nu m_B{}^\nu Q^\nu \quad (14)$$

where $\nu = \nu_+ + \nu_-$, $Q = (\nu_+{}^{\nu+}\nu_-{}^{\nu-})^{1/\nu}$, and $\gamma_\pm$ is the mean ionic molal activity coefficient. A similar relation holds for the concentration activity $y_B c_B$. For the mol fractional activity,

$$f_{x,B}\, x_B = \nu_+{}^{\nu_+} \nu_-{}^{\nu_-} f_\pm{}^\nu x_\pm{}^\nu \quad (15)$$

The quantities x_+ and x_- are the ionic mole fractions (9), which for a single solute are

$$x_+ = \nu_+ x_B/[1+(\nu-1)x_B]; \qquad x_- = \nu_- x_B/[1+(\nu-1)x_B] \quad (16)$$

(ii) *Solvent*, A:

The *osmotic coefficient*, ϕ, of a solvent substance A is defined as (1):

$$\phi = (\mu_A^* - \mu_A)/RT\, M_A \sum_s m_s \quad (17)$$

where μ_A^* is the chemical potential of the pure solvent.

The *rational osmotic coefficient*, ϕ_x, is defined as (1):

$$\phi_x = (\mu_A - \mu_A^*)/RT\ln x_A = \phi M_A \sum_s m_s/\ln(1 + M_A \sum_s m_s) \quad (18)$$

The activity, a_A, or the activity coefficient f_A is often used for the solvent rather than the osmotic coefficient. The activity coefficient is defined relative to pure A, just as for a mixture.

The Liquid Phase

A general thermodynamic differential equation which gives solubility as a function of temperature, pressure and composition can be derived. The approach is that of Kirkwood and Oppenheim (7). Consider a solid mixture containing c' thermodynamic components i. The Gibbs-Duhem equation for this mixture is:

$$\sum_{i=1}^{c'} x_i'(S_i'dT - V_i'dp + d\mu_i) = 0 \quad (19)$$

A liquid mixture in equilibrium with this solid phase contains c thermodynamic components i, where, usually, $c \geq c'$. The Gibbs-Duhem equation for the liquid mixture is:

$$\sum_{i=1}^{c'} x_i(S_i dT - V_i dp + d\mu_i) + \sum_{i=c'+1}^{c} x_i(S_i dT - V_i dp + d\mu_i) = 0 \quad (20)$$

Eliminate $d\mu_1$ by multiplying (19) by x_1 and (20) x_1'. After some algebra, and use of:

$$d\mu_i = \sum_{j=2}^{c} G_{ij}dx_j - S_i dT + V_i dp \quad (21)$$

where (7)

$$G_{ij} = (\partial\mu_i/\partial x_j)_{T,P,x_i \neq x_j} \quad (22)$$

it is found that

$$\sum_{i=2}^{c'} \sum_{j=2}^{c} (x_i' - x_i x_1'/x_1)G_{ij}dx_j - (x_1'/x_1) \sum_{i=c'+1}^{c} \sum_{j=2}^{c} x_i G_{ij} dx_j$$

$$= \sum_{i=1}^{c'} x_i'(H_i - H_i')dT/T - \sum_{i=1}^{c'} x_i'(V_i - V_i')dp \quad (23)$$

where

$$H_i - H_i' = T(S_i - S_i') \tag{24}$$

is the enthalpy of transfer of component i from the solid to the liquid phase, at a given temperature, pressure and composition, and H_i, S_i, V_i are the partial molar enthalpy, entropy, and volume of component i. Several special cases (all with pressure held constant) will be considered. Other cases will appear in individual evaluations.

(a) Solubility as a function of temperature. Consider a binary solid compound A_nB in a single solvent A. There is no fundamental thermodynamic distinction between a binary compound of A and B which dissociates completely or partially on melting and a solid mixture of A and B; the binary compound can be regarded as a solid mixture of constant composition. Thus, with c = 2, c' = 1, $x_A' = n/(n+1)$, $x_B' = 1/(n+1)$, eqn (23) becomes

$$(1/x_B - n/x_A)\{1 + \left(\frac{\partial \ln f_B}{\partial \ln x_B}\right)_{T,P}\}dx_B = (nH_A + H_B - H_{AB}^*)dT/RT^2 \tag{25}$$

where the mole fractional activity coefficient has been introduced. If the mixture is a non-electrolyte, and the activity coefficients are given by the expression for a simple mixture (6):

$$RT \ln f_B = wx_A^2 \tag{26}$$

then it can be shown that, if w is independent of temperature, eqn (25) can be integrated (cf. (5), Chap. XXIII, sect. 5). The enthalpy term becomes

$$nH_A + H_B - H_{AB}^* = \Delta H_{AB} + n(H_A - H_A^*) + (H_B - H_B^*)$$

$$= \Delta H_{AB} + w(nx_B^2 + x_A^2) \tag{27}$$

where ΔH_{AB} is the enthalpy of melting and dissociation of one mole of pure solid A_nB, and H_A^*, H_B^* are the molar enthalpies of pure liquid A and B. The differential equation becomes

$$R\, d \ln\{x_B(1-x_B)^n\} = -\Delta H_{AB}\, d\left(\frac{1}{T}\right) - w\, d\left(\frac{x_A^2 + nx_B^2}{T}\right) \tag{28}$$

Integration from x_B,T to $x_B = 1/(1+n)$, T = T*, the melting point of the pure binary compound, gives:

$$\ln\{x_B(1-x_B)^n\} \simeq \ln\left\{\frac{n^n}{(1+n)^{n+1}}\right\} - \left\{\frac{\Delta H_{AB}^* - T^*\Delta C_p^*}{R}\right\}\left(\frac{1}{T} - \frac{1}{T^*}\right)$$

$$+ \frac{\Delta C_p^*}{R}\ln\left(\frac{T}{T^*}\right) - \frac{w}{R}\left\{\frac{x_A^2 + nx_B^2}{T} - \frac{n}{(n+1)T^*}\right\} \tag{29}$$

where ΔC_p^* is the change in molar heat capacity accompanying fusion plus decomposition of the compound at temperature T*, (assumed here to be independent of temperature and composition), and ΔH_{AB}^* is the corresponding change in enthalpy at T = T*. Equation (29) has the general form

$$\ln\{x_B(1-x_B)^n\} = A_1 + A_2/T + A_3 \ln T + A_4(x_A^2 + nx_B^2)/T \tag{30}$$

If the solid contains only component B, n = 0 in eqn (29) and (30).

If the infinite dilution standard state is used in eqn (25), eqn (26) becomes

$$RT \ln f_{x,B} = w(x_A^2 - 1) \tag{31}$$

and (27) becomes

$$nH_A + H_B - H_{AB} = (nH_A^* + H_B^\infty - H_{AB}^*) + n(H_A - H_A^*) + (H_B - H_B^\infty) = \Delta H_{AB}^\infty + w(nx_B^2 + x_A^2 - 1) \tag{32}$$

where the first term, ΔH_{AB}^∞, is the enthalpy of melting and dissociation of solid compound A_nB to the infinitely dilute state of solute B in solvent A; H_B^∞ is the partial molar enthalpy of the solute at infinite dilution. Clearly, the integral of eqn (25) will have the same form as eqn (29), with $\Delta H_{AB}^\infty(T^*)$, $\Delta C_p^\infty(T^*)$ replacing ΔH_{AB}^* and ΔC_p^* and $x_A^2 - 1$ replacing x_A^2 in the last term.

If the liquid phase is an aqueous electrolyte solution, and the solid is a salt hydrate, the above treatment needs slight modification. Using rational mean activity coefficients, eqn (25) becomes

$$R\nu(1/x_B - n/x_A)\{1+(\partial \ell n f_\pm/\partial \ell n x_\pm)_{T,P}\}dx_B/\{1+(\nu-1)x_B\}$$

$$= \{\Delta H^\infty_{AB} + n(H_A - H_A^*) + (H_B - H_B^\infty)\}d(1/T) \qquad (33)$$

If the terms involving activity coefficients and partial molar enthalpies are negligible, then integration gives (cf. (11)):

$$\ell n\{\frac{x_B^\nu(1-x_B)^n}{1+(\nu-1)x_B{}^{n+\nu}}\} = \ell n\{\frac{n^n}{(n+\nu)^{n+\nu}}\} - \{\frac{\Delta H^\infty_{AB}(T^*)-T^*\Delta C_p^*}{R}\}(\frac{1}{T}-\frac{1}{T^*}) + \frac{\Delta C_p^*}{R}\ell n(T/T^*) \qquad (34)$$

A similar equation (with ν=2 and without the heat capacity terms) has been used to fit solubility data for some $MOH{=}H_2O$ systems, where M is an alkali metal; the enthalpy values obtained agreed well with known values (11). In many cases, data on activity coefficients (9) and partial molal enthalpies (8,10) in concentrated solution indicate that the terms involving these quantities are not negligible, although they may remain roughly constant along the solubility temperature curve.

The above analysis shows clearly that a rational thermodynamic basis exists for functional representation of solubility-temperature curves in two-component systems, but may be difficult to apply because of lack of experimental or theoretical knowledge of activity coefficients and partial molar enthalpies. Other phenomena which are related ultimately to the stoichiometric activity coefficients and which complicate interpretation include ion pairing, formation of complex ions, and hydrolysis. Similar considerations hold for the variation of solubility with pressure, except that the effects are relatively smaller at the pressures used in many investigations of solubility (5).

(b) *Solubility as a function of composition.*
At constant temperature and pressure, the chemical potential of a saturating solid phase is constant:

$$\mu^*_{A_nB} = \mu_{A_nB}(\mathrm{sln}) = n\mu_A + \mu_B \qquad (35)$$

$$= (n\mu_A^* + \nu_+\mu_+^\infty + \nu_-\mu_-^\infty) + nRT\ \ell n f_A x_A$$

$$+ \nu RT\ \ell n \gamma_\pm m_\pm Q_\pm \qquad (36)$$

for a salt hydrate A_nB which dissociates to water, (A), and a salt, B, one mole of which ionizes to give ν_+ cations and ν_- anions in a solution in which other substances (ionized or not) may be present. If the saturated solution is sufficiently dilute, $f_A = x_A = 1$, and the quantity K^0_{s0} in

$$\Delta G^\infty \equiv (\nu_+\mu_+^\infty + \nu_-\mu_-^\infty + n\mu_A^* - \mu_{AB}^*)$$

$$= -RT\ \ell n\ K^0_{s0}$$

$$= -RT\ \ell n\ Q^\nu \gamma_\pm^\nu m_+^{\nu_+} m_-^{\nu_-} \qquad (37)$$

is called the *solubility product* of the salt. (It should be noted that it is not customary to extend this definition to hydrated salts, but there is no reason why they should be excluded.) Values of the solubility product are often given on mole fraction or concentration scales. In dilute solutions, the theoretical behaviour of the activity coefficients as a function of ionic strength is often sufficiently well known that reliable extrapolations to infinite dilution can be made, and values of K^0_{s0} can be determined. In more concentrated solutions, the same problems with activity coefficients that were outlined in the section on variation of solubility with temperature still occur. If these complications do not arise, the solubility of a hydrate salt $C_{\nu_+}A_{\nu_-}\cdot nH_2O$ in the presence of other solutes is given by eqn (36) as

$$\nu\ \ell n\{m_B/m_B(0)\} = -\nu \ell n\{\gamma_\pm/\gamma_\pm(0)\} - n\ \ell n(a_{H_2O}/a_{H_2O}(0)) \qquad (38)$$

where a_{H_2O} is the activity of water in the saturated solution, m_B is the molality of the salt in the saturated solution, and (0) indicates absence of other solutes. Similar considerations hold for non-electrolytes.

The Solid Phase

The definition of solubility permits the occurrence of a single solid phase which may be a pure anhydrous compound, a salt hydrate, a non-stoichiometric compound, or a solid mixture (or solid solution, or "mixed crystals"), and may be stable or metastable. As well, any number of solid phases consistent with the requirements of the phase rule may be present. Metastable solid phases are of widespread occurrence, and may appear as polymorphic (or allotropic) forms or crystal solvates whose rate of transition to more stable forms is very slow. Surface heterogeneity may also give rise to metastability, either when one solid precipitates on the surface of another, or if the size of the solid particles is sufficiently small that surface effects become important. In either case, the solid is not in stable equilibrium with the solution. The stability of a solid may also be affected by the atmosphere in which the system is equilibrated.

Many of these phenomena require very careful, and often prolonged, equilibration for their investigation and elimination. A very general analytical method, the "wet residues" method of Schreinemakers (12) (see a text on physical chemistry) is usually used to investigate the composition of solid phases in equilibrium with salt solutions. In principle, the same method can be used with systems of other types. Many other techniques for examination of solids, in particular X-ray, optical, and thermal analysis methods, are used in conjunction with chemical analyses (including the wet residues method).

COMPILATIONS AND EVALUATIONS

The formats for the compilations and critical evaluations have been standardized for all volumes. A brief description of the data sheets has been given in the FOREWORD; additional explanation is given below.

Guide to the Compilations

The format used for the compilations is, for the most part, self-explanatory. The details presented below are those which are not found in the FOREWORD or which are not self-evident.

Components. Each component is listed according to IUPAC name, formula, and Chemical Abstracts (CA) Registry Number. The formula is given either in terms of the IUPAC or Hill (13) system and the choice of formula is governed by what is usual for most current users: i.e. IUPAC for inorganic compounds, and Hill system for organic compounds. Components are ordered according to:

(a) saturating components;
(b) non-saturating components in alphanumerical order;
(c) solvents in alphanumerical order.

The saturating components are arranged in order according to a 18-column, 2-row periodic table:

Columns 1,2: H, groups IA, IIA;
3,12: transition elements (groups IIIB to VIIB, group VIII, groups IB, IIB);
13-18: groups IIIA-VIIA, noble gases.
Row 1: Ce to Lu;
Row 2: Th to the end of the known elements, in order of atomic number.

Salt hydrates are generally not considered to be saturating components since most solubilities are expressed in terms of the anhydrous salt. The existence of hydrates or solvates is carefully noted in the texts, and CA Registry Numbers are given where available, usually in the critical evaluation. Mineralogical names are also quoted, along with their CA Registry Numbers, again usually in the critical evaluation.

Original Measurements. References are abbreviated in the forms given by *Chemical Abstracts Service Source Index (CASSI)*. Names originally in other than Roman alphabets are given as transliterated by *Chemical Abstracts*.

Experimental Values. Data are reported in the units used in the original publication, with the exception that modern *names* for units and quantities are used; e.g., mass per cent for weight per cent; mol dm^{-3} for molar; etc. Both mass and molar values are given. Usually, only one type of value (e.g., mass per cent) is found in the original paper, and the compiler has added the other type of value (e.g., mole per cent) from computer calculations based on 1976 atomic weights (14). Errors in calculations and fitting equations in original papers have been noted and corrected, by computer calculations where necessary.

Method. Source and Purity of Materials. Abbreviations used in *Chemical Abstracts* are often used here to save space.

Estimated Error. If these data were omitted by the original authors, and if relevant information is available, the compilers have attempted to estimate errors from the internal consistency of data and type of apparatus used. Methods used by the compilers for estimating and reporting errors are based on the papers by Ku and Eisenhart (15).

Comments and/or Additional Data. Many compilations include this section which provides short comments relevant to the general nature of the work or additional experimental and thermodynamic data which are judged by the compiler to be of value to the reader.

References. See the above description for Original Measurements.

Guide to the Evaluations

The evaluator's task is to check whether the compiled data are correct, to assess the reliability and quality of the data, to estimate errors where necessary, and to recommend "best" values. The evaluation takes the form of a summary in which all the data supplied by the compiler have been critically reviewed. A brief description of the evaluation sheets is given below.

Components. See the description for the Compilations.

Evaluator. Name and date up to which the literature was checked.

Critical Evaluation

(a) Critical text. The evaluator produces text evaluating *all* the published data for each given system. Thus, in this section the evaluator review the merits or shortcomings of the various data. Only published data are considered; even published data can be considered only if the experimental data permit an assessment of reliability.

(b) Fitting equations. If the use of a smoothing equation is justifiable, the evaluator may provide an equation representing the solubility as a function of the variables reported on all the compilation sheets.

(c) Graphical summary. In addition to (b) above, graphical summaries are often given.

(d) Recommended values. Data are *recommended* if the results of at least two independent groups are available and they are in good agreement, and if the evaluator has no doubt as to the adequacy and reliability of the applied experimental and computational procedures. Data are reported as *tentative* if only one set of measurements is available, or if the evaluator considers some aspect of the computational or experimental method as mildly undesirable but estimates that it should cause only minor errors. Data are considered as *doubtful* if the evaluator considers some aspect of the computational or experimental method as undesirable but still considers the data to have some value in those instances where the order of magnitude of the solubility is needed. Data determined by an inadequate method or under ill-defined conditions are *rejected.* However references to these data are included in the evaluation together with a comment by the evaluator as to the reason for their rejection.

(e) References. All pertinent references are given here. References to those data which, by virtue of their poor precision, have been rejected and not compiled are also listed in this section.

(f) Units. While the original data may be reported in the units used by the investigators, the final recommended values are reported in S.I. units (1,16) when the data can be accurately converted.

References

1. Whiffen, D. H., ed., *Manual of Symbols and Terminology for Physico-chemical Quantities and Units. Pure Applied Chem.* 1979, *51*, No. 1.
2. McGlashan, M.L. *Physicochemical Quantities and Units.* 2nd ed. Royal Institute of Chemistry. London. 1971.
3. Jänecke, E. *Z. Anorg. Chem.* 1906, *51*, 132.
4. Friedman, H.L. *J. Chem. Phys.* 1960, *32*, 1351.
5. Prigogine, I.; Defay, R. *Chemical Thermodynamics.* D.H. Everett, transl. Longmans, Green. London, New York, Toronto. 1954.
6. Guggenheim, E.A. *Thermodynamics.* North-Holland. Amsterdam. 1959. 4th ed.
7. Kirkwood, J.G.; Oppenheim, I. *Chemical Thermodynamics.* McGraw-Hill, New York, Toronto, London. 1961.
8. Lewis, G.N.; Randall, M. (rev. Pitzer, K.S.; Brewer, L.). *Thermodynamics.* McGraw Hill. New York, Toronto, London. 1961. 2nd ed.
9. Robinson, R.A.; Stokes, R.H. *Electrolyte Solutions.* Butterworths. London. 1959, 2nd ed.
10. Harned, H.S.; Owen, B.B. *The Physical Chemistry of Electrolytic Solutions.* Reinhold. New York. 1958. 3rd ed.
11. Cohen-Adad, R.; Saugier, M.T.; Said, J. *Rev. Chim. Miner.* 1973, *10*, 631.
12. Schreinemakers, F.A.H. *Z. Phys. Chem., stoechiom. Verwandschaftsl.* 1893, *11*, 75.
13. Hill, E.A. *J. Am. Chem. Soc.* 1900, *22*, 478.
14. IUPAC Commission on Atomic Weights. *Pure Appl. Chem.,* 1976, *47*, 75.

15. Ku, H.H., p. 73; Eisenhart, C., p. 69; in Ku, H.H., ed. *Precision Measurement and Calibration*. NBS Special Publication 300. Vol. 1. Washington. 1969.
16. *The International System of Units*. Engl. transl. approved by the BIPM of *Le Système International d'Unités*. H.M.S.O. London. 1970.

R. Cohen-Adad, Villeurbanne, France
J.W. Lorimer, London, Canada
M. Salomon, Fair Haven, New Jersey, U.S.A.

COMPONENTS:	EVALUATOR:
1. Sodium sulfite; Na_2SO_3; [7757-83-7] 2. Water; H_2O; [7732-18-5]	Mary R. Masson, Dept. of Chemistry, University of Aberdeen, Meston Walk, Old Aberdeen, AB9 2UE, Scotland, UK. June 1984

CRITICAL EVALUATION:

There have been 5 studies of the binary system sodium sulfite - water (1 - 5). However, the data given by Efanov *et al.* (5) are expressed in units of g-equivalents per litre, and therefore cannot be compared with the other data. Data are also available from ternary systems (6 - 24). The data show a good deal of scatter, probably because of the difficulty of preventing oxidation of sulfite to sulfate at all stages from preparation to determination.

Data for the equilibrium with ice (269-273 K) come from (1) and (2). Two points from (1) had to be rejected, but otherwise the data are in reasonable agreement. The regression equation is

$$(T - 273.15) = -0.0137 - 0.347y + 0.0026y^2 \qquad s = 0.027 \quad (15\ \text{pts})$$

or alternatively

$$y = -0.0317 - 2.85(T - 273.2) + 0.0809(T - 273.2)^2 \qquad s = 0.088 \quad (15\ \text{pts})$$

where $y = 100w$ is the solubility in mass % of Na_2SO_3, T is the temperature in K, and s is the estimated standard deviation of the dependent variable about the regression line.

For the temperature range 273 - 309 K, the solid phase in equilibrium with the saturated solution is $Na_2SO_3.7H_2O$ [10102-15-5]. Data were available from (1 - 4, 6 - 9, 11, 12, 14 - 16, 19, 21 - 23); there were 79 points in all. All 10 data points from (7) were rejected, because they were totally inconsistent with the other data. Other data points were rejected when regression analysis showed them to lie outwith $\pm 2s$ of the regression line. The process was repeated until a line was obtained with all points lying inside $\pm 2s$. The rejected points came from the following reference: (1) - 1 out of 8 total, (2) - 2/24, (3) - 2/5, (4) - 1/3, (11) - 1/1, (23) - 2/2. The final regression equation is

$$y = 12.03 + 0.377(T - 273.2) + 0.00325(T - 273.2)^2 \qquad s = 0.207 \quad (60\ \text{pts})$$

For the temperature range 307 - 373 K, where the solid phase is the anhydrous salt Na_2SO_3, data were available in (1 - 4, 8, 10, 13, 17 - 20, 22, 23); there were 52 points. All but 1 of the 11 points from (1) were rejected because they were completely lacking in agreement with the other data. Other points rejected because they lay outside $\pm 2s$ came from (2) - 3/19, (3) - 4/5, (4) - 2/4, (8) - 1/1, (13) - 2/3, (18) - 1/1, (19) - 1/2, (23) - 1/1. The final regression equation is

$$y = 34.2 - 0.2024(T - 273.2) + 0.000760(T - 273.2)^2 \qquad s = 0.135 \quad (27\ \text{pts})$$

The temperature of the transition point between the 7-hydrate and the anhydrous salt has

COMPONENTS:

1. Sodium sulfite; Na_2SO_3; [7757-83-7]
2. Water; H_2O; [7732-18-5]

EVALUATOR:

Mary R. Masson,
Dept. of Chemistry, University of Aberdeen,
Meston Walk, Old Aberdeen, AB9 2UE,
Scotland, UK.
June 1984

CRITICAL EVALUATION: (continued)

been variously reported to be 295.2 K (22°C) (1), 306.6 K (33.4°C) (2), 298.2 - 305.2 K (6), 304.7 K (31.5°C) (12) and 306.7 (36.5°C) (31). The point of intersection of the relevant regression equations is 306.57 K (33.41°C). The solubility at that point is 28.25 mass % of Na_2SO_3.

The ice curve intersects the 7-hydrate line at a eutectic temperature of 269.70 K (- 3.46°C); the solubility there is 10.8 mass % of Na_2SO_3.

Rodnyanskii and Galinker (4) actually report solubilities up to 633 K (360°C), but their data were reported only in graphical form, and precision of reading the graph was not good.

With as much data as are available for this system, it would have been hoped to be able to give recommended values. However, because there is such a lot of scatter in the data, it was felt that the values derived from the regression equations should be designated merely as Tentative.

TENTATIVE SOLUBILITIES

T/K	Solubility mass %	Solubility molality mol/kg
	$Na_2SO_3.7H_2O$	
273.2	12.03	1.085
278.2	14.0	1.29
283.2	16.1	1.52
288.2	18.4	1.79
293.2	20.9	2.10
298.2	23.5	2.44
303.2	26.3	2.83
308.2	29.2	3.27
	Na_2SO_3	
308.2	28.0	3.09
313.2	27.3	2.98
318.2	26.6	2.88
323.2	25.95	2.78
328.2	25.3	2.69
333.2	24.8	2.62
338.2	24.2	2.53
343.2	23.7	2.46
348.2	23.3	2.41
353.2	22.8	2.34
358.2	22.5	2.30
363.2	22.1	2.25
368.2	21.8	2.21
373.2	21.5	2.17

COMPONENTS:	EVALUATOR:
1. Sodium sulfite; Na_2SO_3; [7757-83-7] 2. Water; H_2O; [7732-18-5]	Mary R. Masson, Dept. of Chemistry, University of Aberdeen, Meston Walk, Old Aberdeen, AB9 2UE, Scotland, UK. June 1984

CRITICAL EVALUATION: (continued)

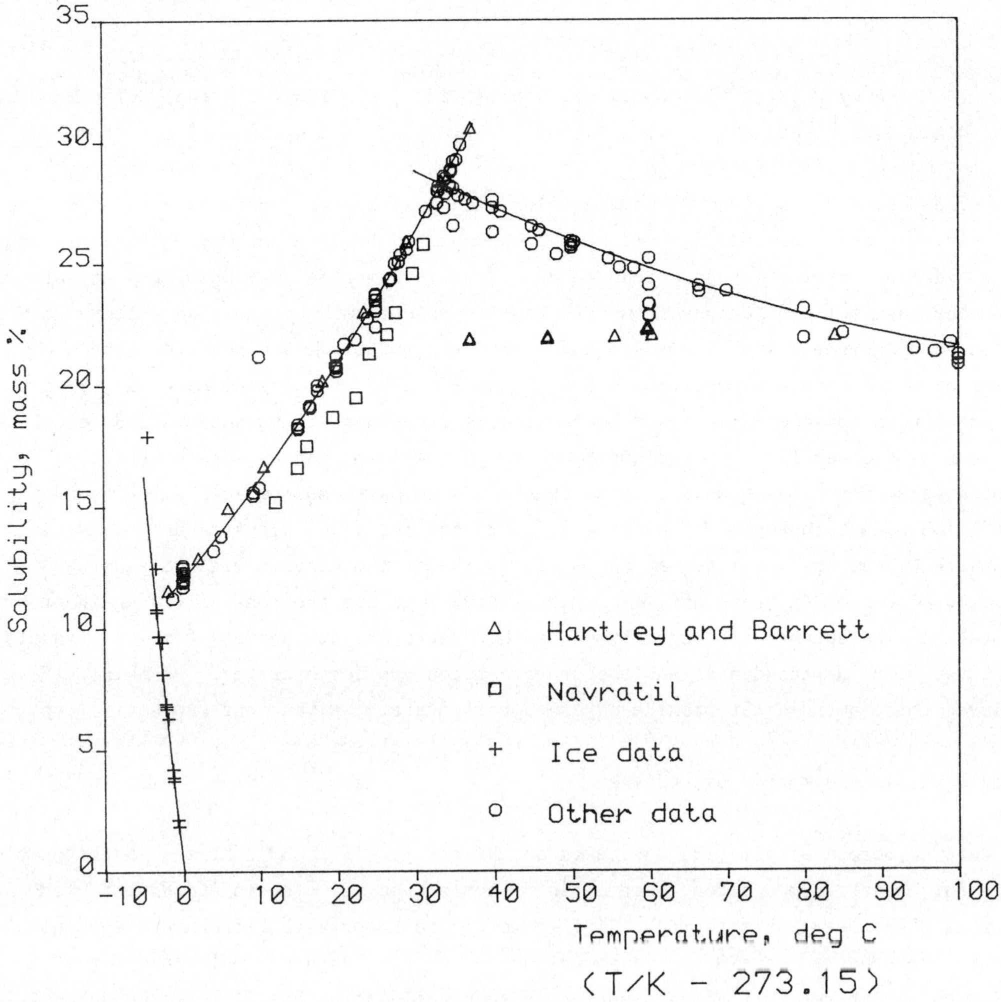

TERNARY SYSTEMS

Sodium sulfite - sodium sulfate - water. This system has been studied by a number of workers (10 - 18), at a number of temperatures. At 273.2 and 273.3 K, the data of Rivett and Lewis (14) and Palkina (15) are in good agreement. The data of Sotova *et al.* (16) are also in agreement, apart from one or two points. Palkina gives no information about the solid phases, and the others differ in their conclusions about the solid phases. A first glance at the phase diagrams would suggest that there were two solid phases, viz. anhydrous Na_2SO_3 and Na_2SO_4. Sotova merely states that the two solid phases are $Na_2SO_3.7H_2O$ and $Na_2SO_4.10H_2O$. Rivett and Lewis say that there are two series of mixed crystals, one between the heptahydrates and the other between the decahydrates. In neither case is convincing evidence presented. It seems certain that some sort of

COMPONENTS:	EVALUATOR:
1. Sodium sulfite; Na_2SO_3; [7757-83-7] 2. Water; H_2O; [7732-18-5]	Mary R. Masson, Dept. of Chemistry, University of Aberdeen, Meston Walk, Old Aberdeen, AB9 2UE, Scotland, UK. June 1984

CRITICAL EVALUATION: (continued)

mixed crystals are indeed formed, because almost certainly it is not the anhydrous salts that are formed.

The data given by Palkina (15) for 278.2, 283.2, 288.2 and 293.2 K appear to be in accord with the data of Rivett and Lewis (14) for 290.7 K and Lewis and Rivett (12) for 294.2 K. The phase diagram for 290.7 K appears to suggest that there are two solid phases, $Na_2SO_3.7H_2O$, and $Na_2SO_4.7H_2O$, forming a simple eutonic system, and this does appear to be possible, although perhaps it would have been expected to find $Na_2SO_4.10H_2O$ rather than the 7-hydrate. Kuznetsova and Sedova (32) report finding $Na_2SO_4.10H_2O$ from sulfate-rich solutions, and solid solutions from sulfite-rich solutions, but their solubility data is not in good agreement with the rest. However, the authors again believed that two series of mixed crystals were formed (14). At 298.2 K, the data of Palkina (15) are in agreement with those of Rivett and Lewis (14), and the data of Kuznetsova and Yaroshenko (11) are also essentially in agreement, although they show rather more scatter. The phase diagram for (11) shows a simple eutonic system, with solids $Na_2SO_3.7H_2O$ and $Na_2SO_4.10H_2O$, but there is not much data. The one for (14) suggests that one solid is $Na_2SO_3.7H_2O$, but the tie-lines at the sulfate side of the diagram meet between the positions for the 7-hydrate and 10-hydrate, which suggests the possibility again of mixed crystals. The authors (14) again propose that there are two series of mixed crystals, one between the heptahydrates and the other between the decahydrates. Palkina (15) also believed this, but did not provide any supporting data. Rivett and Lewis (14) report also a metastable system at 298.2 K; here the solid phases are the anhydrous salts, and mixed crystals are formed extensively.

The data of Sotova *et al.* (18) and Lewis and Rivett (13) for 333.2 K, are in reasonable agreement, apart from a bit of scatter in Sotova's results. Lewis and Rivett also report a short metastable region. The phase diagrams for this temperature show no eutonic, but that a range of solid solutions of the two anhydrous salts are formed. Lewis and Rivett report finding 5 series of such solid solutions. The same sort of phase diagram is seen at 313.2 K (13), 318.2 K (13), 310.7 K (14), and at 373.2 K, where the agreement between Durymanova and Telepneva (10) and Sotova *et al.* (17) is reasonable. Small amounts of data for several other temperatures given by Lewis and Rivett (12) seem in accord with other data. Data given by Wöhler and Dierksen (26) are not in particularly good agreement with the rest.

Sodium sulfite - sodium chloride - water. At 298.2 K, the data of Durymanova and Telepneva (9) are in reasonable agreement with those of Kobe and Hellwig (3), apart from at a part of the curve where Kobe and Hellwig appear to have reported a metastable region. The reported solid phases are NaCl and $Na_2SO_3.7H_2O$, but the phase diagram in (9) suggests that there may be a region where mixed crystals are formed. The data given by Labash and Lusby (8) for 293.2 K seem to be reasonably in accord with (9). At 333.2 K the data of Labash and Lusby (8) agree reasonably with those of Kobe and Hellwig (3),

COMPONENTS:	EVALUATOR:
1. Sodium sulfite; Na_2SO_3; [7757-83-7] 2. Water; H_2O; [7732-18-5]	Mary R. Masson, Dept. of Chemistry, University of Aberdeen, Meston Walk, Old Aberdeen, AB9 2UE, Scotland, UK. June 1984

CRITICAL EVALUATION: (continued)

apart from at one or two points. At this temperature, the phase diagram does appear to show a simple eutonic system, with NaCl and Na_2SO_3 being the solid phases.

Sodium sulfite - sodium pyrosulfite - water. Data for this system are available at 273.2 K (16), 288.2 K (19), 298.2 K (19, 23), 308.2 K (19, 23), 318.2 K (19), 333.2 K (18) and 273.2 K (17). For 298.2 K, the data are in good but not perfect agreement. At 308.2 K, the data show the same trends, but differ by about 1 - 2%. At all temperatures, a simple eutonic system is observed; at 298.2 K and below, the solid phases are $Na_2SO_3.7H_2O$ and $Na_2S_2O_5$, and at 308.2 K and above, they are Na_2SO_3 and $Na_2S_2O_5$.

Sodium sulfite - ammonium sulfite - water. Data are reported by Labash and Lusby (8) for 293.2 and 333.2 K, and by Zil'berman and Ivanov (20) for 333.2 and 358.2 K. The data at 333.2 K are in reasonable agreement.

Sodium sulfite - sodium thiosulfate - water. Data are given by Palkina (15) for 273.2 - 298.2 K, by Mochalov and Monina (21) for 288.2 K, and by Wöhler and Dierksen (26) for 296.2 - 353.2 K. At 288.2 K, the data for the sulfite branch of the solubility curve are in good agreement, but Mochalov and Monina found about 1% more thiosulfate in the thiosulfate branch.

Other ternary systems. The system sodium sulfite - ethanol - water has been studied by Navrátil and Ńývlt (7) and Klebanov and Ostapkevich (22), but the data are not directly comparable. The data in (7) are suspect, because the binary data extracted were found to be unreliable. The system sodium sulfite - sodium hydroxide - water was studied by Hammick and Currie (6) at temperatures from 273.3 K to 305.2 K. The system sodium sulfite - sodium 2-naphtholate - water was studied by Teslo *et al.* (25) at temperatures between 313.2 and 373.2 K.

MORE COMPLEX SYSTEMS

Systems involving sodium sulfite, water, with sodium sulfate and sodium chloride (9, 10); with sodium pyrosulfite and sodium sulfate (11, 27, 33), with ammonium sulfite, ammonium chloride and sodium chloride (20, 30); and with 2-naphthol and sodium hydroxide (29) have been studied. No comparisons were possible, however.

REFERENCES

1. Hartley, H.; Barrett, W.H. *J. Chem. Soc.* 1909, *95*, 1178.
2. Foerster, F.; Brosche, A.; Norberg-Schutz, Chr. *Z. Phys. Chem.* 1924, *10*, 435.
3. Kobe, K.A.; Hellwig, K.C. *Ind. Eng. Chem.* 1955, *47*, 1116.
4. Rodnyanskii, I.M.; Galinker, I.S. *Tf. Khar'kovsk Sel'skokhoz. Inst.* 1961, *35*, 69.

COMPONENTS:	EVALUATOR:
1. Sodium sulfite; Na_2SO_3; [7757-83-7] 2. Water; H_2O; [7732-18-5]	Mary R. Masson, Dept. of Chemistry, University of Aberdeen, Meston Walk, Old Aberdeen, AB9 2UE, Scotland, UK. June 1984

CRITICAL EVALUATION: (continued)

5. Efanov, L.N.; Lel'kin, K.P.;Romashchenko,I.M. *Zh. Neorg. Khim.* 1977, 22, 217;**Russ. J. Inorg. Chem. (Eng. Transl.)* 1977, 22, 120.

6. Hammick, D.L.; Currie, J.A. *J. Chem. Soc.* 1925, *127*, 1623.

7. Navrátil, J.; Nývlt, J. *Chem. Prům.* 1968, *18*, 612.

8. Labash, J.A.; Lusby, G.R. *Can. J. Chem.* 1955, *33*, 774.

9. Durymanova, M.A.; Telepneva, A.E.; Zagrebina, L.A. *Zh. Neorg. Khim.* 1971, *16*, 500; **Russ. J. Inorg. Chem. (Eng. Transl.)* 1971, *16*, 264.

10. Durymanova, M.A.; Telepneva, A.E. *Zh. Priklad. Khim.* 1972, *45*, 1610;**J. Appl. Chem. USSR (Eng. Transl.)* 1972, *45*, 1680.

11. Kuznetsova, A.G.; Yaroshenko, L.B. *Zh. Priklad. Khim.* 1981, *54*, 2197;**J. Appl. Chem. USSR (Eng. Transl.)* 1981, *54*, 1929.

12. Lewis, N.B.; Rivett, A.C.D. *J. Chem. Soc.* 1924, *125*, 1156.

13. Lewis, N.B.; Rivett, A.C.D. *J. Chem. Soc.* 1924, *125*, 1162.

14. Rivett, A.C.D.; Lewis, N.B. *Rec. Trav. Chim. Pays-Bas* 1923, *42*, 954.

15. Palkina, N.A. *Tr. Vornezh. Gos. Univ.* 1950, *17*, 61.

16. Sotova, N.N.; Kuznetsova, A.G.; Torocheshnikov, N.S.; Kononova, I.V. *Zh. Priklad. Khim.* 1978, *51*, 779; *J. Appl. Chem. USSR (Eng. Transl.)* 1978, *51*, 760. Also published in *Mezhvuz. Sb. Altaisk. Politekn. In-t* 1976, 2, 150 and in *Fiz.-Khim. Osn. Tekhnol. Pererab. Khim. Syr'ya* 1976, 2, 150.

17. Sotova, N.N.; Torocheshnikov, N.S.; Kuznetsova, A.G.; Sokolova, E.I. *Khimiya i Tekhnol. Mineral'n Solei i Galurgichesk. Pr.-v, Varnaul* 1978, 53.

18. Sotova, N.N.; Torocheshnikov, N.S.; Kuznetsova, A.G.; Poroshkova, M.A. *Khimiya i Tekhnol. Mineral'n Solei i Galurgichesk. Pr.-v, Varnaul* 1978, 59.

19. Jäger, L.; Rejlek, M.; Klimeček, R.; Machala, J. *Chem. Prům.* 1959, *9*, 361.

20. Zil'berman, Ya.I.; Ivanov, P.T. *Zh. Priklad. Khim.* 1941, *14*, 939-46.

21. Mochalov, K.I.; Monina, S.S. *Uch. Zap. Perm. Univ.* 1970, *229*, 40.

22. Klebanov, G.S.; Ostapkevich, N.A. *Zh. Neorg. Khim.* 1960, 5, 2329;**Russ. J. Inorg. Chem. (Eng. Transl.)* 1960, 5, 1128.

23. Arii, K. *Bull. Inst. Phys. Chem. Research* 1926, *6*, 1065 (in Japanese). *Sci. Rep. Tohoku Imp. Univ.* 1932, *21*, 783 (in English).

24. Klebanov, G.S.; Ostapkevich, N.A. *Zh. Priklad. Khim.* 1966, *39*, 2467.

25. Teslo, S.P.; Gulyamov, Yu.M.; Odarich, V.F. *Vopr. Khim. Khim. Tekhnol.* 1979, *55*, 92.

26. Wöhler, L.; Dierksen, J. *Z. Angew. Chem.* 1926, *39*, 33.

27. Sotova, N.N.; Kuznetsova, A.G.; Torocheshnikov, N.S. *Zh. Priklad. Khim.* 1978, *51*, 940;**J. Appl. Chem. USSR (Eng. Transl.)* 1978, *51*, *905*.

28. Durymanova, M.A.; Telepneva, A.E.; Zagrebina, L.A. *Zh. Neorg. Khim.* 1971, *16*, 500; **Russ. J. Inorg. Chem. (Eng. Transl.)* 1971, *16*, 264.

29. Kogan, I.M.; Planovskii, A.N.; Evdokimov, A.N. *Anilinokrasochnaya Prom.* 1934, *4*, 34.

COMPONENTS:	EVALUATOR:
1. Sodium sulfite; Na_2SO_3; [7757-83-7] 2. Water; H_2O; [7732-18-5]	Mary R. Masson, Dept. of Chemistry, University of Aberdeen, Meston Walk, Old Aberdeen, AB9 2UE, Scotland, UK. June 1984

CRITICAL EVALUATION: (continued)

30. Labash, J.A.; Lusby, G.R. *Can. J. Chem.* 1955, *33*, 787.

31. Arii, K. *Sci. Rep. Tohoku Imp. Univ.* 1932, *21*, 772.

32. Kuznetsova, A.G.; Sedova, V.A. *VINITI Deposited Document* 1981, 5710-81.

33. Kuznetsova, A.G.; Sedova, V.A. *VINITI Deposited Document* 1981, 5711-81.

COMPONENTS:

1. Sodium sulfite; Na_2SO_3; [7757-83-7]
2. Water; H_2O; [7732-18-5]

ORIGINAL MEASUREMENTS:

Hartley, H.; Barrett, W.H.

J. Chem. Soc. 1909, *95*, 1178-85.

VARIABLES:

Temperature: 269 - 357 K

PREPARED BY:

Mary R. Masson

EXPERIMENTAL VALUES:

t/°C	Na_2SO_3 g/100 g water	Na_2SO_3 mass %	Na_2SO_3[a] mol/kg	Solid[b] phase
- 1.9	13.09	11.570	1.038	A
2.0	14.82	12.910	1.176	A
5.9	17.61	14.970	1.397	A
10.6	20.01	16.670	1.587	A
18.2	15.31	20.200	2.008	A
23.5	29.92	23.030	2.374	A
29.0	34.99	25.920	2.776	A
33.5	39.64	28.390	3.145	A
37.2	44.08	30.590	3.497	A
37.0	28.01	21.880	2.222	B
	28.07	21.920	2.227	B
47.0	28.19	21.990	2.236	B
	28.07	21.920	2.227	B
55.6	28.21	22.000	2.238	B
59.8	28.89	22.410	2.292	B
	28.65	22.270	2.273	B
	28.75	22.330	2.281	B
60.4	28.29	22.050	2.244	B
84.0	28.26	22.030	2.242	B

[a] Molalities calculated by the compiler

[b] Solid phases: A - $Na_2SO_3.7H_2O$, B - Na_2SO_3

(continued on next page)

AUXILIARY INFORMATION

METHOD APPARATUS/PROCEDURE:

The solubility of the hydrated salt was determined by the sealed tube method (1). Weighed amounts of solid and water were sealed in small tubes. Two temperatures, differing by about 0.4 K, were found, at the higher of which small crystals were seen to dissolve and at the lower of which small crystals could be seen to grow. The mean of the two temperatures was taken as the temperature at which the weighed amount of solid dissolved in the weighed amount of water.

The solubility of the anhydrous salt was determined by stirring excess of the salt with water under hydrogen, portions of the solutions were withdrawn from time to time and analysed for sulfite by reaction with excess of iodine and back-titration of the excess with standard thiosulfate.

SOURCE AND PURITY OF MATERIALS:

Anhydrous sodium sulfite was prepared as follows. Pure sodium carbonate (Merck, 40g) was dissolved in 120 g of air-free water, and sulfur dioxide was passed into the solution until the gain in weight corresponded to complete conversion into sodium hydrogen sulfite. An equal quantity of sodium carbonate solution was then added. The solution was filtered under hydrogen through glass wool into a vessel maintained at over 100°C. On evaporation in a stream of hydrogen, crystals of the anhydrous salt separated. These were washed with water + alcohol then alcohol, and dried, under hydrogen. This salt was free from sulfate.

ESTIMATED ERROR:

For anhydrous salt, s = 0.15 g/100 g
For hydrate, temperature error = $\pm$ 0.2 K

REFERENCES:

1. Hartley, H.; Thomas
J. Chem. Soc. (Trans.), 1906, *89*, 1016.

COMPONENTS:	ORIGINAL MEASUREMENTS:
1. Sodium sulfite; Na_2SO_3; [7757-83-7]	Hartley, H.; Barrett, W.H.
2. Water; H_2O; [7732-18-5]	*J. Chem. Soc.* 1909, 95, 1178.

EXPERIMENTAL VALUES (continued):

t/°C	Na_2SO_3 g/100 g water	Na_2SO_3 mass %	Na_2SO_3[a] mol/kg	Solid phase
-0.76	2.15	2.10	0.170	ice
-1.37	4.21	4.04	0.334	"
-1.96	6.29	5.92	0.499	"
-2.77	9.44	8.63	0.749	"
-3.51	12.48	11.10	0.991	"
-4.50	17.91	15.19	1.421	"

The freezing points of the solutions of sodium sulfite were found by the "ordinary Beckmann method", under an atmosphere of hydrogen. The salt was added in the anhydrous form.

The ice curve cut the solubility curve at -3.5°C, the eutectic temperature.

[a] Molalities calculated by the compiler.

COMPONENTS:

1. Sodium sulfite; Na_2SO_3; [7757-83-7]

2. Water; H_2O; [7732-18-5]

ORIGINAL MEASUREMENTS:

Foerster, F.; Brosche, A.; Norberg-Schutz, Chr.

Z. Phys. Chem. 1924, *10*, 435-96.

VARIABLES:

Temperature: 270 - 372 K

PREPARED BY:

Mary R. Masson

EXPERIMENTAL VALUES:

t/°C	Na_2SO_3 mass %	Na_2SO_3[a] mol/kg
- 1.3	11.25*	1.006
0.0	12.50	1.133
0.0	12.59	1.143
4.0	13.20*	1.207
9.15	15.47	1.452
9.2	15.60	1.466
16.5	19.14	1.878
16.6	19.07	1.870
19.9	20.82	2.086
22.4	21.89	2.223
24.0	22.76	2.338
26.85	24.32	2.550
27.0	24.39	2.559
28.0	25.07	2.655
28.2	25.36	2.696
29.05	25.59	2.729
33.0	27.99	3.084
33.1	27.98	3.082
33.1	28.15	3.108
33.8	28.58	3.175
34.65	28.78	3.206
34.7	28.86	3.219
35.4	29.27	3.283
35.9	29.89	3.382

*Results considered particularly reliable by the original authors.

Solid phase in equilibrium was $Na_2SO_3.7H_2O$.

[a] Molalities calculated by the compiler (continued on next page)

AUXILIARY INFORMATION

METHOD APPARATUS/PROCEDURE:

Solids were equilibrated with solution under a hydrogen atmosphere, in a vessel maintained in a thermostat. Samples for analysis were withdrawn through a tube plugged with cotton wool.

Samples were reacted with an excess of standard iodine solution, and the excess was back-titrated with thiosulfate.

SOURCE AND PURITY OF MATERIALS:

$Na_2SO_3.7H_2O$ was prepared by passing sulfur dioxide into a 25% solution of sodium hydroxide, taking care not to let the temperature rise above the transition temperature (34°C). When the temperature started to drop rapidly, the supply of sulfur dioxide was shut off. The hydrate crystallized out when then solution was cooled to 0°C.

ESTIMATED ERROR:

Temperature: ± 0.1 K

Analyses: no accurate estimate possible.

REFERENCES:

COMPONENTS:	ORIGINAL MEASUREMENTS:
1. Sodium sulfite; Na_2SO_3; [7757-83-7]	Foerster, F.; Brosche, A.; Norberg-Schutz, Chr.
2. Water; H_2O; [7732-18-5]	*Z. Phys. Chem.* 1924, *10*, 435-96.

EXPERIMENTAL VALUES (continued):

t/°C	Equilibration time	Na_2SO_3 mass %	Na_2SO_3[a] mol/kg	Solid[b] phase
46.0	10 min	26.03	2.792	A
46.0	2 hr	26.35	2.839	A
50.4	15 min	25.85	2.766	A
56.25	10 min	24.80	2.617	A
94.4	10 min	21.44	2.165	A
52.5	16 hr	22.35	2.284	A
48.2	10 min	25.38	2.699	A
66.4	10 min	24.06	2.514	A
97.0	15 min	21.32	2.150	A
58.1	25 hr	25.12	2.662	A
58.1	93 hr	25.78	2.756	A
58.1	177 hr	24.79	2.615	A
54.8	48 hr	25.20*	2.673	B
70.0	100 min	23.85	2.485	B
99.0	24 hr	21.70	2.199	B
34.0	1 hr	26.00	2.788	B
34.0	6.5 hr	27.50	3.009	B
34.0	20 hr	27.70	3.040	B
34.0	92 hr	27.90	3.070	B
37.5	140 min	23.20	2.397	B
37.5	27 hr	26.95	2.927	B
37.5	29 hr	27.50	3.009	B
36.5	51 hr	27.65	3.032	B
34.5	69 hr	28.20	3.116	B
41.0	75 hr	27.35	2.987	B
41.0	78 hr	27.15*	2.957	B
50.0	40 min	23.85	2.485	B
50.0	220 min	24.90	2.631	B
49.9	315 min	25.20	2.673	B
49.7	21 hr	25.45	2.709	B
50.0	29 hr	25.60	2.730	B
50.1	44 hr	25.70	2.744	B
50.0	70 hr	25.90	2.773	B
50.0	93.5 hr	25.75*	2.752	B
66.5	4 hr	22.07	2.247	B
66.5	21 hr	23.82	2.481	B
35.7	30 min	23.05	2.377	C
35.6	150 min	24.40	2.561	C
35.7	20 hr	26.30	2.831	C
35.4	42.5 hr	27.25	2.972	C
35.7	66 hr	27.55	3.017	C
35.7	71 hr	27.60	3.025	C
35.6	90 hr	27.80*	3.055	C
35.6	115 hr	27.70	3.040	C

a Molalities calculated by the compiler.

b Solid phases: A - Na_2SO_3 formed from $Na_2SO_3.7H_2O$ on standing in contact with solution. B - Na_2SO_3 formed on heating under vacuum at 70°C. C - Na_2SO_3 formed on heating at 100°C at atmospheric pressure. The crystals were ground. (continued on next page)

COMPONENTS:

1. Sodium sulfite; Na_2SO_3; [7757-83-7]

2. Water; H_2O; [7732-18-5]

ORIGINAL MEASUREMENTS:

Foerster, F.; Brosche, A.; Norberg-Schutz, Chr.

Z. Phys. Chem. 1924, *10*, 435-96.

EXPERIMENTAL VALUES (continued):

t/°C	Na_2SO_3 mass %	Na_2SO_3[a] mol/kg	Solid phase
-0.667	1.865	0.151	ice
-1.27	3.73	0.307	"
-2.23	6.69	0.569	"
-2.31	6.91	0.589	"
-2.70	8.12	0.701	"
-3.20	9.68	0.850	"
-3.48	10.80	0.961	"
-1.29	3.87	0.319	"
-2.24	6.80	0.579	"
-3.44	10.67	0.948	"

A Beckmann apparatus was used for the determination of freezing points.

[a] Molalities calculated by the compiler.

COMPONENTS:

1. Sodium sulfite; Na_2SO_3; [7757-83-7]
2. Water; H_2O; [7732-18-5]

ORIGINAL MEASUREMENTS:

Kobe, K.A.; Hellwig, K.C.

Ind. Eng. Chem. 1955, 47, 1116-21.

VARIABLES:

Temperature: 273 - 373 K

PREPARED BY:

Mary R. Masson

EXPERIMENTAL VALUES:

t/°C	Na_2SO_3 g/100 g of water	Na_2SO_3[a] mass %	Na_2SO_3[a] mol/kg	Solid[b] phase
0.0	13.3	11.74	1.055	A
25.0	30.7	23.49	2.436	A
29.30	34.9	25.87	2.769	A
32.95	38.0	27.54	3.015	A + B
33.82	37.5	27.27	2.975	B
40.00	35.7	26.31	2.833	B
60.0	31.7	24.07	2.515	B
80.0	28.0	21.87	2.222	B
100.0	26.3	20.82	2.086	B

[a] Molalities and mass % values calculated by the compiler.

[b] Solid phases: A - $Na_2SO_3.7H_2O$, B - Na_2SO_3

AUXILIARY INFORMATION

METHOD APPARATUS/PROCEDURE:

Solids were equilibrated with water in sealed borosilicate-glass tubes for up to 20 days.

Solution was removed from the tube for analysis, under nitrogen. The measured and weighed aliquot was run into excess of standard iodine, the excess of which was back-titrated with sodium thiosulfate.

SOURCE AND PURITY OF MATERIALS:

Sodium sulfite was obtained from J.T. Baker Chemical Co., and was found to assay at 100.1%. Dissolved oxygen was removed from distilled water by boiling under reduced pressure at 60 - 65°C for 1 hr. This water was stored under nitrogen for a maximum of 5 hr.

ESTIMATED ERROR:

No estimates possible.

REFERENCES:

COMPONENTS:

1. Sodium sulfite; Na_2SO_3; [7757-83-7]

2. Water; H_2O; [7732-18-5]

ORIGINAL MEASUREMENTS:

Rodnyanskii, I.M.; Galinker, I.S.

Tf. Khar'kovsk Sel'skokhoz. Inst. 1961, 35, 69-70.

VARIABLES:

Temperature: 273 - 633 K

PREPARED BY:

Mary R. Masson

EXPERIMENTAL VALUES:

t/°C	Na_2SO_3 mass %	Na_2SO_3[a] mol/kg
0	12.3	1.113
10	21.2	2.135
34	28.5	3.162
40	27.3	2.979
60	25.2	2.673
80	23.1	2.383
100	21.1	2.122
200	17.3	1.660
250	14.4	1.335
300	9.2	0.804
320	7.1	0.606
340	5.2	0.435
360	4.6	0.383

Concentrations were read from graph by the compiler.

[a] Molalities calculated by the compiler.

AUXILIARY INFORMATION

METHOD APPARATUS/PROCEDURE:

An isothermal saturation procedure. An autoclave was used to obtain solubilities above 100°C.

SOURCE AND PURITY OF MATERIALS:

ESTIMATED ERROR:

No estimates possible.

Error in reading graph: 2%

REFERENCES:

COMPONENTS:	ORIGINAL MEASUREMENTS:
1. Sodium sulfite; Na_2SO_3; [7757-83-7] 2. Water; H_2O; [7732-18-5]	Efanov, L.N.; Lel'kin, K.P.; Romashchenko, I.M. *Zh. Neorg. Khim.* 1977, 22, 217-9; *Russ. J. Inorg. Chem. (Eng. Transl.)* 1977, 22, 120-1.
VARIABLES:	PREPARED BY:
Temperature: 303 - 308 K	Mary R. Masson

EXPERIMENTAL VALUES:

Solubility in water

t/°C	Na_2SO_3 [a]
30.0	5.05
31.6	5.27
32.6	5.40
33.1	5.60
33.6	5.67
33.7	5.66
33.8	5.65
34.0	5.16
34.1	5.69
34.3	5.61
34.8	5.66

A large dip in the solubility was observed at 34.0°C, which is approximately the value for the transition temperature (between $Na_2SO_3.7H_2O$ and Na_2SO_3) found by other workers.

[a] Expressed in g-equivalents/l., i.e. mol ($\frac{1}{2}Na_2SO_3$)/l.
The values quoted were read from the graph given by the authors.

AUXILIARY INFORMATION

METHOD APPARATUS/PROCEDURE:	SOURCE AND PURITY OF MATERIALS:
Equilibrium was approached from both higher and lower temperatures. Sulfite was determined by reaction with excess of iodine, and titration of the excess with thiosulfate.	Doubly distilled water was used. Analytical-reagent grade salt was used.
	ESTIMATED ERROR: Temperature: ±0.03 K Error in reading from graph: 0.2%
	REFERENCES:

COMPONENTS:	ORIGINAL MEASUREMENTS:
1. Sodium sulfite; Na_2SO_3; [7757-83-7] 2. Sodium sulfate; Na_2SO_4; [7757-82-6] 3. Water; H_2O; [7732-18-5]	Durymanova, M.A.; Telepneva, A.E. *Zh. Priklad. Khim.* 1972, 45, 1610-12; *J. Appl. Chem. USSR (Eng. Transl.)* 1972, 45, 1680-2.

VARIABLES:	PREPARED BY:
Concentrations of the components One temperature: 373 K	Mary R. Masson

EXPERIMENTAL VALUES: Composition of equilibrium solutions at 100°C

Na_2SO_3 mass %	Na_2SO_4 mass %	Na_2SO_3[a] mol/kg	Na_2SO_4[a] mol/kg	Solid[b] phase
20.8	-	2.084	0.	A
19.4	3.4	1.994	0.310	B
19.4	3.7	2.002	0.339	B
19.2	4.2	1.989	0.386	B
18.4	6.4	1.941	0.599	B
18.0	6.5	1.892	0.606	B
17.5	8.6	1.879	0.819	B
16.9	8.4	1.795	0.792	B
15.9	11.0	1.726	1.059	B
15.5	12.5	1.708	1.222	B
15.5	11.3	1.680	1.087	B
15.3	12.0	1.670	1.162	B
15.2	12.9	1.677	1.263	B
14.8	13.5	1.638	1.326	B
14.7	13.5	1.624	1.324	B
14.5	14.7	1.625	1.462	B + C
14.3	14.8	1.600	1.470	B + C
13.1	16.6	1.478	1.662	C
12.4	17.1	1.395	1.708	C
11.0	19.3	1.252	1.949	C
10.4	19.3	1.174	1.933	C
10.3	19.5	1.164	1.956	C
9.5	20.5	1.077	2.062	C
8.5	22.6	0.979	2.309	C

(continued on next page)

AUXILIARY INFORMATION

METHOD APPARATUS/PROCEDURE:	SOURCE AND PURITY OF MATERIALS:
The systems were studied under isothermal conditions in a thermostat filled with glycerol. Equilibrium was reached after 5 days. Sodium sulfite was determined iodometrically, and sodium sulfate by precipitation of the sulfate as barium sulfate and weighing. The identities of the solids were determined by the method of wet residues.	Sodium sulfate was of c.p. grade. Sodium sulfite was prepared by saturating aqueous c.p. sodium carbonate with the stoichiometric amount of sulfur dioxide. ESTIMATED ERROR: Temperature: ± 0.5 K Analyses: no estimate possible. REFERENCES:

COMPONENTS:	ORIGINAL MEASUREMENTS:
1. Sodium sulfite; Na_2SO_3; [7757-83-7]	Durymanova, M.A.; Telepneva, A.E.
2. Sodium sulfate; Na_2SO_4; [7757-82-6]	*Zh. Priklad. Khim.* 1972, *45*, 1610-12; *J. Appl. Chem. USSR (Eng. Transl.)* 1972, *45*, 1680-2.
3. Water; H_2O; [7732-18-5]	

EXPERIMENTAL VALUES (continued):

Na_2SO_3 mass %	Na_2SO_4 mass %	Na_2SO_3[a] mol/kg	Na_2SO_4[a] mol/kg	Solid[b] phase
7.4	23.2	0.846	2.354	C
6.3	24.8	0.725	2.534	C
6.1	25.4	0.707	2.611	D
5.4	25.9	0.624	2.654	D
4.0	27.1	0.461	2.769	D
2.0	28.2	0.227	2.844	D
1.8	29.8	0.209	3.067	D
0.0	29.9	0.000	3.003	E

[a] Molalities calculated by the compiler.

[b] Solid phase: A - Na_2SO_3, B - mixture of Na_2SO_3 with a small amount of Na_2SO_4, C - mixture of 25% Na_2SO_3 and 75% of Na_2SO_4, D - mixture of Na_2SO_4 with a small amount of Na_2SO_3, E - Na_2SO_4.

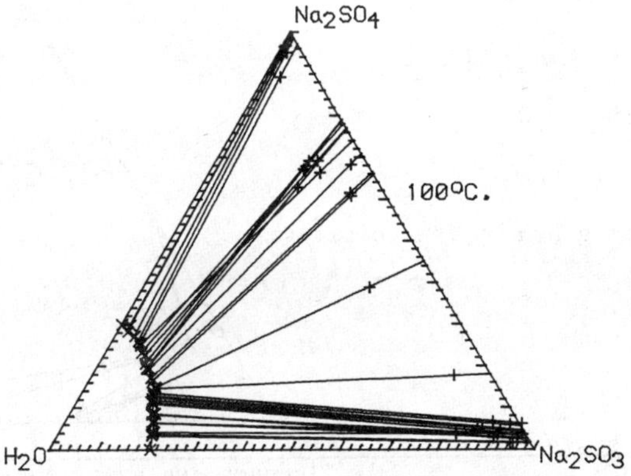

COMPONENTS:

1. Sodium sulfite; Na_2SO_3; [7757-83-7]
2. Sodium sulfate; Na_2SO_4; [7757-82-6]
3. Water; H_2O; [7732-18-5]

ORIGINAL MEASUREMENTS:

Kuznetsova, A.G.; Yaroshenko, L.B.

Zh. Priklad. Khim. 1981, *54*, 2197-2201; *J. Appl. Chem. USSR (Eng. Transl.)* 1981, *54*, 1929-32.

VARIABLES:

Concentrations of the components
One temperature: 298 K

PREPARED BY:

Mary R. Masson

EXPERIMENTAL VALUES:

Compositions of equilibrium solutions at 25°C

Na_2SO_4 mass %	Na_2SO_3 mass %	Na_2SO_3[a] mol/kg	Na_2SO_4[a] mol/kg	Solid[b] phase
0.0	22.4	0.0	2.29	A
2.5	21.7	0.232	2.271	A
8.1	18.3	0.775	1.973	A
12.3	17.9	1.241	2.035	A
14.52	16.07	1.473	1.837	A + B
15.8	14.1	1.587	1.596	B
18.3	8.0	1.748	0.861	B
22.03	0.0	1.989	0.0	B

[a] Molalities calculated by the compiler.

[b] Solid phase: A - $Na_2SO_3.7H_2O$,

B - $Na_2SO_4.10H_2O$

Note: data were read from graph by the compiler.

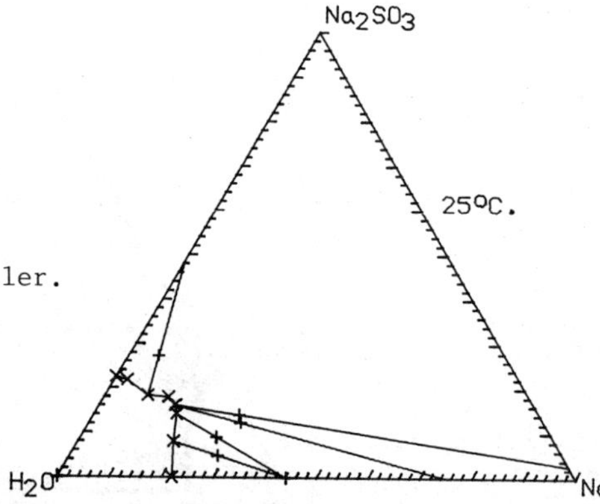

METHOD APPARATUS/PROCEDURE:

An isothermal procedure was used. The desired amounts of salts and water were placed in a vessel and stirred for 6 hr in a thermostat. Sodium sulfite in the solution was determined iodometrically, and the sum of sulfite and sulfate was determined by titration with barium chloride soln. (nitchromazo indicator) after oxidation with hydrogen peroxide.

The solid phases were dried in air after washing with ethyl acetone. The solid phase composition was determined by Schreinemakers' method and checked by chemical analysis.

SOURCE AND PURITY OF MATERIALS:

Sodium sulfate was "chemically pure" and sodium sulfite was special purity grade.

ESTIMATED ERROR:

Temperature: a thermostat was used, but the error was not stated.

Analyses: no estimate possible. There was an error of up to 0.5% in reading the graph.

REFERENCES:

COMPONENTS:	ORIGINAL MEASUREMENTS:
1. Sodium sulfite; Na_2SO_3; [7757-83-7] 2. Sodium sulfate; Na_2SO_4; [7757-82-6] 3. Water; H_2O; [7732-18-5]	Lewis, N.B.; Rivett, A.C.D. *J. Chem. Soc.* 1924, *125*, 1156-67.
VARIABLES: Concentrations of the components Temperature: 273 - 341 K	PREPARED BY: Mary R. Masson

EXPERIMENTAL VALUES:

Systems in equilibrium with hydrated solids

t/°C	Na_2SO_3 mass %	Na_2SO_4 mass %	S_E[a] mass % (mol/kg)[b]
0.10	11.59	1.11	
	11.34	1.76	
	11.37	1.77	
	11.37	1.79	12.3 (1.113)
17.5	18.04	3.79	
	17.09	5.97	
	16.45	7.59	
	16.46	7.66	
	16.48	7.64	20.0 (1.983)
21.0	20.64	1.84	
	19.38	4.67	
	18.09	7.62	21.7 (2.199)
25.0	21.61	4.14	
	19.44	8.97	
	18.27	11.05	
	16.81	14.54	
	16.77	14.50	23.7 (2.464)
27.5	23.33	3.45	
	21.87	6.52	
	20.21	9.99	25.0 (2.645)
31.5	25.64	2.93	
	24.09	6.08	
	22.53	9.23	
	21.00	12.22	27.15 (2.957)
25.0	27.69	2.83	
	27.71	2.74	
	26.41	5.16	
	24.94	7.96	
	23.39	11.30	
	21.83	14.11	29.2 (3.272)

(continued on next page)

AUXILIARY INFORMATION

METHOD APPARATUS/PROCEDURE:

A mixture was prepared of water, excess of $Na_2SO_3.7H_2O$, and varying amounts of sodium sulfate, added either as anhydrous or hydrated form. This was placed in a sealed or well stoppered tube, then sometimes preheated, sometimes not, and placed in a thermostat at the desired temperature, where it was rotated for 5-100 hr. Two samples of the solution were withdrawn for analysis through a cotton wool filter. One was run into excess of iodine, which was back-titrated with thiosulfate to obtain the sulfite concentration. The other was treated with hydrogen peroxide to oxidize sulfite to sulfate, then heated to dryness and weighed to find the total sulfate. Both measured samples were weighed to find the solution density.

SOURCE AND PURITY OF MATERIALS:

Various anhydrous and hydrated salts were used. Consistent results were obtained when the sulfite was BDH commercial anhydrous sodium sulfite, which contains some sodium sulfate as impurity.

ESTIMATED ERROR:

Temperature: ±0.02°, except at 0.1°C, where it was ±0.1°C.
Analyses: about ±0.5%.

REFERENCES:

COMPONENTS:	ORIGINAL MEASUREMETS:
1. Sodium sulfite; Na_2SO_3; [7757-83-7]	Lewis, N.B.; Rivett, A.C.D.
2. Sodium sulfate; Na_2SO_4; [7757-82-6]	*J. Chem. Soc.* 1924, *125*, 1156-67.
3. Water; H_2O; [7732-18-5]	

EXPERIMENTAL VALUES (continued):

Systems in equilibrium with hydrated solids

t/°C	Na_2SO_3 mass %	Na_2SO_4 mass %	S_E[a] mass % (mol/kg)[b]
33.0	25.29	2.30	
	25.33	2.44	
	22.75	6.63	
	21.36	8.87	
	20.10	10.95	26.8 (2.905)
37.5	24.73	2.51	
	24.81	2.54	
	23.54	4.63	
	22.38	6.52	
	21.14	8.59	
	19.94	10.70	26.3 (2.831)
42.5	24.32	2.17	
	24.42	2.16	
	22.88	4.72	
	21.78	6.78	
	20.42	9.06	
	20.42	9.03	
	19.11	11.15	
	19.23	11.07	25.7 (2.744)
47.5	23.89	2.46	
	24.03	2.40	
	22.56	4.77	
	21.22	7.14	25.25 (2.680)

t/°C	Na_2SO_3 mass %	Na_2SO_4 mass %	S_E[a] mass % (mol/kg)[b]
52.5	23.31	2.18	
	23.31	2.24	
	22.54	3.72	
	21.29	5.80	
	20.13	7.73	
	19.26	9.20	24.7 (2.603)
57.5	22.93	2.03	
	22.58	2.69	
	21.50	4.56	
	18.93	9.04	24.1 (2.519)
62.5	22.02	2.95	
	20.82	4.91	
	20.72	5.20	
	19.52	7.52	
	19.47	7.44	
	18.28	9.48	
	17.94	10.16	23.6 (2.451)
67.5	22.23	1.87	
	21.01	4.16	
	19.93	6.27	
	18.85	8.35	23.2 (2.397)

[a] Values obtained for mass % of Na_2SO_3 were plotted against the values of mass % of Na_2SO_4. A series of straight-line plots was obtained. For each temperature, the line was extrapolated to 0.0 mass % of Na_2SO_4. The corresponding solubility of Na_2SO_3, expressed as mass %, is given here as S_E.

The values were found to be in good agreement with solubilities measured in the normal manner.

[b] Molalities calculated by the compiler.

Note: no data on solid phases is given.

COMPONENTS:

1. Sodium sulfite; Na_2SO_3; [7757-83-7]
2. Sodium sulfate; Na_2SO_4; [7757-82-6]
3. Water; H_2O; [7732-18-5]

ORIGINAL MEASUREMENTS:

Lewis, N.B.; Rivett, A.C.D.

J. Chem. Soc. 1924, *125*, 1162-7.

VARIABLES:

Three temperatures: 313, 318, 333 K
Concentrations of the components

PREPARED BY:

Mary R. Masson

EXPERIMENTAL VALUES:

Composition of equilibrium solutions

Na_2SO_3 mass %	Na_2SO_4 mass %	Na_2SO_3[a] mol/kg	Na_2SO_4[a] mol/kg	Solid[b] phase
Temperature = 60°C				
1.01	30.26	0.117	3.100	A
3.90	27.84	0.453	2.871	A
4.24	27.48	0.493	2.833	A
9.76	22.69	1.146	2.365	A
10.92	22.02	1.292	2.312	A + B
13.60	18.59	1.591	1.930	B
15.56	16.59	1.819	1.721	B
17.78	13.66	2.058	1.403	B
19.37	12.52	2.256	1.294	B
20.27	10.62	2.327	1.082	C
20.79	10.35	2.395	1.058	C
20.87	10.13	2.400	1.034	C
21.86	9.56	2.529	0.981	C
24.17	8.05	2.829	0.836	D
25.09	5.70	2.876	0.580	D
25.03	4.94	2.836	0.497	D
20.99	9.25	2.387	0.934	E
21.37	7.20	2.374	0.710	E
21.60	5.26	2.343	0.506	E
22.47	3.96	2.423	0.379	E
22.82	2.83	2.435	0.268	E

(continued on next page)

AUXILIARY INFORMATION

METHOD APPARATUS/PROCEDURE:

For satisfactory results in this system it was necessary to form the solids from homogeneous solutions. In practice, this meant that water had to be evaporated from unsaturated solutions by gentle boiling under reduced pressure. The solutions were contained in stoppered tubes with side-arms for the attachment of pressure tubing. The tubes were rocked in a thermostat. Only a small amount of solid was allowed to form, because its composition would be expected to change continuously.

Sulfite was determined by reaction with excess of iodine and titration of the excess with thiosulfate. Total sulfate was determined after oxidation with hydrogen peroxide by weighing as sodium sulfate.

SOURCE AND PURITY OF MATERIALS:

ESTIMATED ERROR:

Temperature: ±0.2°C
Analyses: about ±0.5%

REFERENCES:

COMPONENTS:	ORIGINAL MEASUREMENTS:
1. Sodium sulfite; Na_2SO_3; [7757-83-7]	Lewis, N.B.; Rivett, A.C.D.
2. Sodium sulfate; Na_2SO_4; [7757-82-6]	*J. Chem. Soc.* 1924, *125*, 1162-7.
3. Water; H_2O; [7732-18-5]	

EXPERIMENTAL VALUES (continued):

Na_2SO_3 mass %	Na_2SO_4 mass %	Na_2SO_3[a] mol/kg	Na_2SO_4[a] mol/kg	Solid[b] phase
Temperature = 45°C				
2.01	30.16	0.235	3.130	A
7.56	25.58	0.897	2.694	A
7.51	25.57	0.890	2.690	A
9.62	24.05	1.151	2.553	A
12.93	20.66	1.545	2.190	B
15.05	18.68	1.802	1.984	B
17.47	15.81	2.077	1.668	B
19.85	13.87	2.376	1.473	B
22.36	10.60	2.646	1.113	U
20.52	12.01	2.413	1.253	C
22.33	9.92	2.615	1.031	C
23.81	8.87	2.806	0.928	C
24.72	7.84	2.908	0.818	C
25.52	7.61	3.028	0.801	C
24.10	8.59	2.841	0.898	C + D
24.75	7.28	2.889	0.754	D
26.35	3.86	2.996	0.389	D
28.06	2.90	3.225	0.296	D
28.19	2.29	3.217	0.232	D
26.36	2.56	2.942	0.254	U
21.33	10.43	2.480	1.076	E
22.71	7.78	2.592	0.788	E
23.48	7.11	2.684	0.721	E
23.91	5.88	2.702	0.590	E
23.82	5.96	2.691	0.598	E
24.61	3.79	2.727	0.373	E
24.64	3.85	2.734	0.379	E
25.16	3.11	2.783	0.305	E
25.89	2.29	2.860	0.224	E
26.53	1.79	2.937	0.176	E
Temperature = 40°C				
1.03	31.6	0.121	3.302	A
3.73	29.3	0.442	3.080	A
6.43	26.8	0.764	2.826	A
8.99	24.5	1.072	2.593	A
12.83	21.5	1.550	2.305	A
13.01	21.2	1.569	2.269	A + B
12.61	21.6	1.521	2.311	B metastable
14.54	20.03	1.763	2.155	B
16.76	17.30	2.017	1.847	B
17.85	16.09	2.144	1.715	B
19.53	14.23	2.339	1.512	B
20.74	12.71	2.473	1.345	B
24.49	9.92	2.962	1.065	B
25.38	8.41	3.041	0.894	B
18.07	16.78	2.201	1.813	C
20.01	14.30	2.417	1.533	C
22.33	11.49	2.677	1.222	C
22.63	11.34	2.719	1.209	C
22.76	10.80	2.718	1.144	C
25.06	8.25	2.981	0.871	C
27.46	7.09	3.329	0.763	?
27.60	7.02	3.349	0.756	?
28.63	6.75	3.515	0.735	?
25.58	7.73	3.043	0.816	D
26.08	5.88	3.041	0.608	D

(continued on next page)

COMPONENTS:	ORIGINAL MEASUREMENTS:
1. Sodium sulfite; Na_2SO_3; [7757-83-7]	Lewis, N.B.; Rivett, A.C.D.
2. Sodium sulfate; Na_2SO_4; [7757-82-6]	*J. Chem. Soc.* 1924, *125*, 1162-7.
3. Water; H_2O; [7732-18-5]	

EXPERIMENTAL VALUES (continued):

Na_2SO_3 mass %	Na_2SO_4 mass %	Na_2SO_3[a] mol/kg	Na_2SO_4[a] mol/kg	Solid[b] phase
Temperature = 40°C				
27.52	4.79	3.226	0.498	D
23.62	8.07	2.743	0.832	E
23.80	6.90	2.725	0.701	E
24.89	5.24	2.826	0.528	E
26.47	2.67	2.964	0.265	E
27.61	0.87	3.063	0.086	E

a Molalities calculated by the compiler.

b Solids A, B, C, D and E are series of mixed crystals corresponding to the five solution curves noted by these authors. U signifies an unstable solid. Several metastable systems were observed.

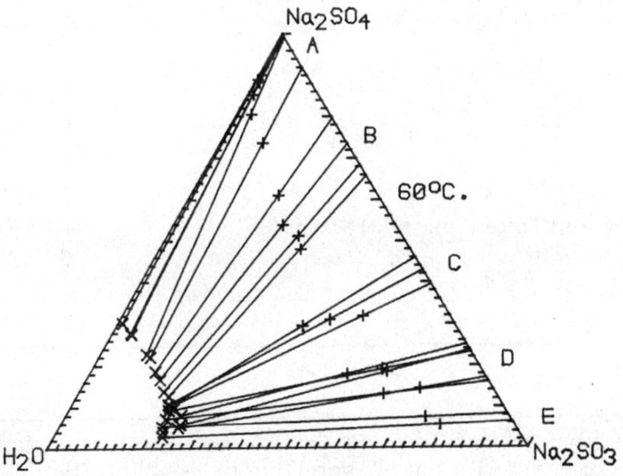

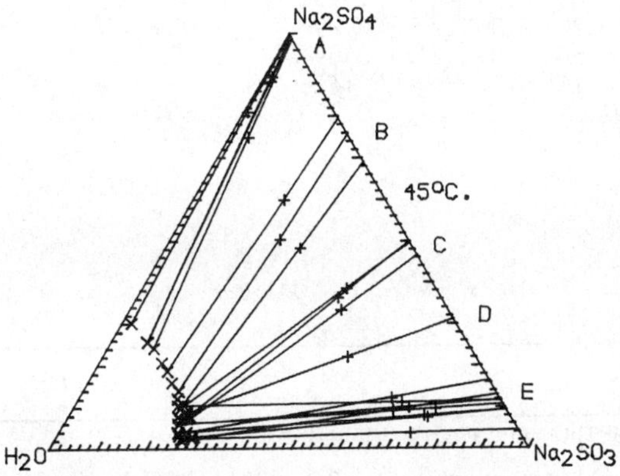

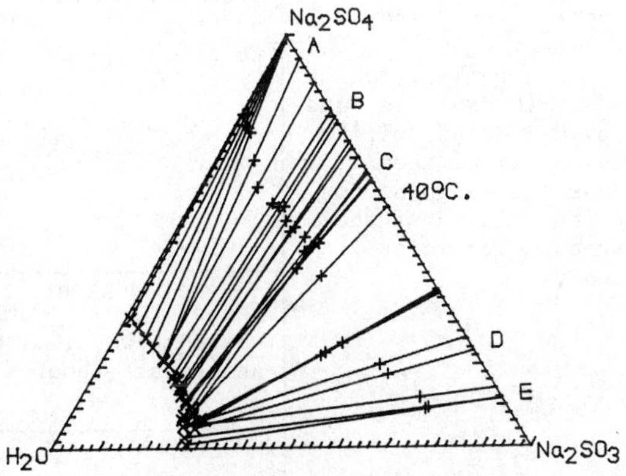

COMPONENTS:

1. Sodium sulfite; Na_2SO_3; [7757-83-7]
2. Sodium sulfate; Na_2SO_4; [7757-82-6]
3. Water; H_2O; [7732-18-5]

ORIGINAL MEASUREMENTS:

Rivett, A.C.D.; Lewis, N.B.

Rec. Trav. Chem. Pays-Bas 1923, *42*, 954-63.

VARIABLES:

Concentrations of the components
Four temperatures: 273 311 K

PREPARED BY:

Mary R. Masson

EXPERIMENTAL VALUES:

Stable system at 25°C

Na_2SO_3 mass %	Na_2SO_4 mass %	Na_2SO_3[a] mol/kg	Na_2SO_4[a] mol/kg
21.61	4.14	2.309	0.393
19.44	8.97	2.154	0.882
18.27	11.05	2.051	1.101
16.81	14.49	1.941	1.485
16.77	14.50	1.936	1.485
16.81	14.54	1.943	1.491
16.81	14.54	1.943	1.491
16.80	14.55	1.942	1.492
16.85	14.52	1.948	1.490
14.61	15.25	1.653	1.531
11.39	16.43	1.252	1.603
9.22	17.31	0.996	1.659
6.59	18.11	0.694	1.693
4.11	19.66	0.428	1.816
2.53	20.4	0.260	1.864

Metastable system at 25°C

Na_2SO_3 mass %	Na_2SO_4 mass %	Na_2SO_3[a] mol/kg	Na_2SO_4[a] mol/kg
27.29	1.69	3.049	0.168
26.57	3.99	3.036	0.405
25.45	6.75	3.010	0.703
26.32	7.47	3.154	0.794
23.40	10.93	2.827	1.172
21.88	12.34	2.639	1.321
20.35	13.82	2.453	1.478
18.52	16.18	2.250	1.744
18.34	16.32	2.227	1.758
16.94	17.82	2.060	1.923
15.43	19.12	1.870	2.057
14.39	20.51	1.754	2.218

(continued on next page)

AUXILIARY INFORMATION

METHOD APPARATUS/PROCEDURE:

Mixtures of solids were equilibrated with water in closed tubes maintained at constant temp. Two samples of the solution phase were drawn off through cotton wool, and analysed for sulfite, by reaction with excess of standard iodine and back-titration of the excess with standard thiosulfate, and for sulfite + sulfate, by oxidizing sulfite to sulfate with hydrogen peroxide, then evaporating all to dryness and weighing to obtain total sulfate. The density of the solution was also measured. The moist solids were similarly analysed.

SOURCE AND PURITY OF MATERIALS:

Commercial "pure" samples of $Na_2SO_3.7H_2O$ and $Na_2SO_4.10H_2O$ were used. The mixtures prepared were heated to about 80°C; the undissolved hydrated crystals broke down to a fine white anhydrous powder.

ESTIMATED ERROR:

Temperature: ±0.02°C, except at 0.1°C, where it was ±0.1°C
Analyses: about ±0.5%

REFERENCES:

COMPONENTS:	ORIGINAL MEASUREMENTS:
1. Sodium sulfite; Na_2SO_3; [7757-83-7]	Rivett, A.C.D.; Lewis, N.B.
2. Sodium sulfate; Na_2SO_4; [7757-82-6]	*Rec. Trav. Chem. Pays-Bas* 1923, *42*, 954-63.
3. Water; H_2O; [7732-18-5]	

EXPERIMENTAL VALUES: (continued):

Stable system at 37.5°C

Na_2SO_3 mass %	Na_2SO_4 mass %	Na_2SO_3[a] mol/kg	Na_2SO_4[a] mol/kg
27.51	1.80	3.088	0.179
26.15	3.35	2.943	0.335
25.29	4.13	2.843	0.412
24.81	5.28	2.816	0.532
26.32	7.26	3.144	0.770
25.78	7.14	3.049	0.749
25.06	7.66	2.955	0.802
25.40	7.76	3.015	0.817
23.23	10.25	2.771	1.085
23.06	9.23	2.702	0.960
21.52	11.61	2.553	1.222
21.05	12.05	2.496	1.268
18.13	15.24	2.159	1.610
17.72	15.43	2.103	1.625
16.47	17.26	1.972	1.834
12.87	20.60	1.535	2.180
8.90	25.16	1.071	2.686
14.47	20.10	1.755	2.163
11.30	22.63	1.357	2.411
9.04	24.66	1.082	2.619
6.13	27.34	0.731	2.893
4.78	28.36	0.567	2.986
2.57	30.07	0.303	3.143

Metastable system at 17.5°C

Na_2SO_3 mass %	Na_2SO_4 mass %	Na_2SO_3[a] mol/kg	Na_2SO_4[a] mol/kg
18.87	1.82	1.888	0.162
18.04	3.79	1.831	0.341
17.09	5.97	1.762	0.546
16.45	7.59	1.718	0.703
16.46	7.66	1.721	0.711
16.48	7.64	1.723	0.709
14.78	7.93	1.517	0.722
11.56	8.54	1.148	0.752
8.48	9.89	0.824	0.853
5.99	10.61	0.570	0.896
2.61	12.15	0.243	1.004

System at 0.1°C

Na_2SO_3 mass %	Na_2SO_4 mass %	Na_2SO_3[a] mol/kg	Na_2SO_4[a] mol/kg
11.59	1.11	1.053	0.090
11.34	1.76	1.035	0.143
11.37	1.77	1.039	0.143
11.37	1.79	1.039	0.145
8.56	1.95	0.759	0.153
5.93	2.42	0.513	0.186
3.07	3.11	0.260	0.233

[a] Molalities calculated by the compiler.

The extrapolated values for the solubility of sodium sulfite in water are:
0.1°C - 12.0 mass % 17.5°C - 19.8 mass % 25°C - 23.75 mass %

The authors conclude as follows:
At 0.1, 17.5 and 25°C the stable systems show two series of mixed crystals (with corresponding solutions curves), one being between the heptahydrates and the other being between the decahydrates of sodium sulfite and sodium sulfate.

At 25°C there is also a metastable system in which there is extensive mixed crystal formation between the anhydrous salts.

At 37.5°C there are three series of mixed crystals formed between the anhydrous salts.

(continued on next page)

COMPONENTS:	ORIGINAL MEASUREMENTS:
1. Sodium sulfite; Na_2SO_3; [7757-83-7] 2. Sodium sulfate; Na_2SO_4; [7757-82-6] 3. Water; H_2O; [7732-18-5]	Rivett, A.C.D.; Lewis, N.B. *Rec. Trav. Chem. Pays-Bas* <u>1923</u>, 42, 954-63.

EXPERIMENTAL VALUES (continued):

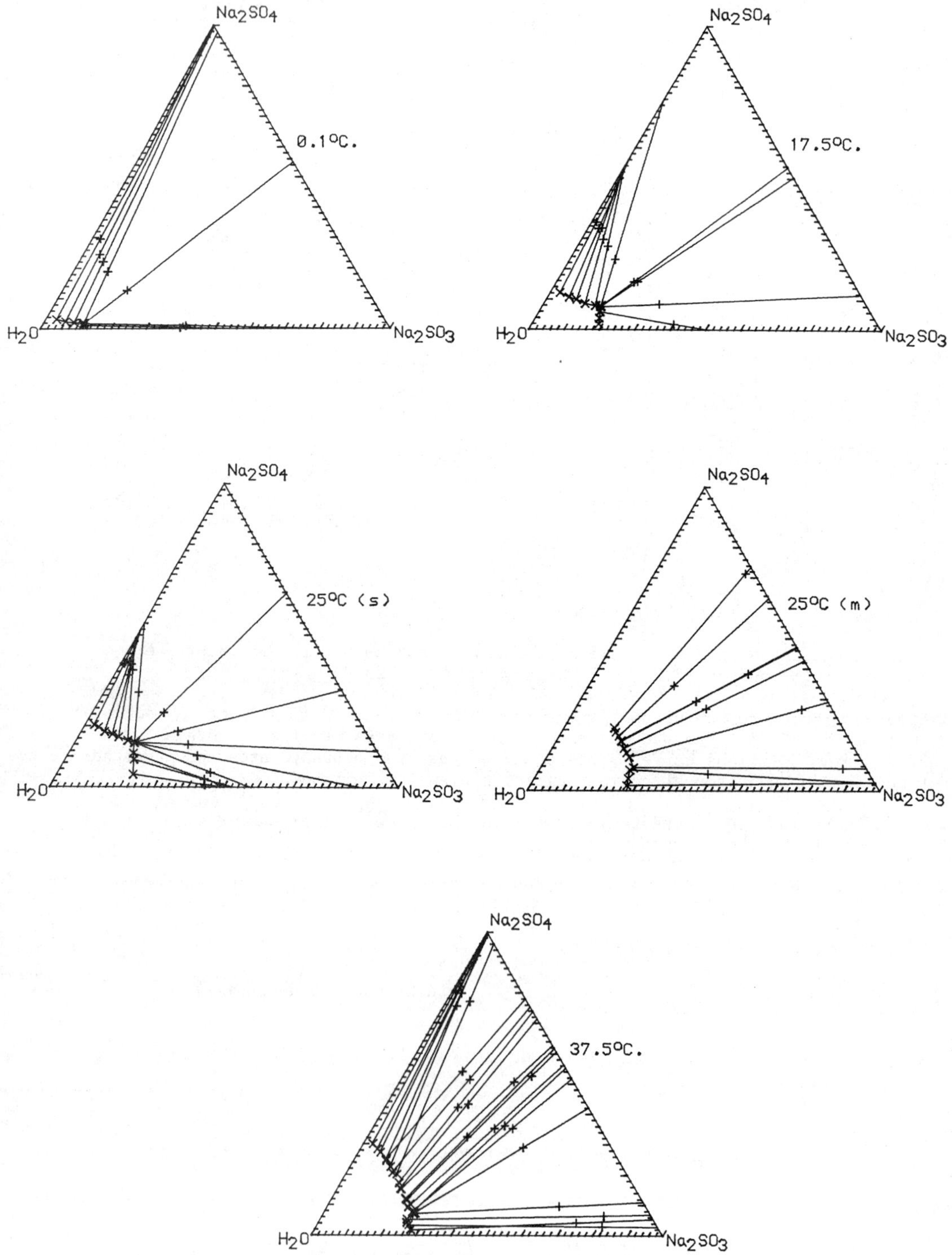

COMPONENTS:

1. Sodium sulfite; Na_2SO_3; [7757-83-5]
2. Sodium sulfate; Na_2SO_4; [7757-82-6]
3. Water; H_2O; [7732-18-5]

ORIGINAL MEASUREMENTS:

Palkina, N.A.

Tr. Vornezh. Gos. Univ. 1950, *17*, 61-88.

VARIABLES:

Six temperatures: 273 - 288 K
Concentrations of the components

PREPARED BY:

Mary R. Masson

EXPERIMENTAL VALUES:

Composition of equilibrium solutions

Na_2SO_4 mass %	Na_2SO_3 mass %	Na_2SO_4[a] mol/kg	Na_2SO_3[a] mol/kg	Solid[b] phase
Temperature = 0°C				
4.22	0.0	0.310	0.	
2.6	4.87	0.198	0.418	A
2.35	6.0	0.181	0.519	A
1.8	9.83	0.143	0.883	A
1.7	11.3	0.138	1.031	*
0.0	11.9	0.	1.072	
Temperature = 5°C				
5.9	0.0	0.441	0.	
4.89	2.1	0.370	0.179	A
4.0	4.75	0.309	0.413	A
2.9	9.71	0.234	0.882	A
2.7	12.4	0.224	1.159	*
2.19	12.5	0.181	1.163	B
0.0	13.8	0.	1.270	
Temperature = 10°C				
8.25	0.0	0.633	0.	
7.35	2.0	0.571	0.175	A
6.2	4.69	0.490	0.418	A
5.0	9.5	0.412	0.882	A
4.4	13.0	0.375	1.249	A
4.3	14.15	0.371	1.377	*
3.5	14.47	0.300	1.400	B
2.12	15.0	0.180	1.436	B
0.0	15.8	0.	1.489	

(continued on next page)

AUXILIARY INFORMATION

METHOD/APPARATUS/PROCEDURE:

A polythermal procedure was used, based on the following systems:

I (5% Na_2SO_3 + 95% water) + Na_2SO_4
II (10% Na_2SO_3 + 90% water) + Na_2SO_4
III (15% Na_2SO_3 + 85% water) + Na_2SO_4
IV (2.5% Na_2SO_4 + 97.5% water) + Na_2SO_3
V (5%Na_2SO_4 + 95% water) + Na_2SO_3
VI (7.5% Na_2SO_4 + 92.5% water)+ Na_2SO_3

SOURCE AND PURITY OF MATERIALS:

ESTIMATED ERROR:

No estimates possible.

REFERENCES:

COMPONENTS:	ORIGINAL MEASUREMENTS:
1. Sodium sulfite; Na_2SO_3; [7757-83-5]	Palkina, N.A.
2. Sodium sulfate; Na_2SO_4; [7757-82-6]	*Tr. Vornezh. Gos. Univ.* 1950, *17*, 61-88.
3. Water; H_2O; [7732-18-5]	

EXPERIMENTAL VALUES (continued):

Composition of equilibrium solutions

Na_2SO_4 mass %	Na_2SO_3 mass %	Na_2SO_4[a] mol/kg	Na_2SO_3[a] mol/kg	Solid[b] phase
Temperature = 15°C				
11.6	0.0	0.924	0.	
9.2	4.54	0.751	0.418	A
7.6	9.24	0.643	0.882	A
6.52	13.1	0.571	1.293	A
6.5	14.12	0.576	1.411	A
6.2	15.5	0.557	1.571	*
4.17	16.5	0.370	1.650	B
2.06	17.5	0.180	1.726	B
0.0	18.2	0.	1.765	
Temperature = 20°C				
16.2	0.0	1.361	0.	
13.6	4.32	1.167	0.418	A
12.0	8.8	1.067	0.882	A
10.5	13.42	0.972	1.400	A
9.4	16.8	0.897	1.806	*
6.13	18.3	0.571	1.921	B
4.03	19.1	0.369	1.971	B
2.0	19.9	0.180	2.022	B
0.0	20.8	0.	2.084	
Temperature = 25°C				
21.9	0.0	1.974	0.	
19.6	4.02	1.807	0.418	A
17.8	8.22	1.694	0.882	A
16.0	12.6	1.578	1.400	A
14.3	17.0	1.465	1.963	*
5.87	20.8	0.564	2.250	A
3.96	21.7	0.375	2.316	A
1.94	22.5	0.181	2.363	A
0.0	23.5	0.	2.437	

[a] Molalties calculated by the compiler.

[b] Solid phases: A - "Solid solution of $Na_2SO_4 \cdot 10H_2O$ and $Na_2SO_3 \cdot 10H_2O$",
B - "Solid solution of $Na_2SO_4 \cdot 7H_2O$ and $Na_2SO_3 \cdot 7H_2O$".
* - A + B, liquid composition determined graphically.

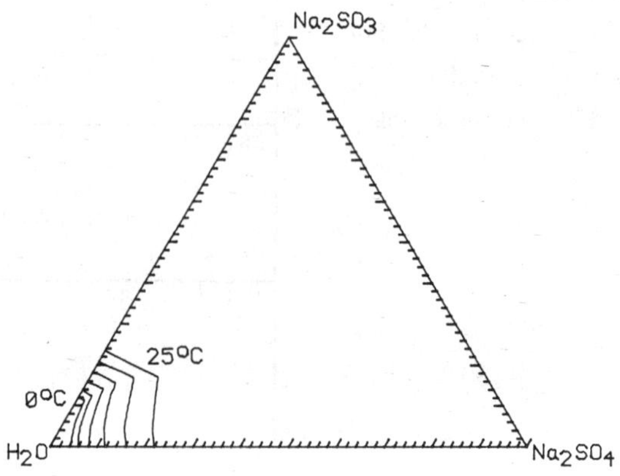

COMPONENTS:

1. Sodium sulfite; Na_2SO_3; [7757-83-7]
2. Sodium sulfate; Na_2SO_4; [7757-82-6]
3. Water; H_2O; [7732-18-5]

ORIGINAL MEASUREMENTS:

Sotova, N.N.; Kuznetsova, A.G.; Torocheshnikov, N.S.
Zh. Priklad. Khim. 1978, *51*, 779-84;**Russ. J. Appl. Chem. (Eng. Transl.)* 1978, *51*, 760-4; also Sotova, N.N.; Kuznetsova, A.G.; Torocheshnikov, N.S.; Kononova, I.V.; **Mezhvuz. Sb. Altaisk. Politekhn. In-t.* 1976, 2(57), 150-5; also**Fiz.-Khim. Osn. Tekhnol. Pererab. Khim. Syr'ya* 1976, 2, 150-5.

VARIABLES:

One temperature: 273 K
Concentrations of the components.

PREPARED BY:

Mary R. Masson

EXPERIMENTAL VALUES:

Composition of equilibrium solutions at 0°C

Na_2SO_4 mass %	Na_2SO_3 mass %	Na_2SO_4[a] mol/kg	Na_2SO_3[a] mol/kg	Solid[b] phase
4.50	0.0	0.332	0.	A
3.00	3.62	0.226	0.308	A
2.98	5.56	0.229	0.482	A
3.20	6.70	0.250	0.590	A
2.19	7.56	0.171	0.665	A
1.85	11.06	0.150	1.008	A
1.60	12.03	0.130	1.105	A + B
1.60	11.85	0.130	1.086	A + B
0.64	12.34	0.052	1.125	B
0.11	12.29	0.009	1.113	B
0.0	12.30	0.	1.113	B

[a] Molality calculated by the compiler.

[b] Solid phases: A - $Na_2SO_4.10H_2O$, B - $Na_2SO_3.7H_2O$

AUXILIARY INFORMATION

METHOD APPARATUS/PROCEDURE:

The isothermal method was used. p-Phenylenediamine was used as anti-oxidant.

Sulfite was determined iodometrically, and sodium sulfate was weighed to obtain total sulfate after a peroxide oxidation.

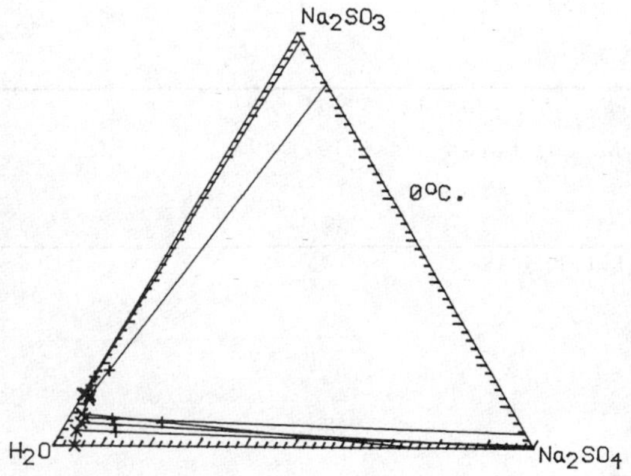

SOURCE AND PURITY OF MATERIALS:

Sodium sulfate was of analytical grade, and sodium sulfite was of high-purity grade.

ESTIMATED ERROR:

Temperature: ±0.2 K
Analyses: no estimate given.

REFERENCES:

COMPONENTS:

1. Sodium sulfite; Na_2SO_3; [7757-83-7]
2. Sodium sulfate; Na_2SO_4; [7757-82-6]
3. Water; H_2O; [7732-18-5]

ORIGINAL MEASUREMENTS:

Sotova, N.N.; Torocheshnikov, N.S.; Kuznetsova, A.G.; Sokolova, E.I.

Khimiya i Tekhnol. Mineral'n. Solei i Galurgichesk. Pr.-v, Varnaul 1978, 59-65.

VARIABLES:

One temperature: 373 K
Concentrations of the components.

PREPARED BY:

Mary R. Masson

EXPERIMENTAL VALUES:

Composition of equilibrium solutions at 100°C

Na_2SO_4 mass %	Na_2SO_3 mass %	Na_2SO_4[a] mol/kg	Na_2SO_3[a] mol/kg	Solid[b] phase
-	21.2	0.	2.135	A
3.2	19.9	0.293	2.053	A
2.79	19.16	0.252	1.948	A + D
13.5	14.8	1.326	1.638	D
15.2	13.7	1.505	1.529	D
18.4	11.5	1.848	1.302	D
19.2	11.2	1.942	1.277	D
22.0	8.8	2.238	1.009	D
25.4	5.6	2.592	0.644	D
27.45	4.55	2.842	0.531	D
27.5	4.9	2.864	0.575	C
29.6	-	2.960	0.	C

[a] Molalities calculated by the compiler.

[b] Solid phases: A - Na_2SO_3, C - Na_2SO_4, D - $\underline{n}Na_2SO_3.\underline{m}Na_2SO_4$

AUXILIARY INFORMATION

METHOD APPARATUS/PROCEDURE:

An isothermal method was used.

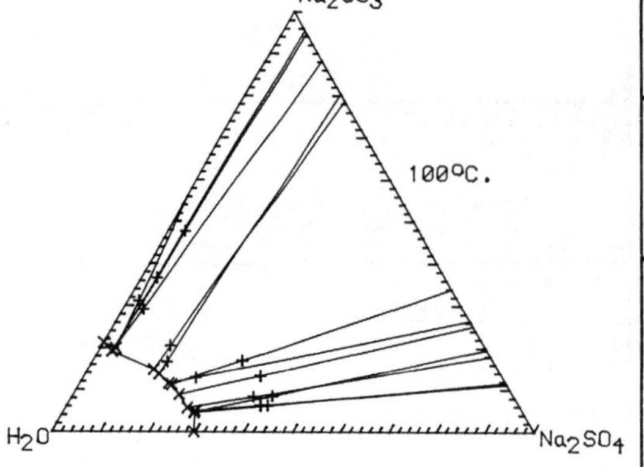

SOURCE AND PURITY OF MATERIALS:

ESTIMATED ERROR:

No estimates possible.

REFERENCES:

COMPONENTS:	ORIGINAL MEASUREMENTS:
1. Sodium sulfite; Na_2SO_3; [7757-83-7] 2. Sodium sulfate; Na_2SO_4; [7757-82-6] 3. Water; H_2O; [7732-18-5]	Sotova, N.N.; Toroheshnikov, N.S.; Kuznetsova, A.G.; Poroshkova, M.A. *Khimiya i Tekhnol. Mineral'n. Solei i Galurgichesk. Pr.-v, Varnaul* 1978, 59-65.
VARIABLES:	PREPARED BY:
One temperature: 333 K Concentrations of the components.	Mary R. Masson

EXPERIMENTAL VALUES:

Composition of equilibrium solutions at 60°C

Na_2SO_4 mass %	Na_2SO_3 mass %	Na_2SO_4[a] mol/kg	Na_2SO_3[a] mol/kg	Solid[b] phase
31.5	0.0	3.237	0.	C
30.0	2.2	3.115	0.257	C
27.7	3.9	2.851	0.452	C
26.0	7.2	2.740	0.855	C
25.6	8.7	2.743	1.051	D
24.2	9.2	2.558	1.096	D
24.0	10.0	2.568	1.230	D
18.0	13.6	1.853	1.578	D
9.4	18.5	0.918	2.036	D
7.5	21.2	0.741	2.359	D
3.9	22.3	0.372	2.397	A
3.4	22.4	0.323	2.395	A
0.0	23.3	0.	2.410	A

a Molalities calculated by the compiler.

b Solid phases: A - Na_2SO_3, C - Na_2SO_4, D - solid solution of sodium sulfite and sodium sulfate

AUXILIARY INFORMATION

METHOD APPARATUS/PROCEDURE:

The isothermal method was used.

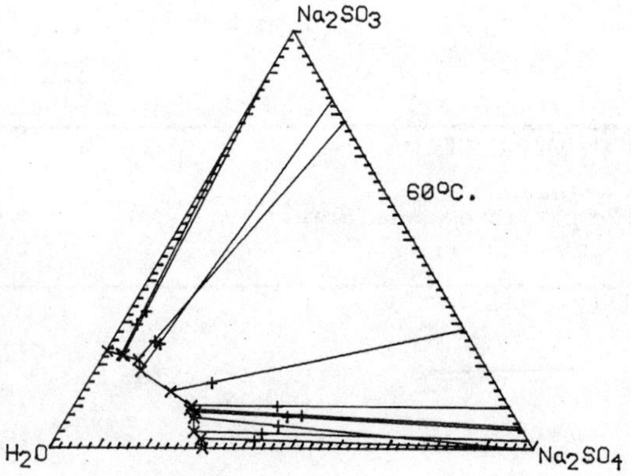

SOURCE AND PURITY OF MATERIALS:

ESTIMATED ERROR:

Temperature: ±0.1 K
Analyses: no estimates possible.

REFERENCES:

COMPONENTS:

1. Sodium sulfite; Na_2SO_3; [7757-83-7]

2. Sodium sulfate; Na_2SO_4; [7757-82-6]

3. Water; H_2O; [7732-18-5]

ORIGINAL MEASUREMENTS:

Wöhler, L.; Dierksen, J.

Z. Angew. Chem. 1926, *39*, 33-36.

VARIABLES:

Temperature: 296 - 353 K
Concentrations of the components.

PREPARED BY:

Mary R. Masson

EXPERIMENTAL VALUES:

Composition of saturated solutions

t/°C	Na_2SO_4 mass %	Na_2SO_3 mass %	Na_2SO_4[a] mol/kg	Na_2SO_3[a] mol/kg
23	9.55	15.25	0.894	1.609
40	18.48	16.26	1.994	1.977
60	14.21	16.59	1.446	1.902
80	14.95	15.15	1.506	1.720

[a] Molalities calculated by the compiler.

AUXILIARY INFORMATION

METHOD APPARATUS/PROCEDURE:

Sulfite was determined iodometrically. Sulfate was determined by difference after oxidation to sodium sulfate with hydrogen peroxide.

SOURCE AND PURITY OF MATERIALS:

ESTIMATED ERROR:

No estimates possible.

REFERENCES:

COMPONENTS:

1. Sodium sulfite; Na_2SO_3; [7757-83-7]
2. Sodium sulfate; Na_2SO_4; [7757-82-6]
3. Water; H_2O; [7732-18-5]

ORIGINAL MEASUREMENTS:

Kuznetsova, A.G.; Sedova, V.A.

VINITI Deposited Document 1981, 5710-81.

VARIABLES:

Temperature: 288 K
Composition

PREPARED BY:

Mary R. Masson

EXPERIMENTAL VALUES:

Composition of saturated solutions

Na_2SO_3 mass %	Na_2SO_4 mass %	Na_2SO_3 [a] mol/kg	Na_2SO_4 [a] mol/kg	Solid [b] phase
0.50	13.50	0.046	1.105	A
2.68	13.42	0.253	1.126	A
3.32	13.11	0.315	1.104	A
5.65	12.67	0.549	1.092	A
11.62	12.07	1.208	1.114	A + B
12.79	10.13	1.316	0.925	B
12.83	8.00	1.286	0.711	B
14.33	5.07	1.411	0.443	B
15.27	5.18	1.523	0.458	B
15.23	2.18	1.463	0.186	B

[a] Molalities calculated by the compiler.

[b] Solid phases: A - $Na_2SO_4 \cdot 10H_2O$, B - solid solution

AUXILIARY INFORMATION

METHOD APPARATUS/PROCEDURE:

Isothermal method.

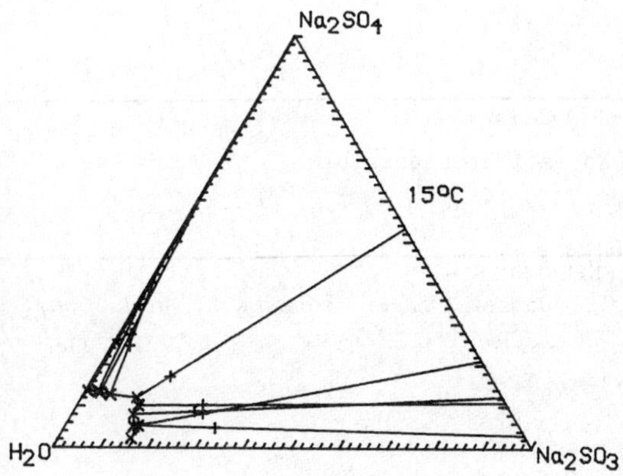

SOURCE AND PURITY OF MATERIALS:

ESTIMATED ERROR:

No estimates possible.

REFERENCES:

COMPONENTS:

1. Sodium sulfite; Na_2SO_3; [7757-83-7]
2. Sodium chloride; NaCl; [7647-14-5]
3. Water; H_2O; [7732-18-5]

ORIGINAL MEASUREMENTS:

Kobe, K.A.; Hellwig, K.C.

Ind. Eng. Chem. 1955, *47*, 1116-21.

VARIABLES:

Concentrations of the components
Temperature: 273 - 373 K

PREPARED BY:

Mary R. Masson

EXPERIMENTAL VALUES:

Composition of saturated solutions

t/°C	NaCl mass %	NaCl g/100g water	NaCl mol/kg	Na_2SO_3 mass %	Na_2SO_3 g/100g water	Na_2SO_3 mol/kg	Solid[c] phase
0.0	0.	0.0	0.	11.70	13.3	1.051	A
	6.21	7.23	1.236	7.85	9.13	0.725	A
	18.10	23.2	3.969	3.87	4.96	0.393	A
	24.44	33.6	5.750	2.83	3.89	0.309	A + C
	26.20	35.5	6.075	0.	0.0	0.	C
25.0	0.	0.0	0.	23.49	30.7	2.436	A
	2.79	3.65	0.624	20.67	27.0	2.143	A
	6.65	8.79	1.504	17.70	23.4	1.856	A
	11.31	14.2	2.629	15.08	20.2	1.625	A
	12.48	16.9	2.891	13.66	18.5	1.467	A
	15.65	21.6	3.695	11.88	16.4	1.301	A
	18.90	26.8	4.586	10.58	15.0	1.190	A
	19.34	28.1	4.809	11.84	17.2	1.365	A + C
	20.17	29.2	4.998	10.77	15.6	1.237	A + C
	20.43	29.4	5.031	10.08	14.5	1.151	A + C
	20.88	29.9	5.117	9.29	13.3	1.056	C
	22.26	31.5	5.390	7.07	10.0	0.794	C
	22.96	32.3	5.527	5.96	8.39	0.665	C
	23.90	33.3	5.699	4.34	6.05	0.480	C
	24.46	33.9	5.800	3.38	4.68	0.372	C
	25.10	34.6	5.921	2.36	3.25	0.258	C
	25.70	35.2	6.018	1.23	1.74	0.134	C
	26.47	36.0	6.160	0.	0.0	0.	C

(continued on next page)

AUXILIARY INFORMATION

METHOD APPARATUS/PROCEDURE:

Solids were equilibrated with water in sealed borosilicate-glass tubes for up to 20 days.

Solution was removed from the tube for analysis, under nitrogen. Separate aliquots were measured and weighed, then analysed for sulfite and chloride. Sulfite was determined by reaction with excess of iodine and back-titration with sodium thiosulfate, and chloride was determined mercurimetrically, after oxidation of sulfite to sulfate with hydrogen peroxide to prevent it from interfering (1).

SOURCE AND PURITY OF MATERIALS:

Sodium sulfite was obtained from J.T.Baker Chemical Co., and was found to assay at 100.1%. Sodium chloride from the same source assayed at 99.8%. Dissolved oxygen was removed from distilled water by boiling under reduced pressure at 60-65°C for 1 hr. This water was stored under nitrogen and used within 5 hr.

ESTIMATED ERROR:

No estimates possible.

REFERENCES:

1. Domask, W.G.; Kobe, K.A. *Anal. Chem.* 1952, *24*, 989.

COMPONENTS:	ORIGINAL MEASUREMENTS:
1. Sodium sulfite; Na_2SO_3; [7757-83-7] 2. Sodium chloride; NaCl; [7647-14-5] 3. Water; H_2O; [7732-18-5]	Kobe, K.A.; Hellwig, K.C. *Ind. Eng. Chem.* 1955, *47*, 1116-21.

EXPERIMENTAL VALUES (continued):

t/°C	NaCl[a] mass %	NaCl[b] g/100g water	NaCl[a] mol/kg	Na_2SO_3[a] mass %	Na_2SO_3[b] g/100g water	Na_2SO_3[a] mol/kg	Solid[c] phase
40.0	0.	0.0	0.	26.25	35.6	2.824	B
	5.00	6.71	1.149	20.55	27.6	2.190	B
	9.82	13.2	2.258	15.77	21.2	1.681	B
	15.49	20.9	3.576	10.38	14.0	1.111	B
	23.14	32.6	5.577	5.86	8.26	0.655	B + C
	24.02	33.6	5.751	4.51	6.31	0.501	B + C
	25.20	34.8	5.955	2.39	3.30	0.262	C
	26.74	36.5	6.246	0.	0.0	0.	C
60.0	0.	0.0	0.	23.78	31.2	2.475	B
	10.29	13.4	2.292	12.90	16.8	1.332	B
	15.76	20.8	3.560	8.48	11.2	0.888	B
	25.16	35.1	6.005	3.14	4.38	0.347	B + C
	25.61	35.6	6.091	2.44	3.39	0.269	C
	26.09	36.1	6.178	1.65	2.28	0.181	C
	27.11	37.2	6.364	0.	0.0	0.	C
80.0	0.	0.0	0.	21.87	28.0	2.222	B
	10.26	13.2	2.258	11.98	15.4	1.222	B
	20.47	27.3	4.672	4.56	6.08	0.483	B
	26.08	36.5	6.244	2.45	3.43	0.272	B + C
	26.77	37.2	6.364	1.25	1.74	0.138	C
	27.59	38.1	6.520	0.	0.0	0.	C
100.0	0.	0.0	20.82	0.	26.3	2.086	B
	10.64	13.5	10.56	2.310	13.4	1.063	B
	20.76	27.6	4.04	4.724	5.37	0.426	B
	26.96	38.0	2.08	6.501	2.93	0.233	B + C
	26.88	·38.9	0.44	6.329	0.61	0.048	C
	28.21	39.3	0.	6.724	0.0	0.	C

[a] Molalities and mass % values calculated by the compiler.

[b] Original data.

[c] Solid phases: A - $Na_2SO_3.7H_2O$, B - Na_2SO_3, C - NaCl

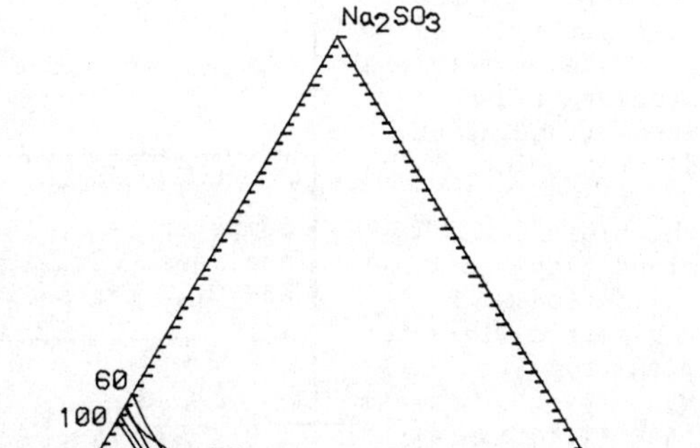

COMPONENTS:	ORIGINAL MEASUREMENTS:
1. Sodium sulfite; Na_2SO_3; [7757-83-7] 2. Sodium chloride; NaCl; [7647-14-5] 3. Water; H_2O; [7732-18-5]	Labash, J.A.; Lusby, G.R. *Can. J. Chem.* 1955, *33*, 774-86
VARIABLES:	**PREPARED BY:**
Two temperatures: 293 and 333 K Concentrations of the components	Mary R. Masson

EXPERIMENTAL VALUES:

Composition of equilibrium solutions

NaCl mass %	Na_2SO_3 mass %	Na_2SO_4 mass %	NaCl[b] mol/kg	Na_2SO_3[b] mol/kg	Na_2SO_4[b] mol/kg	Solid[b] phase
Temperature = 60°C						
26.36	0.0	-	6.125	0.	0.	A
24.22	3.26	0.22	5.732	0.358	0.021	A
23.68	4.35	0.21	5.647	0.481	0.021	u.d.
21.10	7.41	1.24	5.140	0.837	0.124	u.d.
21.19	7.61	1.23	5.182	0.863	0.124	.d.
16.73	9.13	1.14	3.922	0.992	0.110	B
16.89	9.20	0.71	3.948	0.997	0.068	B
10.27	12.46	1.04	2.305	1.297	0.096	B
6.76	14.87	0.86	1.492	1.522	0.078	B
0.0	20.58	0.77	0.	2.076	0.069	B
Temperature = 60°C						
27.03	0.0	-	6.339	0.	0.	A
25.77	1.89	0.30	6.121	0.208	0.029	A + C?
25.14	3.04	0.16	6.003	0.337	0.016	A + C?
18.85	5.98	0.93	4.345	0.639	0.088	C
14.03	9.57	0.70	3.171	1.003	0.065	C
12.75	10.64	1.05	2.887	1.117	0.098	C
11.69	11.46	2.40	2.687	1.221	0.227	C
10.88	11.27	1.06	2.424	1.164	0.097	C
7.55	15.80	2.04	1.732	1.680	0.192	C
4.45	18.90	1.32	1.011	1.991	0.123	C
2.98	21.90	1.02	0.688	2.345	0.097	C
1.29	22.50	0.65	0.292	2.363	0.061	C
0.0	22.86	0.67	0.	2.372	0.062	

(continued on next page)

AUXILIARY INFORMATION

METHOD APPARATUS/PROCEDURE:

Solutions were stirred under nitrogen in a three-necked flask. A saturated solution was prepared of one of the single salts, and portions of the second was added with stirring in the presence of excess of the first salt. Addition of the second salt was continued until the solution composition became constant. Samples were withdrawn, after settling, through a pipette plugged with cotton wool. The weighed sample was diluted to volume in a standard flask.

Sulfite was determined by adding an aliquot of the freshly diluted solution to excess of iodine solution, and back-titrating the excess with thiosulfate. Total sulfate was determined gravimetrically as barium sulfate after oxidation with bromine water, sodium gravimetrically after conversion of all sodium salts to the sulfate, and chloride by addition of excess of silver nitrate, and back-titration with ammonium thiocyanate.

SOURCE AND PURITY OF MATERIALS:

Sodium sulfite heptahydrate was prepared freshly from a saturated solution of anhydrous sodium sulfite by cooling. Anhydrous sodium sulfite and sodium chloride were of analytical grade.

ESTIMATED ERROR:

Temperature: ±1 K
Analyses: 0.2% for sulfite and chloride, and 0.4% for sodium and total sulfate.

REFERENCES:

COMPONENTS:	ORIGINAL MEASUREMENTS:
1. Sodium sulfite; Na_2SO_3; [7757-83-7]	Labash, J.A.; Lusby, G.R.
2. Sodium chloride; NaCl; [7647-14-5]	*Can. J. Chem.* 1955, *33*, 774-86
3. Water; H_2O; [7732-18-5]	

EXPERIMENTAL VALUES (continued):

Composition of equilibrium solutions at 60°C, after equilibration for seven weeks in rotating bottles.

NaCl mass %	Na_2SO_3 mass %	Na_2SO_4 mass %	NaCl[a] mol/kg	Na_2SO_3[a] mol/kg	Na_2SO_4[a] mol/kg	Solid[b] phase
25.77	1.89	0.30	6.121	0.208	0.029	A + C?
25.20	1.97	0.62	5.972	0.216	0.060	A + C?
25.25	2.08	0.70	6.003	0.229	0.068	A + C?
25.55	2.54	0.04??	6.083	0.280	0.004	A + C?
25.14	3.04	0.16	6.003	0.337	0.016	A + C?
25.03	2.93	0.30	5.970	0.324	0.029	A + C?
23.36	2.68	0.56	5.446	0.290	0.054	A + C?
24.72	3.00	0.56	5.898	0.332	0.055	A + C?

[a] Molalities calculated by the compiler.

[b] Solid phases: A - NaCl, B - $Na_2SO_3.7H_2O$, C - Na_2SO_3

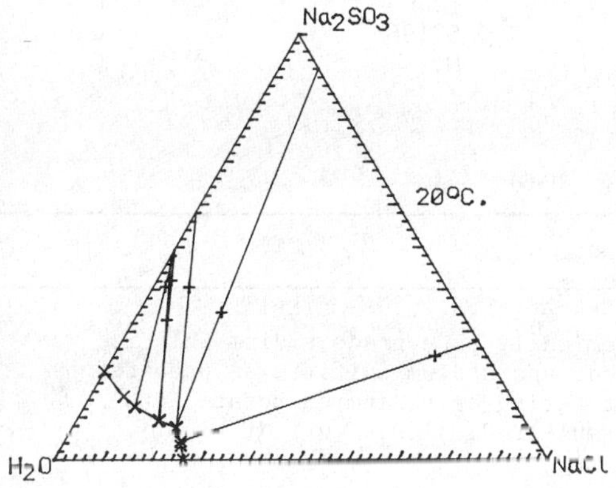

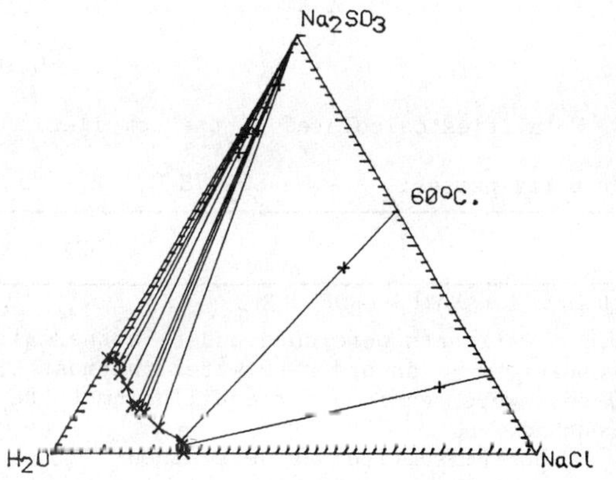

COMPONENTS:	ORIGINAL MEASUREMENTS:
1. Sodium sulfite; Na_2SO_3; [7757-83-77] 2. Sodium chloride; NaCl; [647-14-5] 3. Water; H_2O; [7732-18-5]	Durymanova, M.A.; Telepneva, A.E.; Zagrebina, L.A. *Zh. Neorg. Khim.* 1971, *16*, 500-3; *Russ. J. Inorg. Chem. (Eng. Transl.)* 1971, *16*, 264-6
VARIABLES:	PREPARED BY:
One temperature: 298 K Concentrations of the components	Mary R. Masson

EXPERIMENTAL VALUES: Composition of equilibrium solutions at 25°C

NaCl mass %	Na_2SO_3 mass %	NaCl[a] mol/kg	Na_2SO_3[a] mol/kg	Solid[b] phase
-	23.5	0.	2.437	A
1.2	21.9	0.267	2.259	A
5.1	18.9	1.148	1.973	A
5.6	18.6	1.264	1.947	A
7.2	17.2	1.630	1.805	A
9.5	15.5	2.167	1.640	A
11.1	14.6	2.556	1.559	A + B
11.5	14.7	2.666	1.580	A + B
12.4	13.7	2.871	1.471	B
15.8	10.9	3.688	1.180	B
17.4	9.5	4.073	1.031	B
17.4	9.1	4.051	0.982	B
18.6	9.2	4.408	1.011	B
19.9	7.5	4.690	0.820	B
21.9	6.4	5.227	0.708	B
22.8	5.7	5.457	0.632	B + C
22.9	5.6	5.480	0.621	B + C
23.0	5.5	5.504	0.610	B + C
23.1	5.5	5.536	0.611	B + C
23.5	4.5	5.585	0.496	C
25.2	1.8	5.907	0.196	C
26.5	-	6.169	0.	C

[a] Molalities calculated by the compiler.

[b] Solid phases: A - $Na_2SO_3 . 7H_2O$, B - Na_2SO_3, C - NaCl

AUXILIARY INFORMATION

METHOD APPARATUS/PROCEDURE:

The experiments were done under isothermal conditions in an ordinary water thermostat; 7 days were required for equilibrium to be reached.

Sodium sulfite was determined iodometrically, and sodium chloride by titration with mercury nitrate.

The composition of the solids were determined by Schreinemakers' method, and by chemical and crystal-optical analyses.

SOURCE AND PURITY OF MATERIALS:

Chemically pure grade sodium chloride was used, and sodium sulfite was prepared by saturating cp sodium carbonate with 100% of sulfur dioxide.

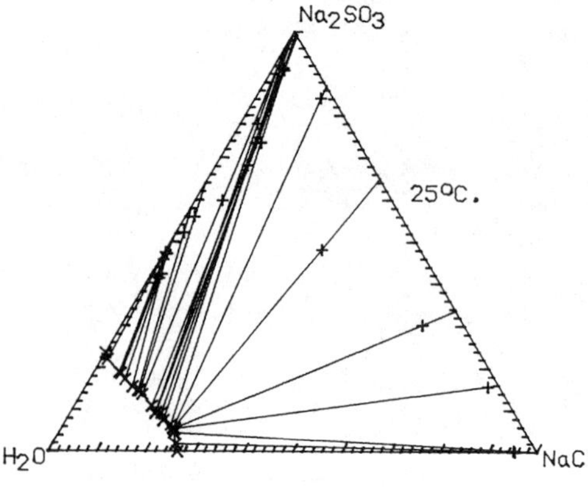

ESTIMATED ERROR:

Temperature: ±0.05 K
Analyses: no estimate possible.

COMPONENTS:	ORIGINAL MEASUREMENTS:
1. Sodium sulfite; Na_2SO_3; [7757-83-7] 2. Sodium pyrosulfite; $Na_2S_2O_5$; [7681-57-4] 3. Water; H_2O; [7732-18-5]	Sotova, N.N.; Kuznetsova, A.G.; Torocheshnikov, N.S. *Zh. Priklad. Khim.* 1978, *51*, 779-84; *J. Appl. Chem. USSR (Eng. Transl.)* 1978, *51*, 760-4.
VARIABLES: One temperature: 273 K Concentrations of the components	PREPARED BY: Mary R. Masson

EXPERIMENTAL VALUES:

Composition of equilibrium solutions at 0°C

Na_2SO_3 mass %	'$NaHSO_3$' mass %	$Na_2S_2O_5$ mass %	Na_2SO_3[a] mol/kg	$NaHSO_3$[a] mol/kg	Na_2SO_3[a] mol/kg	$Na_2S_2O_5$[a] mol/kg	Solid[b] phase
0.0	35.21	32.15	0.	5.222	0.	2.493	A
0.76	35.06	32.03	0.094	5.250	0.090	2.507	A
2.33	33.73	30.81	0.289	5.069	0.276	2.424	A
2.95	33.73	30.81	0.370	5.119	0.353	2.447	A
3.77	32.64	29.82	0.470	4.933	0.450	2.362	A
4.34	32.90	30.05	0.549	5.038	0.525	2.409	A + B
4.71	32.37	29.57	0.594	4.944	0.569	2.367	A + B
4.58	32.66	29.81	0.579	5.001	0.554	2.390	B
4.75	32.88	30.03	0.604	5.066	0.578	2.422	B
6.23	23.30	21.28	0.701	3.177	0.682	1.544	B
8.27	13.66	12.48	0.840	1.681	0.828	0.828	B
8.38	12.21	11.15	0.837	1.478	0.826	0.729	B
12.30	0.0	0.	1.113	0.	1.113	0.	B

[a] Molalities calculated by the compiler.

[b] Solid phases: A - $Na_2S_2O_5.6H_2O$, B - $Na_2SO_3.7H_2O$

AUXILIARY INFORMATION

METHOD APPARATUS/PROCEDURE:

The isothermal method was used.
p-Phenylenediamine was used as anti-oxidant.
Total sulfite was determined iodometrically. Bisulfite was titrated with alkali as bisulfate after oxidation with peroxide. Sodium sulfate was weighed to obtain total sulfur.

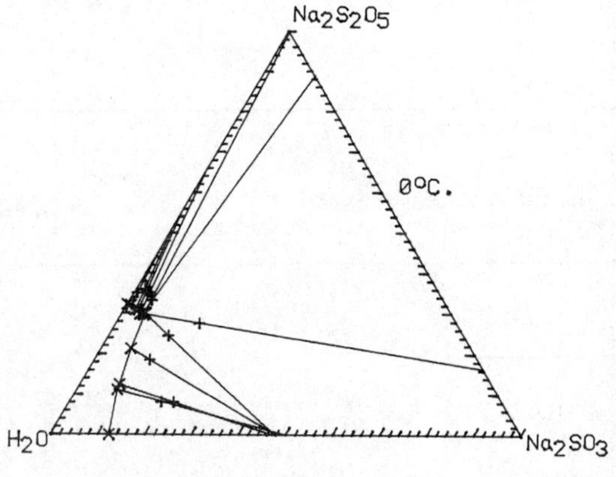

SOURCE AND PURITY OF MATERIALS:

Sodium pyrosulfite was of analytical grade, and sodium sulfite was of high-purity grade.

ESTIMATED ERROR:

Temperature: ±0.2 K
Analyses: no estimate possible.

REFERENCES:

COMPONENTS:	ORIGINAL MEASUREMENTS:
1. Sodium sulfite; Na_2SO_3; [7757-83-7] 2. Sodium pyrosulfite; $Na_2S_2O_5$; [7681-57-4] 3. Water; H_2O; [7732-18-5]	Sotova, N.N.; Torocheshnikov, N.S.; Kuznetsova, A.G.; Sokolova, E.I. *Khimiya i Tekhnol. Mineral'n. Solei i Galurgichesk. Pr.-v, Varnaul* 1978, 53-59.
VARIABLES: One temperature: 373 K Concentrations of the components	PREPARED BY: Mary R. Masson

EXPERIMENTAL VALUES:

Composition of equilibrium solutions at 100°C

'$NaHSO_3$' mass %	$Na_2S_2O_5$ mass %	Na_2SO_3 mass %	$Na_2S_2O_5$[a] mol/kg	Na_2SO_3[a] mol/kg	Solid[b] phase
-	-	21.2	0.0	2.135	A
2.01	1.84	19.4	0.123	1.954	A
2.9	2.65	18.85	0.178	1.905	A
7.5	6.85	17.0	0.473	1.771	A
30.4	27.77	5.8	2.199	0.693	A
46.61	42.58	1.71	4.021	0.244	A
50.83	46.43	2.6	4.792	0.405	A + B
50.83	46.43	2.6	4.792	0.405	A + B
49.95	45.63	2.07	4.590	0.314	B
53.8	49.14	-	5.082	0.	B

[a] Molalities calculated by the compiler.

[b] Solid phases: A - Na_2SO_3, B - $Na_2S_2O_5$

AUXILIARY INFORMATION

METHOD APPARATUS/PROCEDURE:

An isothermal method was used.

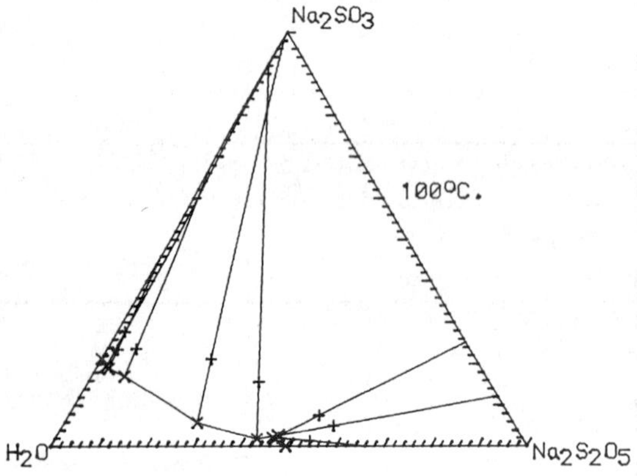

SOURCE AND PURITY OF MATERIALS:

ESTIMATED ERROR:

No estimates possible.

REFERENCES:

COMPONENTS:

1. Sodium sulfite; Na_2SO_3; [7757-83-7]
2. Sodium pyrosulfite; $Na_2S_2O_5$; [7681-57-4]
3. Water; H_2O; [7732-18-5]

ORIGINAL MEASUREMENTS:

Sotova, N.N.; Torocheshnikov, N.S.; Kuznetsova, A.G.; Poroshkova, M.A.

Khimiya i Tekhnol. Mineral'n. Solei i Galurgichesk. Pr.-v, Varnaul 1978, 59-65.

VARIABLES:

One temperature: 333 K
Concentrations of the components

PREPARED BY:

Mary R. Masson

EXPERIMENTAL VALUES:

Composition of equilibrium solutions at 60°C

Na_2SO_3 mass %	$Na_2S_2O_5$ mass %	Na_2SO_3[a] mol/kg	$Na_2S_2O_5$[a] mol/kg	Solid[b] phase
23.3	-	2.410	0.	A
21.4	2.6	2.234	0.180	A
13.9	9.7	1.443	0.668	A
12.4	13.5	1.328	0.958	A
8.3	25.0	0.987	1.972	A
?2.3	41.9	0.327	3.950	A + B
2.9	42.7	0.423	4.129	B
0.0	44.5	0.	4.218	B

[a] Molalities calculated by the compiler.

[b] Solid phases: A - Na_2SO_3, B - $Na_2S_2O_5$

AUXILIARY INFORMATION

METHOD APPARATUS/PROCEDURE:

The isothermal method was used.

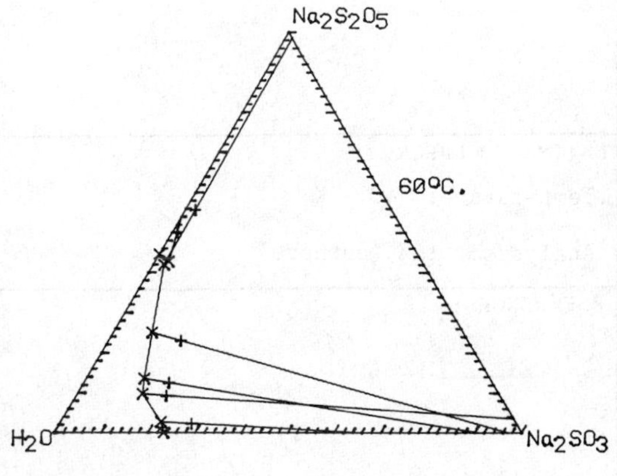

SOURCE AND PURITY OF MATERIALS:

ESTIMATED ERROR:

Temperature: ± 0.1 K
Analyses: no estimate possible.

REFERENCES:

COMPONENTS:

1. Sodium sulfite; Na_2SO_3; [7757-83-7]
2. Sodium pyrosulfite; $Na_2S_2O_5$; [7681-57-4]
3. Water; H_2O; [7732-18-5]

ORIGINAL MEASUREMENTS:

Jäger, L.; Rejlek, M.; Klimeček, R.; Machala, J.

Chem. Prům. 1959, *9*, 361-3.

VARIABLES:

Four temperatures: 288 - 318 K
Concentrations of the components

PREPARED BY:

Mary R. Masson

EXPERIMENTAL VALUES:

Composition of equilibrium solutions

Na_2SO_3 mass %	$Na_2S_2O_5$ mass %	Na_2SO_3[a] mol/kg	$Na_2S_2O_5$[a] mol/kg	Solid[b] phase
Temperature = 15°C				
18.4	-	1.789	0.	A
16.1	4.3	1.605	0.284	A
14.1	9.2	1.459	0.631	A
12.0	14.5	1.295	1.038	A
10.7	20.2	1.229	1.538	A
9.1	26.9	1.128	2.211	A
8.6	29.0	1.093	2.445	A + B
6.5	31.4	0.830	2.660	B
-	37.9	0.	3.210	
Temperature = 25°C				
23.2	-	2.397	0.	A
20.9	4.5	2.223	0.317	A
19.5	7.5	2.119	0.540	A
17.9	10.45	1.982	0.767	A
16.0	16.0	1.867	1.238	A
14.0	23.0	1.763	1.920	A
12.2	27.0	1.592	2.336	A + B
11.2	28.6	1.476	2.499	B
5.5	33.8	0.719	2.929	B
3.6	36.4	0.476	3.191	B
2.6	37.1	0.342	3.236	B
-	39.9	0.	3.492	B

(continued on next page)

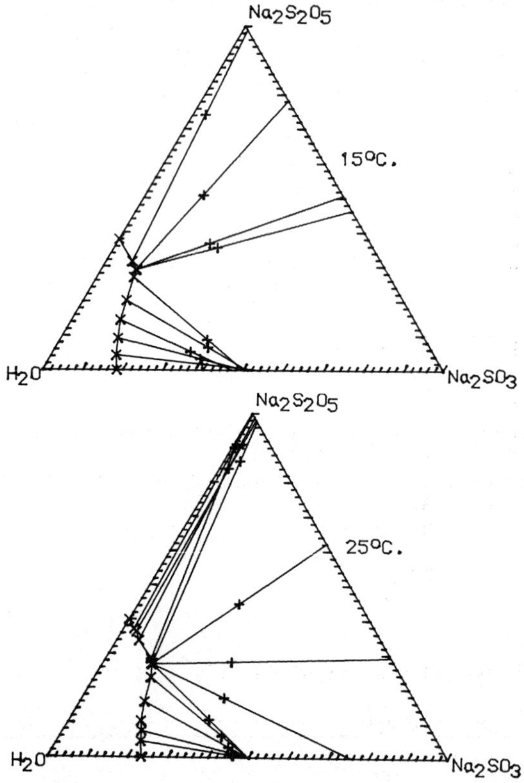

METHOD APPARATUS/PROCEDURE:

An isothermal method. The analysis involved an iodometic titration, and the oxidation of HSO_3^- to SO_4^{2-} with hydrogen peroxide. Solid phases were identified by microscopy and X-ray diffraction.

SOURCE AND PURITY OF MATERIALS:

Commercial sodium pyrosulfite was found to contain 2.2% of sodium sulfite. Sodium sulfite 7-hydrate was recrystallized from water.

ESTIMATED ERROR:

Temperature: ±0.1 K

Analyses: ±2% (authors)

REFERENCES:

COMPONENTS:	ORIGINAL MEASUREMENTS:
1. Sodium sulfite; Na_2SO_3; [7757-83-7] 2. Sodium pyrosulfite; $Na_2S_2O_5$; [7681-57-4] 3. Water; H_2O; [7732-18-5]	Jäger, L.; Rejlek, M.; Klimeček, R.; Machala, J. *Chem. Prům.* 1959, 9, 361-3.

EXPERIMENTAL VALUES (continued):

Composition of equilibrium solutions

Na_2SO_3 mass %	$Na_2S_2O_5$ mass %	Na_2SO_3[a] mol/kg	$Na_2S_2O_5$[a] mol/kg	Solid[b] phase
Temperature =35°C				
28.1	-	3.101	0.	C
25.5	3.5	2.850	0.259	C
22.0	8.7	2.519	0.660	C
16.0	18.9	1.950	1.527	C
13.8	23.7	1.752	1.995	C
11.4	29.3	1.525	2.599	C
10.4	30.7	1.401	2.742	B + C
8.4	33.4	1.145	3.019	B
4.5	37.2	0.612	3.357	B
-	41.0	0.	3.656	B
Temperature = 45°C				
25.8	-	2.759	0.	C
21.5	5.3	2.330	0.381	C
18.7	11.7	2.132	0.884	C
16.9	17.3	2.038	1.383	C
16.0	18.4	1.935	1.475	C
13.0	24.3	1.645	2.039	C
10.6	31.4	1.450	2.848	B + C
6.7	34.1	0.898	3.030	B
3.0	37.5	0.400	3.315	B
-	42.0	0.	3.809	B

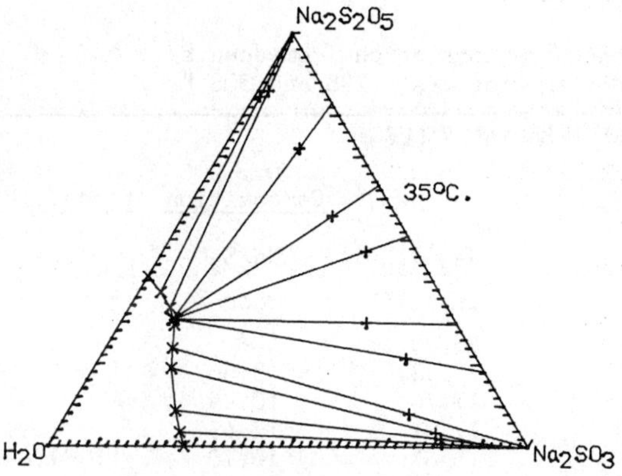

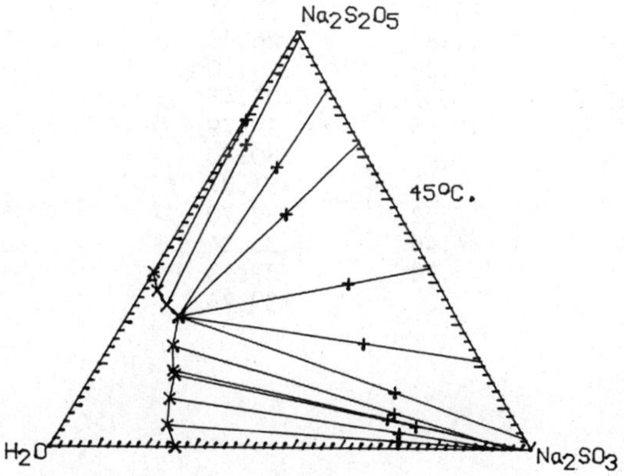

a Molalities calculated by the compiler.

b Solid phases: A - $Na_2SO_3.7H_2O$, B - $Na_2S_2O_5$, C - Na_2SO_3

COMPONENTS:

1. Sodium sulfite; Na_2SO_3; [7757-83-7]
2. Sodium pyrosulfite; $Na_2S_2O_5$; [7681-57-4]
3. Water; H_2O; [7732-18-5]

ORIGINAL MEASUREMENTS:

Arii, K.

Sci. Rep. Tohoku Imp. Univ. 1932, *21*, 783-9; (Original: *Bull. Inst. Phys. and Chem. Research* 1926, *6*, 1065-73.)

VARIABLES:

Concentrations of the components
Two temperatures: 298 and 308 K

PREPARED BY:

Mary R. Masson

EXPERIMENTAL VALUES:

Composition of equilibrium solutions at 25°C

Na_2SO_3 mass %	$Na_2S_2O_5$ mass %	Na_2SO_3[a] mol/kg[t]	$Na_2S_2O_5$[a] mol/kg[t]	Solid[b] phase
23.05	-	2.377	0.	B
20.98	3.66	2.209	0.255	B
17.70	10.39	1.953	0.760	B
17.01	12.49	1.914	0.932	B
15.67	16.70	1.838	1.299	B
15.06	18.65	1.802	1.480	B
13.89	21.91	1.717	1.795	B
11.83	26.73	1.528	2.289	B
11.28	29.10	1.501	2.568	B + C
10.87	29.45	1.445	2.596	C
10.46	29.79	1.389	2.623	C
9.63	30.70	1.280	2.706	C
8.33	31.77	1.103	2.790	C
7.22	32.70	0.953	2.863	C
4.46	35.34	0.588	3.088	C
1.78	37.93	0.234	3.309	C
-	39.75	0.	3.471	C

(continued on next page)

AUXILIARY INFORMATION

METHOD APPARATUS/PROCEDURE:

For systems where Na_2SO_3 was the solid phase, sulfur dioxide gas of the necessary amount was added to 20 ml of 25% sodium hydroxide, under a hydrogen atmosphere, and with cooling.

For systems where the solid was $Na_2S_2O_5$, a suitable amount of sodium hydroxide was added to sodium pyrosulfite solid.

The mixtures were stirred for 50-184 hr until equilibrium was reached. Weighed samples of the solution were analysed for sulfite by reaction with excess of iodine and determination of the excess with thiosulfate. The amount of pyrosulfite was determined by titrating the protons released in the oxidation reaction with alkali to a methyl orange end-point.

The solids were analysed similarly, after filtration under carbon dioxide.

SOURCE AND PURITY OF MATERIALS:

Sodium pyrosulfite was prepared by Foerster's method (2).

ESTIMATED ERROR:

Temperatures: ±0.01 K
Analyses: no estimate possible.

REFERENCES:

1. Arii, K. *Sci. Rep. Tohoku Imp. Univ.* 1932, *21*, 772-8.
2. Foerster, F.; Brosche, A.; Norberg-Schutz, Chr. *Z. Phys. Chem.* 1924, *10*, 435-96.

COMPONENTS:	ORIGINAL MEASUREMENTS:
1. Sodium sulfite; Na_2SO_3; [7757-83-7] 2. Sodium pyrosulfite; $Na_2S_2O_5$; [7681-57-4] 3. Water; H_2O; [7732-18-5]	Arii, K. *Sci. Rep. Tohoku Imp. Univ.* 1932, *21*, 783-9.

EXPERIMENTAL VALUES (continued):

Composition of equilibrium solutions at 35°C

Na_2SO_3 mass %	$Na_2S_2O_5$ mass %	Na_2SO_3[a] mol/kg	$Na_2S_2O_5$[a] mol/kg	Solid[b] phase
26.57	-	2.871	0.	A
25.00	2.13	2.722	0.154	A
24.17	3.27	2.643	0.237	A
21.04	8.00	2.352	0.593	A
19.22	11.54	2.202	0.877	A
16.60	16.33	1.964	1.281	A
14.80	20.40	1.812	1.656	A
11.86	25.84	1.510	2.182	A
8.86	30.84	1.166	2.690	A
6.77	33.03	0.892	2.886	A
5.99	34.91	0.804	3.107	A + C
5.98	34.73	0.800	3.081	C
4.12	36.48	0.550	3.231	C
3.11	37.42	0.415	3.310	C
1.74	38.69	0.232	3.417	C
0.00	40.55	0.	3.588	C

a Molalities calculated by the compiler.

b Solid phases: A - Na_2SO_3, B - $Na_2SO_3.7H_2O$, C - $Na_2S_2O_5$

Notes: The author chose to work at 25°C and 35°C because he had found previously (1) that the transition point between $Na_2SO_3.7H_2O$ and Na_2SO_3 was 33.5°C. He concluded from his results that, when sulfur dioxide passes into sodium hydroxide solution, any solid that separates will be Na_2SO_3, $Na_2SO_3.7H_2O$, or $Na_2S_2O_5$, and not any double salt or solid solution between these salts, at temperatures of 25 and 35°C.

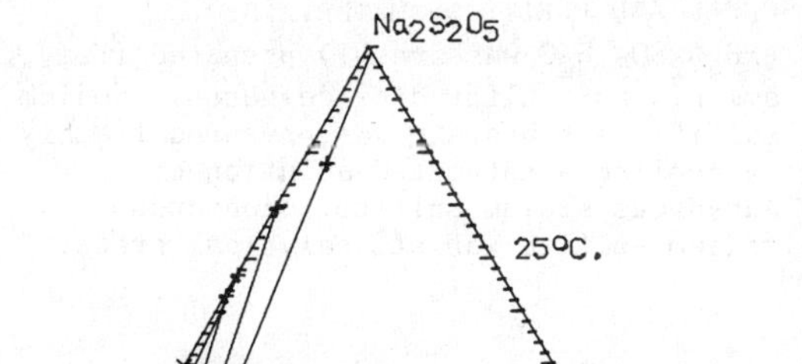

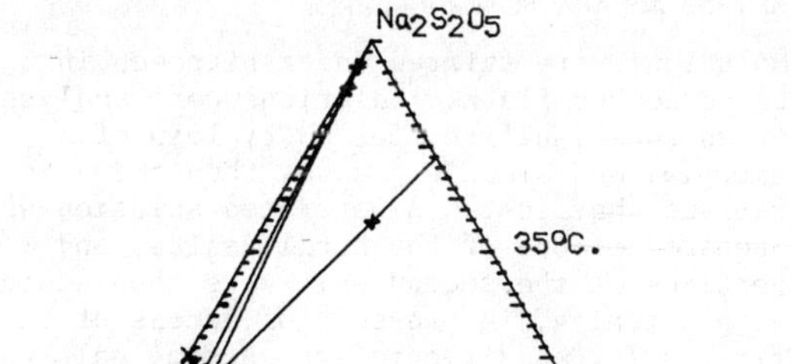

COMPONENTS:

1. Sodium sulfite; Na_2SO_3; [7757-83-7]
2. Ammonium sulfite; $(NH_4)_2SO_3$; [10196-04-0]
3. Water; H_2O; [7732-18-5]

ORIGINAL MEASUREMENTS:

Labash, J.A.; Lusby, G.R.

Can. J. Chem. 1955, *33*, 774-86.

VARIABLES:

Two temperatures: 293 and 333 K
Concentrations of the components

PREPARED BY:

Mary R. Masson

EXPERIMENTAL VALUES:

Composition of equilibrium solutions at 20°C

Na_2SO_3 mass %	$(NH_4)_2SO_3$ mass %	$(NH_4)_2SO_4$ mass %	Na_2SO_3[a] mol/kg	$(NH_4)_2SO_3$[a] mol/kg	$(NH_4)_2SO_4$[a] mol/kg	Solid[b] phase
0.0	37.34	0.49	0.	5.171	0.060	A
4.74	33.52	1.10	0.620	4.760	0.137	A
9.91	30.91	0.62	1.343	0.545	0.080	A
14.79	27.55	1.00	2.071	4.187	0.134	A
14.98	27.43	0.96	2.099	4.171	0.128	A + B
17.26	25.75	1.30	2.459	3.981	0.177	n.d.
17.21	25.70	0.80	2.426	3.931	0.108	C
17.97	20.87	0.88	2.365	2.981	0.110	C
18.65	15.95	0.84	2.292	2.127	0.098	C
19.34	10.44	0.87	2.213	1.296	0.095	C
20.58	-	0.77 (Na)	2.076	0.	0.074	C

(continued on next page)

AUXILIARY INFORMATION

METHOD APPARATUS/PROCEDURE:

Solutions were stirred under nitrogen in a three-necked flask. Solutions were analysed often for bisulfite, formed by loss of ammonia, and ammonia gas was then added to replace that lost. A saturated solution was prepared of one of the single salts, and portions of the second salt were then added, with stirring, in presence of excess of the first salt. Addition of the second salt was continued until the soln composition became constant. Samples were withdrawn, after settling, through a pipette plugged with cotton wool. The weighed sample was diluted to volume in a standard flask.

Bisulfite was determined by acid-base titration as bisulfate after oxidation with neutral hydrogen peroxide. Sulfite was determined by adding an aliquot of the freshly diluted soln to excess of iodine solution. The excess was back-titrated with thiosulfate. Total sulfate was determined as barium sulfate, ammonium by (2) and sodium gravimetrically after conversion of all sodium salts to the sulfate.

SOURCE AND PURITY OF MATERIALS:

$(NH_4)_2SO_3.H_2O$ was freshly prepared from ammonia and sulfur dioxide gases. Sodium sulfite heptahydrate was prepared freshly by cooling a saturated solution of anhydrous sodium sulfite. Anhydrous sodium sulfite was of analytical grade.

ESTIMATED ERROR:

Temperature: $\pm$0.1 K
Analyses: 0.2% for sulfite, ammonium, 0.4% for sodium and total sulfate.

REFERENCES:

1. Zil'berman, Y.I.; Ivanov, P.T. *Zh. Priklad. Khim.* 1941, *14*, 939.
2. Sutton, F. *Volumetric Analysis*, 12th Ed., Blakiston, Philadelphia, 1935. Page 75.

COMPONENTS:	ORIGINAL MEASUREMENTS:
1. Sodium sulfite; Na_2SO_3; [7757-83-7] 2. Ammonium sulfite; $(NH_4)_2SO_3$; [10196-04-0] 3. Water; H_2O; [7732-18-5]	Labash, J.A.; Lusby, G.R. *Can. J. Chem.* 1955, *33*, 774-86.

EXPERIMENTAL VALUES (continued):

Composition of the equilibrium solutions at 60°C expressed as mass %

Na_2SO_3	$(NH_4)_2SO_3$	Na_2SO_4	NH_4HSO_3	NH_3	Solid[b] phase
8.35[c]	45.08	-	-	-	A + D
7.76	44.7	0.74	-	0.08	n.d.
7.69	45.0	0.71	0.08	-	n.d.

Compositions of these solutions expressed as molalities[a], mol/kg (compiler)

1.42[c]	8.33	-	-	-
1.32	8.24	0.11	-	0.10
1.31	8.33	0.11	0.017	

a Molalities calculated by the compiler.

b Solid phases: A - $(NH_4)_2SO_3.H_2O$, B - unknown solid, C - $Na_2SO_3.7H_2O$, D - Na_2SO_3 n.d. - not determined

c From ref. (1)

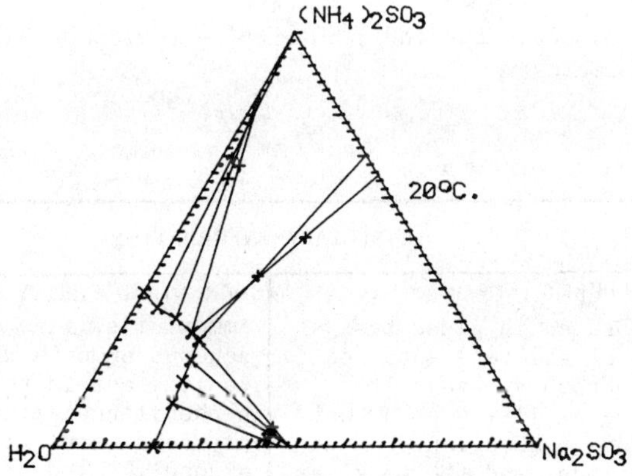

COMPONENTS:

1. Sodium sulfite; Na_2SO_3; [7757-83-7]
2. Ammonium sulfite; $(NH_4)_2SO_3$; [10196-04-0]
3. Water; H_2O; [7732-18-5]

ORIGINAL MEASUREMENTS:

Zil'berman, Ya.I.; Ivanov, P.T.

Zh. Priklad. Khim. 1941, *14*, 939-46.

VARIABLES:

Two temperatures: 333 and 358 K
Concentrations of the components

PREPARED BY:

Mary R. Masson

EXPERIMENTAL VALUES:

Composition of equilibrium solutions

SO_3^{2-} mass %	NH_4^+ mass %	Na_2SO_3 mass %	$(NH_4)_2SO_3$ mass %	Na_2SO_3[a]	$(NH_4)_2SO_3$	Solid[b] phase	
Temperature = 85°C							
-	-	22.1	-	2.250	0.	A	
36.00	14.60	5.52	47.16	0.925	8.581	A	
37.43	15.25	5.25	49.40	0.918	9.379	A	
41.22	16.95	5.51	54.67	1.098	11.821	A + B	
-	-	-	59.53	0.	12.665	B	
Temperature = 60°C							
33.92	11.82	12.20	38.11	1.948	6.604	A	
35.98	13.00	11.06	45.08	2.000	8.850	A	
36.40	13.97	8.35	45.08	1.422	8.335	A + B'	
-	-	(22.0)	-	2.237	0.0	(A)	(1)
-	-	-	(50.94)	0.0	8.940	(B)	(2)

[a] Molalities calculated by the compiler.

[b] Solid phases: A - Na_2SO_3, B - $(NH_4)_2SO_3$, B' - $(NH_4)_2SO_3.H_2O$

(continued on next page)

AUXILIARY INFORMATION

METHOD APPARATUS/PROCEDURE:

Solutions were equilibrated in glass test tubes fitted with spiral stirrers supplied with mercury seals. Freshly made salts were always used for each filling of a test tube. The anti-oxidant p-phenylenediamine was added to all solutions, and the work was done in an atmosphere of nitrogen, but experiments still had to be repeated often because of the formation of unacceptably high concentrations of sulfate.

Sulfite was determined by reaction with iodine solution, ammonium was volatilized by reaction with alkali, and collected in acid, the excess of which was titrated, sodium was weighed as sodium sulfate, and total sulfur as barium sulfate.

SOURCE AND PURITY OF MATERIALS:

Ammonium sulfite was made by saturating aqueous ammonia with sulfur dioxide, with cooling and in the presence of p-phenylenediamine as anti-oxidant. The crystals were filtered off and washed with alcohol. The product usually contained 0.5 - 1% of sulfate.

The sodium sulfite heptahydrate (reagent grade) contained about 1% of sulfate.

ESTIMATED ERROR:

Analyses: 0.2% relative
Temperature: no estimate given (toluene and mercury thermoregulators)

REFERENCES:

1. *Landolt* III (suppl.), Chap. I.
2. Mellor, J.W. *A Comprehensive Treatise on Inorganic and Theoretical Chemistry: Vol. X* Longmans, Green & Co., London, 1930.

COMPONENTS:	ORIGINAL MEASUREMENTS:
1. Sodium sulfite; Na_2SO_3; [7757-83-7]	Zil'berman, Ya.I.; Ivanov, P.T.
2. Ammonium sulfite; $(NH_4)_2SO_3$; [10196-04-0]	*Zh. Priklad. Khim.* 1941, *14*, 939-46.
3. Water; H_2O; [7732-18-5]	

EXPERIMENTAL VALUES (continued):

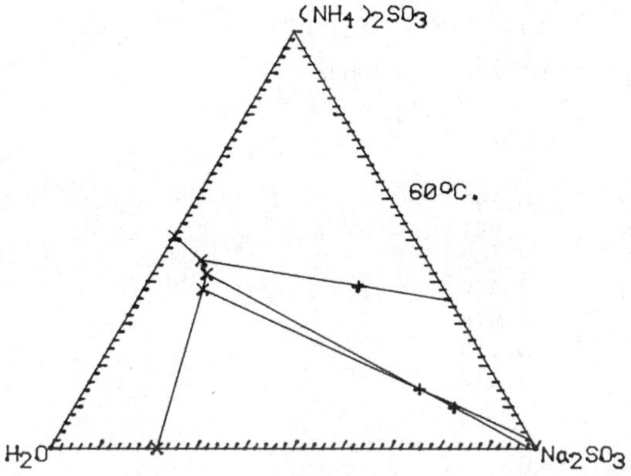

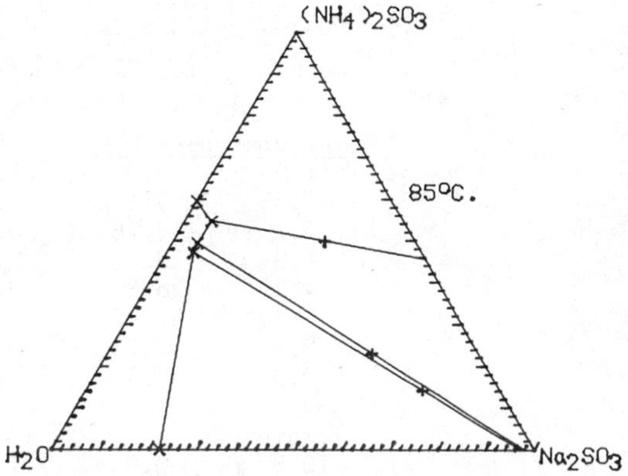

COMPONENTS:

1. Sodium sulfite; Na_2SO_3; [7757-83-5]
2. Sodium thiosulfate; $Na_2S_2O_3$; [7772-04-7]
3. Water; H_2O; [7732-18-5]

ORIGINAL MEASUREMENTS:

Palkina, N.A.

Tr. Vornezh. Gos. Univ. 1950, *17*, 61-88.

VARIABLES:

Six temperatures: 273 - 298 K
Concentrations of the components

PREPARED BY:

Mary R. Masson

EXPERIMENTAL VALUES:

Composition of equilibrium solutions

Na_2SO_3 mass %	$Na_2S_2O_3$ mass %	Na_2SO_3[a] mol/kg	$Na_2S_2O_3$[a] mol/kg	Solid[b] phase
Temperature = 0°C				
11.9	0.0	1.072	0.	A
8.6	9.1	0.829	0.699	A
6.1	18.78	0.644	1.581	A
4.9	23.77	0.545	2.108	A
4.2	26.82	0.483	2.459	A
3.25	30.97	0.392	2.978	*
0.81	32.73	0.097	3.115	B
0.0	33.4	0.	3.172	
Temperature = 5°C				
13.8	0.0	1.270	0.	
9.9	9.0	0.969	0.702	A
7.4	18.52	0.793	1.581	A
6.1	23.47	0.687	2.108	A
5.4	26.48	0.629	2.459	A
4.1	31.54	0.505	3.100	A
4.0	32.1	0.497	3.177	*
0.0	35.2	0.	3.436	

(continued on next page)

AUXILIARY INFORMATION

METHOD APPARATUS/PROCEDURE:

A polythermal procedure was used, based on the following systems:

I (10% $Na_2S_2O_3$ + 90% water) + Na_2SO_3
II (20% $Na_2S_2O_3$ + 80% water) + Na_2SO_3
III (25% $Na_2S_2O_3$ + 75% water) + Na_2SO_3
IV (28% $Na_2S_2O_3$ + 72% water) + Na_2SO_3
V (33% $Na_2S_2O_3$ + 67% water) + Na_2SO_3
VI (36% $Na_2S_2O_3$ + 64% water) + Na_2SO_3
VII (38% $Na_2S_2O_3$ + 62% water) + Na_2SO_3

SOURCE AND PURITY OF MATERIALS:

ESTIMATED ERROR:

No estimates possible.

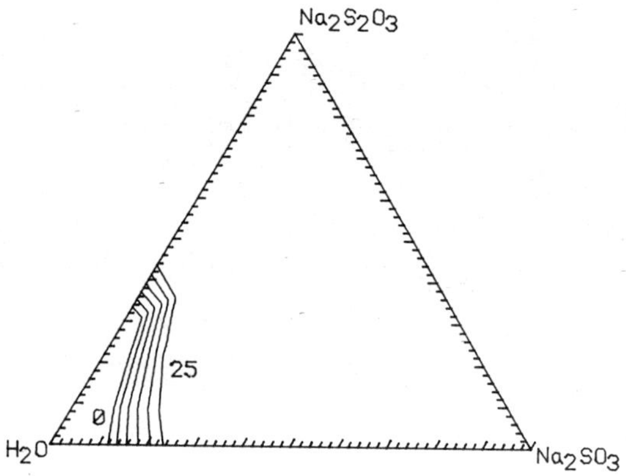

COMPONENTS:	ORIGINAL MEASUREMENTS:
1. Sodium sulfite; Na_2SO_3; [7757-83-5] 2. Sodium thiosulfate; $Na_2S_2O_3$; [7772-04-7] 3. Water; H_2O; [7732-18-5]	Palkina, N.A. *Tr. Vornezh. Gos. Univ.* 1950, *17*, 61-88.

EXPERIMENTAL VALUES (continued):

Composition of equilibrium solutions

Na_2SO_3 mass %	$Na_2S_2O_3$ mass %	Na_2SO_3[a] mol/kg	$Na_2S_2O_3$[a] mol/kg	Solid[b] phase
Temperature = 10°C				
15.8	0.0	1.489	0.	
11.7	8.83	1.168	0.703	A
8.9	18.22	0.969	1.581	A
7.5	23.12	0.858	2.108	A
6.7	26.12	0.791	2.459	A
5.4	31.22	0.676	3.116	A
4.75	33.25	0.608	3.392	*
2.5	35.1	0.318	3.558	B
0.0	37.0	0.	3.715	
Temperature = 15°C				
18.2	0.0	1.765	0.	
13.7	8.63	1.399	0.703	A
10.6	17.88	1.176	1.581	A
9.1	22.47	1.055	2.077	A
8.1	25.73	0.971	2.459	A
6.7	30.79	0.850	3.116	A
5.9	33.87	0.777	3.557	A
5.7	34.3	0.754	3.616	*
1.9	37.32	0.248	3.884	B
0.0	39.1	0.	4.061	
Temperature = 20°C				
20.8	0.0	2.084	0.	
16.0	8.4	1.679	0.703	A
12.5	17.5	1.417	1.581	A
10.7	22.32	1.267	2.108	A
9.7	25.28	1.184	2.459	A
8.1	30.32	1.044	3.114	A
7.3	33.37	0.976	3.558	A
6.6	35.2	0.900	3.825	*
6.0	35.72	0.817	3.877	B
0.0	41.2	0.	4.432	
Temperature = 25°C				
23.5	0.0	2.437	0.	
18.5	8.15	2.001	0.703	A
14.5	17.1	1.682	1.581	A
12.5	21.47	1.502	2.057	A
11.6	24.75	1.446	2.459	A
9.8	29.76	1.286	3.114	A
8.7	32.86	1.181	3.557	A
8.2	34.88	1.143	3.876	A
7.8	35.8	1.097	4.015	*
0.0	43.5	0.	4.870	

[a] Molalities calculated by the compiler.

[b] Solid phases: A - "Solid solution of $Na_2SO_3.7H_2O$ and $Na_2S_2O_3.5H_2O$"
B - $Na_2S_2O_3.5H_2O$
* - A + B, liquid composition determined graphically

COMPONENTS:	ORIGINAL MEASUREMENTS:
1. Sodium sulfite; Na_2SO_3; [7757-83-7] 2. Sodium thiosulfate; $Na_2S_2O_3$; [7772-98-7] 3. Water; H_2O; [7732-18-5]	Mochalov, K.I.; Monina, S.S. *Uch. Zap. Perm. Univ.* 1970, *229*, 40-43.
VARIABLES: One temperature: 288 K Concentrations of the components	PREPARED BY: Mary R. Masson

EXPERIMENTAL VALUES:

Composition of saturated solutions

$Na_2S_2O_3.5H_2O$ mass %	$Na_2SO_3.7H_2O$ mass %	$Na_2S_2O_3$[a] mass %	Na_2SO_3[a] mass %	$Na_2S_2O_3$[b] mol/kg	Na_2SO_3[b] mol/kg	Solid[c] phase
0.0	36.5	0.0	18.2	0.0	1.765	A
14.4	28.0	9.2	14.0	0.758	1.446	A
32.4	19.0	20.6	9.5	1.864	1.078	A
48.8	13.0	31.1	6.5	3.152	0.826	A
55.0	12.5	35.0	6.2	3.765	0.837	A + B
56.5	8.5	36.0	4.2	3.808	0.557	B
60.0	4.0	38.2	2.0	4.040	0.265	B
63.6	0.0	40.5	0.0	4.305	0.0	B

a Values calculated by the compiler.

b Molalities calculated by the compiler.

c Solid phases: A - $Na_2SO_3.7H_2O$, B - $Na_2S_2O_3.5H_2O$

AUXILIARY INFORMATION

METHOD APPARATUS/PROCEDURE:

An isothermal method.

SOURCE AND PURITY OF MATERIALS:

ESTIMATED ERROR:

No estimates possible.

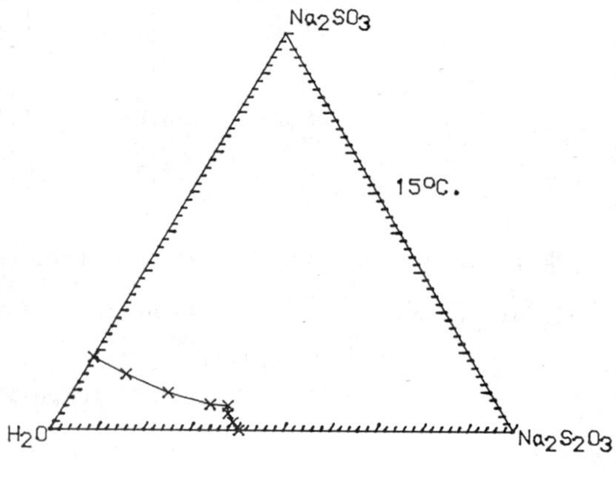

COMPONENTS:

1. Sodium sulfite; Na_2SO_3; [77757-83-5]
2. Sodium thiosulfate; $Na_2S_2O_3$; [7772-03-7]
3. Water; H_2O; [7732-18-5]

ORIGINAL MEASUREMENTS:

Wöhler, L.; Dierksen, J.

Z. Angew. Chem. 1926, *39*, 33-36.

VARIABLES:

Temperature: 296 - 353 K
Concentrations of the components

PREPARED BY:

Mary R. Masson

EXPERIMENTAL VALUES:

Composition of saturated solutions

t/°C	$Na_2S_2O_3$ mass %	Na_2SO_3 mass %	$Na_2S_2O_3$[a] mol/kg	Na_2SO_3[a] mol/kg
23	36.9	5.6	4.059	0.773
40	49.4	1.1	6.312	0.176
60	64.4	0.3	11.539	0.067
80	69.7	0.36	14.724	0.095

[a] Molalities calculated by the compiler.

AUXILIARY INFORMATION

METHOD APPARATUS/PROCEDURE:

Sulfite and thiosulfate were determined by titration with iodine, and then with alkali, to Methyl Orange.

SOURCE AND PURITY OF MATERIALS:

ESTIMATED ERROR:

No estimates possible.

REFERENCES:

COMPONENTS:

1. Sodium sulfite; Na_2SO_3; [7757-83-7]

2. Ethanol; C_2H_5OH; [64-17-5]

3. Water; H_2O; [7732-18-5]

ORIGINAL MEASUREMENTS:

Klebanov, G.S.; Ostapkevich, N.A.

Zh. Neorg. Khim. 1960, 5, 2329-2333; *Russ. J. Inorg. Chem. (Eng. Transl.)* 1960, 5, 1128-9.

VARIABLES:

Two temperatures: 293 - 323 K
Concentrations of the components

PREPARED BY:

Mary R. Masson

EXPERIMENTAL VALUES: Composition of equilibrium solutions

Na_2SO_3 mass %	C_2H_5OH mass %	Na_2SO_3[a] mol/kg	C_2H_5OH[a] mol/kg	Solid[b] phase
Temperature = 20°C				
20.72	-	2.074	0.	A
13.44	7.68	1.352	2.113	A
4.33	25.66	0.491	7.956	A
1.63	36.90	0.210	13.030	A
0.97	41.00	0.133	15.336	A
0.71	45.12	0.104	18.080	A
0.50	48.39	0.078	20.551	A
0.34	51.27	0.056	22.998	A
0.12	58.72	0.023	30.967	A + B
0.10	62.43	0.021	36.165	B
0.08	70.01	0.021	50.807	B
0.02	84.46	0.010	118.125	B
Temperature = 50°C				
25.63	-	2.734	0.	B
17.44	7.28	1.838	2.099	B
12.01	14.61	1.299	4.322	B
6.48	23.87	0.738	7.439	B
2.45	37.20	0.322	13.380	B
1.46	44.44	0.214	17.830	B
0.57	54.63	0.101	26.469	B
0.29	60.28	0.058	33.184	B
0.22	64.31	0.049	39.355	B
0.02	81.10	0.008	93.240	B

[a] Molalities calculated by the compiler.

[b] Solid phases: A - $Na_2SO_3.7H_2O$, B - Na_2SO_3

AUXILIARY INFORMATION

METHOD/APPARATUS/PROCEDURE:

The isothermal method was used; the vessels were glass test tubes fitted with mercury seals, at 20°C, and with reflux condensers at 50°C.

Alcohol was distilled off and determined iodometrically.

SOURCE AND PURITY OF MATERIALS:

"Chemically pure" salts were used. Ethanol and water were redistilled twice.

ESTIMATED ERROR:

Temperature: ±0.1 K
Analyses: no estimate possible.

REFERENCES:

COMPONENTS:	ORIGINAL MEASUREMENTS:
1. Sodium sulfite; Na_2SO_3; [7757-83-7] 2. Ethanol; C_2H_5OH; [64-17-5] 3. Water; H_2O; [7732-18-5]	Navrátil, J.; Nývlt, J. *Chem. Prům.* 1968, *18*, 612-4.
VARIABLES: Temperature: 276 - 310 K Ethanol concentration	PREPARED BY: Mary R. Masson

EXPERIMENTAL VALUES:

Solubility in water

t/°C	Atmos-phere	Na_2SO_3 mass %	Na_2SO_3[a] mol/kg
12.1	air	15.2	1.422
14.9	air	16.6	1.579
19.5	air	18.7	1.825
26.5	air	22.1	2.251
27.6	air	23.0	2.370
29.8	air	24.6	2.589
16.1	N_2	17.5	1.683
22.5	N_2	19.5	1.922
24.2	N_2	21.3	2.147
31.2	N_2	25.8	2.759

Solubility in 10% aqueous ethanol

t/°C	Atmos-phere	Na_2SO_3 mass %	Na_2SO_3[a] mol/kg
8.4	air	4.6	0.383
17.6	air	8.7	0.756
27.6	air	13.8	1.270

Solubility in 30% aqueous ethanol

t/°C	Atmos-phere	Na_2SO_3 mass %	Na_2SO_3[a] mol/kg
7.8	air	0.98	0.079
16.0	air	1.9	0.154
24.5	air	3.7	0.305

Solubility in 20% aqueous ethanol

t/°C	Atmos-phere	Na_2SO_3 mass %	Na_2SO_3[a] mol/kg
14.4	air	4.1	0.339
21.0	air	6.2	0.524
24.2	air	7.3	0.625
28.5	air	9.1	0.794
31.2	air	10.7	0.951
37.5	air	13.5	1.238
3.1	N_2	1.0	0.080
6.5	N_2	1.9	0.154
12.0	N_2	4.0	0.331
18.8	N_2	5.9	0.497
23.0	N_2	7.5	0.643
28.2	N_2	10.0	0.882
32.5	N_2	11.6	1.041
35.5	N_2	13.1	1.196

[a] Molalities calculated by the compiler.

AUXILIARY INFORMATION

METHOD APPARATUS/PROCEDURE:

An isothermal method.

SOURCE AND PURITY OF MATERIALS:

Sodium sulfite heptahydrate was obtained from Lachema, Brno.

ESTIMATED ERROR:

No estimates possible.

REFERENCES:

COMPONENTS:

1. Sodium sulfite; Na_2SO_3; [7757-83-7]
2. Sodium hydroxide; NaOH; [1310-73-2]
3. Water; H_2O; [7732-18-5]

ORIGINAL MEASUREMENTS:

Hammick, D.L.; Currie, J.A.

J. Chem. Soc. 1925, *127*, 1623-8.

VARIABLES:

Concentrations of the components
Temperature

PREPARED BY:

Mary R. Masson

EXPERIMENTAL VALUES:

Composition of equilibrium solutions

NaOH mass %	Na_2SO_3 mass %	NaOH mol/kg	Na_2SO_3 mol/kg	Solid phase
Temperature = 0.15°C				
-	12.2	0.	1.102	A
3.0	9.0	0.852	0.811	A
9.0	5.2	2.622	0.481	A
16.9	2.8	5.262	0.277	A
21.9	2.21	7.214	0.231	A
23.3	2.2	7.819	0.234	A + B
25.4	1.3	8.663	0.141	B
26.6	0.7	9.147	0.076	B
27.1	0.4	9.345	0.044	B
27.9	0.1	9.688	0.011	B
29.1	0.1	10.275	0.011	B + C
29.5	-	10.461	0.	C
Temperature = 20°C				
-	21.2	0.	2.135	A
1.6	18.7	0.502	1.862	A
4.4	15.4	1.372	1.523	A
8.3	12.0	2.604	1.195	A
9.9	11.0	3.129	1.103	A
11.6	9.9	3.694	1.001	A
12.2	9.6	3.900	0.974	A
12.7	9.4	4.076	0.957	A
13.4	9.2	4.328	0.943	A
14.0	9.0	4.545	0.927	A + B

(continued on next page)

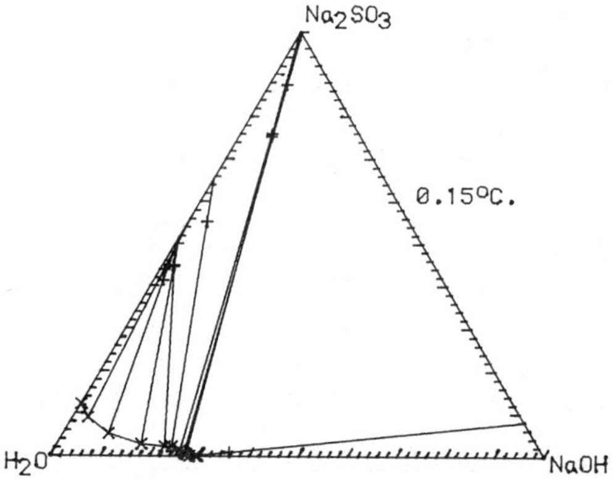

AUXILIARY INFORMATION

METHOD APPARATUS/PROCEDURE:

Mixtures of the components of approx. known composition were made up in resistance-glass test-tubes, or in a silver vessel (for [NaOH] ÷ 25%). The vessels were placed in a thermostat for 24-48 hr, until equilibrium was reached. The solution was separated from the moist solid by filtration, still in the thermostat. The solutions and the solids were analysed for sulfite by oxidation with hydrogen peroxide, decomposition of the excess by boiling, then determination of the sulfate formed as barium sulfate. Hydroxide was determined by titration, and water by difference.

Schreinemakers' method was then used to identify the solids.

SOURCE AND PURITY OF MATERIALS:

Sodium sulfite was prepared by passing sulfur dioxide into a solution of A.R. sodium carbonate (BDH), with exclusion of air. The solution was evaporated under hydrogen until a good crop of crystals was obtained. The washed and dried salt was tested for sulfate, and rejected if any was found. Carbonate-free sodium hydroxide solutions were prepared (1,2).

ESTIMATED ERROR:

Analyses: about ±0.5%
Temperatures: ±0.1°C, and ±0.05°C at 0.15°C

REFERENCES:

1. Cornog, J. *J. Am. Chem. Soc.* 1921, *43*, 2573.
2. Freeth, F.A. *Phil. Trans. A.* 1922, *223*, 35.

COMPONENTS:	ORIGINAL MEASUREMENTS:
1. Sodium sulfite; Na_2SO_3; [7757-83-7] 2. Sodium hydroxide; NaOH; [1310-73-2] 3. Water; H_2O; [7732-18-5]	Hammick, D.L.; Currie, J.A. *J. Chem. Soc.* 1925, *127*, 1623-8.

EXPERIMENTAL VALUES (continued):

Composition of equilibrium solutions

NaOH mass %	Na_2SO_3 mass %	NaOH mol/kg	Na_2SO_3 mol/kg	Solid phase
Temperature = 20°C (continued)				
13.8	8.6	4.446	0.879	A + B
15.5	7.0	5.000	0.717	B
16.0	6.7	5.175	0.688	B
20.2	3.5	6.619	0.364	B
24.9	1.1	8.412	0.118	B
26.8	1.0	9.280	0.110	B
36.3	0.1	14.269	0.012	B
50.1	trace	25.105	0.002	B
52.2**	-	27.301	0.	D
Temperature = 25°C				
-	23.6*	0.	2.451	A
0.7	22.5	0.228	2.324	A
2.5	19.3	0.799	1.958	A
5.0	17.7	1.617	1.817	A
9.7	13.0	3.137	1.334	A
9.8	13.2	3.182	1.360	A + B
9.6	13.8	3.133	1.429	A + B
12.2	10.2	3.930	1.043	B
12.6	10.1	4.075	1.037	B
14.2	8.2	4.575	0.838	B
17.2	5.8	5.584	0.598	B
20.7	2.9	6.774	0.301	B
20.7	2.5	6.738	0.258	B
25.2	1.0	8.537	0.108	B
28.7	0.5	10.134	0.056	B
29.4	0.3	10.455	0.034	B
42.7	0.1	18.663	0.014	B
43.9	trace	10.567	0.001	B
53.3[d]	-	28.533	0.	D
Temperature = 32°C				
-	26.5[c]	0.	2.861	B
0.6	26.4[c]	0.205	2.869	B
2.3	23.1	0.771	2.457	B
5.1	19.1	1.682	1.999	B
7.6	15.7	2.477	1.624	B
11.1	11.2	3.571	1.144	B
16.5	5.4	5.282	0.549	B
20.1	2.9	6.526	0.299	B
27.2	0.6	9.418	0.066	B
36.2	0.1	14.207	0.012	B
51.8	0.001	26.868	0.000	B
54.4[d]	-	29.825	0.	D

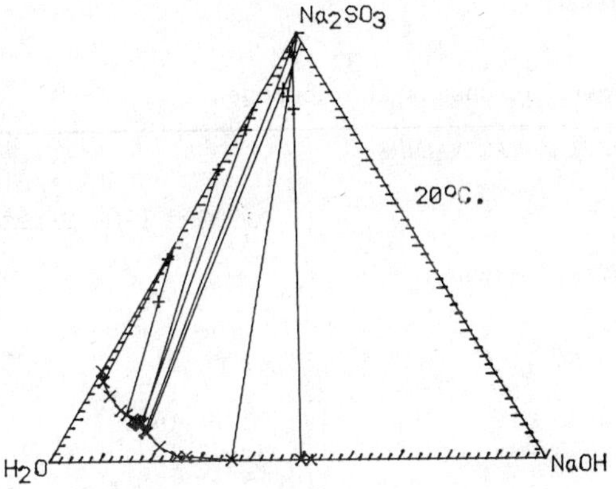

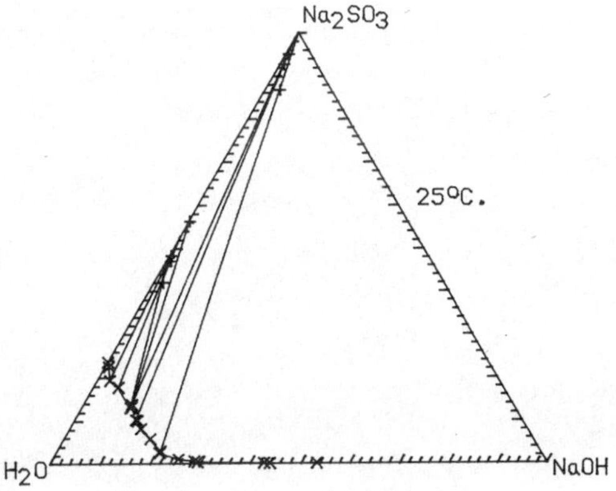

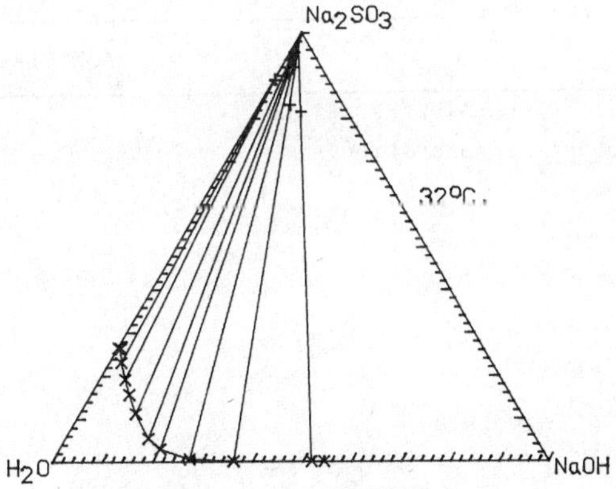

[a] Molalities calculated by the compiler.

[b] Solid phases: A - $Na_2SO_3.7H_2O$, B - Na_2SO_3, C - $NaOH.4H_2O$, D - $NaOH.H_2O$

[c] Values taken from Lewis, N.B.; Rivett, A.C.D. *J. Chem. Soc.* 1924, *125*, 1156-67.

[d] Values taken from Pickering, S.U. *J. Chem. Soc.* 1893, *63*, 890.

COMPONENTS:

1. Sodium sulfite; Na_2SO_3; [7757-83-7]
2. Sodium 2-naphtholate; $NaC_{10}H_7O$; [875-83-2]
3. Water; H_2O; [7732-18-5]

ORIGINAL MEASUREMENTS:

Teslo, S.P.; Gulyamov, Yu.M.; Odarich, V.F.

Vopr. Khim. Khim. Tekhnol. 1979, 55, 92-4.

VARIABLES:

Temperature: 313 - 371 K
Concentrations of the components

PREPARED BY:

Mary R. Masson

EXPERIMENTAL VALUES:

Composition of saturated solutions

[Sodium 2-naphtholate], g/l.	[Sodium sulfite], g/l. 40°C	50°C	60°C	70°C	80°C	90°C	98°C
54.44	235.62	214.83	200.97	189.88	180.18	177.41	172.02
98.71	205.13	177.41	167.71	162.16	155.23	151.07	148.30
134.64	185.72	166.32	153.23	149.69	135.83	130.28	128.90
148.90	171.86	152.46	144.14	138.60	126.12	120.59	117.81
197.42	119.75	125.29	111.99	95.36	87.60	84.55	71.77
259.78	99.79	88.70	76.51	69.85	63.20	56.55	-
321.55	-	-	46.71	45.74	44.54	42.97	-

AUXILIARY INFORMATION

METHOD APPARATUS/PROCEDURE:

Isothermal saturation.

SOURCE AND PURITY OF MATERIALS:

ESTIMATED ERROR:

No estimates possible.

REFERENCES:

COMPONENTS:	ORIGINAL MEASUREMENTS:
1. Sodium sulfite; Na_2SO_3; [7757-833-7] 2. Sodium sulfate; Na_2SO_4; [7757-82-6] 3. Sodium chloride; NaCl; [7647-14-5] 4. Water; H_2O; [7732-18-5]	Durymanova, M.A.; Telepneva, A.E.; Zagrebina, L.A. *Zh. Neorg. Khim.* 1971, *16*, 500-3; *Russ. J. Inorg. Chem. (eng. Transl.)* 1971, *16*, 264-6.
VARIABLES:	PREPARED BY:
One temperature: 298 K Concentrations of the components	Mary R. Masson

EXPERIMENTAL VALUES:

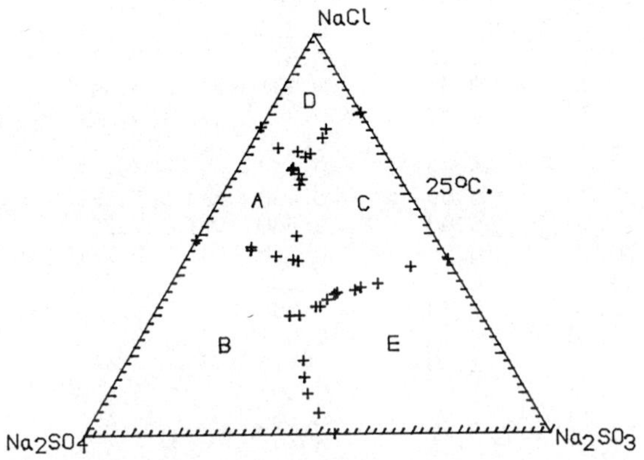

(continued on next page)

AUXILIARY INFORMATION

METHOD APPARATUS/PROCEDURE:

The experiments were done under isothermal conditions in an ordinary water thermostat; 7 days were required for equilibrium to be reached.

Sodium sulfite was determined iodometrically, sodium chloride by titration with mercury nitrate, and sodium sulfate gravimetrically by precipitation of the sulfate with barium chloride.

The compositions of the solids were determined by Schreinemakers' method, and by chemical and crystal-optical analyses.

SOURCE AND PURITY OF MATERIALS:

Chemically pure grade sodium chloride and sodium sulfate were used. Sodium sulfite was prepared by saturating cp sodium carbonate with 100% of sulfur dioxide.

ESTIMATED ERROR:

Temperature: ±0.05 K
Analyses: no estimate possible.

REFERENCES:

COMPONENTS:	ORIGINAL MEASUREMENTS:
1. Sodium sulfite; Na_2SO_3; [7757-83-7] 2. Sodium sulfate; Na_2SO_4; [7757-82-6] 3. Sodium chloride; NaCl; [7647-14-5] 4. Water; H_2O; [7732-18-5]	Durymanova, M.A.; Telepneva, A.E.; Zagrebina, L.A. *Zh. Neorg. Khim.* 1971, *16*, 500-3; *Russ. J. Inorg. Chem. (Eng. Transl.)* 1971, *16*, 264-6.

EXPERIMENTAL VALUES (continued):

	Na_2SO_3 mass %	NaCl mass %	Na_2SO_4 mass %	Na_2SO_3[a] mol/kg	NaCl[a] mol/kg	Na_2SO_4[a] mol/kg	Solid[c] phase
I	-	14.20	15.00	0.	3.432	1.492	A,B
	3.89	14.33	12.40	0.445	3.534	1.258	A,B
	4.03	14.25	12.66	0.463	3.531	1.291	A,B
	6.14	14.15	11.56	0.715	3.553	1.194	A,B
II	7.49	13.95	10.61	0.875	3.513	1.099	A,B,C
	7.94	14.00	10.39	0.931	3.540	1.081	A,B,C
III	-	22.90	6.90	0.	5.582	0.692	A,D
	2.00	22.80	7.05	0.233	5.725	0.728	A,D
	3.70	20.63	6.89	0.427	5.132	0.705	A,D
	3.75	20.84	6.69	0.433	5.189	0.685	A,D
	4.10	26.50	6.90	0.520	7.255	0.777	A,D
	4.43	20.78	6.68	0.516	5.221	0.690	A,C,D
IV	4.97	20.71	6.83	0.584	5.251	0.712	A,C,D
	4.98	20.20	7.20	0.584	5.112	0.750	A,C,D
V	5.60	22.90	-	0.621	5.480	0.	A,C,D
	4.20	22.50	2.80	0.473	5.461	0.280	C,D
	4.43	22.46	3.48	0.505	5.520	0.352	C,D
	4.31	21.78	4.96	0.496	5.405	0.506	C,D
	4.16	21.84	5.53	0.482	5.458	0.569	C,D
	6.85	16.09	9.58	0.805	4.080	0.999	A,C
	12.67	6.16	14.66	1.511	1.585	1.552	B,E
	13.71	4.83	15.57	1.651	1.254	1.664	B,E
	14.49	3.39	15.92	1.737	0.876	1.693	B,E
	15.60	1.73	15.43	1.841	0.440	1.616	B,E
VI	16.80	-	14.50	1.940	0.	1.486	B,E
VII	14.70	11.30	-	1.576	2.613	0.	C,E
	13.60	11.50	2.50	1.490	2.718	0.243	C,E
	13.29	11.27	5.50	1.508	2.757	0.554	C,E
	12.10	10.80	6.60	1.362	2.621	0.659	C,E
	12.61	11.28	7.49	1.458	2.813	0.768	C,E
	11.59	11.19	8.85	1.345	2.801	0.911	C,E
	11.92	11.50	9.45	1.409	2.931	0.991	C,E
	11.55	11.02	10.14	1.362	2.802	1.061	C,E
	10.90	10.00	10.50	1.261	2.494	1.078	C,E
	11.52	10.85	11.70	1.386	2.816	1.249	C,E
VIII	11.20	10.60	13.93	1.383	2.822	1.526	B,C,E
	9.60	9.70	13.50	1.133	2.470	1.414	B,C,E

[a] Molalities calculated by the compiler.

[b] Solid phases: A - solid solutions of Na_2SO_4 with a little Na_2SO_3,
B - solid solutions of $Na_2SO_4.10H_2O$,
C - solid solutions of Na_2SO_3 with a little Na_2SO_4,
D - NaCl,
E - solid solutions of $Na_2SO_3.7H_2O$ and $Na_2SO_4.7H_2O$

COMPONENTS:

1. Sodium sulfite; Na_2SO_3; [7757-83-7]
2. Sodium sulfate; Na_2SO_4; [7757-82-6]
3. Sodium chloride; $NaCl$; [7647-14-5]
4. Water; H_2O; [7732-18-5]

ORIGINAL MEASUREMENTS:

Durymanova, M.A.; Telepneva, A.E.

Zh. Priklad. Khim. 1972, *45*, 1610-12; *J. Appl. Chem. USSR (Eng. Transl.)* 1972, *45*, 1680-2.

VARIABLES:

Concentrations of the components
One temperature: 373 K

PREPARED BY:

Mary R. Masson

EXPERIMENTAL VALUES:

Composition of equilibrium solutions at 100°C

Na_2SO_3 mass %	NaCl mass %	Na_2SO_4 mass %	Na_2SO_3[a] mol/kg	NaCl[a] mol/kg	Na_2SO_4[a] mol/kg	Solid[b] phase
2.1	27.0	-	0.235	6.516	0.	F + B
2.1	26.2	1.6	0.238	6.395	0.543	F + B
2.0	25.0	2.3	0.224	6.051	0.774	F,B,C
2.3	25.8	2.4	0.263	6.352	0.821	F,B,C
2.2	25.7	2.3	0.250	6.300	0.784	F,B,C
1.4	25.6	3.4	0.160	6.294	1.162	F,C
1.3	25.5	3.7	0.148	6.278	1.266	F,C
0.9	25.5	4.0	0.103	6.269	1.367	F,C,D
0.8	25.5	3.9	0.091	6.251	1.329	F,D
-	25.9	4.4	0.	6.359	1.502	F,D
1.1	23.4	4.8	0.123	5.664	1.615	C,D
1.6	20.8	7.1	0.180	5.049	2.396	C,D
1.5	17.8	7.9	0.163	4.184	2.581	C,D
2.6	12.2	12.0	0.282	2.852	3.899	C,D
3.6	9.3	15.5	0.399	2.223	5.149	C,D
4.5	6.0	18.4	0.502	1.444	6.156	C,D
5.1	3.0	22.1	0.580	0.735	7.531	C,D
6.5	-	24.8	0.751	0.	8.587	C,D
2.5	23.8	3.1	0.281	5.768	1.044	B,C
2.6	23.3	3.2	0.291	5.623	1.074	B,C
3.6	19.8	4.5	0.396	4.699	1.485	B,C

(continued on next page)

AUXILIARY INFORMATION

METHOD APPARATUS/PROCEDURE:

The systems were studied under isothermal conditions in a thermostat filled with glycerol. Equilibrium was reached after 5 days. Sodium sulfite was determined iodometrically, sodium chloride by titration with mercuric nitrate solution, and sodium sulfate by precipitation of the sulfate as barium sulfate, and weighing.

The identities of the solids were determined by Schreinemakers' method.

SOURCE AND PURITY OF MATERIALS:

Sodium chloride and sodium sulfate were of c.p. grade. Sodium sulfite was prepared by saturating aqueous c.p. sodium carbonate with the stoichiometric amount of sulfur dioxide.

ESTIMATED ERROR:

Temperature: ±0.5 K
Analyses: no estimate possible.

REFERENCES:

COMPONENTS:	ORIGINAL MEASUREMENTS:
1. Sodium sulfite; Na_2SO_3; [7757-83-7]	Durymanova, M.A.; Telepneva, A.E.
2. Sodium sulfate; Na_2SO_4; [7757-82-6]	*Zh. Priklad. Khim.* 1972, 45, 1610-12; *J. Appl. Chem. USSR (Eng. Transl.)* 1972, 45, 1680-2.
3. Sodium chloride; NaCl; [7647-14-5]	
4. Water; H_2O; [7732-18-5]	

EXPERIMENTAL VALUES (continued):

Composition of equilibrium solutions at 100°C

Na_2SO_3 mass %	NaCl mass %	Na_2SO_4 mass %	Na_2SO_3[a] mol/kg	NaCl[a] mol/kg	Na_2SO_4[a] mol/kg	Solid[b] phase
3.8	19.1	4.6	0.416	4.508	1.509	B,C
5.2	15.7	6.3	0.567	3.690	2.058	B,C
7.6	19.6	8.6	0.939	5.224	3.186	B,C
10.1	6.2	11.0	1.102	1.459	3.599	B,C
12.4	3.0	12.7	1.368	0.714	4.202	B,C
14.4	-	14.75	1.613	0.	4.952	B,C

[a] Molalities calculated by the compiler.

[b] Solid phases: B - mixture of Na_2SO_3 with a small amount of Na_2SO_4
C - mixture of 25% Na_2SO_3 with 75% Na_2SO_4
D - mixture of Na_2SO_4 with a small amount of Na_2SO_3
F - NaCl

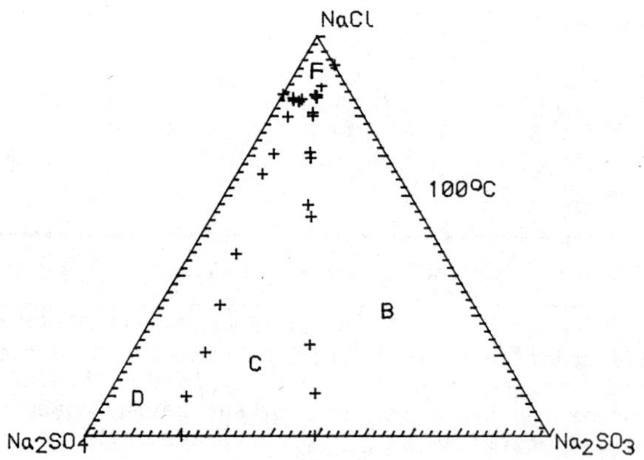

COMPONENTS:

1. Sodium sulfite; Na_2SO_3; [7757-83-7]
2. Sodium pyrosulfite; $Na_2S_2O_5$; [7681-57-4]
3. Sodium sulfate; Na_2SO_4; [7757-82-6]
4. Water; H_2O; [7732-18-5]

ORIGINAL MEASUREMENTS:

Kuznetsova, A.G.; Yaroshenko, L.B.

Zh. Priklad. Khim. 1981, *54*, 2197-2201; *J. Appl. Chem. USSR (Eng. Transl.)* 1981, *54*, 1929-32.

VARIABLES:

Concentrations of the components
One temperature: 298 K

PREPARED BY:

Mary R. Masson

EXPERIMENTAL VALUES: Compositions of equilibrium solutions at 25°C

Na_2SO_3 mass %	$NaHSO_3$ mass %	Na_2SO_4 mass %	Na_2SO_3[a] mol/kg	$NaHSO_3$[a] mol/kg	Na_2SO_4[a] mol/kg
15.150	5.170	13.910	1.828	0.755	1.489
14.180	4.850	13.610	1.670	0.692	1.422
13.900	5.300	13.510	1.641	0.769	1.415
10.320	7.260	11.070	1.148	0.978	1.09
3.140	16.560	12.230	0.366	2.338	1.265
8.060	16.030	13.200	1.020	2.456	1.482
9.620	14.600	12.560	1.207	2.219	1.399
3.580	34.820	7.250	0.523	6.157	0.939
7.560	30.910	6.310	1.086	5.379	0.804
9.930	23.750	4.440	1.273	3.688	0.505
9.300	25.260	4.300	1.207	3.970	0.495
11.670	27.990	5.060	1.675	4.866	0.644
12.050	25.880	1.970	1.591	4.138	0.231
10.950	23.860	3.850	1.416	3.738	0.442
11.010	24.960	7.230	1.538	4.223	0.896
10.500	24.330	7.090	1.434	4.026	0.859
11.720	11.320	12.990	1.454	1.701	1.430
13.120	13.210	12.210	1.694	2.066	1.399

[a] Molalities calculated by the compiler.

Solid phases: A - $Na_2SO_3.7H_2O$,
B - $Na_2SO_4.10H_2O$,
C - $Na_2S_2O_5.6Na_2SO_4$,
D - $Na_2S_2O_5$

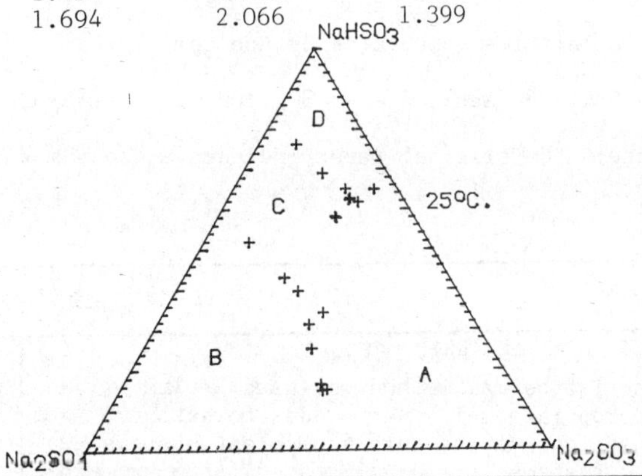

METHOD/APPARATUS/PROCEDURE:

The isothermal procedure was used. The desired amounts of salts and water were loaded into a vessel and stirred for 6 hr in a thermostat. Sodium hydrogen sulfite in the solution was determined by oxidation with hydrogen peroxide and titration of the H_2SO_4 liberated; sulfite plus hydrogen sulfite was determined iodometrically; and sulfite, hydrogen sulfite plus sulfate was determined as sulfate by titration with barium chloride (nitchromazo indicator) after oxidation with hydrogen peroxide. The solid phases were dried in air after washing with ethyl acetone. The solid phase composition was determined by the co-ordinate chain method of Noskov (1), and checked by chemical analysis.

SOURCE AND PURITY OF MATERIALS:

"Chemically pure" grade sodium sulfate and sodium pyrosulfate, and "special-purity" sodium sulfite was used.

ESTIMATED ERROR:

Temperature: a thermostat was used, but the error was not stated.
Analyses: no estimate possible.

REFERENCES:

1. Noskov, N.I. *Zh. Neorg. Khim.* 1959, *3*, 626.

COMPONENTS:

1. Sodium sulfite; Na_2SO_3; [7757-83-7]
2. Sodium pyrosulfite; $Na_2S_2O_5$; [7681-57-4]
3. Sodium sulfate; Na_2SO_4; [7757-82-6]
4. Water; H_2O; [7732-18-5]

ORIGINAL MEASUREMENTS:

Sotova, N.N.; Kuznetsova, A.G. Torocheshnikov, N.S.

Zh. Priklad. Khim. 1978, *51*, 940-3; *J. Appl. Chem. USSR (Eng. Transl.)* 1978, *51*, 905-8.

VARIABLES:

One temperature: 273 K
Concentrations of the components

PREPARED BY:

Mary R. Masson

EXPERIMENTAL VALUES:

Composition of equilibrium solutions at 0°C

Na_2SO_3 mass %	'$NaHSO_3$' mass %	SO_2total mass %	Na_2SO_4 mass %	Na_2SO_3[a] mol/kg	'$NaHSO_3$'[a] mol/kg	Na_2SO_4[a] mol/kg	Solid[b] phase
11.94	0.0	6.11	1.60	1.096	0.	0.130	A,B
11.82	2.39	7.47	1.25	1.109	0.272	0.104	A,B
10.92	4.40	8.25	0.91	1.034	0.505	0.076	A,B
10.46	8.04	10.25	0.65	1.026	0.056	0.057	A,B
9.59	8.16	8.89	0.51	0.931	0.959	0.044	A,B
10.21	12.3	12.74	0.35	1.050	1.532	0.032	A,B
9.09	11.67	11.8	0.42	0.915	1.423	0.038	A,B
9.27	16.19	14.67	0.06	0.987	2.089	0.006	A,B
8.03	16.65	14.32	0.04	0.846	2.125	0.004	A,B
0.0	34.6	21.3	0.61	0.	5.132	0.066	A,C
0.77	35.48	22.21	0.1	0.096	5.357	0.011	A,C
4.54	32.64	22.37	0.0	0.573	4.993	0.	B,C
5.11	34.82	24.0	0.20	0.677	5.589	0.024	B,C
5.03	33.07	22.89	0.23	0.647	5.153	0.026	A,B,C

[a] Molalities calculated by the compiler.

[b] Solid phases: A - $Na_2SO_4.10H_2O$, B - $Na_2SO_3.7H_2O$, C - $Na_2S_2O_5.6H_2O$

Note: the original paper includes a diagram with another dimension (water number).

AUXILIARY INFORMATION

METHOD APPARATUS/PROCEDURE:

The isothermal method was used. The appropriate salts were added to eutonic solutions of the three-component systems. The systems were equilibrated for 6 hr, and the components determined as before (1).

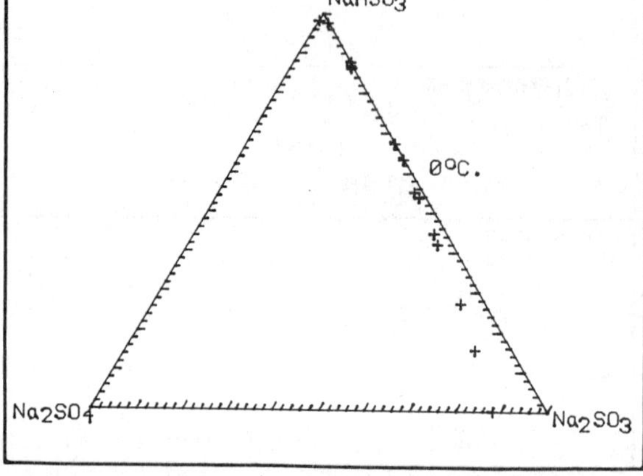

SOURCE AND PURITY OF MATERIALS:

Analytical-grade sodium pyrosulfite and sodium sulfate, and ultrapure sodium sulfite were used.

ESTIMATED ERROR:

Temperature: ±0.2 K

REFERENCES:

1. Sotova, N.N.; Kuznetsova, A.G.; Torocheshnikov, N.S. *Zh. Priklad. Khim.* 1978, *51*, 779.

COMPONENTS:	ORIGINAL MEASUREMENTS:
1. Sodium sulfite; Na_2SO_3; [7757-83-7] 2. Sodium pyrosulfite; $Na_2S_2O_5$; [7681-57-4] 3. Sodium sulfate; Na_2SO_4; [7757-82-6] 4. Water; H_2O; [7732-18-5]	Kuznetsova, A.G.; Sedova, V.A. **VINITI Deposited Document* 1981, 5711-81.

VARIABLES:	PREPARED BY:
Temperature: 288 K Composition	Mary R. Masson

EXPERIMENTAL VALUES:

Composition of saturated solutions

Na_2SO_3 mass %	$Na_2S_2O_5$ mass %	Na_2SO_4 mass %	Na_2SO_3[a] mol/kg	$Na_2S_2O_5$[a] mol/kg	Na_2SO_4[a] mol/kg
8.60	29.00	0.	1.093	2.445	0.
0.	34.10	5.70	0.	2.980	0.667
2.92	2.78	19.70	0.311	0.196	1.859
1.86	3.05	21.92	0.202	0.219	2.109
1.87	12.38	24.44	0.242	1.062	2.806
2.50	12.50	19.05	0.301	0.997	2.034
1.29	14.99	22.12	0.166	1.280	2.528
2.53	15.10	13.42	0.291	1.152	1.370
2.57	18.49	13.28	0.310	1.481	1.424
2.26	20.83	15.09	0.290	1.772	1.719
3.88	19.89	12.15	0.480	1.633	1.335
3.37	23.30	11.41	0.432	1.979	1.297
3.46	27.54	8.89	0.457	2.410	1.041
1.47	31.60	6.62	0.193	2.756	0.773

[a] Molalities calculated by the compiler.

Compositions of remainders and of assumed solid phases are also given in the original paper.

AUXILIARY INFORMATION

METHOD APPARATUS/PROCEDURE:

Isothermal method.

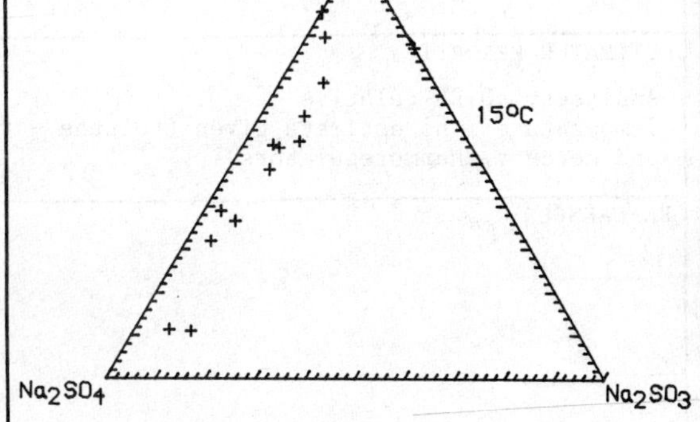

SOURCE AND PURITY OF MATERIALS:

ESTIMATED ERROR:

No estimates possible.

REFERENCES:

COMPONENTS:

1. Sodium sulfite; Na_2SO_3; [7757-83-7]
2. Ammonium sulfite; $(NH_4)_2SO_3$; [10196-04-0]
3. Ammonium chloride; NH_4Cl; [12125-02-9]
4. Sodium chloride; NaCl; [7647-14-5]
5. Water; H_2O; [7732-18-5]

ORIGINAL MEASUREMENTS:

Zil'berman, Ya.I.; Ivanov, P.T.

Zh. Priklad. Khim. 1941, *14*, 939-46.

VARIABLES:

Two temperatures: 333 and 358 K
Concentrations of the components

PREPARED BY:

Mary R. Masson

EXPERIMENTAL VALUES:

Composition of equilibrium solutions

SO_3^{2-}	Cl^-	NH_4^+	Na_2SO_3	$(NH_4)_2SO_3$	NaCl	NH_4Cl	Solid[a] phase
Temperature = 85°C							
1.44	15.82	-	2.27	-	26.08	-	A,C
-	26.88	10.08	-	-	11.57	29.94	C,D
41.22	-	16.95	5.51	54.67	-	-	A,B
29.78	12.24	-	-	43.18	-	18.46	B,D
23.19	13.27	15.65	6.31	27.80	-	20.00	A,D
26.96	13.20	17.42	4.78	34.63	-	19.91	A,D
31.54	12.16	18.96	4.91	41.22	-	18.34	A,B,D
8.51	21.60	11.11	13.40	-	-	32.60	A,C,D
Temperature = 60°C							
-	24.02	8.52	-	-	11.96	25.30	C,D
24.70	11.10	-	-	35.82	-	16.74	B',D
1.50	14.70	-	2.35	-	24.23	-	A,C
36.40	-	13.97	8.36	45.08	-	-	A,B'
8.51	20.45	10.55	12.84	0.52	-	30.84	A,C,D
26.92	10.47	16.20	4.25	35.11	=	15.79	A,B',D

[a] Solid phases: A - Na_2SO_3, B - $(NH_4)_2SO_3$, B' - $(NH_4)_2SO_3.H_2O$, C - NaCl, D - NH_4Cl

(continued on next page)

AUXILIARY INFORMATION

METHOD APPARATUS/PROCEDURE:

Solutions were equilibrated in glass test tubes fitted with spiral stirrers supplied with mercury seals. Freshly made salts were always used for each filling of a test tube. The anti-oxidant p-phenylenediamine was added to all solutions, and the work was done in an atmosphere of nitrogen, but experiments still had to be repeated often because of the formation of unacceptably high concentrations of sulfate.

Sulfite was determined by reaction with iodine solution, ammonium was volatilized by reaction with alkali, and collected in acid, the excess of which was titrated, sodium was weighed as sodium sulfate, total sulfur was weighed as barium sulfate, and chloride was titrated by the Volhard method.

If too much ammonia was found to have been lost, the experiment had to be repeated.

SOURCE AND PURITY OF MATERIALS:

Ammonium chloride, sodium chloride and sodium sulfite heptahydrate were commercial reagents. The $Na_2SO_3.7H_2O$ contained about 1% of sulfate.

Ammonium sulfite was made by saturating aqueous ammonia with sulfur dioxide, with cooling and in the presence of p-phenylenediamine as anti-oxidant. The product usually contained 0.5 - 1% of sulfate.

ESTIMATED ERROR:

Analyses: 0.2% relative
Temperature: no estimate given (toluene and mercury thermoregulators).

REFERENCES:

COMPONENTS:	ORIGINAL MEASUREMENTS:
1. Sodium sulfite; Na_2SO_3; [7757-83-7] 2. Ammonium sulfite; $(NH_4)_2SO_3$; [10196-04-0] 3. Ammonium chloride; NH_4Cl; [12125-02-9] 4. Sodium chloride; NaCl; [7647-14-5] 5. Water; H_2O; [7732-18-5]	Zil'berman, Ya.I.; Ivanov, P.T. *Zh. Priklad. Khim.* 1941, *14*, 939-46.

EXPERIMENTAL VALUES (continued):

Compositions of equilibrium solutions expressed as molalities[b], mol/kg

Na_2SO_3	$(NH_4)_2SO_3$	NaCl	NH_4Cl
Temperature = 85°C			
0.251	0.	6.228	0.
0.	0.	3.385	9.570
1.098	11.821	0.	0.
0.	9.692	0.	8.997
1.091	5.216	0.	8.148
0.932	7.330	0.	9.150
1.096	9.989	0.	9.650
1.969	0.	0.	11.286
Temperature = 60°C			
0.	0.	3.262	7.539
0.	6.501	0.	6.597
0.254	0.	5.647	0.
1.423	8.335	0.	0.
1.826	0.080	0.	10.333
0.752	6.740	0.	6.582

[b] Molalities calculated by the compiler.

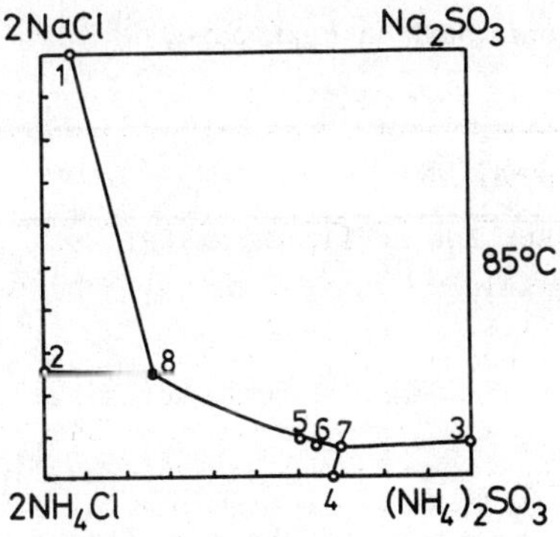

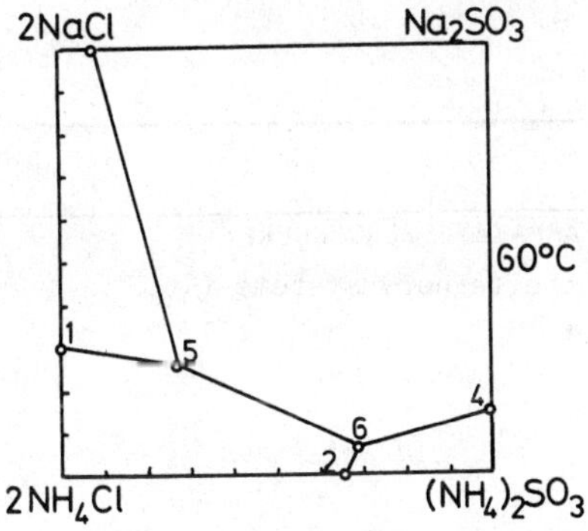

COMPONENTS:

1. Sodium sulfite; Na_2SO_3; [7757-83-7]
2. Ammonium sulfite; $(NH_4)_2SO_3$; [10196-04-0]
3. Sodium chloride; NaCl; [7647-14-5]
4. Ammonium chloride; NH_4Cl; [12125-02-9]
5. Water; H_2O; [7732-18-5]

ORIGINAL MEASUREMENTS:

Labash, J.A.; Lusby, G.R.

Can. J. Chem. 1955, *33*, 787-796.

VARIABLES:

Two temperatures: 293 and 333 K
Concentrations of the components

PREPARED BY:

Mary R. Masson

EXPERIMENTAL VALUES:

Composition of equilibrium solutions, expressed as mass %

NaCl	NH_4Cl	$(NH_4)_2SO_3$	Na_2SO_3	Na_2SO_4	NH_4HSO_3	Solid[a] phase
Temperature = 20°C						
17.63	14.87	-	-	-	-	A,D
10.24	16.54	-	8.20	0.65	-	A,B,D
10.47	16.50	-	8.51	0.46	-	A,B,D
23.68	0.0	-	4.35	0.21	-	u.d.
0.35	21.52	-	16.61	0.44	-	B,D
3.86	18.55	-	16.36	0.59	-	A,B,D
0.0	16.59	23.61	0.0	1.17[b]	-	D,E
0.0	0.0	27.43	14.98	0.96[b]	-	u.d.
0.0	0.0	25.75	17.26	1.30[b]	-	u.d.
0.0	14.58	17.23	11.60	1.51	-	B,D,E
0.0	13.88	17.88	10.71	2.91	-	B,D,E
0.0	13.80	15.78	15.18	0.89	-	B,D,E
0.0	17.33	1.68	22.25	0.71	-	A,B,C
21.43	0.0	0.0	7.44	0.86	-	u.d.
0.0	19.95	0.0	18.92	1.45	-	B,D
26.36	0.0	0.0	0.0	0.0	-	A
0.0	0.0	0.0	20.58	0.77	-	C
0.0	0.0	37.34	0.0	0.49	-	E
0.0	27.26	0.0	0.0	0.0	-	D

(continued on next page)

AUXILIARY INFORMATION

METHOD APPARATUS/PROCEDURE:

As for the ternary systems (1).

SOURCE AND PURITY OF MATERIALS:

As (1).

ESTIMATED ERROR:

Temperature: ±0.1 K
Analyses: 0.2% for sulfite, ammonium, chloride, 0.4% for sodium and sulfate.

REFERENCES:

1. Labash, J.A.; Lusby, G.R. *Can. J. Chem.* 1955, *33*, 774-86.

COMPONENTS:	ORIGINAL MEASUREMENTS:
1. Sodium sulfite; Na_2SO_3; [7757-83-7] 2. Ammonium sulfite; $(NH_4)_2SO_3$; [10196-04-0] 3. Sodium chloride; NaCl; [7647-14-5] 4. Ammonium chloride; NH_4Cl; [12125-02-9] 5. Water; H_2O; [7732-18-5]	Labash, J.A.; Lusby, G.R. *Can. J. Chem.* 1955, *33*, 787-796.

EXPERIMENTAL VALUES (continued):

Composition of equilibrium solutions, expressed as mass %

NaCl	NH_4Cl	$(NH_4)_2SO_3$	Na_2SO_3	Na_2SO_4	NH_4HSO_3	Solid[a] phase
Temperature = 60°C						
13.60	24.70	-	-	-	-	A,D
5.96	27.17	-	7.24	1.73	-	A,B,D
6.00	27.43	-	7.64	0.80	0.16[c]	A,B,D
25.77	0.0	-	1.89	0.30	-	A,B
0.0	30.48	-	12.42	0.76	-	B,D
0.0	29.20	0.41	14.77	0.62	0.07	A,B,D
0.0	17.84	37.72	0.0	0.91[b]	-	D,E
0.0	0.0	45.0	7.69	0.71	0.08	B,E
0.0	16.87	34.5	5.23	1.26	0.37	B,D,E
0.0	16.87	34.4	5.55	0.64	0.79	B,D,E
0.0	15.89	34.3	6.24	2.29	-	B,D,E
25.14	0.0	0.0	3.04	0.16	0.0	A,B
0.0	22.18	20.22	6.68	1.48	-	B,D
27.03	0.0	0.0	0.0	0.0	0.0	A
0.0	0.0	0.0	22.86	0.67	0.0	B
0.0	0.0	50.48	0.0	-	0.13	E
0.0	35.37	0.0	0.0	0.0	0.0	D

Composition of equilibrium solutions, expressed as mass % after equilibration for 6 - 7 weeks in rotating bottles.

NaCl	NH_4Cl	$(NH_4)_2SO_3$	Na_2SO_3	Na_2SO_4	NH_4HSO_3	Solid[a] phase
Temperature = 60°C						
0.0	29.18	0.41	14.90	0.80	0.02	B,D?
0.0	29.04	0.52	14.66	0.75	-	B,D?
0.0	29.15	0.29	14.31	0.80	0.04	B,D?
0.0	28.80	0.87	14.32	0.84	-	B,D?
0.0	29.08	0.29	14.75	0.59	-	B,D?
0.0	28.98	0.41	14.59	0.98	-	B,D?
0.0	30.04	0.0	12.57	0.71	-	B,D?
0.0	29.35	0.0	12.31	0.86	-	B,D?
0.0	30.20	0.0	12.88	0.56	-	B,D?
0.23	29.98	0.0	12.59	0.84	0.04	B,D?
0.0	29.40	0.0	12.84	0.49	-	B,D?
0.0	29.82	0.17	12.36	1.35	-	B,D?

[a] Solid phases: A - NaCl, B - Na_2SO_3, C $Na_2SO_3.7H_2O$, D - NH_4Cl, E - $(NH_4)_2SO_3.H_2O$

[b] $(NH_4)_2SO_4$ rather than Na_2SO_4

[c] $NaHSO_3$ rather than NH_4HSO_4

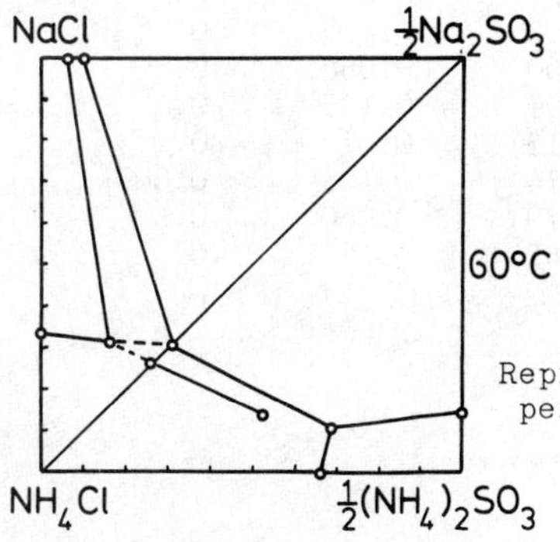

Reprinted by permission

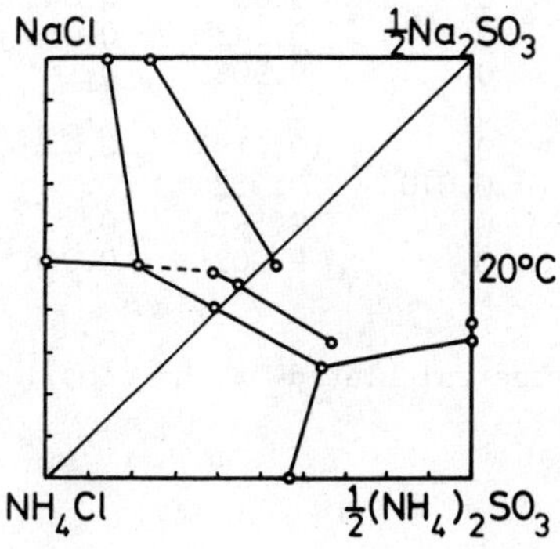

COMPONENTS:
1. Sodium sulfite; Na_2SO_3; [7757-83-7]
2. Ammonium sulfite; $(NH_4)_2SO_3$; [10196-04-0]
3. Sodium chloride; NaCl; [7647-14-5]
4. Ammonium chloride; NH_4Cl; [12125-02-9]
5. Water; H_2O; [7732-18-5]

ORIGINAL MEASUREMENTS:
Labash, J.A.; Lusby, G.R.

Can. J. Chem. 1955, *33*, 787-796.

EXPERIMENTAL VALUES (continued):

Compositions of equilibrium solutions expressed as molalities[d], mol/kg

NaCl	NH_4Cl	$(NH_4)_2SO_3$	Na_2SO_3	Na_2SO_4	NH_4HSO_3
Temperature = 20°C					
4.469	4.118	0.	0.	0.	
2.722	4.804	0.	1.011	0.071	
2.797	4.815	0.	1.054	0.051	
5.647	0.	0.	0.481	0.021	
0.098	6.587	0.	2.158	0.051	
1.089	5.719	0.	2.141	0.068	
0.	5.290	3.467	0.	0.151*	
0.	0.	4.171	2.099	0.128*	
0.	0.	3.981	2.459	0.177*	
0.	4.949	2.693	1.671	0.193	
0.	4.751	2.819	1.556	0.375	
0.	4.747	2.500	2.216	0.115	
0.	5.583	0.249	3.042	0.086	
5.218	0.	0.	0.840	0.086	
0.	6.249	0.	2.515	0.171	
6.125	0.	0.	0.	0.	
0.	0.	0.	2.076	0.069	
0.	0.	5.171	0.	0.055	
0.	7.006	0.	0.	0.	
Temperature = 60°C					
3.772	7.484	0.	0.	0.	0.
1.761	8.773	0.	0.992	0.210	0.
1.771	8.846	0.	1.046	0.097	0.027
6.121	0.	0.	0.208	0.029	0.
0.	10.114	0.	1.749	0.095	0.
0.	9.938	0.064	2.133	0.079	0.013
0.	7.662	7.461	0.	0.158	0.
0.	0.	8.329	1.312	0.107	0.017
0.	7.551	7.112	0.993	0.212	0.089
0.	7.554	7.094	1.055	0.108	0.191
0.	7.196	7.154	1.199	0.391	0.
6.003	0.	0.	0.337	0.016	0.
0.	8.387	3.521	1.072	0.211	0.
6.339	0.	0.	0.	0.	0.
0.	0.	0.	2.372	0.062	0.
0.	0.	8.800	0.	0.	0.027
0.	10.231	0.	0.	0.	0.

Compositions of solutions from rotating bottle experiments, expressed as molalities[d], mol/kg

0.	9.975	0.065	2.162	0.103	0.004
0.	9.866	0.081	2.114	0.096	0.
0.	9.835	0.045	2.049	0.102	0.007
0.	9.759	0.136	2.059	0.107	0.
0.	9.833	0.045	2.117	0.075	0.
0.	9.843	0.064	2.103	0.125	0.
0.	9.908	0.	1.760	0.088	0.
0.	9.546	0.	1.699	0.105	0.
0.	10.018	0.	1.813	0.070	0.
0.070	9.952	0.	1.774	0.105	0.007
0.	9.597	0.	1.779	0.060	0.
0.	9.902	0.026	1.742	0.169	0.

[d] Molalities calculated by the compiler.

COMPONENTS:

1. Sodium sulfite; Na_2SO_3; [7757-83-7]
2. 2-Naphthol; $C_{10}H_7O$; [135-19-3]
3. Sodium hydroxide; NaOH; [1310-73-2]
4. Water; H_2O; [7732-18-5]

ORIGINAL MEASUREMENTS:

Kogan, I.M.; Planovskii, A.N.; Evdokimov, A.N.

Anilinokrasochnaya Prom. 1934, 4, 34-7.

VARIABLES:

Three temperatures: 363, 373 and 379 K
Concentrations of the components

PREPARED BY:

Mary R. Masson

EXPERIMENTAL VALUES:

Composition of saturated solutions

Solvent	$[Na_2SO_3]$, g/100 ml solution			Data for 100°C soln.	
	90°C	100°C	106°C	Na_2SO_3[a] mol/kg	Na_2SO_3[a] mass %
Water	23.6	23.3	22.5	1.897	19.3
1% NaOH	22.7	21.8	21.3	1.765	18.2
3% NaOH	21.3	20.7	20.2	1.683	17.5
5% NaOH	19.6	18.8	18.5	1.500	15.9
3.6% 2-Naphthol in 1% NaOH	21.3	20.9	18.1	1.660	17.3
10.8% 2-Naphthol in 3% NaOH	17.4	16.6	15.7	1.082	12.0
18% 2-Naphthol in 5% NaOH	14.4	11.4	12.7	-	-

[a] Molalities and mass % values calculated by the compiler.

Compiler's comment: The authors also give titration volumes which do not seem to correspond with the concentrations given in the tables: comparison with other work suggests that they made an error when reporting the titrant concentration or sample volume, and that the values listed here are the correct ones.

AUXILIARY INFORMATION

METHOD APPARATUS/PROCEDURE:

An isothermal procedure. Sulfite was determined iodometrically.

SOURCE AND PURITY OF MATERIALS:

ESTIMATED ERROR:

Temperature: ±0.5°C for 90 and 100°C
±1°C for 106°C

Analyses: no estimate possible.

REFERENCES:

COMPONENTS:	EVALUATOR:
1. Sodium pyrosulfite; $Na_2S_2O_5$; [7681-57-4] 2. Water; H_2O; [7732-18-5]	Mary R. Masson, Dept. of Chemistry, University of Aberdeen, Meston Walk, Old Aberdeen, AB9 2UE, Scotland, UK. June 1984.

CRITICAL EVALUATION:

The binary system sodium pyrosulfite - water has been studied by only one group of workers (1), but data are also available from ternary systems (2 -8). Above 278.7 K the anhydrous salt is in equilibrium with solution. At lower temperatures, the stable solid phase is $Na_2S_2O_5.7H_2O$ [91498-96-3] (264 - 279 K), and there is also a metastable solid phase, $Na_2S_2O_5.6H_2O$ [91498-97-4] (263 - 277 K), according to Foerster *et al.* (1). Sotova *et al.* claim to have found, at 273.2 K, $Na_2S_2O_5$ (5) and $Na_2S_2O_5.6H_2O$ (4,13). One of the points given in (4), which is also given in (5), namely 31.31 mass % of $Na_2S_2O_5$ is in agreement with the data given by Foerster *et al.* for the 7-hydrate. The other point from (4) agrees neither with the 6-hydrate nor 7-hydrate data in (1).

There are four equations, for the equilibria with ice and with the three solids.

(1) ice, for 263.5 - 273.2 K (data from (1))

$$(T - 273.15) = -0.0150 - 0.365y + 0.00292y^2 - 0.000161y^3 \qquad s = 0.015 \text{ (11 pts)}$$

or

$$y = -0.0764 - 2.83(T - 273.2) + 0.0187(T - 273.2)^2 + 0.00488(T - 273.2)^3 \qquad s = 0.055 \text{ (11 pts)}$$

(2) $Na_2S_2O_5.6H_2O$, for 263 - 277 K (data from (1))

$$y = 33.0 + 1.29(T - 273.2) + 0.0226(T - 273.2)^2 - 0.0019(T - 273.2)^3 \qquad s = 0.149 \text{ (19 pts)}$$

(3) $Na_2S_2O_5.7H_2O$, for 264 - 27 K (data from (,4))

$$y = 31.2 + 1.068(T - 273.2) - 0.00280(T - 273.2)^3 \qquad s = 0.137 \text{ (9 pts)}$$

(4) $Na_2S_2O_5$, for 278 - 373 K (data from (1,2,3,6,7,8)

$$y = 37.4 + 0.1026(T - 273.2) + 0.000186(T - 273.2)^2 \qquad s = 0.288 \text{ (35 pts)}$$

In all the equations, T is the temperature in K, $y = 100w$ is the concentration expressed as mass % of $Na_2S_2O_5$, and s is the estimated standard deviation of the dependent variable about the regression line. In all cases, some, but not many, points were rejected.

COMPONENTS:	EVALUATOR:
1. Sodium pyrosulfite; $Na_2S_2O_5$; [7681-57-4] 2. Water; H_2O; [7732-18-5]	Mary R. Masson, Dept. of Chemistry, University of Aberdeen, Meston Walk, Old Aberdeen, AB9 2UE, Scotland, UK. June 1984.

CRITICAL EVALUATION: (continued)

TENTATIVE SOLUBILITIES

Although the data for the equilibrium with the anhydrous salt are essentially in agreement, there is a good deal of scatter, so the calculated solubilities are regarded as tentative. The following values were calculated from regression equation (4).

T/K	Solubility mass %	Solubility molality mol/kg
278.2	37.9	3.21
283.2	38.4	3.28
293.2	39.5	3.43
303.2	40.6	3.60
313.2	41.8	3.78
323.2	43.0	3.97
333.2	44.2	4.17
343.2	45.5	4.39
353.2	46.8	4.63
363.2	48.1	4.88
373.2	49.5	5.16

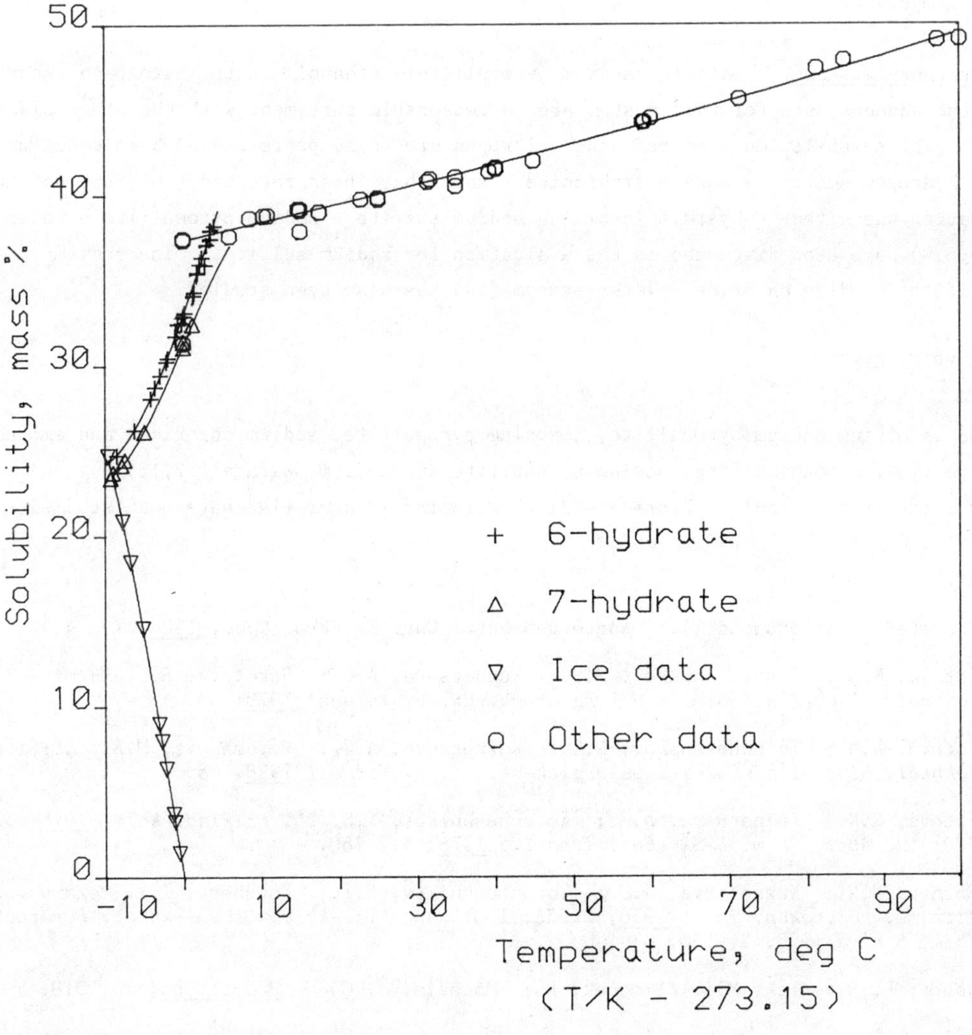

COMPONENTS:	EVALUATOR:
1. Sodium pyrosulfite; $Na_2S_2O_5$; [7681-57-4] 2. Water; H_2O; [7732-18-5]	Mary R. Masson, Dept. of Chemistry, University of Aberdeen, Meston Walk, Old Aberdeen, AB9 2UE, Scotland, UK. June 1984.

CRITICAL EVALUATION: (continued)

TERNARY SYSTEMS

Sodium pyrosulfite - sodium sulfate - water. Data for this system have been reported for 273.2 K (4,5), 288.2, 298.2, 308.2 and 318.2 K (6), 333.2 K (3) and 373.2 K (2). At temperatures up to 298.2 K the sulfate solid phase is $Na_2SO_4.10H_2O$; at higher temperatures it is the anhydrous salt. The pyrosulfite solid phase at 273.2 K has already been discussed. At higher temperatures, it is the anhydrous salt. At temperatures of 298.2 K and above, an additional phase is formed. Jäger (6) reports this as $Na_2S_2O_5.6Na_2SO_4$, which is in accordance with the data presented. Sotova, on the other hand (2,3) reports it as $Na_2S_2O_5.8Na_2SO_4$. The Sotova data show much more scatter, so it is possible that Sotova should also have found $Na_2S_2O_5.6Na_2SO_4$. However, Jäger's data for 318.2 K also show a bit of scatter, so it may be that the salt does not have a definite composition, or that both $Na_2S_2O_5.6Na_2SO_4$ and $Na_2S_2O_5.8Na_2SO_4$ are formed at the higher temperatures.

Other ternary systems. Data for sodium pyrosulfite - ethanol - water have been reported (7); the aqueous data for this system are in reasonable agreement with the other binary data. The solubilities measured under nitrogen are to be preferred. Some data for sodium hydrogen sulfite - sodium dithionite - water have been reported (10), but not much information was given. Systems involving sodium sulfite - sodium pyrosulfite - water (2,4,6,8,9) have been discussed in the evaluation for sodium sulfite. The sodium pyrosulfite - sodium chloride - water system (14) has also been studied.

MORE COMPLEX SYSTEMS

Systems involving sodium pyrosulfite, ammonium pyrosulfite, sodium chloride, and ammonium chloride (11), sodium sulfite, sodium pyrosulfite and sodium sulfate (12,13,15), and sodium pyrosulfite - sodium chloride - diethanolamine - water (14) have been studied.

REFERENCES

1. Foerster, F.; Brosche, A.; Norberg-Schutz, Chr. *Z. Phys. Chem.* 1924, *10*, 435.

2. Sotova, N.N.; Torocheshnikov, N.S.; Kuznetsova, A.G.; Sokolova, E.I. *Khimiya i Tekhnol. Mineral'n. Solei i Galurgichesk. Pr.-v, Varnaul* 1978, 53.

3. Sotova, N.N.; Torocheshnikov, N.S.; Kuznetsova, A.G.; Poroshkova, M.A. *Khimiya i Tekhnol. Mineral'n. Solei i Galurgichesk. Pr.-v, Varnaul* 1978, 65.

4. Sotova, N.N.; Kuznetsova, A.G.; Torocheshnikov, N.S. *Zh. Priklad. Khim.* 1978, *51*, 779; *J. Appl. Chem. USSR (Eng. Transl.)* 1978, *51*, 760.

5. Sotova, N.N.; Kuznetsova, A.G.; Torocheshnikov, N.S.; Kononova, I.V. *Mezhvuz. Sb. Altaisk. Politekhn. In.-t* 1976, 2(57), 150: and *Fiz.-Khim. Osn. Tekhnol. Pererab. Khim. Syr'ya* 1976, 2, 150.

6. Jäger, L.; Rejlek, M.; Klimeček, R.; Machala, J. *Chem. Prům.* 1960, *10*, 518.

COMPONENTS:	EVALUATOR:
1. Sodium pyrosulfite; $Na_2S_2O_5$; [7681-57-4] 2. Water; H_2O; [7732-18-5]	Mary R. Masson, Dept. of Chemistry, University of Aberdeen, Meston Walk, Old Aberdeen, AB9 2UE, Scotland, UK. June 1984.

CRITICAL EVALUATION: (continued)

7. Navrátil, J.; Nývlt, J. *Chem. Prům.* 1968, *18*, 612.

8. Arii, K. *Bull. Inst. Phys. Chem. Research* 1926, *6*, 1065 (in Japanese); *Sci. Rep. Tohoku Imp. Univ.* 1926, *6*, 1065.

9. Sotova, N.N.; Torocheshnikov, N.S.; Kuznetsova, A.G.; Poroshkova, M.A. *Khimiya i Tekhnol. Mineral'n. Solei i Galurgichesk. Pr.-v, Varnaul* 1978, 59.

10. Iijima, T.; Kageyama, T. *Kanto Gakuin Daigaku Kogakubu Kenkyu Hokuku* 1972, *16*, 69.

11. Zil'berman, Ya.I.; Ivanov, P.T. *Zh. Priklad. Khim.* 1941, *14*, 939.

12. Kuznetsova, A.G.; Yaroshenko, L.B. *Zh. Priklad. Khim.* 1981, 54, 2197.

13. Sotova, N.N.; Kuznetsova, A.G.; Torocheshnikov, N.S. *Zh. Priklad. Khim.* 1978, *51*, 940.

14. Yavorskii, V.T.; Perekupko, T.V.; Matsyk, L.V. *Zh. Priklad. Khim.* 1984, 57, 3; *J. Appl. Chem. USSR (Eng. Transl.)* 1984, 57, 1.

15. Kuznetsova, A.G.; Sedova, V.A. *VINITI Deposited Document* 1981, 5711-81.

COMPONENTS:

1. Sodium pyrosulfite; $Na_2S_2O_5$; [7681-57-4]

2. Water; H_2O; [7732-18-5]

ORIGINAL MEASUREMENTS:

Foerster, F.; Brosche, A.; Norberg-Schutz, Chr.

Z. Phys. Chem. 1924, *10*, 435-96.

VARIABLES:

Temperature: 263 - 370 K

PREPARED BY:

Mary R. Masson

EXPERIMENTAL VALUES:

Temp.	$Na_2S_2O_5$ mass %	$Na_2S_2O_5$[a] mol/kg
0.0	37.47	3.152
0.0	37.36	3.137
6.0	37.60	3.170
8.6	38.65[b]	3.314
10.4	38.79	3.334
10.9	38.80	3.335
14.8	39.20	3.392
15.0	39.12	3.380
15.0	39.20	3.392
17.5	39.02	3.366
17.5	39.01	3.365
22.8	39.77	3.473
31.4	40.79	3.624
31.7	40.94	3.646
39.3	41.39	3.715
40.2	41.60	3.747
40.2	41.52	3.735
59.0	44.27	4.179
59.2	44.29	4.182
71.4	45.62	4.413
81.4	47.40	4.740
85.0	47.89	4.834
97.2	49.06	5.066

Solid phase: $Na_2S_2O_5$

[b] Result considered particularly reliable by the authors.

(continued on next page)

AUXILIARY INFORMATION

METHOD APPARATUS/PROCEDURE:

Solids were equilibrated with solution under a hydrogen atmosphere, in a vessel maintained in a thermostat. Samples for analyses were withdrawn through a tube plugged with cotton wool.

Samples were reacted with an excess of standard iodine solution, and the excess was back-titrated with thiosulfate.

A Beckman apparatus was used for the determination of freezing points (1).

SOURCE AND PURITY OF MATERIALS:

A 30% solution of sodium hydroxide or a hot saturated soda solution was saturated with sulfur dioxide. This solution was allowed to cool to not lower than 25°C; crystals of $Na_2S_2O_5$ were obtained.

(Below 25°C, a hydrate containing less than 1 water molecule is formed.)

ESTIMATED ERROR:

Temperature: ±0.1 K
Analyses: no accurate estimate possible.

REFERENCES:

1. Ostwald, W.; Luther, R. *Hand-und Hilfsbuch zur Ausfuhrung physicochemischer Messungen* 5th Ed., Akademische Verlag., Leipzig, 1931.

COMPONENTS:	ORIGINAL MEASUREMENTS:
1. Sodium pyrosulfite; $Na_2S_2O_5$; [7681-57-4]	Foerster, F.; Brosche, A.; Norberg-Schutz, Chr.
2. Water; H_2O; [7732-18-5]	*Z. Phys. Chem.* <u>1924</u>, *10*, 435-96.

EXPERIMENTAL VALUES (continued)):

Temp.	$Na_2S_2O_5$ mass %	$Na_2S_2O_5$[a] mol/kg	
-8.5	24.75	1.730	
-6.3	26.22*	1.869	
-4.2	28.10*	2.056	
-3.65	28.75	2.123	
-3.0	29.45	2.196	
-2.2	30.25*	2.281	
-2.0	30.45	2.303	
-1.0	31.75	2.447	
-0.7	32.45	2.527	
-0.2	32.50	2.533	
0.0	32.90	2.579	Solid phase: $Na_2S_2O_5.6H_2O$
0.2	33.10	2.603	
1.3	34.10	2.722	
1.4	34.30	2.746	
1.8	35.40	2.883	
1.8	35.45	2.889	
2.1	35.95	2.953	
2.4	36.35	3.004	
2.8	35.90	2.946	
3.0	37.10	3.103	
3.5	37.45	3.150	
3.5	37.95	3.127	
4.0	38.20	3.252	
-9.3	23.45	1.611	
-9.0	23.77	1.640	
-7.9	24.20	1.679	
-7.5	24.50	1.707	Solid phase: $Na_2S_2O_5.7H_2O$
-5.0	26.15	1.863	
0.0	31.10	2.374	
0.0	31.45	2.413	
+1.2	32.45*	2.527	
-0.56	1.435	0.0766	
-1.18	3.275	0.178	
-2.28	6.40	0.360	
-3.19	9.00	0.520	
-1.35	3.75	0.205	Solid phase: ice
-2.82	8.00	0.457	
-5.24	14.64	0.902	
-6.81	18.50	1.194	
-7.84	20.92	1.392	
-9.7	24.73	1.728	

[a] Molalities calculated by the compiler.

COMPONENTS:	ORIGINAL MEASUREMENTS:
1. Sodium pyrosulfite; $Na_2S_2O_5$; [7681-57-4] 2. Sodium sulfate; Na_2SO_4; [7757-82-6] 3. Water; H_2O; [7732-18-5]	Sotova, N.N.; Torocheshnikov, N.S.; Kuznetsova, A.G.; Sokolova, E.I. *Khimiya i Tekhnol. Mineral'n. Solei i Galurgichesk. Pr.-v, Varnaul* 1978, 53-59.
VARIABLES:	PREPARED BY:
One temperature: 373 K Concentrations of the components	Mary R. Masson

EXPERIMENTAL VALUES:

Composition of equilibrium solutions at 100°C

Na_2SO_4 mass %	'$NaHSO_3$' mass %	$Na_2S_2O_5$ mass %	Na_2SO_4[a] mol/kg	$Na_2S_2O_5$[a] mol/kg	Solid[b] phase
-	53.8	49.15	0.	5.082	B
0.89	53.19	48.59	0.124	5.059	B
1.99	53.5	48.9	0.285	5.238	B + C
2.25	52.1	47.57	0.316	4.987	C
2.7	43.75	39.96	0.332	3.666	C
4.4	38.04	35.57	0.516	3.117	C
10.0	27.02	24.7	1.078	1.990	C
12.9	23.3	21.3	1.380	1.703	C
18.1	15.2	14.2	1.882	1.103	C
21.1	12.6	11.5	2.204	0.898	C
22.9	11.82	10.8	2.432	0.857	C
22.3	12.0	10.96	2.352	0.864	C + D
22.2	11.58	10.78	2.332	0.846	D
25.6	6.27	5.85	2.629	0.449	D
28.1	2.14	1.96	2.829	0.147	D
29.0	0.79	0.72	2.905	0.054	D
29.6	-	-	2.960	0.	D

[a] Molalities calculated by the compiler.

[b] Solid phases: B - $Na_2S_2O_5$, C - $Na_2S_2O_5.8Na_2SO_4$, D - Na_2SO_4

AUXILIARY INFORMATION

METHOD APPARATUS/PROCEDURE:

An isothermal method was used.

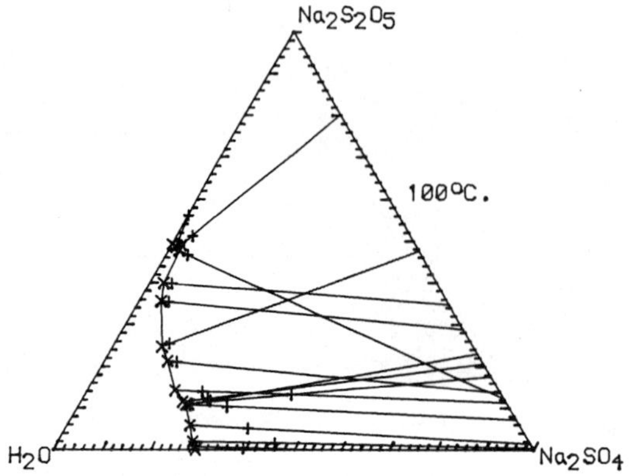

SOURCE AND PURITY OF MATERIALS:

ESTIMATED ERROR:

No estimates possible.

REFERENCES:

COMPONENTS:	ORIGINAL MEASUREMENTS:
1. Sodium pyrosulfite; $Na_2S_2O_5$; [7681-57-4] 2. Sodium sulfate; Na_2SO_4; [7757-82-6] 3. Water; H_2O; [7732-18-5]	Sotova, N.N.; Torocheshnikov, N.S.; Kuznetsova, A.G.; Poroshkova, M.A.; *Khimiya i Tekhnol. Mineral'n. Solei i Galurgichesk. Pr.-v, Varnaul* 1978, 65-69.
VARIABLES: One temperature: 333 K Concentrations of the components	PREPARED BY: Mary R. Masson

EXPERIMENTAL VALUES:

Composition of equilibrium solutions at 60°C

Na_2SO_4 mass %	$Na_2S_2O_5$ mass %	Na_2SO_4[a] mol/kg	$Na_2S_2O_5$[a] mol/kg	Solid[b] phase
31.5	0.0	3.237	0.	A
31.2	0.73	3.227	0.056	A
30	1.8	3.097	0.139	A
29.6	3.0	3.092	0.234	A
29.9	3.4	3.156	0.268	A + B
27.8	3.5	2.849	0.268	B
26.2	5.5	2.701	0.424	B
25.4	6.1	2.611	0.468	B
19.1	10.9	1.921	0.819	B
17.6	13.7	1.804	1.049	B
14.6	17.4	1.512	1.346	B
3.8	32.2	0.418	2.647	B
3.6	34.3	0.408	2.905	B
3.8	40.8	0.483	3.874	B + C
3.0	41.5	0.381	3.933	C
1.0	45.5	0.132	4.474	C
0.0	44.5	0.	4.218	C

[a] Molalities calculated by the compiler.

[b] Solid phases: A - Na_2SO_4, B - $Na_2S_2O_5 . 8Na_2SO_4$, C - $Na_2S_2O_5$

AUXILIARY INFORMATION

METHOD APPARATUS/PROCEDURE:

The isothermal method was used.

Note: Na_2SO_3 appears erroneously in the table-heading in the original, in place of Na_2SO_4.

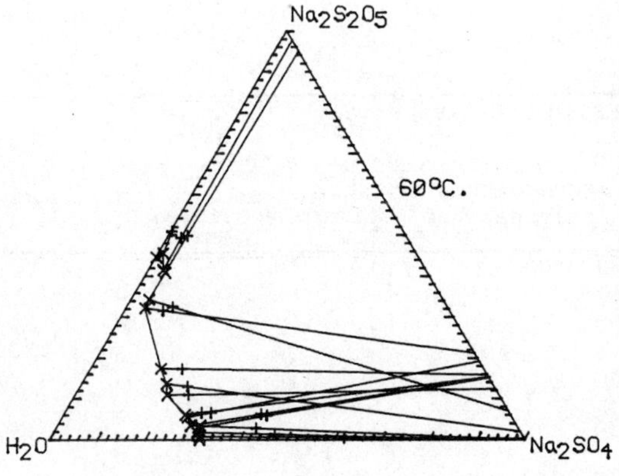

SOURCE AND PURITY OF MATERIALS:

ESTIMATED ERROR:

Temperature: ±0.1 K
Analyses: no estimate possible.

REFERENCES:

COMPONENTS:

1. Sodium pyrosulfite; $Na_2S_2O_5$; [7681-57-4]
2. Sodium sulfate; Na_2SO_4; [7757-82-6]
3. Water; H_2O; [7732-18-5]

ORIGINAL MEASUREMENTS:

Sotova, N.N.; Kuznetsova, A.G.; Torocheshnikov, N.S.

Zh. Priklad. Khim. 1978, *51*, 779-84; *J. Appl. Chem. USSR (Eng. Transl.)* 1978, *51*, 760-4.

VARIABLES:

One temperature: 273 K
Concentrations of the components

PREPARED BY:

Mary R. Masson

EXPERIMENTAL VALUES: Composition of equilibrium solutions at 0°C

Na_2SO_4 mass %	'$NaHSO_3$' mass %	$Na_2S_2O_5$ mass %	Na_2SO_4[a] mol/kg	'$NaHSO_3$'[a] mol/kg	Na_2SO_4[a] mol/kg	$Na_2S_2O_5$[a] mol/kg	Solid[b] phase
4.50	0.0	0.	0.332	0.	0.332	0.	A
3.28	4.86	4.44	0.251	0.508	0.250	0.253	A
2.75	6.50	5.94	0.213	0.688	0.212	0.342	A
2.09	12.05	11.01	0.171	1.349	0.169	0.666	A
1.85	19.40	17.72	0.165	2.367	0.162	1.159	A
1.83	24.56	22.43	0.175	3.206	0.170	1.558	A
1.31	31.42	28.70	0.137	4.488	0.132	2.157	A
0.64	34.65	31.65	0.070	5.146	0.067	2.459	A + B
0.58	34.57	31.57	0.063	5.123	0.060	2.448	A + B
0.62	34.62	31.62	0.067	5.137	0.064	2.455	A + B
0.48	34.38	31.34	0.052	5.072	0.050	2.418	B
0.32	34.23	31.29	0.034	5.026	0.033	2.407	B
0.0	34.31	31.31	0.	5.019	0.	2.398	B

[a] Molalities calculated by the compiler.
[b] Solid phases: A - $Na_2SO_4.10H_2O$, B - $Na_2S_2O_5.6H_2O$

AUXILIARY INFORMATION

METHOD/APPARATUS/PROCEDURE:

The isothermal method was used. p-Phenylenediamine was used as anti-oxidant. Total sulfite was determined iodometrically. Bisulfite was titrated with alkali as bisulfate after oxidation with peroxide. Sodium sulfate was then weighed to obtain total sulfate.

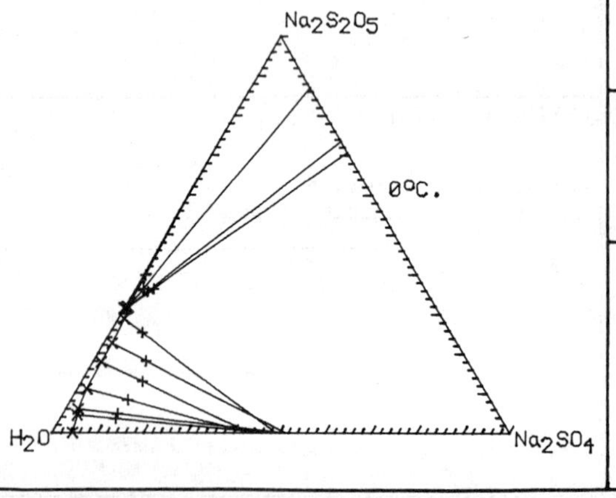

SOURCE AND PURITY OF MATERIALS:

Sodium sulfate and sodium pyrosulfite were of analytical grade.

ESTIMATED ERROR:

Temperature: ±0.2 K
Analyses: no estimate possible.

REFERENCES:

COMPONENTS:

1. Sodium pyrosulfite; $Na_2S_2O_5$; [7681-57-4]
2. Sodium sulfate; Na_2SO_4; [7757-82-6]
3. Water; H_2O; [7732-18-5]

ORIGINAL MEASUREMENTS:

Sotova, N.N.; Kuznetsova, A.G.; Torocheshnikov, N.S.; Kononova, I.V.

Mezhvuz. Sb. Altaisk. Politekhn. In.-t. 1976, 2(57), 150-5; also *Fiz.-Khim. Osn. Tekhnol. Pererab. Khim. Syr'ya* 1976, 2, 150-5.

VARIABLES:

One temperature: 273 K
Concentration of the components

PREPARED BY:

Mary R. Masson

EXPERIMENTAL VALUES: Composition of equilibrium solutions at 0°C

Na_2SO_4 mass %	'NaHSO₃' mass %	Na_2SO_4[a] mol/kg	$NaHSO_3$[a] mol/kg	$Na_2S_2O_5$[b] mass %	Na_2SO_3[a] mol/kg	$Na_2S_2O_5$[a] mol/kg	Solid[c] phase
4.50	0.0	0.332	0.	0.	0.332	0.	A
3.28	4.86	0.251	0.508	4.44	0.250	0.253	A
2.75	6.50	0.213	0.688	5.94	0.212	0.342	A
2.09	12.05	0.171	1.349	11.01	0.169	0.666	A
1.65	19.40	0.165	2.367	17.72	0.162	1.159	A
1.83	24.56	0.175	3.206	22.43	0.170	1.558	A
1.31	31.42	0.137	4.488	28.70	0.132	2.157	A
0.64	34.65	0.070	5.146	31.65	0.067	2.459	A + C
0.58	34.60	0.063	5.130	31.60	0.060	2.451	A + C
0.66	34.57	0.072	5.129	31.58	0.069	2.452	A + C
0.65	34.62	0.071	5.140	31.62	0.068	2.456	A + C
0.32	34.23	0.034	5.026	31.27	0.033	2.405	C
0.0	34.31	0.	5.019	31.34	0.	2.401	C

[a] Molalities calculated by the compiler.

[b] Mass % of $Na_2S_2O_5$ calculated by the compiler.

[c] Solid phases: A - $Na_2SO_4.10H_2O$, C - $Na_2S_2O_5$

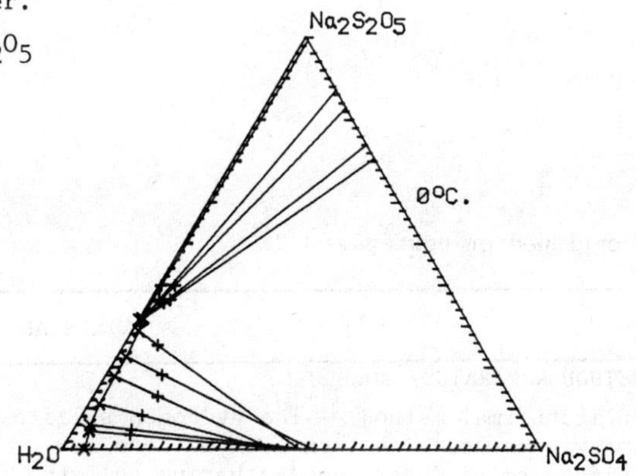

METHOD/APPARATUS/PROCEDURE:

The isothermal method was used.

SOURCE AND PURITY OF MATERIALS:

ESTIMATED ERROR:

No estimates possible.

REFERENCES:

COMPONENTS:	ORIGINAL MEASUREMENTS:
1. Sodium pyrosulfite; $Na_2S_2O_5$; [7681-57-4]	Jäger, L.; Rejlek, M.; Klimeček, R.; Machala, J.
2. Sodium sulfate; Na_2SO_4; [7757-82-6]	*Chem. Prům.* 1960, *10*, 518-20.
3. Water; H_2O; [7732-18-5]	

VARIABLES:	PREPARED BY:
Four temperatures: 288 - 318 K Concentrations of the components	Mary R. Masson

EXPERIMENTAL VALUES:

Composition of equilibrium solutions

Na_2SO_4 mass %	$Na_2S_2O_5$ mass %	Na_2SO_4[a] mol/kg	$Na_2S_2O_5$[a] mol/kg	Solid[b] phase
Temperature = 15°C				
11.8	-	0.942	0.	A
10.4	4.7	0.862	0.291	A
8.4	11.1	0.735	0.725	A
7.1	16.9	0.658	1.170	A
6.1	25.3	0.626	1.940	A
5.7	34.2	0.668	2.993	A
5.7	34.1	0.667	2.980	A + B
3.6	35.5	0.416	3.066	B
-	37.9	0.	3.210	B

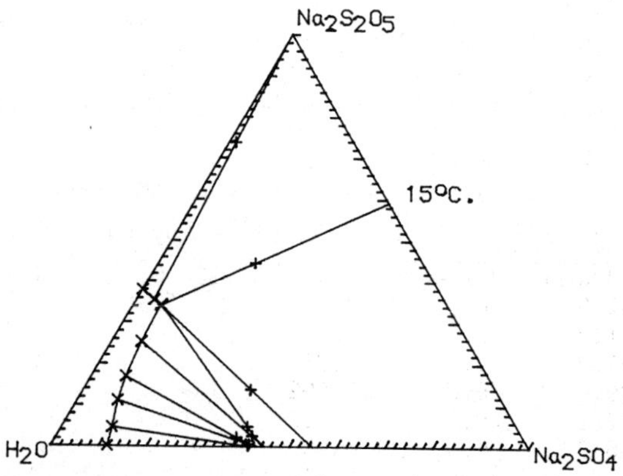

(continued on next page)

AUXILIARY INFORMATION

METHOD APPARATUS/PROCEDURE:	SOURCE AND PURITY OF MATERIALS:
An isothermal method. The hydrogen sulfite ion was masked with formaldehyde to allow sulfate to be determined as barium sulfate. Solid phases were identified by Schreinemakers' method, and by microscopy and X-ray diffraction.	Sodium sulfate was of analytical grade.

ESTIMATED ERROR:

Temperature: ±0.1 K
Analyses: no estimate possible.

REFERENCES:

COMPONENTS:	ORIGINAL MEASUREMENTS:
1. Sodium pyrosulfite; $Na_2S_2O_5$ [7681-57-4]	Jäger, L.; Rejlek, M.; Klimeček, R.; Machala, J.
2. Sodium sulfate; Na_2SO_4; [7757-82-6]	*Chem. Prům.* 1960, *10*, 518-20.
3. Water; H_2O; [7732-18-5]	

EXPERIMENTAL VALUES (continued):

Composition of equilibrium solutions

Na_2SO_4 mass %	$Na_2S_2O_5$ mass %	Na_2SO_4[a] mol/kg	$Na_2S_2O_5$[a] mol/kg	Solid[b] phase
Temperature = 25°C				
21.7	-	1.951	0.	A
20.0	5.0	1.877	0.351	A
18.8	8.1	1.811	0.583	A
18.5	9.1	1.799	0.661	A
17.7	11.1	1.750	0.820	A
17.3	12.9	1.745	0.972	A
16.6	14.9	1.706	1.144	A
16.7	17.4	1.784	1.389	A + D
12.4	21.7	1.325	1.732	D
9.7	26.2	1.065	2.150	D
8.4	28.4	0.936	2.364	D
7.0	31.1	0.796	2.643	D
5.2	35.6	0.618	3.163	B + D
5.1	35.7	0.607	3.172	B
3.6	36.9	0.426	3.262	B
1.9	38.1	0.223	3.340	B
-	39.8	0.	3.478	B
Temperature = 35°C				
33.0	-	3.468	0.	C
32.3	0.8	3.384	0.063	C
31.7	1.6	3.346	0.126	C
31.2	2.5	3.313	0.198	C + D
26.0	6.0	2.692	0.464	D
22.2	9.3	2.282	0.714	D
16.3	16.0	1.695	1.243	D
11.4	22.6	1.216	1.801	D
8.2	28.6	0.913	2.380	D
7.3	30.3	0.824	2.554	D
4.3	37.7	0.522	3.419	B + D
-	41.1	0.	3.671	B
Temperature = 45°C				
32.2	-	3.344	0.	C
31.6	0.8	3.291	0.062	C
30.8	1.2	3.189	0.093	C
30.9	1.7	3.228	0.133	C
30.0	2.5	3.129	0.195	C
29.6	2.7	3.078	0.210	D
27.8	3.8	2.861	0.292	D
26.3	5.3	2.707	0.408	D
21.7	9.2	2.211	0.700	D
18.6	13.0	1.914	1.000	D
15.8	15.8	1.626	1.215	D
13.9	19.3	1.465	1.520	D
9.7	24.8	1.043	1.992	D
9.0	25.9	0.973	2.093	D
7.8	28.3	0.859	2.330	D
5.7	32.9	0.654	2.819	D
3.8	38.8	0.466	3.556	B + D
2.8	39.4	0.341	3.586	B
-	42.0	0.	3.809	B

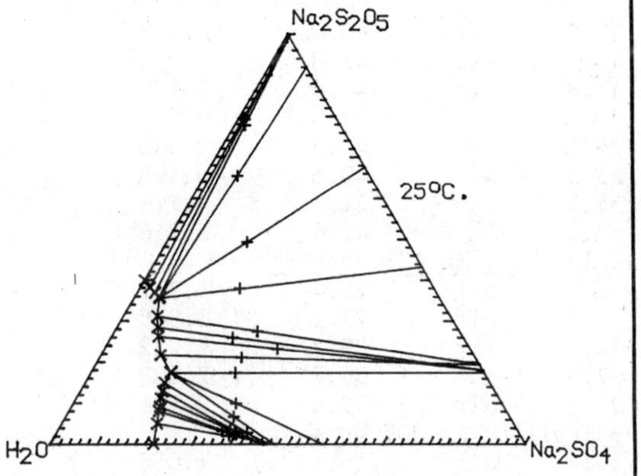

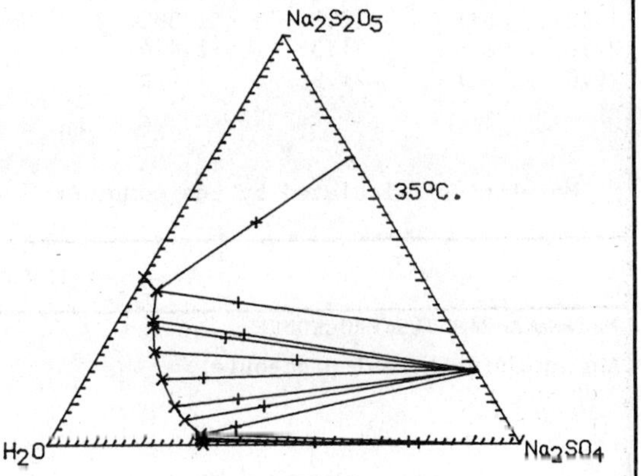

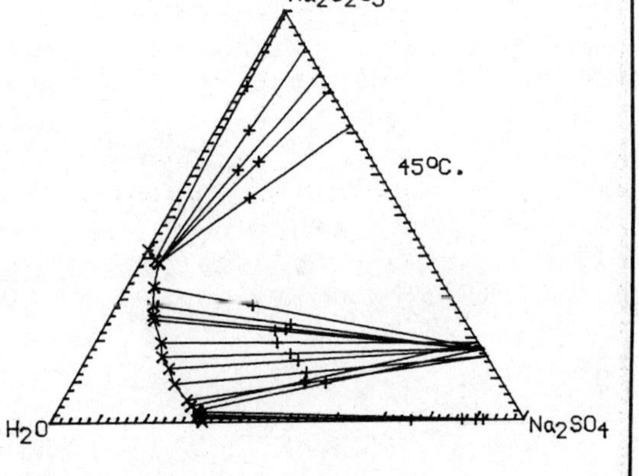

[a] Molalities calculated by the compiler.

[b] Solid phases: A - $Na_2SO_4.10H_2O$, B - $Na_2S_2O_5$, C - Na_2SO_4, D - $Na_2S_2O_5.6Na_2SO_4$

COMPONENTS:	ORIGINAL MEASUREMENTS:
1. Sodium pyrosulfite; $Na_2S_2O_5$; [7681-57-4]	Navrátil, J.; Nývlt, J.
2. Ethanol; C_2H_5OH; [64-17-5]	*Chem. Prům.* 1968, *18*, 612-4.
3. Water; H_2O; [7732-18-5]	

VARIABLES:	PREPARED BY:
Temperature: 276-308 K Ethanol concentration	Mary R. Masson

EXPERIMENTAL VALUES:

Solubility in water

t/°C	Atmos-phere	$Na_2S_2O_5$ mass %	$Na_2S_2O_5$[a] mol/kg
26.5	air	40.0	3.507
42.5	air	42.8	3.936
43.8	air	42.5	3.888
24.1	N_2	40.5	3.581
25.3	N_2	40.6	3.595
34.5	N_2	41.4	3.716
35.1	N_2	41.4	3.716

Solubility in 10% ethanol

t/°C	Atmos-phere	$Na_2S_2O_5$ mass %	$Na_2S_2O_5$[a] mol/kg
3.0	air	25.4	1.791
9.1	air	27.1	1.956
16.5	air	29.7	2.222
22.5	air	30.7	2.330
27.5	air	32.4	2.521

Solubility in 30% ethanol

t/°C	Atmos-phere	$Na_2S_2O_5$ mass %	$Na_2S_2O_5$[a] mol/kg
9.4	air	16.7	1.055
11.1	air	18.1	1.163
16.6	air	20.0	1.315
19.0	air	20.8	1.382
22.8	air	21.3	1.424
26.0	air	21.9	1.475

Solubility in 20% ethanol

t/°C	Atmos-phere	$Na_2S_2O_5$ mass %	$Na_2S_2O_5$[a] mol/kg
2.6	N_2	21.5	1.441
6.0	N_2	22.5	1.527
10.1	N_2	23.2	1.589
11.5	N_2	23.7	1.634
13.8	N_2	24.3	1.689
16.1	N_2	24.9	1.744
18.5	N_2	25.5	1.801
21.9	N_2	26.1	1.858
24.9	N_2	26.6	1.906
27.2	N_2	26.8	1.926
27.8	N_2	27.2	1.965
29.1	N_2	27.1	2.066
32.1	N_2	28.2	2.180
34.2	N_2	29.3	2.180
6.5	air	24.1	1.670
9.5	air	24.5	1.707
15.5	air	25.5	1.801
16.5	air	25.8	1.829
19.0	air	26.5	1.897
21.1	air	27.0	1.946
25.5	air	27.8	2.025
30.5	air	28.6	2.107
33.3	air	30.0	2.254
35.4	air	29.8	2.233

[a] Molalities calculated by the compiler.

AUXILIARY INFORMATION

METHOD APPARATUS/PROCEDURE:	SOURCE AND PURITY OF MATERIALS:
A simple isothermal procedure.	Analytical grade $Na_2S_2O_5$ was obtained from Carlo Erba, Milan.
	ESTIMATED ERROR: No estimates possible.
	REFERENCES:

COMPONENTS:

1. Sodium hydrogen sulfite; $NaHSO_3$; [7631-90-5]
2. Sodium dithionite; $Na_2S_2O_4$; [7775-14-6]
3. Water; H_2O; [7732-18-5]

ORIGINAL MEASUREMENTS:

Iijima, T.; Kageyama, T.

Kanto Gakuin Daigaku Kogakubu Kenkyu Hokoku 1972, *16*, 69-74.

VARIABLES:

Concentrations of the components
One temperature: 278 K

PREPARED BY:

Mary R. Masson

EXPERIMENTAL VALUES:

Extreme points: $Na_2S_2O_4$ = 13.0 g/100 g of solution (0.858[a] mol/kg)
$NaHSO_3$ = 69.3 g/100 g of solution (21.693[a] mol/kg)

[a] Molalities calculated by the compiler.

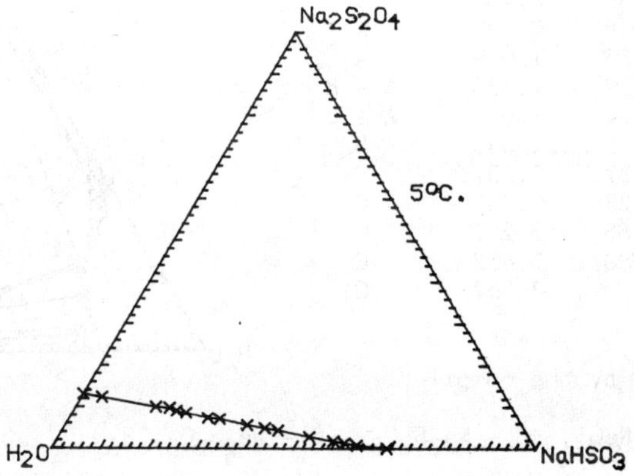

AUXILIARY INFORMATION

METHOD APPARATUS/PROCEDURE:

Experiments were done in a nitrogen atmosphere of 0.5 kg/cm^2.

SOURCE AND PURITY OF MATERIALS:

Not stated.

ESTIMATED ERROR:

Temperature: ±0.5 K

REFERENCES:

COMPONENTS:

1. Sodium pyrosulfite; $Na_2S_2O_5$; [7681-57-4]
2. Sodium chloride; NaCl; [7647-14-5]
3. Water; H_2O; [7732-18-5]

ORIGINAL MEASUREMENTS:

Yavorskii, V.T.; Perekupko, T.V.; Matsyk, L.V.

Zh. Priklad. Khim. 1984, 57, 3-7; *J. Appl. Chem. USSR (Eng. Transl.)* 1984, 57, 1-4.

VARIABLES:

Temperature: 298 K
Composition

PREPARED BY:

Mary R. Masson

EXPERIMENTAL VALUES:

Composition of saturated solutions

NaCl mass %	$Na_2S_2O_5$ mass %	NaCl[a] mol/kg	$Na_2S_2O_5$[a] mol/kg	Solid[b] phase
26.60	-	6.201	0.0	A
24.92	1.81	5.820	0.130	A
24.87	3.46	5.938	0.254	A
24.49	4.05	5.864	0.298	A
22.39	4.82	5.263	0.348	A
22.09	5.50	5.220	0.400	A
20.62	6.35	4.831	0.457	A
20.33	9.22	4.938	0.688	A
18.77	13.40	4.735	1.039	A
16.35	20.91	4.459	1.753	A + B
16.45	20.06	4.434	1.662	A + B
15.71	20.06	4.185	1.643	B
10.09	26.83	2.737	2.237	B
4.84	32.79	1.328	2.766	C
2.01	37.08	0.565	3.202	C
0.63	39.37	0.180	3.452	C
0.00	40.34	0.0	3.557	C

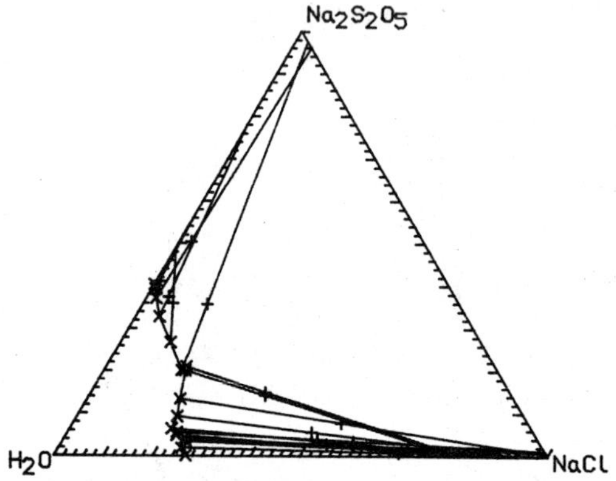

[a] Molality calculated by the compiler.

[b] Solid phases: A - NaCl, B - $Na_2S_2O_5$, C - $Na_2S_2O_5 \cdot 7H_2O$

AUXILIARY INFORMATION

METHOD APPARATUS/PROCEDURE:

Isothermal method. Compositions of saturated solutions and solid phases were determined by chemical analysis, and solid-phase compositions were established by Schreinemakers' method.

Prevention of oxidation by use of an inert gas or an anti-oxidant is not mentioned.

SOURCE AND PURITY OF MATERIALS:

"Pure" grade salts and distilled water were used.

ESTIMATED ERROR:

Temperature: ±0.1 K
Analyses: nothing stated.

REFERENCES:

COMPONENTS:	ORIGINAL MEASUREMENTS:
1. Sodium pyrosulfite; $Na_2S_2O_5$; [7681-57-4] 2. Ammonium pyrosulfite; $(NH_4)_2S_2O_5$; [32736-64-4] 3. Sodium chloride; NaCl; [7647-14-5] 4. Ammonium chloride; NH_4Cl; [12125-02-9] 5. Water; H_2O; [7732-18-5]	Zil'berman, Ya.I.; Ivanov, P.T. *Zh. Priklad. Khim.* 1941, *14*, 939-946.
VARIABLES:	**PREPARED BY:**
Two temperatures: 298 and 333 K Concentrations of the components	Mary R. Masson

EXPERIMENTAL VALUES:

Composition of equilibrium solutions, expressed as mass %

HSO_3^-	Cl^-	NH_4^+	$NaHSO_3$	NH_4HSO_3	NaCl	NH_4Cl	Solid[a] phase
Temperature = 60°C							
21.20	9.37	-	27.56	-	15.44	-	C,E
50.42	7.45	-	-	61.80	-	12.28	D,F
64.80	-	13.54	6.34	74.50	-	-	E,F
30.05	12.87	9.16	23.34	14.40	-	19.41	C,D,E
45.30	8.37	13.38	5.43	50.19	-	12.62	D,E,F
-	24.02	8.52	-	-	11.96	25.30	C,D
Temperature = 25°C							
16.44	10.28	-	21.10	-	16.94	-	C,E
58.67	-	11.92	6.43	65.57	-	-	E,F
55.05	2.52	-	-	67.47	-	3.81	D,F
21.35	12.58	6.00	27.48	-	1.24	17.84	C,D,E
56.69	1.98	12.28	7.61	62.04	-	2.98	D,E,F
-	-	-	-	-	16.82	16.16	C,D ()

[a] Solid phases: C - NaCl, D - $NH_4$1, E - "$NaHSO_3$", F - "NH_4HSO_3"

(continued on next page)

AUXILIARY INFORMATION

METHOD APPARATUS/PROCEDURE:

Solutions were equilibrated in glass test tubes fitted with spiral stirrers supplied with mercury seals. Freshly made salts were always used for each filling of a test tube. The anti-oxidant p-phenylenediamine was added to all solutions, and the work was done in an atmosphere of nitrogen, but experiments still had to be repeated often because of the formation of unacceptably high concentrations of sulfate.

Bisulfate was determined by reaction with iodine solution, ammonia was volatilized by reaction with alkali, and collected in acid, the excess of which was titrated, sodium was weighed as sodium sulfate, total sulfur was weighed as barium sulfate, and chloride was titrated by the Volhard method.

If too much ammonia was found to have been lost, the experiment had to be repeated.

SOURCE AND PURITY OF MATERIALS:

Ammonium chloride and sodium chloride were commercial reagents.

Sodium pyrosulfite was made by saturating soda with sulfur dioxide, and ammonium pyrosulfite by saturating aqueous ammonia with sulfur dioxide, both in the presence of p-phenylenediamine as anti-oxidant.

ESTIMATED ERROR:

Analyses: 0.2% relative
Temperature: no estimate given (toluene and mercury thermoregulators).

REFERENCES:

1. *Techn. Enc.* VI.

COMPONENTS:	ORIGINAL MEASUREMENTS:
1. Sodium pyrosulfite; $Na_2S_2O_5$; [7681-57-4]	Zil'berman, Ya.I.; Ivanov, P.T.
2. Ammonium pyrosulfite; $(NH_4)_2S_2O_5$; [32736-64-4]	*Zh. Priklad. Khim.* 1941, *14*, 939-946.
3. Sodium chloride; NaCl; [7647-14-5]	
4. Ammonium chloride; NH_4Cl; [12125-02-9]	
5. Water; H_2O; [7732-18-5]	

EXPERIMENTAL VALUES (continued):

Compositions of equilibrium solutions expressed as molalities[b], mol/kg

$NaHSO_3$	NH_4HSO_3	NaCl	NH_4Cl
Temperature = 60°C			
4.646	0.	4.635	0.
0.	24.057	0.	8.857
3.180	39.232	0.	0.
5.234	3.391	0.	8.468
1.643	15.945	0.	7.429
0.	0.	3.262	7.539
Temperature = 25°C			
3.273	0.	4.678	0.
2.207	23.628	0.	0.
0.	23.703	0.	2.480
4.942	0.	0.397	6.241
2.672	22.871	0.	2.035
0.	0.	4.294	4.508

b Molalities calculated by the compiler.

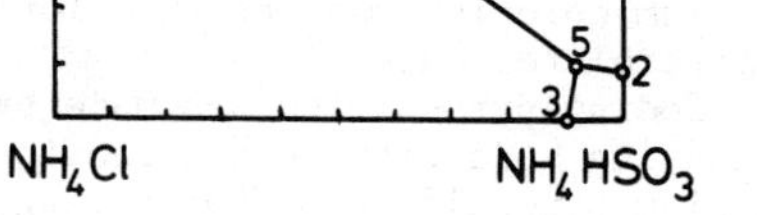

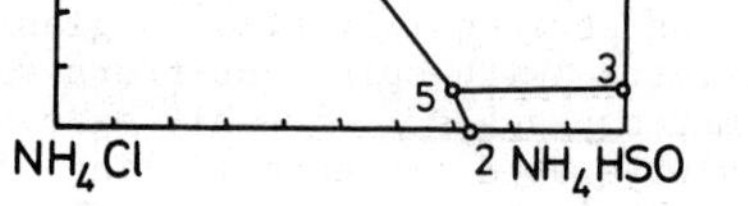

COMPONENTS:

1. Sodium pyrosulfite; $Na_2S_2O_5$; [7681-57-4]
2. Sodium chloride; NaCl; [7647-14-5]
3. Water; H_2O; [7732-18-5]
4. Diethanolamine; $[CH_2(OH)CH_2]_2NH$; [111-42-2]

ORIGINAL MEASUREMENTS:

Yavorskii, V.T.; Perekupko, T.V.; Matsyk, L.V.

Zh. Priklad. Khim. 1984, *57*, 3-7; *J. Appl. Chem. USSR (Eng. Transl.)* 1984, *57*, 1-4.

VARIABLES:

Temperature: 298 K
Composition

PREPARED BY:

Mary R. Masson

EXPERIMENTAL VALUES:

Composition of saturated solutions prepared with 30% aqueous diethanolamine (DEA) as the solvent.

NaCl mass %	$Na_2S_2O_5$ mass %	NaCl[a] mol/kg	$Na_2S_2O_5$[a] mol/kg	Solid[b] phase
0.	26.80	0.	1.926	A
0.98	25.58	0.228	1.832	A
1.64	25.43	0.385	1.834	A
3.62	23.09	0.845	1.657	A
3.75	21.00	0.853	1.468	A
7.78	19.62	1.834	1.422	A
11.12	17.35	2.660	1.276	A
14.96	15.75	3.694	1.196	A
15.01	14.60	3.649	1.091	A + B
14.88	14.80	3.621	1.107	A + B
15.08	13.75	3.626	1.016	B
15.32	12.05	3.609	0.873	B
15.51	7.16	3.432	0.487	B
15.32	3.93	3.246	0.256	B
15.44	2.88	3.235	0.185	B
16.51	2.18	3.475	0.141	B
16.66	1.60	3.488	0.103	B
17.21	1.07	3.604	0.069	B
18.00	0.	3.756	0.	B

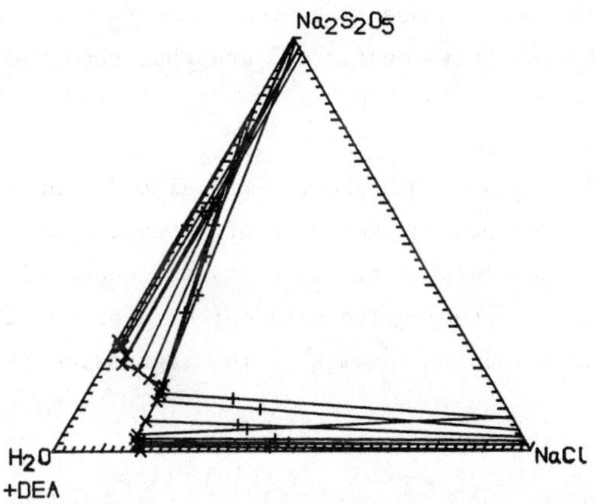

[a] Molalities calculated by the compiler.
[b] Solid phases: A - $Na_2S_2O_5$, B - NaCl

AUXILIARY INFORMATION

METHOD APPARATUS/PROCEDURE:

Isothermal method. Compositions of saturated solutions and solid phases were determined by chemical analysis, and solid-phase compositions were established by Schreinemakers'' method.

Prevention of oxidation by use of an inert gas or an anti-oxidant is not mentioned.

Additional Data

Solubility isotherms (T/K = 298, 313, 323) for the system $Na_2S_2O_5$-DEA-H_2O are given, but only in graphical form.

SOURCE AND PURITY OF MATERIALS:

"Pure" grade salts and distilled water were used. Purity of the DEA is not mentioned.

ESTIMATED ERROR:

Temperature: ±0.1 K
Analyses: nothing stated.

REFERENCES:

COMPONENTS:	EVALUATOR:
1. Potassium sulfite; K_2SO_3; [10117-38-1] 2. Water; H_2O; [7732-18-5]	Mary R. Masson, Dept. of Chemistry, University of Aberdeen, Meston Walk, Old Aberdeen, AB9 2UE, Scotland, UK. March 1984

CRITICAL EVALUATION:

Only Foerster *et al.* (1) have studied the binary system potassium sulfite - water, but data are also available from studies of ternary systems (2 - 6). The solid phase reported in all the solubility studies was the anhydrous salt. (In some early work reported by Mellor, a monohydrate and a dihydrate were believed to have been synthesized, but no solubility data for these salts is available.)

The data of Hölzl (6) differ markedly from the rest; there is no obvious reason for the difference, since the work appears to have been done under oxygen-free conditions, and the solid phase is stated to be K_2SO_3. It does appear possible that the solid phase was incorrectly identified, and that some hydrate was present in this sytem. These data were all rejected.

The data of Babenko (3 - 5) also differ somewhat from the other data (1,2). This may have arisen because the work of Babenko appears to have been done with the solutions in contact with air, since it was concerned with the properties of fertilizer formulations: thus contamination with carbon dioxide or oxidation of sulfite to sulfate may have affected the results. The regression line found for Babenko's results was

$$y = 52.0 + 0.021(T - 273.2) \qquad s = 0.03$$

All of Babenko's results were, therefore, also omitted from the data set used for calculation of the general regression line, which is

$$y = 51.3 + 0.00756(T - 273.2) + 0.00010(T - 273.2)^2 \qquad s = 0.16$$

where $y = 100w$ is the solubility in mass %, T is the temperature in K, and s is the estimated standard deviation of the dependent variable about the regression line.

TENTATIVE SOLUBILITIES

The following tentative solubility values were calculated from the regression equation.

T/K	Solubility	
	mass %	molality mol/kg
243.2	51.16	6.62
253.2	51.19	6.63
263.2	51.23	6.64
273.2	51.30	6.66
283.2	51.39	6.68
293.2	51.49	6.71
298.2	51.55	6.72
303.2	51.62	6.74
313.2	51.76	6.78
323.2	51.93	6.83
333.2	52.11	6.88
343.2	52.32	6.933
353.2	52.54	6.99
363.2	52.79	7.06
373.2	53.06	7.14

COMPONENTS:	EVALUATOR:
1. Potassium sulfite; K_2SO_3; [10117-38-1] 2. Water; H_2O; [7732-18-5]	Mary R. Masson, Dept. of Chemistry, University of Aberdeen, Meston Walk, Old Aberdeen, AB9 2UE, Scotland, UK. March 1984

CRITICAL EVALUATION: (continued)

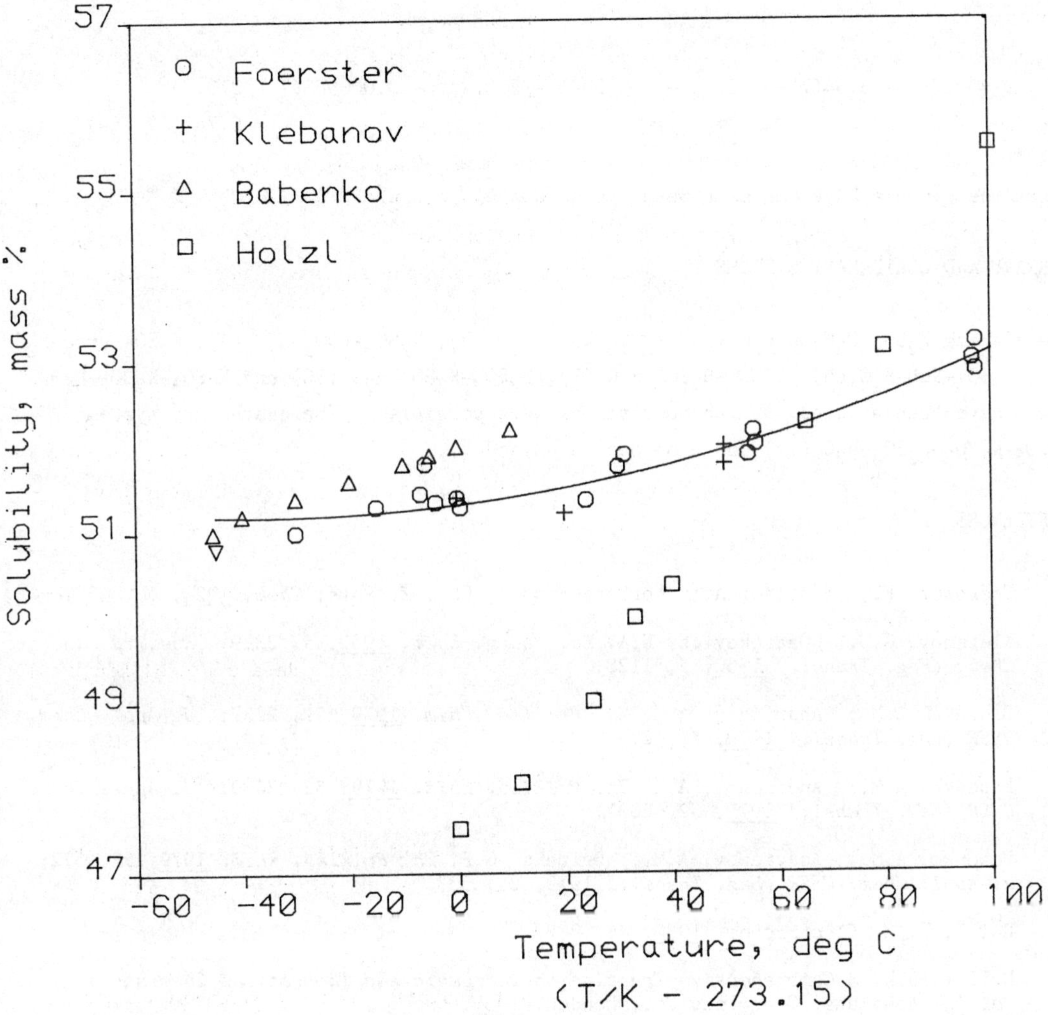

Data for the solubility of potassium sulfite in aqueous t-butanol, said to come from a paper by Ginnings and Robbins (8), have been reported (9). However, inspection reveals that the original paper gives no data for K_2SO_3, but "K_2SO_3" does appear in a table as a misprint for "K_2CO_3".

Foerster *et al*. (1) also give data for the equilibrium with ice. The regression equation for this data is:

COMPONENTS:

1. Potassium sulfite; K_2SO_3; [10117-38-1]
2. Water; H_2O; [7732-18-5]

EVALUATOR:

Mary R. Masson,
Dept. of Chemistry,
University of Aberdeen,
Meston Walk, Old Aberdeen, AB9 2UE,
Scotland, UK.
March 1984

CRITICAL EVALUATION: (continued)

$$(T - 273.15) = 0.0623 - 0.338y + 0.0646y^2 - 0.000339y^3 \qquad s = 0.096 \text{ (12 pts)}$$

or

$$y = 0.763 - 3.34(T - 273.2) - 0.0939(T - 273.2)^2 - 0.000988(T - 273.2)^3 \qquad s = 0.74 \text{ (12 pts)}$$

where the symbols have the same meanings as above.

TERNARY AND QUATERNARY SYSTEMS

The systems K_2SO_3-C_2H_5OH-H_2O (2), K_2SO_3-KNO_3-H_2O (3), K_2SO_3-KNO_2-H_2O (4), K_2SO_3-urea-H_2O (5), K_2SO_3-KOH-H_2O (6), K_2SO_3-H_2SO_3-H_2O (7), K_2SO_3-K_2SO_4-H_2O (10) and K_2SO_3-K_2CO_3-H_2O (11), have been studied, but no comparisons were possible. The quaternary system K_2SO_3-K_2CO_3-K_2SO_4-H_2O (11) has also been examined.

REFERENCES

1. Foerster, F.; Brosche, A.; Norberg-Schutz, Chr. *Z. Phys. Chem.* 1924, *10*, 435.

2. Klebanov, G.S.; Ostapkevich, N.A. *Zh. Neorg. Khim.* 1960, 5, 2329; *Russ. J. Inorg. Chem. (Eng. Transl.)* 1960, 5, 1128.

3. Babenko, A.M.; Andrianov, A.M. *Zh. Priklad. Khim.* 1979, 52, 2237; *J. Appl. Chem. USSR (Eng. Transl.)* 1979, 52, 2.

4. Babenko, A.M.; Andrianov, A.M. *Zh. Priklad. Khim.* 1979, 52, 2483; *J. Appl. Chem. USSR (Eng. Transl.)* 1979, 52, 2351.

5. Babenko, A.M.; Andrianov, A.M.; Deineka, G.F. *Zh. Priklad. Khim.* 1979, 52, 572; *J. Appl. Chem. USSR (Eng. Transl.)* 1979, 25, 533.

6. Hölzl, F. *Z. Electrochem.* 1937, *43*, 302.

7. Mellor, J.W. *A Comprehensive Treatise on Inorganic and Theoretical Chemistry: Vol. X.* Longmans, Green and Co., London, 1930.

8. Ginnings, P.M.; Robbins, D. *J. Am. Chem. Soc.* 1930, 52, 2282.

9. Linke, W.F. *Solubilities of Inorganic and Metal-Organic Compounds* (originated by A. Seidell) Vol. 2. American Chemical Society, Washington, D.C. 1965. p.295.

10. Bishimbaev, V.K.; Shokin, I.N.; Kuznetsova, A.G. *Khim. Khim. Tekhnol. (Alma-Ata)* 1971 *12*, 203.

11. Kuznetsova, A.G.; Trukhanova, E.A. *VINITI Deposited Document* 1983, 6890-83.

COMPONENTS:

1. Potassium sulfite; K_2SO_3; [10117-38-1]

2. Water; H_2O; [7732-18-5]

ORIGINAL MEASUREMENTS:

Foerster, F.; Brosche, A.; Norberg-Schutz, Chr.

Z. Phys. Chem. 1924, *10*, 435-96.

VARIABLES:

Temperature: 228 - 370 K

PREPARED BY:

Mary R. Masson

EXPERIMENTAL VALUES:

t/°C	K_2SO_3 mass %	K_2SO_3[a] mol/kg
-30.0	51.0	6.577
-15.0	51.30	6.656
- 6.7	51.45[b]	6.696
- 5.8	51.80	6.791
- 3.9	51.35	6.669
+ 0.1	51.40	6.683
0.7	51.29	6.653
24.0	51.37	6.675
30.0	51.76	6.780
31.2	51.90	6.818
54.4	51.90	6.818
55.4	52.18	6.895
55.8	52.02	6.851
96.6	53.01	7.128
97.2	53.22	7.189
97.2	52.88	7.091

Solid phase: K_2SO_3

[b] Result considered particularly reliable by the authors.

(continued on next page)

AUXILIARY INFORMATION

METHOD APPARATUS/PROCEDURE:

Solids were equilibrated with solution under a hydrogen atmosphere, in a vessel maintained in a thermostat. Samples for analysis were withdrawn through a tube plugged with cotton wool.

Samples were reacted with excess of standard iodine solution, and the excess was back-titrated with thiosulfate.

A Beckman apparatus (1) was used for the determination of freezing points.

SOURCE AND PURITY OF MATERIALS:

Potassium hydroxide (100 g) was dissolved in 200 g of nitrogen-flushed water, then the solution was saturated with sulfur dioxide. Another 100 g of potassium hydroxide was dissolved in the minimum of water, and added. The solution was evaporated under nitrogen to obtain crystals of potassium sulfite.

ESTIMATED ERROR:

Temperature: ±0.1 K
Analyses: no accurate estimate possible.

REFERENCES:

1. Ostwald, W.; Luther, R. *Hand-und Hilfsbuch zur Ausfuhrung physicochemischer Messungen* 5th Ed., Akademische Verlag., Leipzig, 1931.

COMPONENTS:	ORIGINAL MEASUREMENTS:
1. Potassium sulfite; K_2SO_3; [10117-38-1] 2. Water; H_2O; [7732-18-5]	Foerster, F.; Brosche, A.; Norberg-Schutz, Chr. *Z. Phys. Chem.* 1924, *10*, 435-96.

EXPERIMENTAL VALUES (continued):

t/°C	K_2SO_3 mass %	K_2SO_3[a] mol/kg
- 1.69	5.78	0.388
- 2.71	9.20	0.640
- 4.10	13.37	0.975
- 5.27	16.47	1.246
- 5.74	17.57	1.347
- 6.59	19.51	1.532
- 6.84	20.02	1.582
-10.88	26.70	2.302
-14.06	30.6	2.786
-31.0	44.0	4.96
-45.0	50.8	6.52

Solid phase: ice

[a] Molalities calculated by the compiler.

COMPONENTS:	ORIGINAL MEASUREMENTS:
1. Potassium sulfite; K_2SO_3; [10117-38-1] 2. Ethanol; C_2H_5OH; [64-17-5] 3. Water; H_2O; [7732-18-5]	Klebanov, G.S.; Ostapkevich, N.A. *Zh. Neorg. Khim.* 1960, 5, 2329-2333; *Russ. J. Inorg. Chem. (Eng. Transl.)* 1960, 5, 1128-9.

VARIABLES:	PREPARED BY:
Two temperatures: 293 - 323 K Concentrations of the components	Mary R. Masson

EXPERIMENTAL VALUES:

Composition of equilibrium solutions

Lower phase				Upper phase				Initial
K_2SO_3 mass %	C_2H_5OH mass %	K_2SO_3[a] mol/kg	C_2H_5OH[a] mol/kg	K_2SO_3 mass %	C_2H_5OH mass %	K_2SO_3[a] mol/kg	C_2H_5OH[a] mol/kg	C_2H_5OH mass %
Temperature = 20°C								
51.22	0.0	6.635	0.	-	-	-	-	0.0
41.20	3.82	4.735	1.508	-	-	-	-	10.0
41.20	3.80	4.733	1.500	0.26	76.30	0.070	70.656	30.0
41.10	3.85	4.718	1.518	0.27	76.30	0.073	70.686	50.0
41.12	3.85	4.722	1.519	0.26	76.25	0.070	70.459	60.0
41.20	3.83	4.736	1.512	0.26	76.30	0.070	70.656	70.0
-	-			0.07	78.90	0.021	81.437	80.0
-	-			0.02	84.40	0.008	117.586	90.0
Temperature = 50°C								
51.79	0.0	6.788	0.	-	-	-	-	0.0
42.70	3.75	5.038	1.520	-	-	-	-	10.0
42.70	3.75	5.038	1.520	0.35	77.10	0.098	74.215	30.0
42.72	3.73	5.041	1.512	0.36	77.20	0.101	74.675	40.0
42.75	3.72	5.046	1.508	0.35	77.30	0.099	75.073	50.0
42.70	3.73	5.037	1.511	0.36	77.30	0.102	75.107	60.0
42.72	3.70	5.038	1.499	0.35	77.20	0.099	74.642	70.0
-	-			0.06	84.71	0.025	120.730	90.0

[a] Molalities calculated by the compiler.

AUXILIARY INFORMATION

METHOD APPARATUS/PROCEDURE:

The isothermal method was used; the vessels were glass test-tubes fitted with mercury seals at 20°C, and with reflux condensers at 50°C.

Two liquid phases formed for ethanol concentrations between 3.8 and 76.3% at 20°C, and between 3.75 and 77.3% at 50°C. At both temperatures the solid phase was anhydrous potassium sulfite.

Alcohol was distilled off and determined iodometrically.

SOURCE AND PURITY OF MATERIALS:

"Chemically pure" salts were used. Ethanol and water were redistilled twice.

ESTIMATED ERROR:

Temperature: ±0.1 K
Analyses: no estimate possible.

REFERENCES:

COMPONENTS:

1. Potassium sulfite; K_2SO_3; [10431-47-7]
2. Potassium nitrate; KNO_3; [7757-79-1]
3. Water; H_2O; [7732-18-5]

ORIGINAL MEASUREMENTS:

Babenko, A.M.; Andrianov, A.M.

Zh. Priklad. Khim. 1979, 52, 2237-2240; *J. Appl. Chem. USSR (Eng. Transl.)* 1979, 52, 2114-17.

VARIABLES:

Concentrations of the components
Temperature: 228 - 283 K

PREPARED BY:

Mary R. Masson

EXPERIMENTAL VALUES: Solubility at various temperatures and compositions

t/°C	KNO_3 mass %	K_2SO_3 mass %	KNO_3[a] mol/kg	K_2SO_3[a] mol/kg	Solid[b] phase
- 2.9	10.0	-	1.099	0.	A + B
- 5.4	7.5	9.25	0.891	0.702	A + B
- 9.2	4.5	19.1	0.583	1.580	A + B
-14.6	2.0	29.4	0.288	2.708	A + B
-25.0	1.0	39.6	0.167	4.212	A + B
-35.0	1.0	43.56	0.178	4.965	A + B
+ 1.0	0.7	50.643	0.142	6.577	B + C
-45.5	-	51.0	0.	6.577	A + C
-38.0	0.9	51.75	0.188	6.906	A,B,C
-10.0	-	24.4	0.	2.039	
-10.0	2.0	39.2	0.336	4.212	
-10.0	2.5	42.9	0.453	4.965	
-10.0	-	51.8	0.	6.791	
0.0	11.6	-	1.298	0.	
0.0	9.0	9.1	1.087	0.702	
0.0	6.0	18.8	0.789	1.580	
0.0	3.8	28.86	0.558	2.708	
0.0	2.4	39.04	0.405	4.212	
0.0	3.1	42.635	0.565	4.964	
0.0	-	52.0	0.	6.845	

(continued on next page)

AUXILIARY INFORMATION

METHOD APPARATUS/PROCEDURE:

A polythermal technique was used (1).

SOURCE AND PURITY OF MATERIALS:

Potassium nitrate was twice recrystallized from a chemically or analytically pure grade of reagent. "Pure" potassium sulfite was further purified (2). A 51% solution of potassium sulfite was prepared.

ESTIMATED ERROR:

Temperature: ±0.4 K
Sulfite analyses: ±1-3%

REFERENCES:

1. Éraizer, L.N.; Kaganskii, I.M. *Zavod. Lab.* 1967, *33*, 119.
2. Klebanov, G.S.; Ostapkevich, N.A. *Zh. Neorg. Khim.* 1960, 5, 2331.

COMPONENTS:	ORIGINAL MEASUREMENTS:
1. Potassium sulfite; K_2SO_3; [10431-47-7] 2. Potassium nitrate; KNO_3; [7757-79-1] 3. Water; H_2O; [7732-18-5]	Babenko, A.M.; Andrianov, A.M. *Zh. Priklad. Khim.* 1979, *52*, 2237-2240; *J. Appl. Chem. USSR (Eng. Transl.)* 1979, *52*, 2114-17.

EXPERIMENTAL VALUES (continued):

t/°C	KNO_3 mass %	K_2SO_3 mass %	KNO_3[a] mol/kg	K_2SO_3[a] mol/kg
+10.0	17.5	-	2.098	0.
+10.0	12.8	8.72	1.613	0.702
+10.0	8.6	18.28	1.163	1.580
+10.0	5.2	28.44	0.775	2.708
+10.0	3.8	38.48	0.651	4.212
+10.0	3.7	42.372	0.679	4.965
+10.0	1.4	50.286	0.287	6.577
+10.0	-	52.2	0.	6.900

[a] Molalities calculated by the compiler.

[b] Solid phases: A - ice, B - KNO_3, C - K_2SO_3

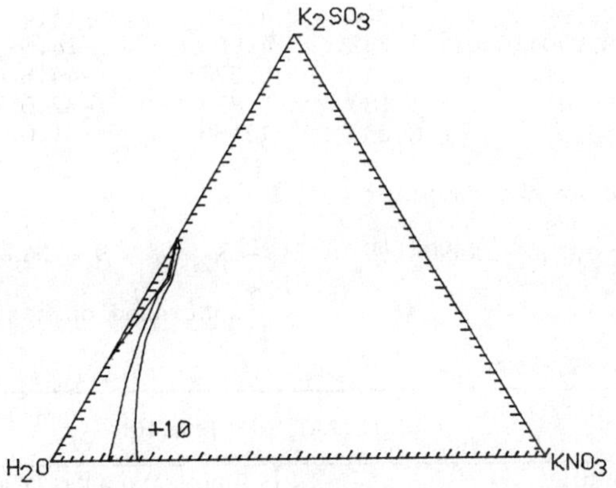

COMPONENTS:

1. Potassium sulfite; K_2SO_3; [10117-38-1]
2. Potassium nitrite; KNO_2; [7758-09-0]
3. Water; H_2O; [7732-18-5]

ORIGINAL MEASUREMENTS:

Babenko, A.M.; Andrianov, A.M.

Zh. Priklad. Khim. 1979, 52, 2483-6; *J. Appl. Chem. USSR (Eng. Transl.)* 1979, 52, 2351-4.

VARIABLES:

Temperature: 228-283 K
Concentrations of the components

PREPARED BY:

Mary R. Masson

EXPERIMENTAL VALUES:

Composition of liquid phases

KNO_2 mass %	K_2SO_3 mass %	KNO_2[a] mol/kg	K_2SO_3[a] mol/kg	Cryst. temp. °C	Solid[b] phase
46.0	0.0	10.010	0.	-22.5	A,B
41.0	5.9	9.073	0.702	-23.4	A,B
32.0	13.6	6.912	1.580	-27.6	A,B
24.0	22.8	5.301	2.708	-34.0	A,B
16.4	33.4	3.839	4.204	-40.0	A,B
9.0	40.04	2.075	4.965	-44.2	A,B
0.0	51.0	0.	6.577	-45.5	A,C
73.0	0.0	31.771	0.	- 3.0	B,C
59.8	4.02	19.422	0.702	- 6.4	B,C
49.0	10.2	14.113	1.580	- 2.2	B,C
38.0	18.6	10.289	2.708	0.0	B,C
27.0	29.2	7.244	4.212	-11.4	B,C
18.4	35.904	4.732	4.965	-14.0	B,C
2.6	49.674	0.640	6.577	-44.8	A,B,C
5.0	45.6	1.189	5.833	-32.0	B,C,D
1.0	53.5	0.258	7.430	+ 1.0	C,D

[a] Molalities calculated by the compiler.

[b] Solid phases: A - ice, B - $NaNO_2.\frac{1}{2}H_2O$, C - K_2SO_3, D - $NaNO_2$

(continued on next page)

AUXILIARY INFORMATION

METHOD APPARATUS/PROCEDURE:

An improved polythermal technique was used (1).

Potassium sulfite was determined by an iodometric method.

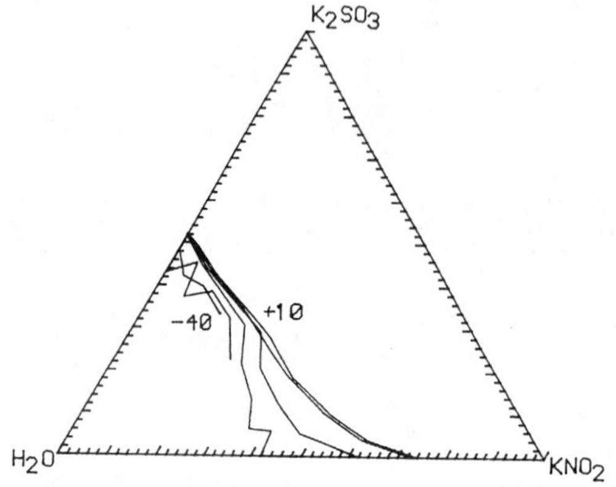

SOURCE AND PURITY OF MATERIALS:

Pure grade potassium sulfite was purified as in (2). Potassium nitrite was recrystallized twice from the general-purpose or analytical grade reagent, keeping the temperature above 0°C to avoid obtaining the hydrate $KNO_2.\frac{1}{2}H_2O$.

ESTIMATED ERROR:

Crystallization temperatures: ±0.4 K
Analyses: 1 - 3% relative

REFERENCES:

1. Éraizer, L.N.; Kaganskii, N.M. *Zavod. Lab.* 1967, *33*, 119.
2. Klebanov, G.S.; Ostapkevich, N.A. *Zh. Neorg. Khim.* 1960, 5(10), 2331.

COMPONENTS:	ORIGINAL MEASUREMENTS:
1. Potassium sulfite; K_2SO_3; [10117-38-1] 2. Potassium nitrite; KNO_2; [7758-09-0] 3. Water; H_2O; [7732-18-5]	Babenko, A.M.; Andrianov, A.M. *Zh. Priklad. Khim.* 1979, 52, 2483-6; *J. Appl. Chem. USSR (Eng. Transl.)* 1979, 52, 2351-4.

EXPERIMENTAL VALUES (continued):

KNO_2 mass %	K_2SO_3 mass %	KNO_2[a] mol/kg	K_2SO_3[a] mol/kg
Temperature = -40°C			
16.4	33.44	3.842	4.212
16.2	33.52	3.786	4.212
9.4	39.86	2.177	4.964
4.0	42.24	0.874	4.965
0.0	48.4	0.	5.927
0.0	51.2	0.	6.629
Temperature = -30°C			
24.0	22.8	5.301	2.708
19.0	32.4	4.594	4.212
18.8	24.36	3.887	2.708
12.0	38.72	2.861	4.965
6.6	37.36	1.384	4.212
5.4	45.408	1.290	5.833
0.0	43.1	0.	4.786
0.0	51.4	0.	6.683
Temperature = -20°C			
49.0	0.0	11.290	0.
42.0	0.0	8.509	0.
41.2	5.88	9.148	0.702
36.4	6.36	7.473	0.702
33.4	13.32	7.366	1.580
26.8	21.96	6.146	2.708
23.0	30.8	5.850	4.212
22.0	15.6	4.143	1.580
15.7	37.09	3.908	4.964
8.4	27.4	1.537	2.697
6.8	44.736	1.649	5.833
0.0	51.6	0.	6.736
0.0	36.4	0.	3.616
Temperature = -10°C			
62.0	0.0	19.172	0.
47.0	5.3	11.578	0.702
39.4	12.13	9.552	1.581
31.4	20.58	7.684	2.708
27.2	29.12	7.317	4.212
20.6	7.94	3.387	0.702
19.0	35.64	4.922	4.965
8.0	44.16	1.965	5.833
4.2	19.6	0.648	1.625
0.0	24.4	0.	2.039
0.0	51.8	0.	6.791
Temperature = 0°C			
73.0	0.0	31.771	0.
60.6	3.94	20.082	0.702
49.4	10.12	14.340	1.580
38.0	18.6	10.289	2.708
20.2	35.112	5.312	4.965
9.4	43.48	2.344	5.831
0.0	52.0	0.	6.845
Temperature = +10°C			
74.0	0.0	33.445	0.
62.0	3.8	21.303	0.702
51.0	9.8	15.288	1.580
39.1	18.27	10.778	2.708
29.4	28.24	8.156	4.212
21.6	34.496	5.781	4.965
10.8	42.72	2.730	5.808
3.6	49.164	0.896	6.577
0.0	52.0	0.	6.845

COMPONENTS:

1. Potassium sulfite; K_2SO_3; [10117-38-1]

2. Urea; $CO(NH_2)_2$; [57-13-6]

3. Water; H_2O; [7732-18-5]

ORIGINAL MEASUREMENTS:

Babenko, A.M.; Andrianov, A.M. Deineka, G.F.

Zh. Priklad. Khim. 1979, 52, 572-6; *J. Appl. Chem. USSR (Eng. Transl.)* 1979, 52, 533-7.

VARIABLES:

Temperature: 228-283 K
Concentrations of the components

PREPARED BY:

Mary R. Masson

EXPERIMENTAL VALUES:

Composition of liquid phases

urea mass %	K_2SO_3 mass %	urea[a] mol/kg	K_2SO_3[a] mol/kg	Cryst. temp. °C	Solid[b] phase
33.0	0.0	8.201	0.	-10.6	A,B
31.6	6.84	8.547	0.702	-12.8	A,B
25.2	14.96	7.012	1.580	-17.4	A,B
15.80	25.26	4.463	2.708	-20.6	A,B
7.0	37.2	2.089	4.212	-29.4	A,B
3.2	42.592	0.983	4.965	-39.6	A,B
4.4	43.976	1.419	5.383	-33.4	A,B
2.0	50.0	0.694	6.582	-46.0	A,B,C
0.0	51.0	0.	6.577	-45.5	A,C
3.0	49.47	1.051	6.577	+ 1.0	B,C

[a] Molalities calculated by the compiler.

[b] Solid phases: A - ice, B - urea, D - potassium sulfite

(continued on next page)

AUXILIARY INFORMATION

METHOD APPARATUS/PROCEDURE:

An improved polythermal technique was used (1).

Potassium sulfite was determined by an iodometric method.

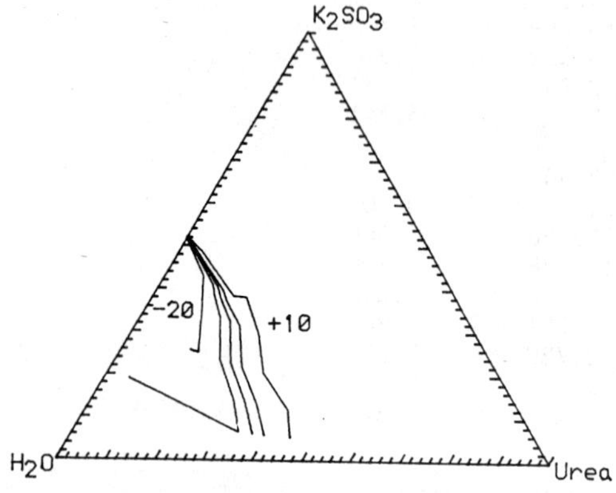

SOURCE AND PURITY OF MATERIALS:

Analytical grade urea was recrystallized twice and dried at 60°C. Pure grade potassium sulfite was purified as in (2).

ESTIMATED ERROR:

Crystallization temperatures: ±0.4 K
Analyses: 1 - 3% relative

REFERENCES:

1. Éraizer, L.N.; Kaganskii, N.M. *Zavod. Lab.* 1967, *33*(1), 119.
2. Klebanov, G.S.; Ostapkevich, N.A. *Zh. Neorg. Khim.* 1960, 5(10), 2331.

COMPONENTS:

1. Potassium sulfite; K_2SO_3; [10117-38-1]

2. Urea; $CO(NH_2)_2$; [57-13-6]

3. Water; H_2O; [7732-18-5]

ORIGINAL MEASUREMENTS:

Babenko, A.M.; Andrianov, A.M. Deineka, G.F.

Zh. Priklad. Khim. 1979, *52*, 572-6; *J. Appl. Chem. USSR (Eng. Transl.)* 1979, *52*, 533-7.

EXPERIMENTAL VALUES (continued):

urea mass %	K_2SO_3 mass %	urea[a] mol/kg	K_2SO_3[a] mol/kg
Temperature = -20°C			
14.0	25.8	3.872	2.708
16.0	25.2	4.531	2.708
0.0	36.4	0.	3.616
11.6	35.36	3.641	4.212
9.4	39.86	3.085	4.964
8.0	42.92	2.714	5.526
0.0	51.6	0.	6.736
Temperature = -10°C			
5.0	19.0	1.095	1.580
34.0	6.58	9.527	0.700
21.4	23.58	6.476	2.708
29.0	14.2	8.501	1.580
16.4	33.44	5.444	4.212
12.6	38.45	4.286	4.963
11.0	40.94	3.811	5.383
0.0	51.8	0.	6.791
Temperature = -5°C			
36.8	6.32	10.772	0.702
31.6	13.68	9.615	1.580
19.0	32.4	6.509	4.212
24.0	22.8	7.511	2.708
14.4	37.66	5.001	4.964
12.4	40.30	4.365	5.384
0.0	51.9	0.	6.818
Temperature = 0°C			
39.4	6.06	12.028	0.702
34.2	13.16	10.817	1.580
21.4	31.44	7.555	4.212
26.6	22.02	8.620	2.708
16.0	36.96	5.663	4.965
13.8	39.65	4.936	5.382
0.0	52.0	0.	6.845
Temperature = +10°C			
44.8	5.52	15.015	0.702
41.2	11.76	14.583	1.580
26.2	29.52	9.852	4.212
31.8	20.46	11.091	2.708
19.4	38.10	7.600	5.665
16.6	38.36	6.137	5.382
5.2	48.34	1.864	6.574
0.0	52.0	0.	6.845

[a] Molalities calculated by the compiler.

COMPONENTS:	ORIGINAL MEASUREMENTS:
1. Potassium sulfite; K_2SO_3; [10117-38-1] 2. Potassium hydroxide; KOH; [1310-58-3] 3. Water; H_2O; [7732-18-5]	Hölzl, F. *Z. Electrochem.* 1937, *43*, 302-4.
VARIABLES:	PREPARED BY:
Temperature: 273 - 373 K Concentrations of the components	Mary R. Masson

EXPERIMENTAL VALUES: Composition of equilibrium solutions

t/°C	K_2O mass %	K_2SO_3 mass %	K_2O[a] mol/kg	K_2SO_3[a] mol/kg	Solid[b] phase
0.0	41.01	0.00	7.380	0.000	A
	40.71	0.43	7.342	0.046	A,D
	0.00	47.52	0.000	5.721	D
11.5	42.87	0.00	7.966	0.000	A
	42.41	0.48	7.883	0.053	A,D
	0.00	48.06	0.	5.847	D
25.0	45.51	0.00	8.866	0.	A
	45.03	0.47	8.771	0.054	A,D
	0.00	49.01	0.	6.073	D
33.0	48.29	0.00	9.914	0.	A,C
	48.16	0.00	9.862	0.	B,C
	47.81	0.37	9.794	0.045	C,D
	0.00	49.99	0.	6.316	D
40.0	48.71	0.00	10.082	0.	C
	48.32	0.40	10.003	0.049	C,D
	0.00	50.37	0.	6.413	D
65.3	50.65	0.00	10.895	0.	C
	50.14	0.45	10.773	0.058	C,D
	0.00	52.27	0.	6.920	D
80.0	51.81	0.00	11.413	0.	C
	51.50	0.39	11.364	0.051	C,D
	0.00	53.15	0.	7.168	D
100.0	54.67	0.00	12.803	0.	C
	54.29	0.40	12.720	0.056	C,D
	0.00	55.53	0.	7.890	D

[a] Molalities calculated by the compiler.

[b] Solid phases:

A - $KOH.2H_2O$,

B - $KOH.1\frac{1}{2}H_2O$,

C - $KOH.H_2O$,

D - K_2SO_3

AUXILIARY INFORMATION

METHOD/APPARATUS/PROCEDURE:

All work was done under oxygen-free nitrogen, and in sealed vessels whenever possible.

Sulfite was determined titrimetrically by a method involving iodate and a Landolt reaction. K_2O was determined gravimetrically as KCl or K_2SO_4.

SOURCE AND PURITY OF MATERIALS:

Potassium sulfite was prepared by saturation of a potassium hydroxide solution with sulfur dioxide. The water was free from dissolved oxygen and carbon dioxide.

ESTIMATED ERROR:

No estimates possible.

REFERENCES:

COMPONENTS:	ORIGINAL MEASUREMENTS:
1. Potassium sulfite; K_2SO_3; [10117-38-1] 2. Sulfurous acid; H_2SO_3; [7782-99-2] 3. Water; H_2O; [7732-18-5]	Hölzl, F. *Z. Electrochem.* 1937, *43*, 302-4.
VARIABLES:	PREPARED BY:
Temperature: 273 - 373 K Concentrations of the components	Mary R. Masson

EXPERIMENTAL VALUES:

Composition of equilibrium solutions

t/°C	K_2SO_3 mass %	H_2SO_3 mass %	K_2SO_3 mol/kg	H_2SO_3 mol/kg	Solid phase
0.0	47.52	0.0	5.722	0.	A
	50.36	0.44	6.468	1.109	A,B
	14.06	7.49	1.132	1.163	B
11.5	48.06	0.0	5.847	0.	A
	50.96	1.11	6.718	0.282	A,C
25.0	49.01	0.0	6.073	0.	A
	51.92	1.83	7.093	0.482	A,C
40.0	50.37	0.0	6.413	0.	A
	53.01	2.64	7.553	0.725	A,C
65.3	52.27	0.0	6.920	0.	A
	55.12	4.37	8.598	1.314	A,C
80.0	53.15	0.0	7.168	0.	A
	55.82	5.48	9.114	1.725	A,C
100.0	55.53	0.0	7.890	0.	A
	56.60	7.0	9.825	2.343	A,C

[a] Molalities calculated by the compiler.
[b] Solid phases: A - K_2SO_3, B - $K_2S_2O_5 \cdot 2/3H_2O$, C - $K_2S_2O_5$

AUXILIARY INFORMATION

METHOD APPARATUS/PROCEDURE:

All work was done under oxygen-free nitrogen, and in sealed vessels whenever possible.

Sulfite was determined titrimetrically by a method involving iodate and a Landlolt reaction. K_2O was determined gravimetrically as KCl or K_2SO_4.

SOURCE AND PURITY OF MATERIALS:

Potassium sulfite was prepared by saturation of potassium hydroxide solution with sulfur dioxide. The water used was free from dissolved oxygen and carbon dioxide.

ESTIMATED ERROR:

No estimates given.

REFERENCES:

COMPONENTS:

1. Potassium sulfite; K_2SO_3; [10117-38-1]
2. Potassium sulfate; K_2SO_4; [7778-80-5]
3. Water; H_2O; [7732-18-5]

ORIGINAL MEASUREMENTS:

Bishimbaev, V.K.; Shokin, I.N.; Kuznetsova, A.G.

Khim. Khim. Tekhnol. (Alma-Ata) 1971, *12*, 203-5.

VARIABLES:

Two temperatures: 293 - 333 K
Concentrations of the components

PREPARED BY:

Mary R. Masson

EXPERIMENTAL VALUES:

Composition of equilibrium solutions at 20°C

K_2SO_4 mass %	K_2SO_3 mass %	K_2SO_4[a] mol/kg	K_2SO_3[a] mol/kg	Solid[b] phase
10.5	0.0	0.673	0.	A
10.5	0.96	0.681	0.069	A
9.5	1.85	0.615	0.132	A
7.25	3.96	0.469	0.282	A
7.52	4.94	0.493	0.357	A
6.42	5.76	0.420	0.414	A
4.64	8.13	0.305	0.589	A
4.46	10.80	0.302	0.805	A
4.32	14.85	0.307	1.161	A
2.85	20.75	0.214	1.716	A
1.45	22.10	0.109	1.827	A
1.83	27.50	0.149	2.459	A
1.45	32.10	0.125	3.052	A
1.43	40.90	0.142	4.481	A + B

(continued on next page)

AUXILIARY INFORMATION

METHOD APPARATUS/PROCEDURE:

A saturation method was used. Sulfite was determined iodometrically, and sulfate (total) was weighed as barium sulfate after conversion of sulfite into sulfate with hydrogen peroxide.

SOURCE AND PURITY OF MATERIALS:

ESTIMATED ERROR:

Temperature: ±0.1 K
Analyses: no estimate possible.

REFERENCES:

COMPONENTS:	ORIGINAL MEASUREMENTS:
1. Potassium sulfite; K_2SO_3; [10117-38-1]	Bishimbaev, V.K.; Shokin, I.N.; Kuznetsova, A.G.
2. Potassium sulfate; K_2SO_4; [7778-80-5]	*Khim. Khim. Tekhnol. (Alma-Ata)* 1971, *12*, 203-5.
3. Water; H_2O; [7732-18-5]	

EXPERIMENTAL VALUES (continued):

Composition of equilibrium solutions at 60°C

K_2SO_4 mass %	K_2SO_3 mass %	K_2SO_4[a] mol/kg	K_2SO_3[a] mol/kg	Solid[b] phase
15.13	0.0	1.023	0.	A
14.60	1.09	0.994	0.082	A
12.35	3.49	0.842	0.262	A
15.90	5.97	1.168	0.483	A
9.30	8.65	0.650	0.666	A
8.06	12.02	0.579	0.950	A
3.90	19.63	0.293	1.622	A
3.40	25.00	0.273	2.206	A
2.18	29.20	0.182	2.689	A
1.83	38.30	0.175	4.042	A
1.08	41.60	0.108	4.586	A + B

[a] Molalities calculated by the compiler.

[b] Solid phases: A - K_2SO_4, B - K_2SO_3

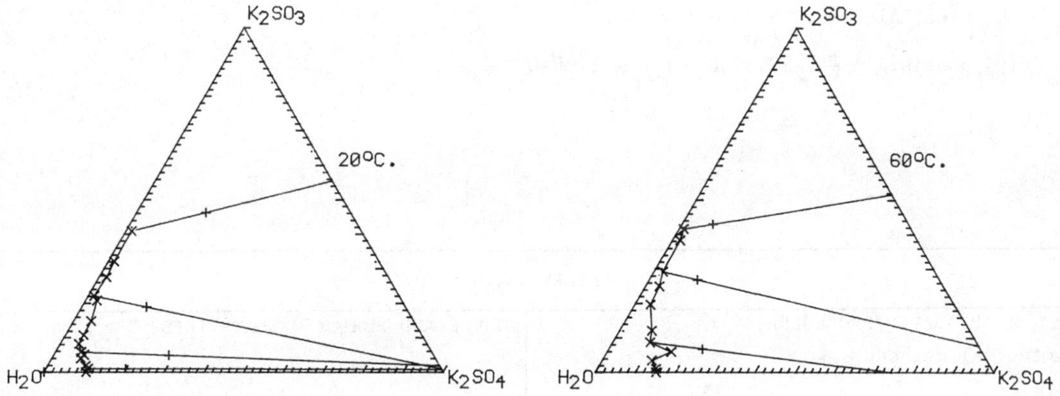

COMPONENTS:

1. Potassium sulfite; K_2SO_3; [10117-38-1]
2. Potassium carbonate; K_2CO_3; [584-08-7]
3. Water; H_2O; [7732-18-5]

ORIGINAL MEASUREMENTS:

Kuznetsova, A.G.; Trukhanova, E.A.

**VINITI Deposited Document* 1983, 6890-83.

VARIABLES:

Temperature: 293 K
Composition

PREPARED BY:

Mary R. Masson

EXPERIMENTAL VALUES:

Composition of saturated solutions

K_2SO_3 mass %	K_2CO_3 mass %	K_2SO_3[a] mol/kg	K_2CO_3[a] mol/kg	Solid[b] phase
33.45	13.15	3.958	1.782	A
27.30	24.90	3.609	3.769	A
19.80	30.00	2.492	4.324	A
19.25	36.20	2.730	5.879	A
15.10	38.00	2.034	5.862	A
15.15	38.80	2.079	6.096	A
18.10	38.40	2.629	6.387	A
16.81	39.50	2.431	6.541	A
11.90	43.87	1.700	7.176	A
8.50	47.70	1.226	7.880	A
6.18	51.00	0.912	8.618	A + B

a Molalities calculated by the compiler.

b Solid phases: A - K_2SO_3, B - $K_2CO_3.1.5H_2O$

AUXILIARY INFORMATION

METHOD APPARATUS/PROCEDURE:

Isothermal method.

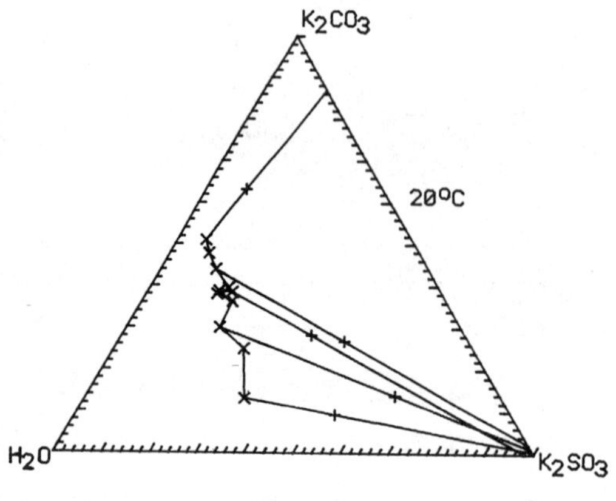

SOURCE AND PURITY OF MATERIALS:

ESTIMATED ERROR:

No estimates possible.

REFERENCES:

COMPONENTS:

1. Potassium sulfite; K_2SO_3; [10117-38-1]
2. Potassium carbonate; K_2CO_3; [584-08-7]
3. Potassium sulfate; K_2SO_4; [7778-80-5]
4. Water; H_2O; [7732-18-5]

ORIGINAL MEASUREMENTS:

Kuznetsova, A.G.; Trukhanova, E.A.

**VINITI Deposited Document* 1983, 6890-83.

VARIABLES:

Temperature: 293 K
Composition

PREPARED BY:

Mary R. Masson

EXPERIMENTAL VALUES:

Composition of saturated solutions

K_2SO_3 mass %	K_2CO_3 mass %	K_2SO_4 mass %	K_2SO_3[a] mol/kg	K_2CO_3[a] mol/kg	K_2SO_4[a] mol/kg
41.2	13.0	1.90	5.930	2.143	0.248
32.4	15.1	2.17	4.068	2.171	0.247
30.3	30.3	1.09	4.998	5.723	0.163
25.5	30.0	1.27	3.727	5.021	0.169
15.3	38.9	1.00	2.158	6.282	0.128
13.9	41.4	1.63	2.039	6.955	0.217
6.02	48.0	1.46	0.854	7.801	0.188
40.9	0.	1.43	4.481	0.	0.142
0.	53.2	0.023	0.	8.229	0.003
6.16	51.0	0.	0.909	8.614	0.

[a] Molalities calculated by the compiler.

Note: compositions of "wet residues" are also given in the original paper.

AUXILIARY INFORMATION

METHOD APPARATUS/PROCEDURE:

Isothermal method.

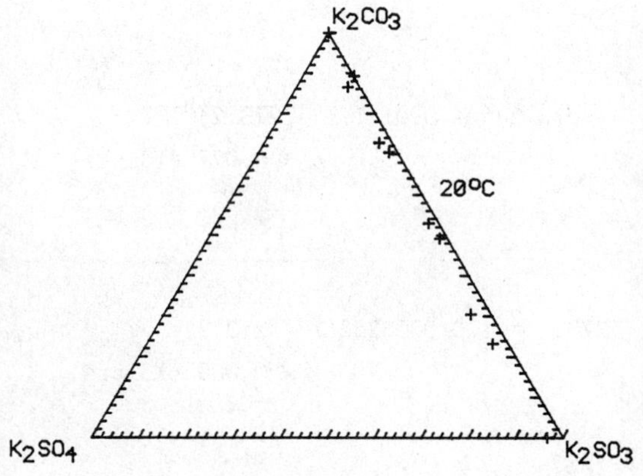

SOURCE AND PURITY OF MATERIALS:

ESTIMATED ERROR:

No estimates possible.

REFERENCES:

COMPONENTS:	EVALUATOR:
1. Potassium pyrosulfite; $K_2S_2O_5$; [16731-55-8] 2. Water; H_2O: [7732-18-5]	Mary R. Masson, Dept. of Chemistry, University of Aberdeen, Meston Walk, Old Aberdeen, AB9 2UE, Scotland, UK. March 1984.

CRITICAL EVALUATION:

There are two sets of data for this system (1,2), although one set is reported in terms of $KHSO_3$, and had to be converted for purposes of comparison (2). However, the two sets of data are not in agreement. The first set (1) gives the regression equation

$$y = 22.12 + 0.472(T - 273.2) - 0.00124(T - 273.2)^2 \qquad s = 0.14 \text{ (25 pts)}$$

and the second set (2) gives the equation

$$y = 25.1 + 0.255(T - 273.2) \qquad s = 0.72 \text{ (11 pts)}$$

where $y = 100w$ is the solubility expressed as mass % of $K_2S_2O_5$, T is the temperature in K and s is the standard deviation of the dependent variable about the regression line.

There is no obvious reason for the differences; the data of Foerster *et al.* are more self-consistent, and therefore probably to be preferred.

TENTATIVE SOLUBILITIES

The following tentative solubility values were calculated from the regression equation.

T/K	Solubility mass%	Solubility molality mol/kg
273.2	22.12	1.278
298.2	33.15	2.230
323.2	42.62	3.341
348.2	50.55	4.598
363.2	54.6	5.409

Foerster (1) also reports data for equilibrium with ice and with $K_2S_2O_5 \cdot 2/3H_2O$ [91498-98-5]. The regression equations are as follows:

Ice -

$$(T - 273.15) = -0.00051 - 0.294y + 0.00223y^2 - 0.0000972y^3 \qquad s = 0.023 \text{ (13 pts)}$$

or

$$y = -0.00230 - 3.40(T - 273.2) + 0.0923(T - 273.2)^2 + 0.0141(T - 273.2)^3$$
$$s = 0.077 \text{ (13 pts)}$$

$K_2S_2O_5 \cdot 2/3H_2O$ -

$$y = 21.8 + 0.600(T - 273.2) - 0.00806(T - 273.2)^2 + 0.0003252(T - 273.2)^3$$
$$s = 0.089 \text{ (13 pts)}$$

where the symbols have the same meanings as above.

COMPONENTS:	EVALUATOR:
1. Potassium pyrosulfite; $K_2S_2O_5$; [16731-55-8] 2. Water; H_2O: [7732-18-5]	Mary R. Masson, Dept. of Chemistry, University of Aberdeen, Meston Walk, Old Aberdeen, AB9 2UE, Scotland, UK. March 1984.

CRITICAL EVALUATION: (continued)

REFERENCES

1. Foerster, F.; Brosche, A.; Norberg-Schutz, Chr. *Z. Phys. Chem.* 1924, *10*, 435.

2. Platt, J.H.; Hudson, D. *J. Soc. Dyers Colourists* 1926, *42*, 348.

COMPONENTS:

1. Potassium pyrosulfite; $K_2S_2O_5$; [16731-55-8]
2. Water; H_2O; [7732-18-5]

ORIGINAL MEASUREMENTS:

Foerster, F.; Brosche, A.; Norberg-Schutz, Chr.

Z. Phys. Chem. 1924, *10*, 435-96.

VARIABLES:

Temperature: 267 - 367 K

PREPARED BY:

Mary R. Masson

EXPERIMENTAL VALUES:

Composition of equilibrium solutions

t/°C	$K_2S_2O_5$ mass %	$K_2S_2O_5$[a] mol/kg
- 6.0	19.30	1.076
- 6.0	19.25	1.072
- 5.6	19.75	1.107
- 5.5	19.52[b]	1.091
- 3.1	20.52	1.161
- 3.0	20.65[b]	1.171
0.0	22.20	1.283
+ 0.2	22.21	1.284
0.2	22.14	1.279
10.0	26.50	1.622
18.9	30.53	1.977
22.0	31.82	2.099
28.5	34.61	2.381
33.7	36.46	2.581
41.0	39.35	2.918
46.2	41.39	3.176
46.4	41.60	3.204
50.2	42.79	3.364
60.1	46.11	3.849
60.5	46.65	3.933
60.7	46.15	3.855
70.2	49.27	4.369
72.0	49.64	4.434
82.8	52.42	4.956
90.4	53.70	5.217
93.6	55.50	5.610
94.0	55.51	5.612

Solid phase: $K_2S_2O_5$

(continued on next page)

AUXILIARY INFORMATION

METHOD/APPARATUS/PROCEDURE:

Solids were equilibrated with solution under a hydrogen atmosphere, in a vessel maintained in a themostat. Samples for analysis were withdrawn through a tube plugged with cotton wool.

Samples were reacted with excess of standard iodine solution, then the excess was back-titrated with thiosulfate.

A Beckman apparatus (1) was used for the determination of freezing points.

SOURCE AND PURITY OF MATERIALS:

Prepared by saturating a concentrated solution of potassium hydroxide or potassium carbonate with sulfur dioxide, and allowing the salt to crystallize.

ESTIMATED ERROR:

Temperature: ±0.1 K
Analyses: no estimate possible.

REFERENCES:

1. Ostwald, W.; Luther, R. *Hand-und Hilfsbuch zur Ausfuhrung physicochemischer Messungen* 5th Ed., Akademische Verlag., Leipzig, 1931.

COMPONENTS:	ORIGINAL MEASUREMENTS:
1. Potassium pyrosulfite; $K_2S_2O_5$; [16731-55-8] 2. Water; H_2O; [7732-18-5]	Foerster, F.; Brosche, A.; Norberg-Schutz, Chr. *Z. Phys. Chem.* 1924 *10*, 435-96.

EXPERIMENTAL VALUES (continued):

t/°C	$K_2S_2O_5$ mass %	$K_2S_2O_5$[a] mol/kg	
- 0.4	21.50*	1.232	
+ 1.0	22.75	1.325	
1.2	22.50	1.306	
1.8	22.85	1.332	
4.0	24.15	1.432	
5.0	23.75	1.401	
6.1	25.10	1.507	
7.7	26.10	1.589	
8.9	26.75	1.643	Solid phase: $K_2S_2O_5.2/3H_2O$
10.1	27.25	1.685	
11.6	28.30	1.775	
12.2	28.37[b]	1.782	
15.0	30.05	1.932	
16.4	30.95	2.016	
18.0	31.30	2.049	
20.1	33.20	2.236	
- 1.07	3.73	0.174	
- 1.93	6.75	0.326	
- 3.73	13.15	0.681	Solid phase: ice
- 4.87	17.19	0.934	
- 5.63	19.47	1.087	$K_2S_2O_5$ dissolved
- 1.94	6.95	0.336	
- 2.93	10.50	0.528	
- 3.93	13.90	0.726	
- 1.34	4.64	0.219	
- 2.39	8.41	0.413	Solid phase: ice
- 3.48	12.32	0.632	
- 5.50	19.20	1.069	$K_2S_2O_5.2/3H_2O$ dissolved

[a] Molalities calculated by the compiler.

[b] Results considered particularly reliable by the authors.

COMPONENTS:

1. Potassium hydrogen sulfite; $KHSO_3$; [7773-08-4]
2. Water; H_2O; [7732-18-5]

ORIGINAL MEASUREMENTS:

Platt, J.H.; Hudson, D.

J. Soc. Dyers Colourists 1926, *42*, 348-9.

VARIABLES:

Temperature: 287 - 373 K

PREPARED BY:

Mary R. Masson

EXPERIMENTAL VALUES:

t/°C	$KHSO_3$ [a] g/100 g of water	$KHSO_3$ [b] mass %	$KHSO_3$ [c] mol/kg	$K_2S_2O_5$ [b] mass %
100	115.3	53.55	9.594	49.54
90	109.0	52.15	9.069	48.24
83	107.0	51.69	8.904	47.82
73	89.39	47.20	7.439	43.66
69	85.01	45.95	7.074	42.51
60	76.57	43.37	6.373	40.12
50	66.67	40.00	5.548	37.00
40	62.86	38.60	5.231	35.71
31	54.67	35.35	4.550	32.70
20	49.00	32.89	4.078	30.43
14	44.72	30.90	3.721	28.58

[a] Original data.

[b] Calculated by the compiler.

[c] Molalities calculated by the compiler.

AUXILIARY INFORMATION

METHOD APPARATUS/PROCEDURE:

The hydrogen sulfite was dissolved in distilled water. The solution, with excess of solid, was slowly heated to 100°C, and stirred there for 15 min. After settling for 15 min, about 1 g of solution was withdrawn, weighed, and analysed by reaction with excess of iodine and back-titration with thiosulfate. The solution was allowed to cool about 10°C, kept at constant temperature for about 15 min, then another sample was removed. This was repeated.

Sampling was difficult at temperatures above 75°C, owing to the rapid crystallization.

SOURCE AND PURITY OF MATERIALS:

The hydrogen sulfite used was recrystallized from commercial potassium metabisulfite. The crystals were washed and well drained on the filter (reduced pressure) but were not dried (to prevent oxidation to sulfate).

ESTIMATED ERROR:

No estimates possible. Replicates said to be "in close agreement".

REFERENCES:

COMPONENTS:	EVALUATOR:
1. Ammonium sulfite; $(NH_4)_2SO_3$; [10196-04-0] 2. Water; H_2O; [7732-18-5]	Mary R. Masson, Dept. of Chemistry, University of Aberdeen, Meston Walk, Old Aberdeen, AB9 2UE, Scotland, UK. March 1984.

CRITICAL EVALUATION:

Three studies have been made of the binary system ammonium sulfite - water (1 -3) and further data are available from studies of ternary systems (3 - 8). Yasuda (1) reported his results as amount-of-substance concentrations, and these cannot be converted to mass % or molality; thus, the data cannot be compared with other work. Yasuda reported that the solid phase in equilibrium with solutions at temperatures from 298 - 333 K (25 - 60°C) was $2(NH_4)_2SO_3.3H_2O$, but this does not seem to be confirmed in any other work; Mellor (9) reports some early work that postulated formation of $(NH_4)_2SO_3.1\frac{1}{3}H_2O$, but says that this was probably imperfectly dried monohydrate.

The other data all say that the solid phase, up to 353.2 K, is the monohydrate, $(NH_4)_2SO_3.H_2O$ [7783-11-1]. Above that, it is the anhydrous salt.

The data of Terres and co-workers (6 - 8) are very imprecise, and they are not in agreement with the other data, so all the points are rejected. The rest of the data are in reasonable agreement, with the exception of a few points, which were rejected before the final regression equations were derived.

The regression equations are (1) for 258.2 - 273.2 K, solid phase = ice

$$(T - 273.15) = -0.0467 - 0.331y - 0.000147y^3 \qquad s = 0.103 \text{ (20 pts)}$$

or

$$y = -0.095 - 3.11(T - 273.2) - 0.0698(T - 273.2)^2 \qquad s = 0.206 \text{ (20 pts)}$$

and (2) for 258.2 - 353.2 K, solid phase = $(NH_4)_2SO_3.H_2O$

$$y = 32.3 + 0.273(T - 273.2) + 0.00000945(T - 273.2)^3 \qquad s = 0.203 \text{ (33 pts)}$$

where $y = 100w$ is the solubility in mass % of $(NH_4)_2SO_3$, T is the temperature in K, and s is the standard deviation of the dependent variable about the regression line.

At temperatures above 354.0 K there is a different solubility line, for the solution in equilibrium with $(NH_4)_2SO_3$, but there were insufficient points available to justify calculation of a smoothing equation.

COMPONENTS:	EVALUATOR:
1. Ammonium sulfite; $(NH_4)_2SO_3$; [10196-04-0] 2. Water; H_2O; [7732-18-5]	Mary R. Masson, Dept. of Chemistry, University of Aberdeen, Meston Walk, Old Aberdeen, AB9 2UE, Scotland, UK. March 1984.

CRITICAL EVALUATION: (continued)

TENTATIVE SOLUBILITIES

The following tentative solubility values for $(NH_4)_2SO_3$ in water were calculated from the second regression equation.

T/K	Solubility mass %	molality, mol/kg
263.2	29.5	3.60
273.2	32.2	4.09
283.2	34.9	4.62
293.2	37.7	5.21
303.2	40.6	5.89
313.2	43.7	6.68
323.2	47.0	7.64
333.2	50.6	8.82
343.2	54.5	10.31
353.2	58.9	12.34

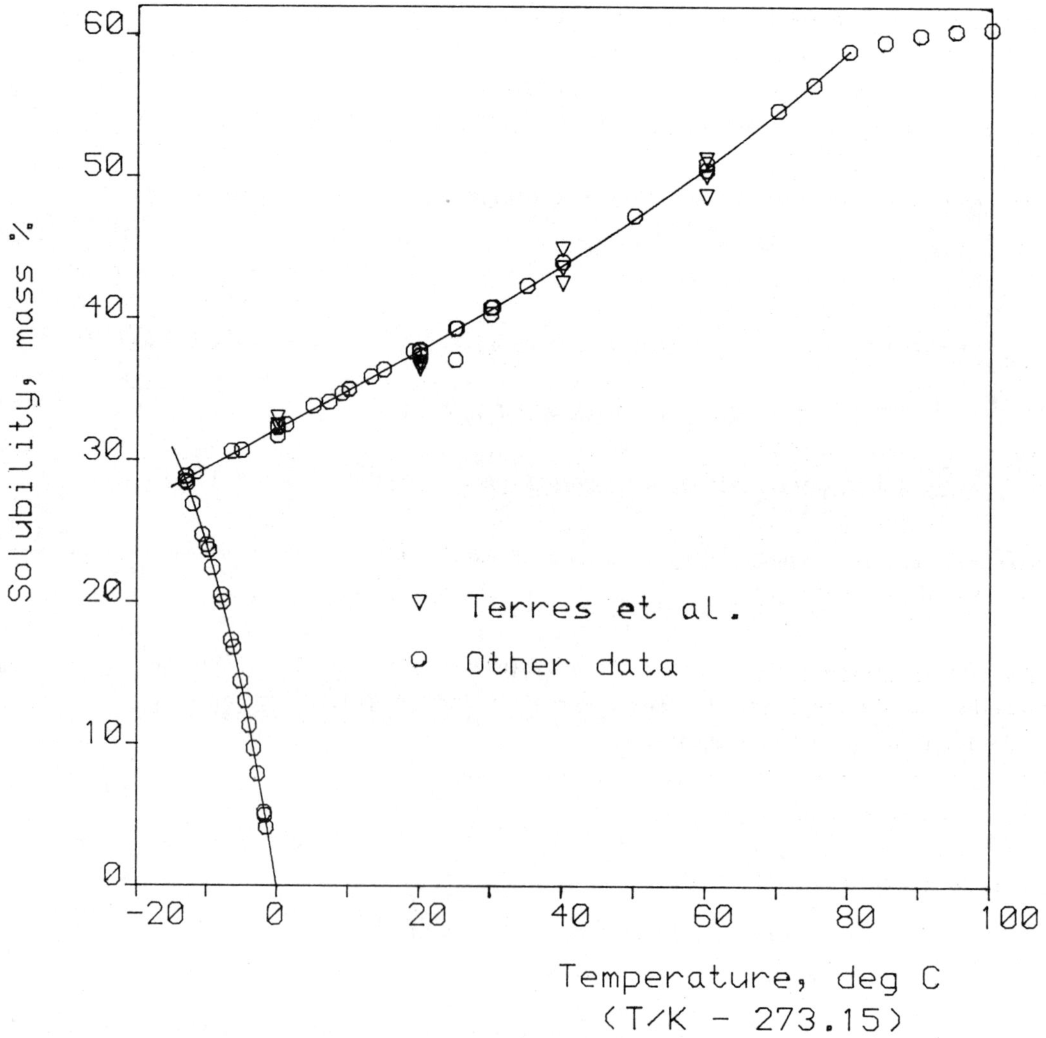

COMPONENTS:	EVALUATOR:
1. Ammonium sulfite; $(NH_4)_2SO_3$; [10196-04-0] 2. Water; H_2O; [7732-18-5]	Mary R. Masson, Dept. of Chemistry, University of Aberdeen, Meston Walk, Old Aberdeen, AB9 2UE, Scotland, UK. March 1984.

CRITICAL EVALUATION: (continued)

Ammonium sulfite - ammonia - water. The solubilities measured by Hill (10) for this system are rather lower than those of Ishikawa and Hiroshi (11), despite being measured at the same temperature. The experiments differed in that Hill worked with SO_2 and NH_3, whereas Ishikawa and Hiroshi used $(NH_4)_2SO_3$ and NH_3. The trends observed are similar, and there is no clear indication of which results should be preferred; I am inclined to favour those of Ishikawa and Hiroshi, because their work was more directly concerned with measurement of solubilities. The solid phase is the monohydrate $(NH_4)_2SO_3.H_2O$ [7783-11-1].

The results of Terres and Hahn (6) are not in agreement with the other data, except at one or two points. Examination of the data shows it to be highly erratic. Although a small number of the points do appear reasonable, many more obviously contain gross errors. This work is therefore rejected.

Ammonium sulfite - ammonium pyrosulfite - water. This system has also been described in terms of ammonium sulfite - sulfur dioxide - water and ammonium sulfite - ammonium hydrogen sulfite - water: to facilitate comparison, all the data were converted so as to express the system in terms of ammonium sulfite - ammonium pyrosulfite - water.

The results of Vasilenko (3) for 273.2, 293.2 and 303.2 K (0, 20 and 30°C) appear to be in reasonable agreement with those of Ishikawa and Hiroshi (11) for 298.2 K (25°C). The results of Hill (10) for 298.2 K are rather lower - they lie below Vasilenko's 293.2 K line. The results of Terres and Hahn (6) for 293.2 K agree with Vasilenko's data for the same temperature when the ammonium pyrosulfite content is below 15%. At higher pyrosulfite contents, they approach Vasilenko's data for 303 K.

Ammonium sulfite - ammonium sulfate - water. The results of Ishikawa and Murooka (12) at 288.2 and 303.2 K are in good agreement with those of Vasilenko (3) at 283.2 and 293.2 K. The rather limited amount of data provided by Ishikawa and Murooka for higher temperatures also appears to be in agreement with Vasilenko's more detailed work. The data of Terres and Heinsen (8) are in poor agreement with these.

Ammonium sulfite - sodium sulfite - water. Ternary systems (5,13) and more complex ones (13,14) are discussed under sodium sulfite.

Other ternary systems. Ammonium sulfite - ammonium chloride - water (5), and ammonium sulfite - ammonium thiosulfate - water (7) have also been studied.

QUATERNARY SYSTEMS

Two studies of the quaternary system ammonium sulfite - ammonium hydrogen sulfite - ammonium sulfate - water have been described (15,16).

COMPONENTS:	EVALUATOR:
1. Ammonium sulfite; $(NH_4)_2SO_3$; [10196-04-0] 2. Water; H_2O; [7732-18-5]	Mary R. Masson, Dept. of Chemistry, University of Aberdeen, Meston Walk, Old Aberdeen, AB9 2UE, Scotland, UK. March 1984.

CRITICAL EVALUATION: (continued)

REFERENCES

1. Yasuda, M. *Bull. Inst. Phys. Chem. Research (Tokyo)* 1924, *3*, 43.

2. Ishikawa, F.; Murooka, T. *Bull. Inst. Phys. Chem. Research (Tokyo)* 1928, *7*, 1160 (in Japanese); *Sci. Repts. Tohoku Imp. University* 1933, *22*, 202 (in English).

3. Vasilenko, N.A. *Zh. Priklad. Khim.* 1950, *23*, 472.

4. Vasilenko, N.A. *Zh. Priklad. Khim.* 1949, *22*, 338.

5. Labash, J.A.; Lusby, G.R. *Can. J. Chem.* 1955, *33*, 774.

6. Terres, E.; Hahn, E. *Das Gas- und Wasserfach* 1927, *70*, 363.

7. Terres, E.; Overdick, F. *Das Gas- und Wasserfach* 1928, *71*, 106.

8. Terres, E.; Heinsen, A. *Das Gas- und Wasserfach* 1927, *70*, 1157.

9. Mellor, J.W. *A Comprehensive Treatise on Inorganic and Theoretical Chemistry: Vol. X.* Longmans, Green and Co., London, 1930.

10. Hill, L.M. *J. Chem. Soc.* 1948, 76.

11. Ishikawa, F.; Hiroshi, H. *Bull. Inst. Phys. Chem. Research (Tokyo)* 1931, *10*, 166 (in Japanese); *Sci. Repts. Tohoku Imp. University* 1933, *22*, 235 (in English).

12. Ishikawa, F.; Murooka, T. *Bull. Inst. Phys. Chem. Research (Tokyo)* 1929, *8*, 75 (in Japanese); *Sci. Repts. Tohoku Imp. University* 1933, *22*, 220 (in English).

13. Zil'berman, Ya.I.; Ivanov, P.T. *Zh. Priklad. Khim.* 1941, *14*, 939.

14. Labash, J.A.; Lusby, G.R. *Can. J. Chem.* 1955, *33*, 787.

15. Vasilenko, N.A. *Zh. Priklad. Khim.* 1953, *26*, 650-2.

16. Vasilenko, N.A. *Nauch. -Tekh. Inform. Byull. Nauch. Inst. po Udobren i Insektofungisidam* 1957 (5 - 6) 105.

COMPONENTS:

1. Ammonium sulfite; $(NH_4)_2SO_3$; [10196-18-5]

2. Water; H_2O; [7732-18-5]

ORIGINAL MEASUREMENTS:

Yasuda, M.

Bull. Inst. Phys. Chem. Research (Tokyo) 1924, *3*, 43-50.

VARIABLES:

Temperature: 285 - 333 K

PREPARED BY:

Mary R. Masson

EXPERIMENTAL VALUES:

Composition of equilibrium solutions

t/°C	SO_2 mol dm^{-3}	NH_3 mol dm^{-3}	$(NH_4)_2SO_3$ g dm^{-3}	Solid[a] phase
12	3.463	6.899	403.2	B
15	3.675	7.450	426.5	B
20	3.874	7.728	451.5	B
25	4.060	8.092	470.96	A
30	4.189	8.406	485.9	A
40	4.328	8.859	502.4	A
50	4.919	9.641	570.6	A
60	54.64	10.81	633.7	A

[a] Solid phases: A - $2(NH_4)_2SO_3.3H_2O$, B - $(NH_4)_2SO_3.H_2O$

AUXILIARY INFORMATION

METHOD APPARATUS/PROCEDURE:

Equilibrium solutions were analysed by "standard methods" (1).

SOURCE AND PURITY OF MATERIALS:

Not stated.

ESTIMATED ERROR:

No estimates possible.

REFERENCES:

1. Treadwell, *Analytical Chemistry*, 5th Ed., Vol. II, 560, 692.

COMPONENTS:

1. Ammonium sulfite; $(NH_4)_2SO_3$; [10196-04-0]
2. Water; H_2O; [7732-18-5]

ORIGINAL MEASUREMENTS:

Ishikawa, F.; Murooka, T.

Bull. Inst. Phys. Chem. Research (Tokyo) 1928, *7*, 1160-76. (In Japanese); *Sci. Repts. Tohoku Imp. University* 1933, *22*, 201-219. (In English).

VARIABLES:

Temperature: 260 - 373 K

PREPARED BY:

Mary R. Masson

EXPERIMENTAL VALUES:

t/°C	Time hr	$(NH_4)_2SO_3$ mass %	$(NH_4)_2SO_3$ mean mass %	$(NH_4)_2SO_3$ g/100 ml soln.	$(NH_4)_2SO_3$[a] mol/kg
0	4.5	32.52			
	5	32.29			
	27	32.46			
	53	32.31	32.40	38.21	4.127
5	4.5	33.75			
	4.5	33.75			
	28	33.79			
		33.87			
		33.89	33.81	40.05	4.398
10	5	35.05			
	30	35.05	35.05	41.69	4.647
15	4.5	36.39			
	5.5	36.40			
	28	36.40			
	51	36.41	36.40	43.47	4.928
20	5	37.78			
	24	37.82	37.80	45.34	5.233
25	24	39.30			
	24	39.28	39.29	47.31	5.572
30	5	40.75			
	16	40.82			
	18	40.73	40.77	49.32	5.927

(continued on next page)

AUXILIARY INFORMATION

METHOD APPARATUS/PROCEDURE:

A simple saturation technique was used. An atmosphere of ammonia-satd. nitrogen was used, in an open vessel for temperatures up to 80°C, and in a closed vessel at the higher temps. Solution samples were removed through a filtering tube containing cotton wool into a pipette with stopcocks at both ends. The solution removed was weighed and analysed.

Sulfur dioxide was determined by reaction of the sample with excess of acidified iodine solution, followed by back-titration with thiosulfate. Ammonia was determined after addition of sodium hydroxide, collection of the evolved ammonia in standard sulfuric acid solution, and titration of the excess of acid with standard sodium hydroxide to a sodium alizarin sulfonate end-point.

The solubility was calculated from the mean of the values corresponding to these two analyses.

SOURCE AND PURITY OF MATERIALS:

Sulfur dioxide gas from a cylinder was passed into an ammonia solution under an atmosphere of hydrogen, until only a little free ammonia remained. The small crystals obtained were redissolved in the solution by heating, then the solution was cooled very slowly to allow large crystals to separate. These were filtered off under N_2 and kept in a special desiccator.

ESTIMATED ERROR:

Temperature: ±0.02°C (up to 80°C)
±0.05°C (above 80°C)

Analyses: r.s.d. generally < 0.2%.

REFERENCES:

COMPONENTS:	ORIGINAL MEASUREMENTS:
1. Ammonium sulfite; $(NH_4)_2SO_3$; [10196-04-0] 2. Water; H_2O; [7732-18-5]	Ishikawa, F.; Murooka, T. *Bull. Inst. Phys. Chem. Research (Tokyo)* 1928, *7*, 1160-76. (In Japanese); *Sci. Repts. Tohoku Imp. University* 1933, *22*, 201-219. (In English).

EXPERIMENTAL VALUES (continued):

t/°C	Time hr	$(NH_4)_2SO_3$ mass %	$(NH_4)_2SO_3$ mean mass %	$(NH_4)_2SO_3$ g/100 ml soln.	$(NH_4)_2SO_3$[a] mol/kg
35	5	42.35			
	24	42.31			
	24	42.31	42.32	51.44	6.317
40	5	43.93			
	29	43.98			
	29	43.96	43.96	53.64	6.754
50	2	47.27			
	4	47.25	47.26	58.16	7.716
60	2	50.90			
	4.5	50.97	50.94	63.31	8.940
70	2.5	54.72			
	3	54.70			
	4.5	54.72	54.71	68.70	10.401
75	2	56.51			
	4.5	56.54	56.52	71.38	11.193
80	1.5	58.88			
	1.5	58.90	58.89	74.88	12.334
85[b]	1	59.42			
	2	59.50			
	2.5	59.68	59.53		12.665
90[b]	1.5	60.10			
	3.5	59.90	60.00		12.915
95[b]	1	60.27			
	1.5	60.34	60.30		13.078
100[b]	1	60.63			
	1.5	60.26			
	1.5	60.44	60.44		13.155
- 6.55	2	30.69			
	2	30.55	30.62		3.800
-11.52	2	29.14			
	3	29.19	29.16		3.544
-12.96		28.87			
		28.84	28.85		3.491
- 1.73[c]		4.961			0.449
- 1.82[c]		5.162			0.469
- 3.35[c]		9.698			0.925
- 4.61[c]		13.044			1.292
- 6.27[c]		16.817			1.741
- 7.97[c]		20.505			2.221
- 9.69[c]		23.652			2.667
-12.74[c]		28.418			3.418
-12.96		28.855			3.492

[a] Molalities calculated by the compiler.

[b] Solid phases: $(NH_4)_2SO_3$, [c] solid phase ice, otherwise $(NH_4)_2SO_3.H_2O$.

The transition temperature between the monohydrate and the anhydrous salt was found to be 80.8± 0.2°C.

COMPONENTS:

1. Ammonium sulfite; $(NH_4)_2SO_3$; [10196-04-0]

2. Water; H_2O; [7732-18-5]

ORIGINAL MEASUREMENTS:

Vasilenko, N.A.

Zh. Priklad. Khim. 1950, *23*, 472-81.

VARIABLES:

Temperature: 261 - 303 K

PREPARED BY:

Mary R. Masson

EXPERIMENTAL VALUES:

Composition of equilibrium solutions

t/°C	$(NH_4)_2SO_3$ mass %	$(NH_4)_2SO_3$ [a] mol/kg	Solid [b] phase
-12.0	26.9	3.168	A
-10.6	24.7	2.824	A
- 9.2	22.4	2.485	A
- 7.8	20.0	2.153	A
- 6.6	17.3	1.801	A
- 5.3	14.4	1.448	A
- 4.0	11.3	1.097	A
- 2.8	7.9	0.739	A
- 1.5	4.1	0.368	A
- 5.2	30.7	3.814	B
+ 1.2	32.5	4.146	B
7.2	34.1	4.455	B
9.0	34.7	4.575	B
13.2	35.9	4.822	B
19.1	37.7	5.211	B
25.2	39.3	5.575	B
30.3	40.8	5.934	B

[a] Molalities calculated by the compiler.

[b] Solid phases: A - ice, B - $(NH_4)_2SO_3.H_2O$

AUXILIARY INFORMATION

METHOD APPARATUS/PROCEDURE:

A simple saturation procedure.

SOURCE AND PURITY OF MATERIALS:

ESTIMATED ERROR:

No estimates possible.

REFERENCES:

COMPONENTS:

1. Ammonium sulfite; $(NH_4)_2SO_3$; [10196-04-0]
2. Ammonia; NH_3; [7664-41-7]
3. Water; H_2O; [7732-18-5]

ORIGINAL MEASUREMENTS:

Terres, E.; Hahn, E.

Das Gas- und Wasserfach 1927, *70*, 363-367.

VARIABLES:

Four temperatures: 273 - 333 K
Concentrations of the components

PREPARED BY:

Mary R. Masson

EXPERIMENTAL VALUES: Composition of equilibrium solutions

$(NH_4)_2SO_3$ mass %	NH_3 mass %	$(NH_4)_2SO_3$[a] mol/kg	NH_3[a] mol/kg
Temperature = 0°C			
32.93	0.0	4.227	0.
26.11	3.97	3.215	3.334
24.78	7.52	3.152	6.523
24.34	8.41	3.116	7.343
32.70	8.57	4.794	8.569
40.81	10.26	7.181	12.313
33.08	10.40	5.039	10.805
26.91	10.43	3.698	9.774
34.41	10.93	5.420	11.742
31.89	12.11	4.903	12.698
102.87?	13.94	—	—
26.93	16.80	4.121	17.531
33.47	14.38	5.526	16.192
20.80	17.98	2.925	17.246
32.28	18.08	5.599	21.387
66.02	18.92	37.746	73.770
71.08	19.02	61.820	112.813
26.70	19.57	4.279	21.387
47.96	20.20	12.970	37.253
43.97	20.40	10.626	33.620
38.09	20.76	7.970	29.624
27.19	21.15	4.532	24.040

(continued on next page)

AUXILIARY INFORMATION

METHOD APPARATUS/PROCEDURE:

To a saturated solution of ammonium sulfite was added ammonia gas. After separation of the solid, the solution was analysed for NH_3 and SO_2, then the mass % of ammonium sulfite and the excess amount of ammonia were calculated.

SOURCE AND PURITY OF MATERIALS:

ESTIMATED ERROR:

Temperature: ±0.1 K
Analyses: no estimate possible.

REFERENCES:

COMPONENTS:	ORIGINAL MEASUREMENTS:
1. Ammonium sulfite; $(NH_4)_2SO_3$; [10196-04-0] 2. Ammonia; NH_3; [7664-41-7] 3. Water; H_2O; [7732-18-5]	Terres, E.; Hahn, E. *Das Gas- und Wasserfach* 1927, *70*, 363-367.

EXPERIMENTAL VALUES (continued):

$(NH_4)_2SO_3$ mass %	NH_3 mass %	$(NH_4)_2SO_3$[a] mol/kg	NH_3[a] mol/kg
Temperature = 20°C			
36.60	0.0	4.971	0.
36.49	0.52	4.988	0.485
32.09	2.59	4.230	2.328
30.28	4.94	4.025	4.478
29.84	5.65	3.983	5.143
32.14	8.21	4.644	8.085
33.19	8.43	4.895	8.479
30.15	8.69	4.245	8.343
30.37	8.77	4.297	8.462
33.38	8.98	4.986	9.148
38.90	9.07	6.437	10.236
34.37	9.20	5.244	9.573
38.92	9.35	6.478	10.613
36.26	10.57	5.872	11.673
34.26	12.86	5.578	14.280
36.63	13.07	6.270	15.258
56.55	14.71	16.942	30.055
32.18	14.98	5.244	16.647
32.36	16.54	5.453	19.006
34.00	17.24	6.004	20.762
38.00	17.39	7.334	22.890
54.70	17.47	16.924	36.961
43.28	18.09	9.647	27.498
46.05	18.15	11.076	29.770
72.71	18.18	68.722	117.182
66.10	18.59	37.174	71.300
64.93	19.09	34.985	70.148
Temperature = 40°C			
42.50	1.05	6.483	1.092
40.01	2.28	5.969	2.320
37.24	4.92	5.544	4.995
37.70	5.74	5.739	5.959
36.00	6.27	5.369	6.378
32.55	7.93	4.709	7.823
55.60	9.06	13.546	15.054
31.95	9.53	4.701	9.563
32.42	9.79	4.830	9.948
37.08	9.81	6.011	10.846
31.56	9.84	4.637	9.860
36.20	10.03	5.797	10.953
39.90	10.16	6.879	11.946
35.30	10.41	5.599	11.259
38.06	10.30	6.346	11.712
37.03	10.49	6.075	11.737
31.82	10.54	4.753	10.737
38.37	11.50	6.590	13.471
39.30	11.84	6.926	14.229
39.33	12.00	6.958	14.478
55.23?	12.24	14.619	22.094
37.00	12.70	6.334	14.826
36.60	13.26	6.285	15.529
51.70?	13.34	12.733	22.406
37.27	14.07	6.595	16.979
37.86	14.26	6.808	17.488
36.12	14.58	6.308	17.366
36.05	15.03	6.345	18.041
44.33	15.44	9.488	22.536

(continued on next page)

COMPONENTS:	ORIGINAL MEASUREMENTS:
1. Ammonium sulfite; $(NH_4)_2SO_3$; [10196-04-0]	Terres, E.; Hahn, E.
2. Ammonia; NH_3; [7664-41-7]	*Das Gas- und Wasserfach* 1927, *70*, 363-367.
3. Water; H_2O; [7732-18-5]	

EXPERIMENTAL VALUES (continued):

$(NH_4)_2SO_3$ mass %	NH_3 mass %	$(NH_4)_2SO_3$[a] mol/kg	NH_3[a] mol/kg
Temperature = 60°C			
50.07	0.53	8.727	0.630
47.58	2.18	8.154	2.548
46.68	4.05	8.158	4.827
45.25	4.77	7.795	5.604
45.60	4.78	7.913	5.657
39.01	11.68	6.812	13.909
44.33	5.55	7.616	6.502
41.87	7.11	7.066	8.183
39.80	9.03	6.697	10.362
38.73	9.79	6.478	11.167
38.86	10.16	6.563	11.703
38.46	10.23	6.454	11.707
38.43	10.49	6.478	12.059
39.14	10.82	6.735	12.697
56.92	10.87	15.216	19.816
52.95	12.29	13.116	20.761
39.54	12.53	7.103	15.351
40.68	12.54	7.480	15.741
47.34	12.66	10.190	18.585
47.68	12.90	10.414	19.216
38.99	13.36	7.045	16.464

[a] Molalities calculated by the compiler

[b] The solid phase was $(NH_4)_2SO_3.H_2O$ throughout.

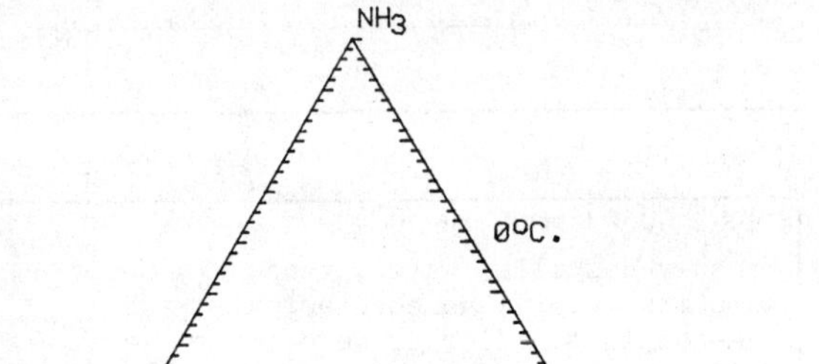

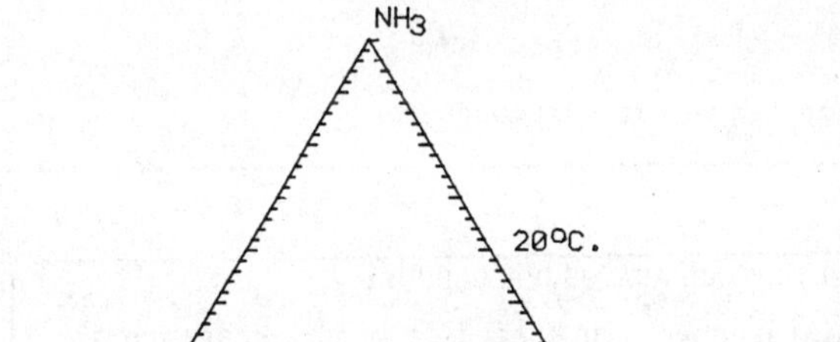

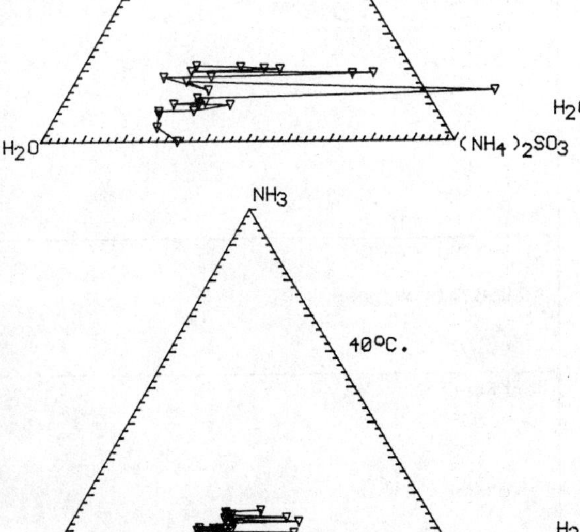

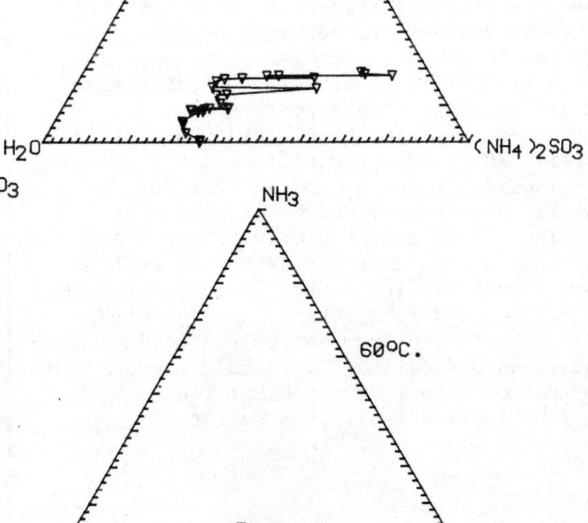

COMPONENTS:	ORIGINAL MEASUREMENTS:
1. Ammonia; NH_3; [7664-41-7] 2. Sulfur dioxide; SO_2; [7446-09-5] 3. Water; H_2O; [7732-18-5]	Hill, L.M. *J. Chem. Soc.* 1948, 76-8.
VARIABLES:	PREPARED BY:
Concentrations of the components One temperature: 298 K	Mary R. Masson

EXPERIMENTAL VALUES:

Composition of equilibrium solutions at 25°C

NH_3 mass %	SO_2 mass %	SO_3 mass %	NH_3[a] mol/kgt	SO_2[a] mol/kg
23.8	6.5	0.2	20.051	1.456
19.9	9.3	-	16.505	2.051
17.4	13.5	-	14.786	3.050
16.1	15.9	0.3	13.903	3.650
15.8	16.4	0.1	13.684	3.776
13.8	19.0	-	12.059	4.414
13.8	19.2	0.1	12.095	4.473
11.9	20.7	0.7	10.367	4.794
12.1	21.0	-	10.620	4.900
11.5	22.0	0.3	10.155	5.164
11.8	27.3	0.2	11.378	6.998
12.4	33.3	0.3	13.409	9.573
13.4	41.4	0.8	17.408	14.298
13.8	43.2	0.2	18.845	15.683
14.1	44.9	0.4	20.194	17.095
14.8	49.2	0.5	24.140	21.334

[a] Molalities calculated by the compiler.
Solid phase: $2NH_3.SO_2.2H_2O$ [= $(NH_4)_2SO_3.H_2O$]

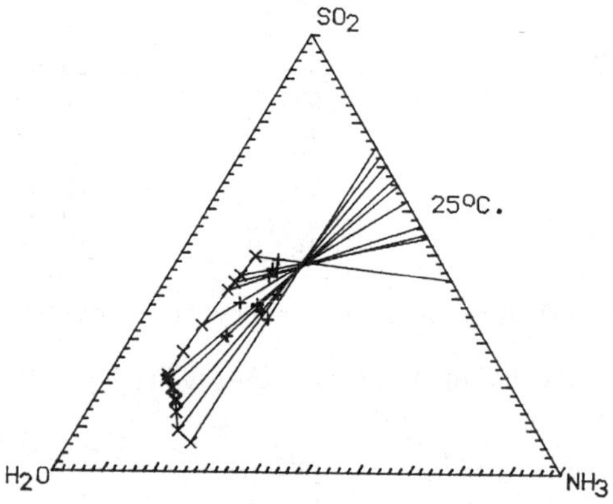

(continued on next page)

AUXILIARY INFORMATION

METHOD APPARATUS/PROCEDURE:

Mixtures were prepared from the reagents in an 8-oz bottle fitted with a glass stirrer. The mixture was stirred at constant temperature under coal gas (but some oxidation still occurred). Seeding with solid was done when supersaturation occurred. Overnight (17 hr) was usually sufficient time for equilibration. Samples of liquid were separated and weighed. Sulfite was determined by titration with iodine, total sulfur gravimetrically as barium sulfate after oxidation of sulfite with bromine, alkalinity by titration with sulfuric acid to Methyl Orange, and ammonia by distillation from sodium hydroxide solution and collection in sulfuric acid, or by calculation from the other analyses (NH_3 = alkalinity + $\frac{1}{2}$sulfite + sulfate).

SOURCE AND PURITY OF MATERIALS:

Freshly distilled water, reagent ammonia solution (0.88), gaseous ammonia, and gaseous sulfur dioxide were the reagents used in this study.

ESTIMATED ERROR:

No estimates possible.

REFERENCES:

COMPONENTS:	ORIGINAL MEASUREMENTS:
1. Ammonia; NH_3; [7664-41-7]	Hill, L.M.
2. Sulfur dioxide; SO_2; [7446-09-5]	*J. Chem. Soc.* 1948, 76-8.
3. Water; H_2O; [7732-18-5]	

ADDITIONAL MATERIAL:

The original data have been recalculated by the compiler in order to facilitate comparison with other data.

$(NH_4)_2SO_3$ mol/kg	$(NH_4)_2SO_3$ mass %	NH_3 mol/kg	NH_3 mass %
1.456	11.58	17.133	19.97
2.051	16.43	12.403	14.57
3.050	23.58	8.686	9.85
3.650	27.59	6.603	7.32
3.776	28.42	6.132	6.77
4.414	32.70	3.231	3.51
4.473	33.02	3.149	3.41
4.794	35.46	0.779	0.85
4.900	35.95	0.820	0.88

$(NH_4)_2SO_3$ mol/kg	$(NH_4)_2SO_3$ mass %	SO_2 mol/kg	SO_2 mass %
5.078	36.97	0.086	0.35
5.689	37.87	1.309	4.81
6.704	39.68	2.869	9.37
8.704	42.67	5.594	15.13
9.423	43.86	6.260	16.07
10.097	44.74	6.998	17.10
12.07	46.80	9.264	19.81

OR

$(NH_4)_2SO_3$ mol/kg	$(NH_4)_2SO_3$ mass %	$(NH_4)_2S_2O_5$ mol/kg	$(NH_4)_2S_2O_5$ mass %
4.992	36.32	0.086	0.97
4.380	29.14	1.309	13.52
3.835	22.68	2.869	26.35
3.110	15.23	5.594	42.55
3.163	14.71	6.260	45.21
3.099	13.72	6.998	48.12
2.806	10.87	9.264	55.74

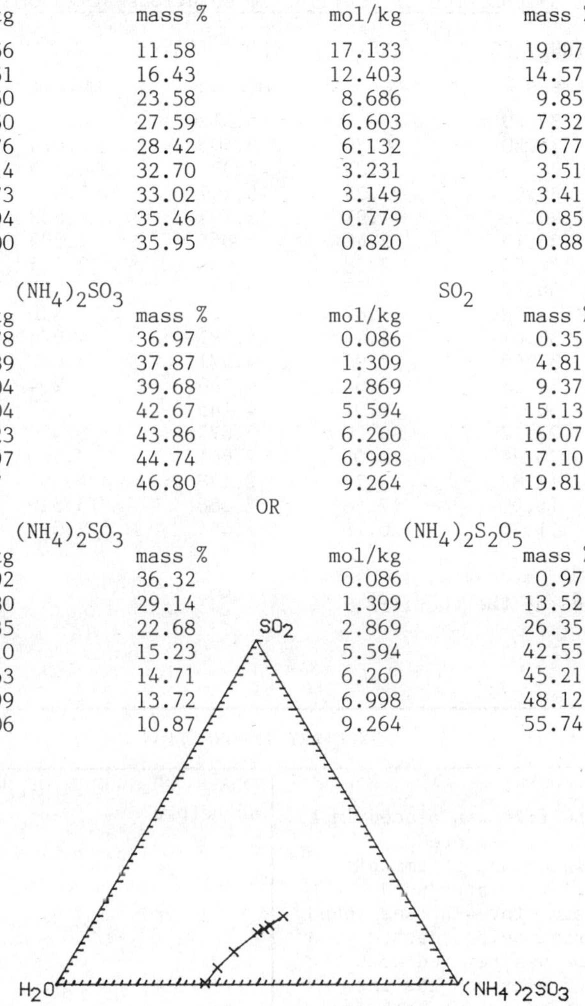

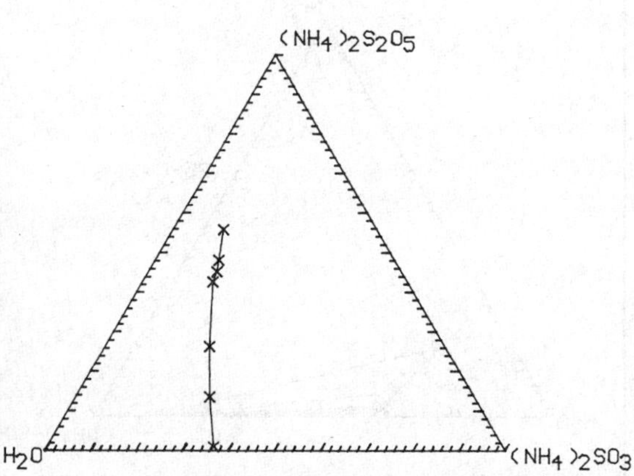

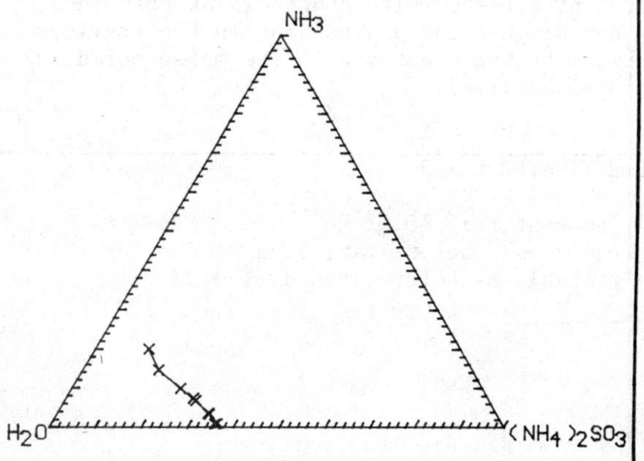

COMPONENTS:

1. Ammonium sulfite; $(NH_4)_2SO_3$; [10196-04-0]

2. Ammonia; NH_3; [7664-41-7]

3. Water; H_2O; [7732-18-5]

ORIGINAL MEASUREMENTS:

Ishikawa, F.; Hiroshi, H.

Bull. Inst. Phys. Chem. Research (Tokyo) 1931, *10*, 166-72 (in Japanese); *Sci. Repts. Tohoku Imp. University* 1933, 22, 235-243 (in English).

VARIABLES:

One temperature: 298 K
Concentrations of the components

PREPARED BY:

Mary R. Masson

EXPERIMENTAL VALUES:

Composition of equilibrium solutions at 25°C

$(NH_4)_2SO_3$ mass %	NH_3 mass %	$(NH_4)_2SO_3$[a] mol/kg	NH_3[a] mol/kg
39.29	-	5.572	0.
38.30	0.70	5.406	0.674
38.15	0.82	5.382	0.789
36.98	1.75	5.197	1.677
36.90	1.91	5.192	1.833
36.15	2.40	5.065	2.293
36.05	2.58	5.058	2.469
35.97	2.67	5.047	2.555
34.58	3.67	4.822	3.490
33.01	4.94	4.581	4.675
31.66	5.97	4.371	5.621
30.51	6.64	4.180	6.204
30.25	6.91	4.145	6.457
27.17	9.20	3.677	8.490
20.29	14.05	2.661	12.565
16.82	16.70	2.178	14.751
15.90	17.58	2.058	15.519
11.51	20.78	1.464	18.021

[a] Molality calculated by the compiler.

AUXILIARY INFORMATION

METHOD APPARATUS/PROCEDURE:

A sample of ammonia sulfite was placed in a closed vessel under an atmosphere of nitrogen, then a known amount of ammonia solution (oxygen-free) was introduced. Stirring in a thermostat for 4 hr was found to result in equilibrium being reached. A sample of the solution was removed with the aid of the pressure of nitrogen gas into a pipette fitted with stopcocks at both ends, and weighed and analysed as in the previous work by these authors. The moist solid was also analysed.

SOURCE AND PURITY OF MATERIALS:

As before.

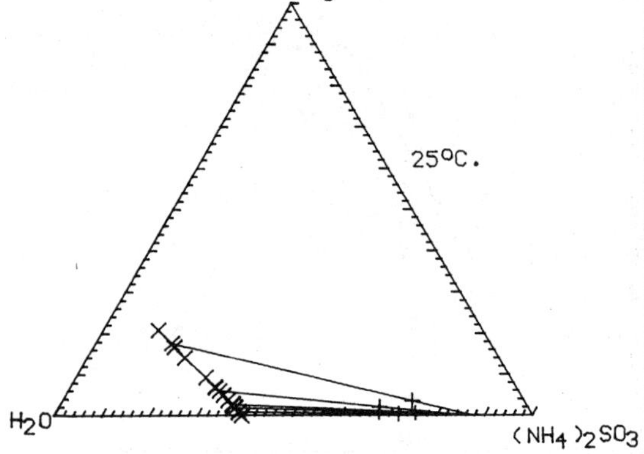

ESTIMATED ERROR:

Temperature: ±0.01 K
Analyses: no estimate given (compiler - probably as before, r.s.d. < 0.2%).

COMPONENTS:	ORIGINAL MEASUREMENTS:
1. Ammonium sulfite; $(NH_4)_2SO_3$; [10196-04-0] 2. Ammonium pyrosulfite; $(NH_4)_2S_2O_5$; [32736-64-4] 3. Water; H_2O; [7732-18-5]	Vasilenko, N.A. *Zh. Priklad. Khim.* 1949, 22, 338-41.
VARIABLES:	PREPARED BY:
Three temperatures: 273 - 303 K Concentrations of the components	Mary R. Masson

EXPERIMENTAL VALUES: Composition of equilibrium solutions

$(NH_4)_2SO_3$ mass %	NH_4HSO_3 mass %	$(NH_4)_2SO_3$[a] mol/kg	NH_4HSO_3[a] mol/kg	$(NH_4)_2S_2O_5$[b] mass %	Solid[c] phase
Temperature = 0°C					
31.71	0.0	3.998	0.	0.0	A
22.58	19.50	3.357	3.397	17.73	A
15.40	39.17	2.919	8.699	35.61	A
10.78	56.29	2.819	17.247	51.17	A
9.06	60.76	2.585	20.313	55.24	A + B
7.20	63.99	2.152	22.410	58.17	B
2.48	69.18	0.753	24.630	62.89	B
0.0	72.38	0.	26.441	65.80	B
Temperature = 20°C					
37.40	0.0	5.144	0.	0.0	A
30.24	13.99	4.669	2.531	12.72	A
24.26	27.75	4.353	5.834	25.23	A
18.19	42.50	3.984	10.909	38.64	A
14.79	54.54	4.152	17.943	49.58	A
11.31	65.04	4.118	27.748	59.13	A + B
9.73	67.56	3.689	30.016	61.42	B
6.48	70.57	2.431	31.026	64.16	B
4.61	71.27	1.646	29.813	64.79	B
3.54	73.84	1.348	32.937	67.13	B
1.71	74.26	0.613	31.181	67.51	B
0.0	75.95	0.	31.864	69.05	B

(continued on next page)

AUXILIARY INFORMATION

METHOD APPARATUS/PROCEDURE:

An isothermal procedure was used.

Data expressed in terms of NH_3 and SO_2 were also presented in the paper.

SOURCE AND PURITY OF MATERIALS:

ESTIMATED ERROR:

No estimates possible.

REFERENCES:

COMPONENTS:	ORIGINAL MEASUREMENTS:
1. Ammonium sulfite; $(NH_4)_2SO_3$; [10196-04-0] 2. Ammonium pyrosulfite; $(NH_4)_2S_2O_5$; [32736-64-4] 3. Water; H_2O; [7732-18-5]	Vasilenko, N.A. *Zh. Priklad. Khim.* 1949, 22, 338-41.

EXPERIMENTAL VALUES (continued):

$(NH_4)_2SO_3$ mass %	NH_4HSO_3 mass %	$(NH_4)_2SO_3$[a] mol/kg	NH_4HSO_3[a] mol/kg	$(NH_4)_2S_2O_5$[b] mass %	Solid[c] phase
Temperature = 30°C					
40.30	0.0	5.812	0.	0.0	A
30.77	18.81	5.255	3.764	17.10	A
21.79	38.26	4.696	9.663	34.78	A
16.13	55.94	4.973	20.208	50.86	A
12.75	66.73	5.350	32.812	60.67	A + B
8.21	70.98	3.397	34.415	64.53	B
4.09	74.45	1.641	35.004	67.68	B
1.72	76.85	0.691	36.183	69.87	B
0.0	79.31	0.	38.677	72.10	B

[a] Molalities calculated by the compiler.

[b] Calculated by the compiler.

[c] Solid phases: A - $(NH_4)_2SO_3.H_2O$, B - $(NH_4)_2S_2O_5$

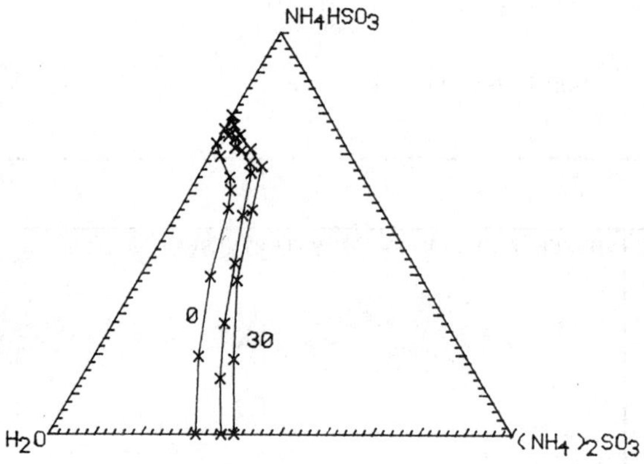

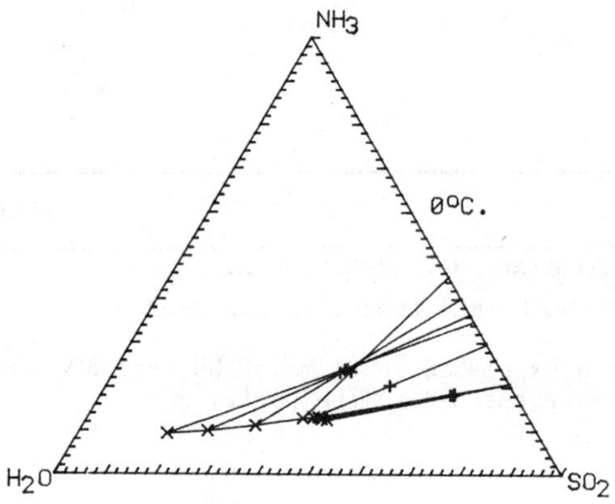

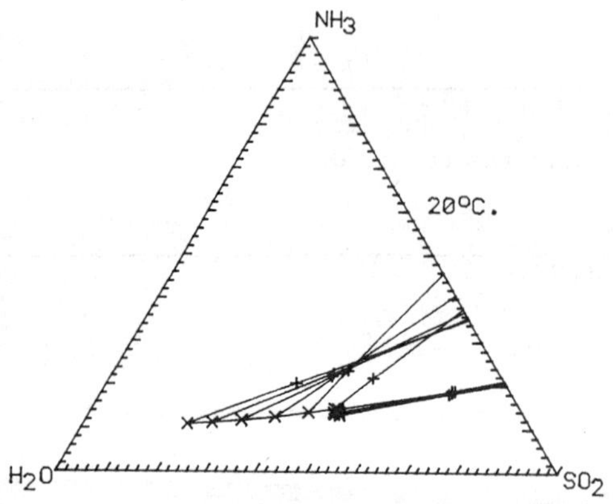

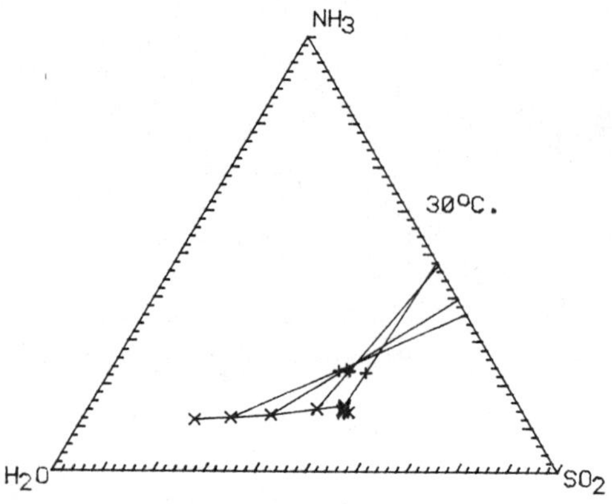

COMPONENTS:	ORIGINAL MEASUREMENTS:
1. Ammonium sulfite; $(NH_4)_2SO_3$; [10196-04-0] 2. Sulfur dioxide; SO_2; [7446-09-5] 3. Water; H_2O; [7732-18-5]	Terres, E.; Hahn, E. *Das Gas- und Wasserfach* 1927, *70*, 363-367.
VARIABLES:	PREPARED BY:
Concentrations of the components Four temperatures: 273 - 333 K	Mary R. Masson

EXPERIMENTAL VALUES:

Composition of equilibrium solutions

$(NH_4)_2SO_3$ mass %	SO_2 mass %	$(NH_4)_2SO_3$[a] mol/kg	SO_2[a] mol/kg	Solid[b] phase
Temperature = 0°C				
32.93	0.0	4.227	0.	A
34.64	0.50	4.599	0.120	A
33.57	1.30	4.438	0.312	A
35.04	3.44	4.904	0.873	A
36.63	6.99	5.594	1.935	A
39.77	13.60	7.344	4.553	A
40.55	14.37	7.745	4.976	A
42.78	15.86	8.906	5.986	A
45.83	16.50	10.475	6.838	A
45.28	17.50	10.475	7.340	A + B
43.17	22.87	10.945	10.513	B
42.35	26.34	11.646	13.132	B
39.14	31.87	11.625	17.161	B

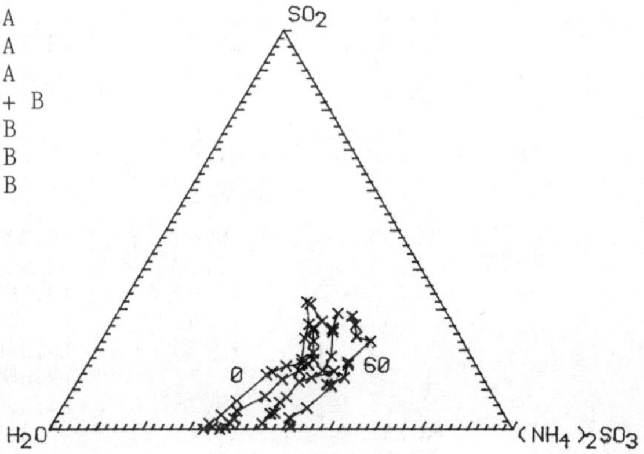

(continued on next page)

AUXILIARY INFORMATION

METHOD APPARATUS/PROCEDURE:

To a saturated solution of ammonium sulfite was added sulfur dioxide gas. After separation of the solid, the solution was analysed for NH_3 and SO_2, then the mass % of ammonium sulfite and the excess amount of sulfur dioxide were calculated.

SOURCE AND PURITY OF MATERIALS:

ESTIMATED ERROR:

Temperature: 0.1 K
Analyses: no estimate possible.

REFERENCES:

COMPONENTS:

1. Ammonium sulfite; $(NH_4)_2SO_3$; [10196-04-0]

2. Sulfur dioxide; SO_2; [7446-09-5]

3. Water; H_2O; [7732-18-5]

ORIGINAL MEASUREMENTS:

Terres, E.; Hahn, E.

Das Gas- und Wasserfach 1927, *70*, 363-367.

EXPERIMENTAL VALUES (continued):

$(NH_4)_2SO_3$ mass %	SO_2 mass %	$(NH_4)_2SO_3$[a] mol/kg	SO_2[a] mol/kg	Solid[b] phase
Temperature = 20°C				
36.50	0.0	4.949	0.	A
37.56	0.33	5.207	0.083	A
37.66	1.66	5.344	0.427	A
38.36	2.49	5.584	0.657	A
37.97	4.44	5.677	1.204	A
42.18	8.24	7.325	2.594	A
43.61	12.43	8.542	4.414	A
46.10	14.67	10.118	5.837	A
47.13	16.27	11.088	6.939	A
47.77	17.45	11.826	7.832	A + B
46.88	19.14	11.879	8.793	B
45.49	22.09	12.082	10.636	B
43.86	24.91	12.092	12.451	B
43.78	25.38	12.223	12.847	B
44.74	27.09	13.675	15.012	B
40.11	31.43	12.135	17.239	B
Temperature = 40°C				
44.92	1.43	7.209	0.416	A
45.43	3.79	7.703	1.165	A
46.34	3.93	8.023	1.234	A
46.03	6.69	8.383	2.209	A
46.80	8.00	8.915	2.763	A
47.53	11.90	10.087	4.579	A
49.59	12.91	11.386	5.374	A
53.70	14.50	14.540	7.118	A + B
51.08	14.72	12.860	6.719	A
51.18	18.28	14.429	9.344	B
48.39	24.44	15.335	14.042	B
47.92	25.13	15.310	14.556	B
47.97	25.25	15.423	14.719	B
47.25	29.02	17.144	19.090	B
Temperature = 60°C				
51.25	0.66	9.176	0.214	A
50.02	2.74	9.117	0.905	A
52.52	5.18	10.691	1.912	A
54.98	10.77	13.822	4.909	A
53.71	11.29	13.213	5.035	A
56.64	12.92	16.021	6.626	A
55.87	16.19	17.218	9.046	B
55.57	16.82	17.330	9.510	B
57.83	22.12	24.835	17.222	A + B
54.22	23.22	20.694	16.067	B
52.50	26.12	21.143	19.071	B
50.67	28.32	20.766	21.042	B

[a] Molalities calculated by the compiler.

[b] Solid phases: A - $(NH_4)_2SO_3.H_2O$, B - NH_4HSO_3

COMPONENTS:	ORIGINAL MEASUREMENTS:
1. Ammonium sulfite; $(NH_4)_2SO_3$; [10196-04-0] 2. Sulfur dioxide; SO_2; [7446-099-5] 3. Water; H_2O; [7732-18-5]	Ishikawa, F.; Hiroshi, H. *Bull. Inst. Phys. Chem. Research (Tokyo)* 1931, *10*, 166-72 (in Japanese); *Sci. Repts. Tohoku Imp. University* 1933, 22, 235-243 (in English).
VARIABLES:	PREPARED BY:
One temperature: 298 K Concentrations of the components	Mary R. Masson

EXPERIMENTAL VALUES: Composition of equilibrium solutions at 25°C

(a). $(NH_4)_2SO_3$ - $(NH_4)_2S_2O_5$ - H_2O system

$(NH_4)_2SO_3$ mass %	$(NH_4)_2S_2O_5$ mass %	$(NH_4)_2SO_3$[a] mol/kg	$(NH_4)_2S_2O_5$[a] mol/kg
39.29	-	5.572	0.
31.75	13.44	4.988	1.361
29.12	18.38	4.776	1.943
26.66	23.24	4.582	2.574
21.63	34.22	4.218	4.301
16.49	46.79	3.867	7.071
11.70	60.63	3.641	12.160
11.49	60.94	3.588	12.266
9.74	62.36	3.006	12.404
3.54	67.55	1.054	12.966
1.99	69.13	0.593	13.284
0.33	70.87	0.099	13.656
-	71.33	0.	13.807

(b). $(NH_4)_2S_2O_5$ - SO_2 - H_2O system

$(NH_4)_2S_2O_5$ mass %	SO_2 mass %	$(NH_4)_2S_2O_5$[a] mol/kg	SO_2[a] mol/kg
71.33	-	13.807	0.
70.57	1.64	14.092	0.921
69.86	2.82	14.190	1.611
69.26	3.86	14.299	2.242

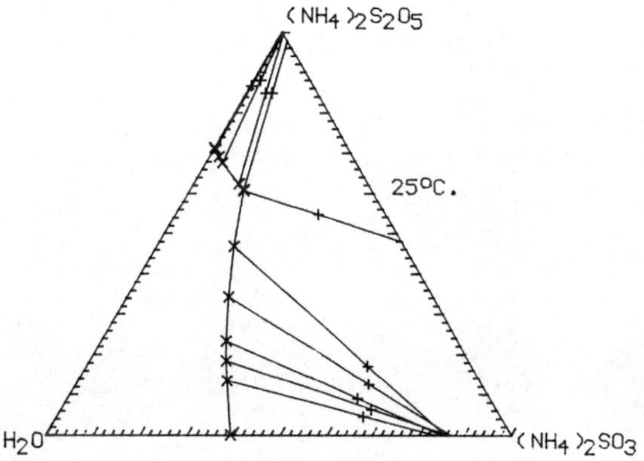

[a] Molalities calculated by the compiler.

AUXILIARY INFORMATION

METHOD APPARATUS/PROCEDURE:	SOURCE AND PURITY OF MATERIALS:
A sample of ammonium sulfite was placed in a closed vessel under an atmosphere of nitrogen, then a known amount of oxygen-free water was added. A known volume of dried sulfur dioxide gas (from a cylinder) was introduced through a tube. Stirring in a thermostat for 4 hr was found to result in equilibrium being reached. A sample of the solution was removed with the aid of the pressure of nitrogen gas, into a pipette fitted with a stopcock at both ends, and weighed and analysed as in the previous work by these authors. The moist solid was also analysed.	As before
ESTIMATED ERROR:	REFERENCES:
Temperature: ±0.01 K Analyses: no estimate given (compiler - probably as before, r.s.d. < 0.2%).	

COMPONENTS:

1. Ammonium sulfite; $(NH_4)_2SO_3$; [10196-04-0]
2. Ammonium sulfate; $(NH_4)_2SO_4$; [7783-20-2]
3. Water; H_2O; [7732-18-5]

ORIGINAL MEASUREMENTS:

Vasilenko, N.A.

Zh. Priklad. Khim. 1950, *23*, 472-81.

VARIABLES:

Six temperatures: 254 - 293 K
Concentrations of the components

PREPARED BY:

Mary R. Masson

EXPERIMENTAL VALUES: Composition of equilibrium solutions

$(NH_4)_2SO_4$ mass %	$(NH_4)_2SO_3$ mass %	$(NH_4)_2SO_4$[a] mol/kg	$(NH_4)_2SO_3$[a] mol/kg	Solid[b] phase
Temperature = -19°C				
39.3	0.0	4.900	0.	A + C
36.6	3.1	4.593	0.443	A
33.6	6.6	4.252	0.950	A
30.8	10.0	3.937	1.454	A
29.8	11.0	3.809	1.600	A + B
35.8	3.2	4.441	0.452	C
32.0	6.8	3.957	0.957	C
28.3	10.4	3.494	1.461	C
24.6	13.8	3.022	1.929	B + C
Temperature = -13°C				
39.8	0.0	5.003	0.	A
37.3	3.1	4.736	0.448	A
34.3	6.6	4.392	0.962	A
31.5	9.9	4.068	1.455	A
29.5	12.5	3.849	1.856	A + B
28.3	13.2	3.661	1.943	B
21.3	16.8	2.604	2.337	B
18.8	17.8	2.244	2.417	B
11.5	21.9	1.307	2.831	B
32.8	0.0	3.694	0.	C
28.0	5.5	3.186	0.712	C
23.3	7.6	2.552	0.947	C
18.8	11.7	2.047	1.450	C
17.2	13.6	1.881	1.692	C
14.2	15.8	1.535	1.943	C
10.1	19.8	1.090	2.432	C
5.5	23.4	0.585	2.834	C
0.0	28.5	0.	3.432	B + C

(continued on next page)

AUXILIARY INFORMATION

METHOD/APPARATUS/PROCEDURE:

A polythermal procedure was used, based on the following systems:

I (4.9% aq. Y) + Z
II (10.0% aq. Y) + Z
III (14.4% aq. Y) + Z
IV (18.4% aq. Y) + Z
V (22.0% aq. Y) + Z
VI (24.7% aq. Y) + Z
VII (32.8% aq. Y) + Z
VIII (22.3% Y + 28.1% Z) + water

where Y = $(NH_3)_2SO_3$ and
Z = $(NH_4)_2SO_4$

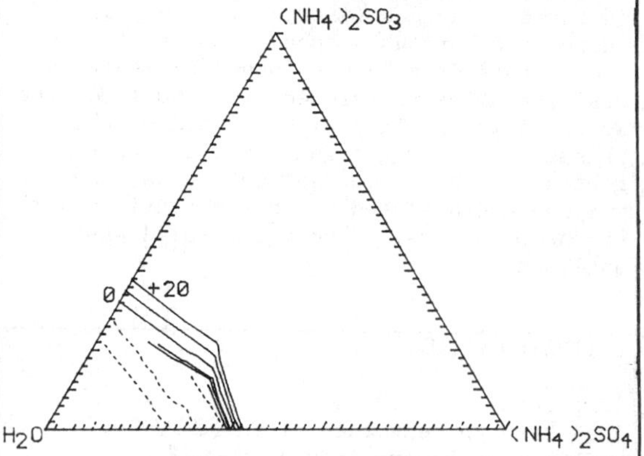

——— Equilibrium with $(NH_4)_2SO_4$

.... Equilibrium with ice

COMPONENTS:	ORIGINAL MEASUREMENTS:
1. Ammonium sulfite; $(NH_4)_2SO_3$; [10196-04-0] 2. Ammonium sulfate; $(NH_4)_2SO_3$; [7783-20-2] 3. Water; H_2O; [7732-18-5]	Vasilenko, N.A. *Zh. Priklad. Khim.* 1950, *23*, 472-81.

EXPERIMENTAL VALUES (continued):

$(NH_4)_2SO_4$ mass %	$(NH_4)_2SO_3$ mass %	$(NH_4)_2SO_4$[a] mol/kg	$(NH_4)_2SO_3$[a] mol/kg	Solid[b] phase
Temperature = -10°C				
40.0	0.0	5.045	0.	A
37.5	3.1	4.778	0.449	A
34.5	6.5	4.425	0.949	A
32.0	9.8	4.161	1.450	A
29.8	12.9	3.936	1.938	A
29.2	13.0	3.823	1.937	A + B
21.8	17.2	2.705	2.428	B
21.0	17.4	2.580	2.432	B
14.0	21.3	1.638	2.835	B
28.0	0.0	2.943	0.	C
22.5	3.8	2.310	0.444	C
18.0	8.3	1.848	0.970	C
14.3	11.3	1.455	1.308	C
13.0	12.6	1.322	1.458	C
7.8	17.0	0.785	1.946	C
2.5	21.4	0.249	2.421	C
0.0	24.0	0.	2.719	C
Temperature = 0°C				
41.0	0.0	5.259	0.	A
38.5	3.1	4.989	0.457	A
35.8	6.4	4.687	0.953	A
33.0	9.7	4.358	1.458	A
30.8	12.8	4.133	1.954	A
28.5	15.5	3.851	2.383	A + B
27.8	15.9	3.737	2.432	B
23.5	18.6	3.072	2.766	B
21.3	19.5	2.723	2.836	B
0.0	32.3	0.	4.108	B
Temperature = +10°C				
42.0	0.0	5.480	0.	A
39.3	3.0	5.154	0.448	A
36.8	6.3	4.894	0.953	A
34.3	9.5	4.619	1.455	A
32.0	12.5	4.363	1.939	A
29.6	15.5	4.080	2.431	A
28.0	17.8	3.910	2.828	A
27.5	18.5	3.854	2.950	A + B
25.3	20.0	3.500	3.148	B
7.3	30.5	0.888	4.222	B
0.0	35.0	0.	4.636	B
Temperature = +20°C				
42.8	0.0	5.663	0.	A
40.3	2.9	5.369	0.440	A
37.8	6.2	5.108	0.953	A
35.3	9.3	4.822	1.445	A
33.0	12.3	4.566	1.936	A
30.8	15.2	4.316	2.424	A
29.0	17.6	4.110	2.838	A
26.8	21.2	3.900	3.510	A
26.3	21.8	3.835	3.617	A + B
16.6	27.4	2.243	4.213	B
0.0	37.7	0.	5.210	B

[a] Molalities calculated by the compiler.

[b] Solid phases: A - $(NH_4)_2SO_4$, B - $(NH_4)_2SO_3.H_2O$, C - ice

COMPONENTS:

1. Ammonium sulfite; $(NH_4)_2SO_3$; [10196-04-0]

2. Ammonium sulfate; $(NH_4)_2SO_4$; [7783-20-2]

3. Water; H_2O; [7732-18-5]

ORIGINAL MEASUREMENTS:

Terres, E.; Heinsen, A.

Das Gas- und Wasserfach 1927, *70*, 1157-61.

VARIABLES:

Concentrations of the components
Four temperatures: 273 - 333 K

PREPARED BY:

Mary R. Masson

EXPERIMENTAL VALUES:

$(NH_4)_2SO_4$ mass %	$(NH_4)_2SO_3$ mass %	$(NH_4)_2SO_4$[a] mol/kg	$(NH_4)_2SO_3$[a] mol/kg	Solid[b] phase
Temperature = 0°C				
42.8	0.0	5.663	0.	A
37.7	6.92	5.152	1.076	A
32.4	13.54	4.536	2.157	A
30.83	15.12	4.317	2.409	A
30.50	15.97	4.312	2.569	A
30.17	16.13	4.252	2.586	A + B
22.25	20.23	2.927	3.028	B
15.50	24.14	1.943	3.444	B
11.08	26.57	1.345	3.669	B
0.0	32.4	0.	4.127	B
Temperature = 20°C				
44.5	0.0	6.068	0.	A
41.45	4.32	5.784	0.686	A
40.84	5.29	5.737	0.846	A
36.73	9.91	5.209	1.599	A
36.47	10.21	5.176	1.649	A
34.22	14.4	5.040	2.413	A
32.08	15.92	4.669	2.636	A
28.54	21.57	4.329	3.723	A
28.27	21.58	4.266	3.705	A
27.67	22.08	4.167	3.783	A
27.75	22.46	4.218	3.884	A + B

(continued on next page)

AUXILIARY INFORMATION

METHOD APPARATUS/PROCEDURE:

To the saturated solution of one salt was added various amounts of the other. After equilibrium was reached, ammonia was determined by the Kjeldahl method, sulfite by titrimetry, and sulfate gravimetrically.

SOURCE AND PURITY OF MATERIALS:

ESTIMATED ERROR:

Temperature: $\pm$0.1 K
Analyses: no estimates possible.

REFERENCES:

COMPONENTS:	ORIGINAL MEASUREMENTS:
1. Ammonium sulfite; $(NH_4)_2SO_3$; [10196-04-0] 2. Ammonium sulfate; $(NH_4)_2SO_4$; [7783-20-2] 3. Water; H_2O; [7732-18-5]	Terres, E.; Heinsen, A. *Das Gas- und Wasserfach* 1927, *70*, 1157-61.

EXPERIMENTAL VALUES (continued):

$(NH_4)_2SO_4$ mass %	$(NH_4)_2SO_3$ mass %	$(NH_4)_2SO_4$[a] mol/kg	$(NH_4)_2SO_3$[a] mol/kg	Solid[b] phase
Temperature = 20°C (continued)				
20.58	26.32	2.933	4.268	B
17.74	27.8	2.465	4.395	B
12.00	30.77	1.587	4.629	B
0.0	36.93	0.	5.042	B
Temperature = 40°C				
47.17	0.0	6.757	0.	A
40.45	8.83	6.035	1.499	A
39.75	9.5	5.927	1.612	A
34.86	16.16	5.386	2.841	A
28.46	24.36	4.565	4.446	A
26.01	28.24	4.302	5.315	A
25.60	28.37	4.209	5.307	A
25.07	28.77	4.110	5.367	A
25.09	29.04	4.139	5.451	A + B
23.19	30.29	3.772	5.606	B
16.69	34.08	2.566	5.961	B
15.53	34.8	2.366	6.033	B
12.71	36.93	1.910	6.314	B
9.91	38.05	1.441	6.296	B
8.39	39.01	1.207	6.386	B
0.0	43.56	0.	6.645	B
Temperature = 60°C				
48.40	0.0	7.098	0.	A
44.44	6.14	6.805	1.070	A
41.63	8.86	6.363	1.541	A
39.36	11.98	6.121	2.120	A
38.19	14.12	6.060	2.549	A
31.64	23.22	5.304	4.429	A
30.22	25.19	5.129	4.864	A
24.03	35.37	4.479	7.501	A
23.48	36.02	4.387	7.658	A
22.51	36.85	4.192	7.807	A + B
16.01	40.54	2.788	8.034	B
14.51	41.57	2.500	8.150	B
0.0	48.67	0.	8.164	B

[a] Molalities calculated by the compiler.
[b] Solid phases: A - ammonium sulfate, B - ammonium sulfite

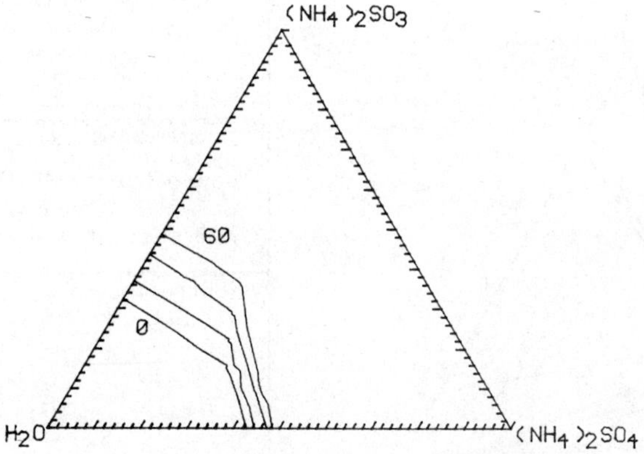

COMPONENTS:

1. Ammonium sulfite; $(NH_4)_2SO_3$; [10196-04-0]
2. Ammonium sulfate; $(NH_4)_2SO_4$; [7783-20-2]
3. Water; H_2O; [7732-18-5]

ORIGINAL MEASUREMENTS:

Ishikawa, F.; Murooka, T.

Bull. Inst. Phys. Chem. Research (Tokyo) 1929, *8*, 75-88 (in Japanese); *Sci. Repts. Tohoku Imp. University* 1933, 22, 220-234 (in English).

VARIABLES:

Temperature: 251 - 373 K
Concentrations of the components

PREPARED BY:

Mary R. Masson

EXPERIMENTAL VALUES:

Compositions of equilibrium solutions at 15°C

$(NH_4)_2SO_3$ mass %	$(NH_4)_2SO_4$ mass %	$(NH_4)_2SO_3$[a] mol/kg	$(NH_4)_2SO_4$[a] mol/kg	Solid[b] phase
-	42.52	0.	5.598	A
5.45	38.08	0.831	5.103	A
9.60	34.82	1.487	4.741	A
13.63	31.68	2.146	4.384	A
17.15	28.80	2.732	4.032	A
19.70	27.22	3.196	3.881	A
19.70	27.23	3.196	3.883	A + B
19.74	27.23	3.205	3.886	A + B
19.75	27.13	3.201	3.865	A + B
21.92	23.38	3.450	3.235	B
22.81	21.76	3.543	2.971	B
25.65	16.94	3.847	2.233	B
27.11	14.19	3.977	1.829	B
29.73	10.31	4.269	1.301	B
31.22	7.97	4.421	0.992	B
34.10	3.64	4.716	0.442	B
36.40	-	4.928	0.	B

(continued on next page)

AUXILIARY INFORMATION

METHOD APPARATUS/PROCEDURE:

A simple saturation technique was used - details similar to other work by these authors.

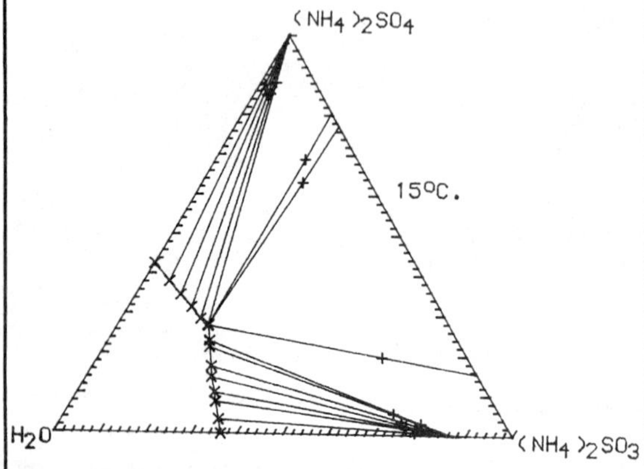

SOURCE AND PURITY OF MATERIALS:

Ammonium sulfite - prepared as in previous work.
Ammonium sulfate (Kahlbaum, zur Analyse) was recrystallized from an ammoniacal solution. The crystals were dried over phosphorus pentoxide then by heating for several hr at 120°C.

ESTIMATED ERROR:

Temperature: ±0.02°C (up to 80°C)
±0.05°C (above 80°C)
Analyses: r.s.d. approx. 0.2% max.

REFERENCES:

COMPONENTS:	ORIGINAL MEASUREMENTS:
1. Ammonium sulfite; $(NH_4)_2SO_3$; [10196-04-0]	Ishikawa, F.; Murooka, T.
2. Ammonium sulfate; $(NH_4)_2SO_4$; [7783-20-2]	*Bull. Inst. Phys. Chem. Research (Tokyo)* <u>1929</u>, *8*, 75-88 (in Japanese); *Sci. Repts. Tohoku Imp. University* <u>1933</u>, *22*, 220-234 (in English).
3. Water; H_2O; [7732-18-5]	

EXPERIMENTAL VALUES (continued):

<u>Compositions of equilibrium solutions at 15°C</u>

$(NH_4)_2SO_3$ mass %	$(NH_4)_2SO_4$ mass %	$(NH_4)_2SO_3$[a] mol/kg	$(NH_4)_2SO_4$[a] mol/kg	Solid[b] phase
-	43.87	0.	5.915	A
1.23	42.91	0.190	5.813	A
2.23	42.05	0.345	5.711	A
2.74	41.68	0.424	5.675	A
4.00	40.65	0.622	5.558	A
6.70	38.55	1.054	5.329	A
8.41	37.13	1.330	5.160	A
12.66	33.88	2.039	4.796	A
15.83	31.46	2.586	4.517	A
18.41	29.60	3.049	4.309	A
21.93	27.12	3.706	4.028	A
22.58	26.61	3.826	3.963	A
24.62	25.23	4.227	3.807	A + B
24.66	25.17	4.232	3.797	A + B
24.64	25.20	4.230	3.802	A + B
24.63	25.23	4.230	3.808	A + B
25.87	23.31	4.383	3.471	B
28.04	19.61	4.612	2.835	B
30.03	16.41	4.828	2.319	B
31.33	14.30	4.962	1.990	B
33.33	11.19	5.173	1.526	B
36.19	6.99	5.484	0.931	B
38.86	2.83	5.738	0.367	B
40.43	0.52	5.895	0.067	B
40.77	-	5.927	0.	B

a Molalities calculated by the compiler.
b Solid phases: A - $(NH_4)_2SO_4$, B - $(NH_4)_2SO_3 \cdot H_2O$

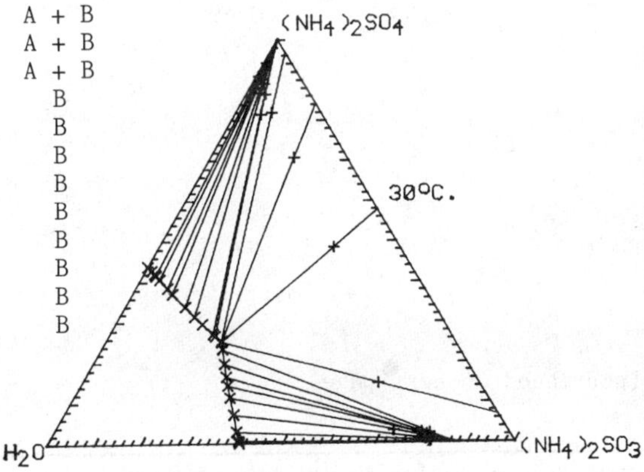

Compositions of solutions saturated with ammonium sulfite and ammonium sulfate at various temperatures.

t/°C	$(NH_4)_2SO_3$ mass %	$(NH_4)_2SO_4$ mass %	$(NH_4)_2SO_3$[a] mol/kg	$(NH_4)_2SO_4$[a] mol/kg
-21.5	10.89	30.57	1.602	3.952
-21.2	10.95	30.55	1.612	3.952
-11.52	12.88	29.85	1.936	3.944
0	15.41	29.05	2.389	3.958
15	19.72	27.20	3.199	3.878
30	24.64	25.21	4.230	3.804
50	32.75	21.89	6.217	3.652
70	42.28	17.71	9.099	3.350
80	48.26	15.13	11.350	3.128
90	49.14	15.01	11.802	3.169
100	50.53	14.80	12.549	3.231

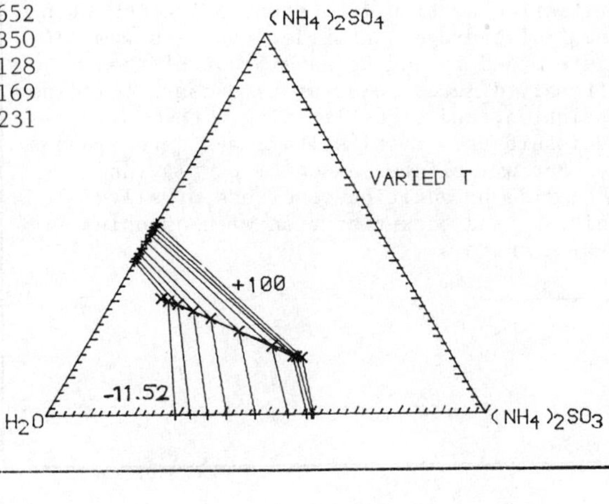

COMPONENTS:

1. Ammonium sulfite; $(NH_4)_2SO_3$; [10196-04-0]
2. Ammonium chloride; NH_4Cl; [12125-02-9]
3. Water; H_2O; [7732-18-5]

ORIGINAL MEASUREMENTS:

Labash, J.A.; Lusby, G.R.

Can.J. Chem. 1955, *33*, 774-86.

VARIABLES:

Concentrations of the components
Two temperatures: 293 - 333 K

PREPARED BY:

Mary R. Masson

EXPERIMENTAL VALUES: Compositions of equilibrium solutions at 20°C

NH_4Cl mass %	$(NH_4)_2SO_3$ mass %	$(NH_4)_2SO_4$ mass %	NH_4Cl[a] mol/kg	$(NH_4)_2SO_3$[a] mol/kg	$(NH_4)_2SO_4$[a] mol/kg	Solid[b] phase
27.26	0.	0.	7.006	0.	0.	A
24.70	5.25	0.45	6.635	0.649	0.049	A
22.45	10.00	1.06	6.312	1.295	0.121	A
20.08	15.52	0.	5.829	2.075	0.	A
18.39	19.13	0.	5.503	2.636	0.	A
16.61	23.57	1.20	5.297	3.462	0.155	A + B
16.57	23.64	1.15	5.283	3.471	0.148	A + B
15.36	24.55	0.90	4.851	3.571	0.115	B
9.66	28.85	0.	2.937	4.040	0.	B
5.13	32.63	0.	1.541	4.514	0.	B
0.	37.34	0.49	0.	5.171	0.060	B

(continued on next page)

AUXILIARY INFORMATION

METHOD APPARATUS/PROCEDURE:

Solutions were stirred under nitrogen in a three-necked flask. Solutions were analysed often for bisulfite, formed by loss of ammonia, and ammonia gas was then added to replace that lost. Samples were withdrawn, after settling, through a pipette plugged with cotton wool. Weighed samples were diluted to volume in a standard flask.

Bisulfite was determined by acid-base titration as bisulfate after oxidation with neutral hydrogen peroxide. Sulfite was determined by adding an aliquot of the freshly diluted solution to excess of iodine solution, and back-titrating with thiosulfate. Total sulfate was determined as barium sulfate, ammonium by (1), and chloride by addition of excess of silver nitrate and back-titration with ammonium thiocyanate.

SOURCE AND PURITY OF MATERIALS:

$(NH_4)_2SO_3$ was freshly prepared from ammonia and sulfur dioxide gases. Ammonium chloride was of analytical reagent grade.

ESTIMATED ERROR:

Temperature: ± 0.1 K
Analyses: 0.2% for sulfite, ammonium and chloride, 0.4% for total sulfate.

REFERENCES:

1. Sutton, F. *Volumetric Analysis*, 12th Ed., Blakiston, Philadelphia, 1935, 75.

COMPONENTS:
1. Ammonium sulfite; $(NH_4)_2SO_3$; [10196-04-0]
2. Ammonium chloride; NH_4Cl; [12125-02-9]
3. Water; H_2O; [7732-18-5]

ORIGINAL MEASUREMENTS:

Labash, J.A.; Lusby, G.R.

Can. J. Chem. 1955, *33*, 774-86.

EXPERIMENTAL VALUES (continued):

Compositions of equilibrium solutions at 60°C, expressed as mass %.

NH_4Cl	$(NH_4)_2SO_3$	$(NH_4)_2SO_4$	$(NH_4)HSO_3$	NH_3	Solid[b] phase
35.37	-	-	-	-	A
32.04	6.95	-	-	-	A
28.56	13.97	-	0.23	-	A
25.35	20.75	-	0.09	-	A
22.98	25.75	-	0.22	-	A
19.82	32.06	-	0.43	-	A
17.84	37.72	0.91	-	-	A + B
17.82	37.86	0.99	-	-	A + B
17.71	37.7	-	0.17	-	B
16.90	37.6	-	-	0.22	B
13.82	39.6	-	-	0.02	B
15.64	38.5	-	-	-	B
12.29	40.7	0.89	-	-	B
9.05	43.1	-	-	0.02	B
5.35	46.2	-	-	0.05	B
3.14	48.6	-	0.51	-	B
0.0	50.48	-	0.13	-	B

Compositions of equilibrium solutions expressed as molalities, mol/kg[a]

10.23	-	-	-	-
9.82	0.98	-	-	-
9.33	2.10	-	0.041	-
8.81	3.32	-	0.017	-
8.42	4.34	-	0.043	-
7.77	5.79	-	0.091	-
7.66	7.46	0.158	-	-
7.69	7.52	0.173	-	-
7.45	7.31	-	0.039	-
6.98	7.15	-	-	0.29
5.55	7.32	-	-	0.025
6.38	7.23	-	-	-
4.98	7.60	0.146	-	-
3.54	7.76	-	-	0.025
2.07	8.22	-	-	0.06
1.23	8.76	-	0.108	-
0.0	8.80	-	0.027	-

a Molalities calculated by the compiler.

b Solid phases: A - NH_4Cl,
B - $(NH_4)_2SO_3.H_2O$

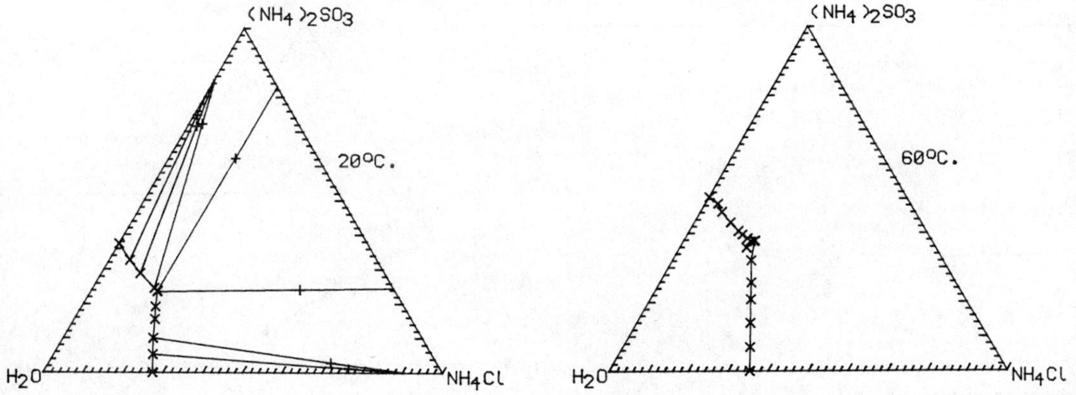

COMPONENTS:

1. Ammonium sulfite; $(NH_4)_2SO_3$; [10196-04-0]
2. Ammonium thiosulfate; $(NH_4)_2S_2O_3$; [7783-18-8]
3. Water; H_2O; [7732-18-5]

ORIGINAL MEASUREMENTS:

Terres, E.; Overdick, F.;

Das Gas- und Wasserfach 1928, *71*, 106-110.

VARIABLES:

Concentrations of the components
One temperature: 298 K

PREPARED BY:

Mary R. Masson

EXPERIMENTAL VALUES:

$(NH_4)_2S_2O_3$ mass %	$(NH_4)_2SO_3$ mass %	$(NH_4)_2S_2O_3$ [a] mol/kg	$(NH_4)_2SO_3$ [a] mol/kg	Solid [b] phase
64.10	0.0	12.048	0.	A
58.25	7.60	11.510	1.916	A
55.80	9.15	10.742	2.248	A
55.63	10.78	11.175	2.763	A + B
48.75	13.80	8.784	3.173	B
44.55	15.65	7.553	3.386	B
44.20	15.80	7.456	3.401	B
40.40	18.05	6.561	3.740	B
35.70	20.15	5.456	3.930	B
35.30	20.70	5.413	4.051	B
32.00	22.20	4.715	4.174	B
30.30	23.50	4.425	4.380	B
26.45	24.75	3.657	4.367	B
22.25	26.90	2.953	4.555	B
17.05	28.40	2.109	4.483	B
16.20	30.40	2.047	4.902	B
12.10	32.15	1.465	4.965	B
4.90	34.50	0.546	4.902	B
0.00	37.10	0.	5.079	B

[a] Molalities calculated by the compiler.
[b] Solid phases: A - ammonium thiosulfate, B - ammonium sulfite

Compiler's note: some other tables of data are given, but no temperature is stated, and it is not clear whether they refer to solubility studies or not.

AUXILIARY INFORMATION

METHOD APPARATUS/PROCEDURE:

To the saturated solution of one salt was added various amounts of the other. After equilibrium was reached, ammonia was determined by the Kjeldahl method, sulfite by titrimetry, and sulfate gravimetrically.

SOURCE AND PURITY OF MATERIALS:

ESTIMATED ERROR:

Temperature: ±0.1 K

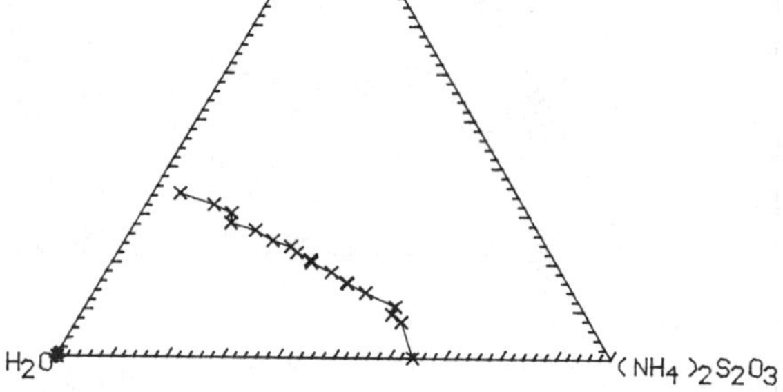

COMPONENTS:

1. Ammonium sulfite; $(NH_4)_2SO_3$; [10196-04-0]
2. Ammonium hydrogen sulfite; NH_4HSO_3; [10192-30-0]
3. Ammonium sulfate; $(NH_4)_2SO_4$; [7783-20-2]
4. Water; H_2O; [7732-18-5]

ORIGINAL MEASUREMENTS:

Vasilenko, N.A.

Zh. Priklad. Khim. 1953, *26*, 650-2.

VARIABLES:

One temperature: 303 K
Concentrations of the components

PREPARED BY:

Mary R. Masson

EXPERIMENTAL VALUES:

Composition of equilibrium solutions expressed as mass %, at 30°C

NH_4HSO_3	$(NH_4)_2SO_3$	$(NH_4)_2SO_4$	H_2SO_3	Solid[a] phase
0.0	24.51	25.92	-	A + B
0.0	24.14	25.56	-	A + B
5.01	23.15	22.27	-	A + B
11.05	22.68	20.36	-	A + B
20.52	20.69	16.37	-	A + B
27.80	19.13	14.00	-	A + B
35.40	17.86	11.44	-	A + B
43.50	15.64	9.78	-	A + B
48.12	15.07	7.87	-	A + B
48.57	15.99	6.37	-	A + B
64.03	11.19	5.51	-	A,B,C
63.45	11.65	5.39	-	A,B,C
64.03	11.86	5.10	-	A,B,C
65.88	11.68	3.46	-	B + C
66.41	11.86	2.60	-	B + C
66.56	12.24	1.68	-	B + C
67.08	12.84	0.36	-	B + C
66.63	12.65	u.d.	-	B + C
66.47	12.79	u.d.	-	B + C
64.51	10.84	5.18	-	A + C
65.80	9.10	5.50	-	A + C
68.27	6.29	5.67	-	A + C
70.11	3.30	5.93	-	A + C
72.14	1.87	6.36	-	A + C
73.29	0.73	6.17	-	A + C

(continued on next page)

AUXILIARY INFORMATION

METHOD/APPARATUS/PROCEDURE:

An isothermal procedure.

SOURCE AND PURITY OF MATERIALS:

ESTIMATED ERROR:

No estimates possible.

REFERENCES:

COMPONENTS:
1. Ammonium sulfite; $(NH_4)_2SO_3$; [10196-04-0]
2. Ammonium hydrogen sulfite; NH_4HSO_3; [10192-30-0]
3. Ammonium sulfate; $(NH_4)_2SO_4$; [7783-20-2]
4. Water; H_2O; [7732-18-5]

ORIGINAL MEASUREMENTS:

Vasilenko, N.A.

Zh. Priklad. Khim. 1953, *26*, 650-2.

EXPERIMENTAL VALUES (continued):

NH_4HSO_3	$(NH_4)_2SO_3$	$(NH_4)_2SO_4$	H_2SO_3	Solid[a] phase
74.57	0.0	5.97	0.54	A + C
73.82	0.0	6.98	0.47	A + C
73.34	0.0	6.19	0.74	A + C
73.58	0.0	6.14	0.64	A + C

Compositions of equilibrium solutions expressed as molalities[b], mol/kg

NH_4HSO_3	$(NH_4)_2SO_3$	$(NH_4)_2SO_4$	H_2SO_3
0.	4.257	3.957	0.
0.	4.132	3.846	0.
1.020	4.021	3.400	0.
2.428	4.254	3.356	0.
4.881	4.200	2.920	0.
7.179	4.216	2.712	0.
10.118	4.356	2.453	0.
14.122	4.333	2.381	0.
16.777	4.484	2.058	0.
16.858	4.736	1.658	0.
33.526	5.000	2.164	0.
32.814	5.141	2.091	0.
33.985	5.372	2.030	0.
35.022	5.299	1.380	0.
35.027	5.338	1.029	0.
34.405	5.399	0.651	0.
34.322	5.606	0.138	0.
32.446	5.257	0.	0.
32.337	5.310	0.	0.
33.431	4.794	2.013	0.
33.873	3.998	2.124	0.
34.842	2.739	2.170	0.
34.240	1.375	2.172	0.
37.080	0.820	2.452	0.
37.329	0.317	2.357	0.
39.767	0.	2.388	0.348
39.767	0.	2.820	0.306
37.506	0.	2.374	0.457
37.801	0.	2.366	0.397

[a] Solid phases: A - $(NH_4)_2SO_4$, B - $(NH_4)_2SO_3.H_2O$
C - $(NH_4)_2S_2O_5$.

[b] Molalities calculated by the compiler.

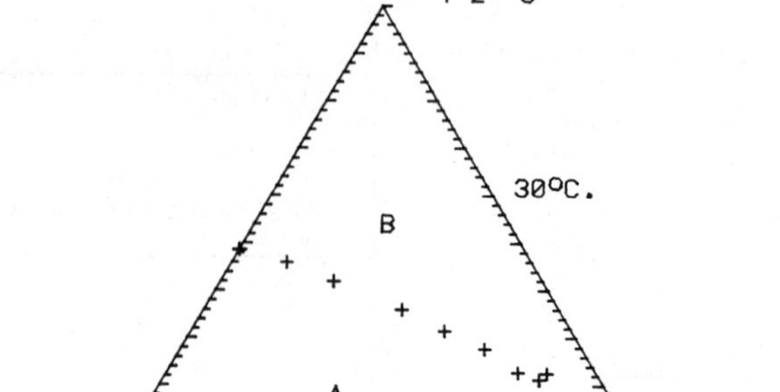

COMPONENTS:	ORIGINAL MEASUREMENTS:
1. Ammonium sulfite; $(NH_4)_2SO_3$; [10196-04-0] 2. Ammonium hydrogen sulfite; NH_4HSO_3; [10192-30-0] 3. Ammonium sulfate; $(NH_4)_2SO_4$; [7783-20-2] 4. Water; H_2O; [7732-18-5]	Vasilenko, N.A. *Nauch.-Tekh. Inform. Byull. Nauch. Inst. po Udobren i Insektofungisidam* 1957, (5-6), 105-10.
VARIABLES:	PREPARED BY:
One temperature: 283 K Concentrations of the components	Mary R. Masson

EXPERIMENTAL VALUES:

Composition of equilibrium solutions at 10°C

NH_4HSO_3 mass %	$(NH_4)_2SO_3$ mass %	$(NH_4)_2SO_4$ mass %	NH_4HSO_3[a] mol/kg	$(NH_4)_2SO_3$[a] mol/kg	$(NH_4)_2SO_4$[a] mol/kg	Solid[b] phase
0.0	19.0	27.0	0.	3.030	3.784	A + B
1.20	18.41	27.89	0.231	3.019	4.020	A + B
6.65	17.18	25.68	1.329	2.930	3.849	A + B
13.99	16.40	21.58	2.939	2.940	3.400	A + B
22.49	15.26	18.03	5.132	2.971	3.086	A + B
29.97	14.92	14.76	7.494	3.184	2.768	A + B
38.86	13.00	11.91	10.822	3.090	2.488	A + B
49.54	11.28	9.09	16.612	3.228	2.286	A + B
58.20	8.93	7.41	23.065	3.020	2.203	A + B
60.12	9.37	7.47	26.328	3.502	2.454	A + B
61.20	7.91	7.11	25.967	2.864	2.263	A,B,C
61.81	6.97	6.97	25.718	2.475	2.175	A + C
64.21	4.50	7.25	26.950	1.612	2.282	A + C
68.08	0.0	7.39	28.003	0.	2.280	A + C
69.00	0.0	6.50	28.416	0.	2.008	A + C
62.66	10.73	2.86	26.620	3.890	0.911	B + C
63.00	10.00	0.0	23.543	3.189	0.	B + C

[a] Molalities calculated by the compiler.

[b] Solid phases: A - $(NH_4)_2SO_4$, B - $(NH_4)_2SO_3.H_2O$, C - $(NH_4)_2S_2O_5$.

AUXILIARY INFORMATION

METHOD APPARATUS/PROCEDURE:	SOURCE AND PURITY OF MATERIALS:
An isothermal procedure.	
	ESTIMATED ERROR: No estimates possible.
	REFERENCES:

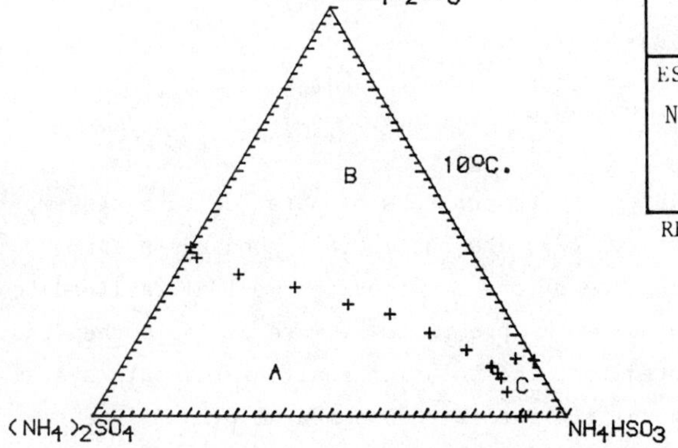

COMPONENTS:	EVALUATOR:
1. Ammonium pyrosulfite; $(NH_4)_2S_2O_5$; [32736-64-4] 2. Water; H_2O; [7732-18-5]	Mary R. Masson, Dept. of Chemistry, University of Aberdeen, Meston Walk, Old Aberdeen, AB9 2UE, Scotland, UK. March, 1984.

CRITICAL EVALUATION:

The binary system ammonium pyrosulfite - water was studied by Vasilenko (1), and data are also available from studies of ternary systems (1 - 3). The solubilities are reported in terms of mass % of ammonium hydrogen sulfite, rather than of ammonium pyrosulfite.

Apart from one or two points, which were rejected, the data are in reasonable agreement. There are two regression equations. The first applies to the equilibrium with ice as solid phase (at 243 - 273 K).

$$(T - 273.15) = 0.101 - 0.412y + 0.00344y^2 - 0.0000865y^3 \qquad s = 0.193 \text{ (15 pts)}$$

or

$$y = -0.240 - 3.01(T - 273.2) - 0.0350(T - 273.2)^2 \qquad s = 0.360 \text{ (15 pts)}$$

and the second to the equilibrium with solid $(NH_4)_2S_2O_5$ (at 243 - 333 K)

$$y = 72.07 + 0.2825(T - 273.2) - 0.002423(T - 273.2)^2 + 0.00001996(T - 273.2)^3 \qquad s = 0.376 \text{ (33 pts)}$$

where $y = 100w$ is the solubility in mass % of NH_4HSO_3, T is the temperature in K, and s is the estimated standard deviation of the dependent variable about the regression line.

TENTATIVE SOLUBILITIES

The following tentative solubility values for $(NH_4)_2S_2O_5$ in water were calculated from the second regression equation.

T/K	Solubility		
	mass % of NH_4HSO_3	mass % of $(NH_4)_2S_2O_5$	molality of $(NH_4)_2S_2O_5$ mol/kg
253.2	65.5	59.5	8.15
263.2	69.0	62.7	9.33
273.2	72.1	65.5	10.5
283.2	74.7	67.9	11.7
293.2	76.8	69.8	12.8
303.2	78.4	71.3	13.8
313.2	79.5	72.3	14.5
323.2	80.2	72.9	14.9
333.2	80.4	73.1	15.1

Ammonium pyrosulfite - ammonium sulfate - water. The two sets of data for this system, measured by Vasilenko (1) and Terres and Heinsen (2), are not in very good agreement, although the trends are similar. In the absence of any other evidence, I am inclined to favour the data of Vasilenko, because his other work appears to be more reliable than that of Terres and co-workers. It should be noted that the solution analyses for this system are expressed in terms of ammonium hydrogen sulfite rather than ammonium pyrosulfite.

COMPONENTS:	EVALUATOR:
1. Ammonium pyrosulfite; $(NH_4)_2S_2O_5$; [32736-64-4] 2. Water; H_2O; [7732-18-5]	Mary R. Masson, Dept. of Chemistry, University of Aberdeen, Meston Walk, Old Aberdeen, AB9 2UE, Scotland, UK. March, 1984.

CRITICAL EVALUATION: (continued)

REFERENCES

1. Vasilenko, N.A. *Zh. Priklad. Khim.* 1948, *21*, 917.

2. Vasilenko, N.A. *Zh. Priklad. Khim.* 1949, *22*, 338.

3. Terres, E.; Heinsen, A. *Das Gas- und Wasserfach* 1927, *70*, 1157.

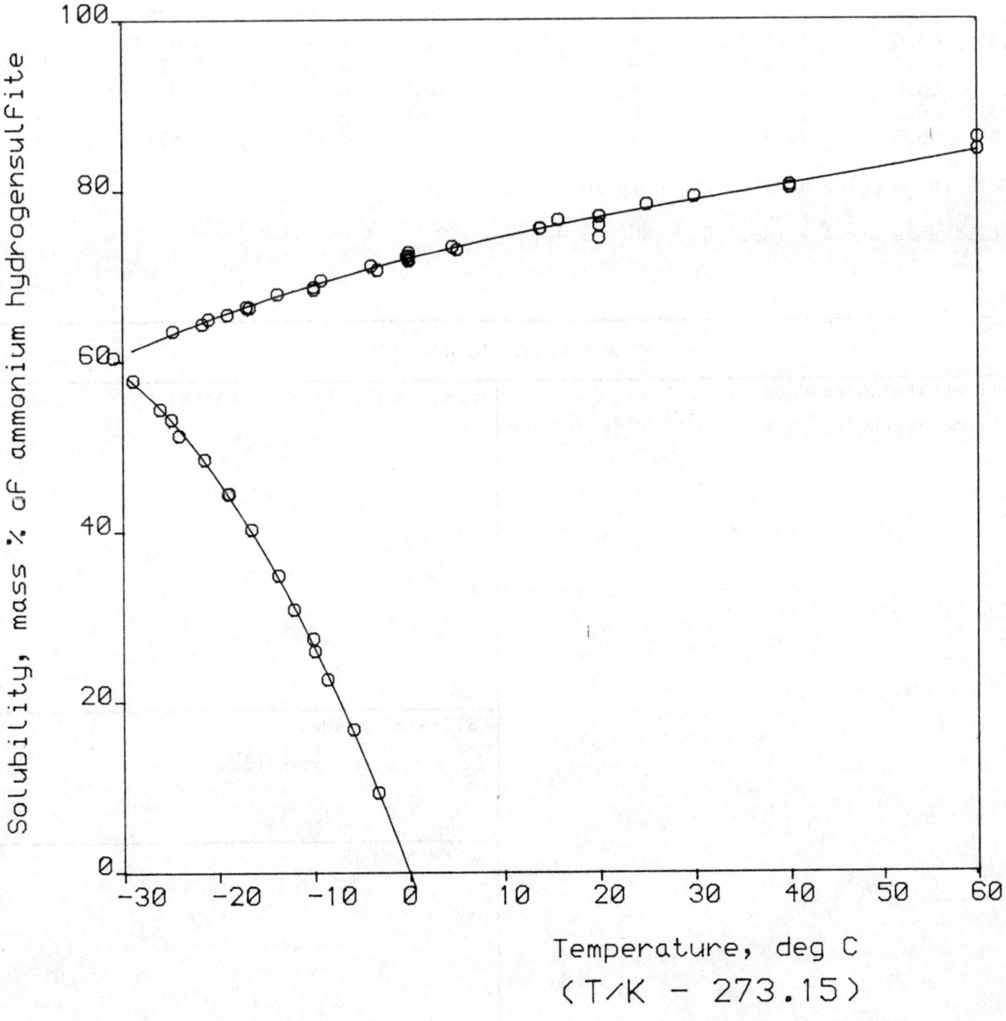

COMPONENTS:

1. Ammonium pyrosulfite; $(NH_4)_2S_2O_5$; [32736-64-4]
2. Water; H_2O; [7732-18-5]

ORIGINAL MEASUREMENTS:

Vasilenko, N.A.

Zh. Priklad. Khim. 1948, *21*, 917-26.

VARIABLES:

Temperature: 256 - 333 K

PREPARED BY:

Mary R. Masson

EXPERIMENTAL VALUES:

t/°C	NH_4HSO_3 mass %	NH_4HSO_3[a] mol/kg	Solid[b] phase	t/°C	NH_4HSO_3 mass %	NH_4HSO_3[a] mol/kg	Solid[b] phase
-17.0	66.4	.1h8					
- 3.4	9.4	1.047	A	-13.8	67.9	21.343	B
- 6.0	16.8	2.037	A	-10.0	68.4	21.840	B
- 8.7	22.7	2.963	A	- 9.2	69.5	22.992	B
-10.2	27.5	3.827	A	- 3.3	70.7	24.346	B
-12.2	30.9	4.512	A	- 4.0	71.2	24.944	B
-13.8	34.9	5.409	A	- 0.2	72.3	26.335	B
-16.6	40.3	6.811	A	0.0	72.8	27.005	B
-18.9	44.5	8.090	A	0.0	71.8	25.690	B
-21.5	48.5	9.502	A	+ 4.6	73.5	27.985	B
-24.2	51.3	10.628	A	5.1	73.1	27.419	B
-25.0	53.2	11.470	A	13.8	75.6	31.262	B
-26.2	54.4	12.037	A	15.7	76.6	33.029	B
-29.0	57.8	13.820	A	20.0	74.5	29.478	B
-24.8	63.6	17.629	B	25.0	78.41	36.644	B
-21.7	64.4	18.252	B	40.0	80.3	41.127	B
-21.0	65.0	18.738	B	40.0	80.64	42.027	B
-16.7	66.3	19.850	B	60.0	86.1	62.499	B
-17.0	66.4	19.939	B	60.0	84.7	55.857	B

[a] Molalities calculated by the compiler.

[b] Solid phases: A - ice, B - $(NH_4)_2S_2O_5$

AUXILIARY INFORMATION

METHOD APPARATUS/PROCEDURE:

Isothermal method.

SOURCE AND PURITY OF MATERIALS:

ESTIMATED ERROR:

No estimates possible.

REFERENCES:

COMPONENTS:	ORIGINAL MEASUREMENTS:
1. Ammonium pyrosulfite; $(NH_4)_2S_2O_5$; [32736-64-4] 2. Ammonium sulfate; $(NH_4)_2SO_4$; [7783-20-2] 3. Water; H_2O; [7732-18-5]	Vasilenko, N.A. *Zh. Priklad. Khim.* 1948, *21*, 917-26.
VARIABLES:	PREPARED BY:
Five temperatures: 242 - 293 K Concentrations of the components	Mary R. Masson

EXPERIMENTAL VALUES:

Composition of equilibrium solutions

$(NH_4)_2SO_4$ mass %	NH_4HSO_3 mass %	$(NH_4)_2SO_4$[a] mol/kg	NH_4HSO_3[a] mol/kg	Solid[b] phase
Temperature = -31°C				
0.0	60.5	0.	15.454	A + B
2.6	59.0	0.583	15.503	B
5.7	56.2	1.288	14.883	B
6.7	54.8	1.498	14.362	B
9.4	52.9	2.147	14.158	B
9.8	52.0	2.209	13.735	B + C
12.5	45.5	2.563	10.931	A + C
8.7	49.0	1.771	11.688	A
6.3	51.3	1.279	12.208	A
5.3	53.0	1.094	12.824	A
2.5	56.6	0.526	13.963	A

(continued on next page)

AUXILIARY INFORMATION

METHOD APPARATUS/PROCEDURE:

A polythermal procedure was used, based on the following systems:

I (15% aq. Y) + Z
II (30% aq. Y) + Z
III (43.6% aq. Y) + Z
IV (52.6% Y + 13.4% Z) + water
V (69.6% Y + 3.0% Z) + water
VI (75.6% aq. Y) + Z
VII (70.1% Y + 7.0% Z) + water
VIII (72% aq. Y) + Z
IX (66.5% Y + 8.1% Z) + water
X (61.6% Y + 9.5% Z) + water
XI (64.5% aq. Y) + Z
XII (60.2% Y + 10.7% Z) + water

where Y = NH_4HSO_3 and Z = $(NH_4)_2SO_4$

SOURCE AND PURITY OF MATERIALS:

ESTIMATED ERROR:

No estimates possible.

REFERENCES:

COMPONENTS:	ORIGINAL MEASUREMENTS:
1. Ammonium pyrosulfite; $(NH_4)_2S_2O_5$; [32736-64-4]	Vasilenko, N.A.
2. Ammonium sulfate; $(NH_4)_2SO_4$; [7783-20-2]	*Zh. Priklad. Khim.* 1948, *21*, 917-26.
3. Water; H_2O; [7732-18-5]	

EXPERIMENTAL VALUES (continued):

$(NH_4)_2SO_4$ mass %	NH_4HSO_3 mass %	$(NH_4)_2SO_4$[a] mol/kg	NH_4HSO_3[a] mol/kg	Solid[b] phase
Temperature = -19°C				
39.25	0.0	5.563	0.	A + C
32.5	10.2	4.884	1.796	C
24.5	22.7	3.995	4.338	C
17.7	35.9	3.285	7.807	C
12.4	48.7	2.745	12.632	C
9.7	54.6	2.339	15.431	C
8.8	57.2	2.229	16.975	C
8.8	57.1	2.222	16.895	B + C
7.2	58.9	1.829	17.531	B
6.0	60.0	1.519	17.806	B
3.8	62.1	0.960	18.375	B
2.8	63.6	0.718	19.099	B
0.0	65.5	0.	19.156	B
0.0	44.5	0.	8.090	A
1.0	43.1	0.154	7.779	A
8.4	33.0	1.234	5.682	A
16.3	25.1	2.395	4.322	A
30.0	10.5	4.341	1.781	A
Temperature = -10°C				
40.0	0.0	5.740	0.	C
33.5	10.0	5.105	1.786	C
25.5	22.4	4.214	4.338	C
18.5	35.5	3.463	7.787	C
12.5	49.1	2.803	12.901	C
9.9	55.4	2.457	16.109	C
9.0	58.1	2.355	17.818	C
8.4	59.2	2.232	18.436	C
8.0	60.8	2.208	19.662	B + C
7.6	61.8	2.139	20.377	B
6.3	62.5	1.739	20.212	B
2.9	66.3	0.811	21.719	B
0.0	68.7	0.	22.146	B
13.5	13.1	1.584	1.801	A
28.0	0.0	3.348	0.	A
0.0	26.0	0.	3.545	A
Temperature = 0°C				
41.0	0.	5.983	0.	C
34.5	9.9	5.343	1.797	C
26.5	22.1	4.439	4.338	C
19.5	35.1	3.698	7.801	C
12.8	50.3	2.987	13.754	C
10.0	56.3	2.555	16.856	C
9.1	59.1	2.464	18.752	C
8.9	58.8	2.372	18.368	C
7.9	64.2	2.438	23.217	C
7.4	65.3	2.334	24.134	B + C
6.6	65.6	2.044	23.809	B
3.0	69.3	0.933	25.243	B
2.0	70.6	0.628	25.998	B
0.	72.0	0.	25.945	B

(continued on next page)

COMPONENTS:	ORIGINAL MEASUREMENTS:
1. Ammonium pyrosulfite; $(NH_4)_2S_2O_5$; [32736-64-4] 2. Ammonium sulfate; $(NH_4)_2SO_4$; [7783-20-2] 3. Water; H_2O; [7732-18-5]	Vasilenko, N.A. *Zh. Priklad. Khim.* 1948, *21*, 917-26.

EXPERIMENTAL VALUES (continued):

$(NH_4)_2SO_4$ mass %	NH_4HSO_3 mass %	$(NH_4)_2SO_4$[a] mol/kg	NH_4HSO_3[a] mol/kg	Solid[b] phase
Temperature = 20°C				
36.5	9.6	5.831	1.797	C
29.0	21.3	5.024	4.324	C
21.5	34.2	4.179	7.789	C
13.3	52.0	3.300	15.120	C
5.9	71.2	2.218	31.371	C
6.9	68.6	2.425	28.251	C
7.3	66.8	2.427	26.023	C
8.0	65.5	2.599	24.939	C
9.4	60.7	2.707	20.483	C
9.9	58.1	2.664	18.319	C
10.3	58.1	2.807	18.551	C
42.8	0.	6.443	0.	C
6.0	71.8	2.327	32.633	B + C
0.	77.0	0.	33.779	B

[a] Molalities calculated by the compiler.

[b] Solid phases: A - ice, B - $(NH_4)_2S_2O_5$, C - $(NH_4)_2SO_4$

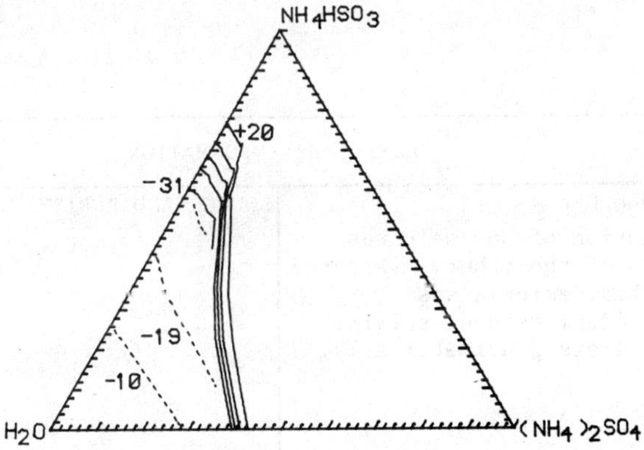

——— Equilibrium with salt(s)

------ Equilibrium with ice

COMPONENTS:

1. Ammonium hydrogen sulfite; NH_4HSO_3; [10192-30-0]
2. Ammonium sulfate; $(NH_4)_2SO_4$; [7783-20-2]
3. Water; H_2O; [7732-18-5]

ORIGINAL MEASUREMENTS:

Terres, E.; Heinsen, A.

Das Gas- und Wasserfach 1927, *70*, 1157-61.

VARIABLES:

Concentrations of the components
Four temperatures: 273 - 333 K

PREPARED BY:

Mary R. Masson

EXPERIMENTAL VALUES:

$(NH_4)_2SO_4$ mass %	NH_4HSO_3 mass %	$(NH_4)_2SO_4$[a] mol/kg	NH_4HSO_3[a] mol/kg	Solid[b] phase
Temperature = 0°C				
42.8	0.0	5.663	0.	A
37.95	7.30	5.246	1.345	A
29.80	19.62	4.459	3.914	A
22.22	30.95	3.591	6.668	A
14.84	45.67	2.844	11.669	A
7.71	62.40	1.952	21.064	A
5.85	66.92	1.626	24.797	A
4.04	69.85	1.171	26.992	B
3.39	70.33	0.976	27.002	B
2.02	71.07	0.568	26.647	B
0.0	71.8	0.	25.690	B

(continued on next page)

AUXILIARY INFORMATION

METHOD APPARATUS/PROCEDURE:

To the saturated solution of one salt was added various amounts of the other. After equilibrium was reached, ammonia was determined by the Kjeldahl method, sulfite by titrimetry, and sulfate gravimetrically.

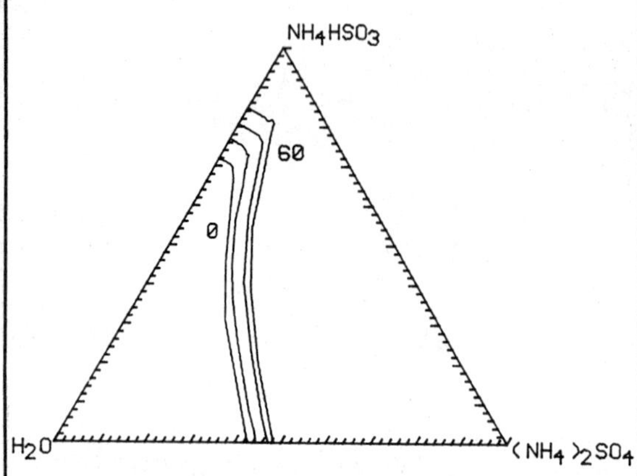

SOURCE AND PURITY OF MATERIALS:

ESTIMATED ERROR:

Temperature: 0.1 K
Analyses: no estimate possible.

REFERENCES:

COMPONENTS:	ORIGINAL MEASUREMENTS:
1. Ammonium hydrogen sulfite; NH_4HSO_3; [10192-30-0]	Terres, E.; Heinsen, A.
2. Ammonium sulfate; $(NH_4)_2SO_4$; [7783-20-2]	*Das Gas- und Wasserfach* 1927, *70*, 1157-61.
3. Water; H_2O; [7732-18-5]	

EXPERIMENTAL VALUES (continued):

$(NH_4)_2SO_4$ mass %	NH_4HSO_3 mass %	$(NH_4)_2SO_4$[a] mol/kg	NH_4HSO_3[a] mol/kg	Solid[b] phase
Temperature = 20°C				
44.5	0.0	6.068	0.	A
43.0	2.68	5.991	0.498	A
38.96	8.18	5.578	1.561	A
26.93	26.78	4.403	5.837	A
21.25	36.25	3.784	8.606	A
16.66	44.91	3.281	11.791	A
11.58	56.50	2.745	17.859	A
8.86	65.11	2.576	25.238	A
7.31	69.08	2.343	29.522	A
7.21	69.65	2.358	30.370	A
6.39	72.11	2.249	33.841	A
6.33	72.75	2.290	35.088	A
5.81	72.87	2.062	34.486	A
4.29	74.57	1.536	35.591	B
3.46	75.22	1.228	35.598	B
0.0	76.95	0.	33.684	B
Temperature = 40°C				
47.17	0.0	6.757	0.	A
42.68	6.47	6.352	1.284	A
32.38	22.43	5.423	5.008	A
21.69	40.24	4.312	10.665	A
14.02	57.52	3.693	20.106	A
11.98	63.00	3.624	25.406	A
9.54	70.42	3.603	35.445	A
8.27	74.34	3.599	43.133	A
7.72	75.73	3.530	46.169	A
7.55	76.09	3.492	46.927	A
6.35	77.28	2.936	47.632	B
6.09	77.72	2.847	48.436	B
4.91	78.29	2.212	47.020	B
2.72	79.43	1.153	44.898	B
0.0	80.62	0.	41.973	B
Temperature = 60°C				
48.4	0.0	7.098	0.	A
34.02	20.21	5.890	4.545	A
23.93	38.71	4.847	10.454	A
16.27	54.45	4.205	18.763	A
11.55	68.7	4.426	35.097	A
8.66	77.21	4.638	55.133	A
7.90	79.82	4.868	65.584	A
7.70	80.79	5.063	70.821	A
6.71	81.79	4.416	71.760	B
6.04	81.38	3.633	65.271	B
4.93	82.30	2.922	65.027	B
0.0	84.70	0.	55.857	B

[a] Molalities calculated by the compiler.

[b] Solid phases: A - ammonium sulfate, B - ammonium hydrogen sulfate

COMPONENTS:	EVALUATOR:
1. Beryllium sulfite; $BeSO_3$; [25454-04-0] 2. Water; H_2O; [7732-18-5]	H.D. Lutz, Dept. of Chemistry, University of Siegen, FR Germany. April 1983.

CRITICAL EVALUATION:

Only a small amount of work has been done on the $BeSO_3$-H_2O system, mostly in the 19th century. It is not clear what compounds, e.g. hydrates or possible basic salts, exist. From aqueous solutions prepared by dissolving beryllium hydroxide in sulfurous acid, no neutral sulfite can be crystallized (1,2). The solubility of beryllium sulfites, especially the hydrogen sulfite, in water is obviously very high (1 - 5). Numerical data are not available.

REFERENCES

1. Krusz, G.; Moraht, H. *Justus Liebigs Ann. Chem.* 1890, *260*, 178; *Ber. Dtsch. Chem. Ges.* 1890, *23*, 734.

2. Atterberg, A. *Svenska Akad. Handl.* 1873, *12*, Nr. 5, 27.

3. Earl, Ch.B.; Hughes, F. *Chem. Abstr. 78*, 151072m; *P.S. African* 7200045.

4. Schneider, R.Th.; Taylor, J.A.; Willis, W.D. *Chem. Abstr. 74*, 79271y; *P. Ger. Offen.* 2 034 453.

5. Terrana, J.D.; Miller, L.A.; Taylor, J.A. *Chem. Abstr. 71*, 72477p; *P. Ger. Offen.* 1 807 926.

COMPONENTS:	EVALUATOR:
1. Magnesium sulfite; $MgSO_3$; [7757-88-2] 2. Water; H_2O; [7732-18-5]	H.D. Lutz, Dept. of Chemistry, University of Siegen, FR Germany. January 1985.

CRITICAL EVALUATION:

Magnesium sulfite crystallizes from aqueous solutions in the form of the hexahydrate, $MgSO_3.6H_2O$ [13446-29-2], at temperatures below 313 K and as the trihydrate, $MgSO_3.3H_2O$ [19086-20-5], above 313 K (1-4). Furthermore the hydrates $MgSO_3.3.5H_2O$ [85017-92-1] (5), $MgSO_3.2.5H_2O$ [85017-92-1] (6), $MgSO_3.2H_2O$ [40854-09-9] (2,6,7), $MgSO_3.1H_2O$ [72860-77-6] (6,7), $MgSO_3.xH_2O$ (8), and $MgSO_3.x'H_2O$ (8) have been reported. The existence of the hydrate $MgSO_3.3.5H_2O$ (5) could not be confirmed (6). It is unknown whether $MgSO_3.xH_2O$ and $MgSO_3.x'H_2O$ are identical with $MgSO_3.2.5H_2O$ or $MgSO_3.2H_2O$, respectively. Solubility data of magnesium sulfite are available for the systems $MgSO_3$-H_2O, $MgSO_3$-SO_2-H_2O, $MgSO_3$-$MgSO_4$-H_2O, $MgSO_3$-$MgCl_2$-H_2O, $MgSO_3$-Na_2SO_3-H_2O, and $MgSO_3$-sucrose-H_2O.

SOLUBILITY OF MAGNESIUM SULFITE IN PURE WATER

$MgSO_3.6H_2O$

Numerical data on the solubility of $MgSO_3.6H_2O$ in water have been given by several authors (1,8-15). The published data are in relatively good agreement. For ambient temperature the following saturation concentrations have been reported: 0.046 (14) and 0.0573 (15) mol kg^{-1} (molality scale) at 298.2 and 303.2 K respectively, and 0.058 (9) and 0.0501 (10) mol dm^{-3} (molarity scale) at 298 and 291 K, respectively. The most reliable solubility data seem to be those of Trendafelov *et al.* (14) and Nývlt *et al.* (15).

RECOMMENDED VALUES

The solubility of $MgSO_3.6H_2O$ in water at 298.2 K (25°C) on the molality scale is 0.050 (±0.005) mol kg^{-1} (5.2 g $MgSO_3$/kg H_2O).

The solubility of $MgSO_3.6H_2O$ increases strongly with increasing temperature (1,5,8, 12-15). The following equation, fitted by evaluator to the data given by Hagisawa (1), Markant *et al.* (8), Kovachev, Trendafelov and Bakalov (13,14), and Nývlt *et al.* (15), which are in relatively good agreement, is recommended.

$$\log x = -237.382 + 9474.30/T + 81.8616 \log T \qquad (1)$$

with x = mole fraction of $MgSO_3$, and T = temperature (K). The correlation coefficient is 0.988. The given equation is valid for the range 273 - 363 K. A graph derived from this equation is shown in Fig. 1.

COMPONENTS:	EVALUATOR:
1. Magnesium sulfite; $MgSO_3$; [7757-88-2] 2. Water; H_2O; [7732-18-5]	H.D. Lutz, Dept. of Chemistry, University of Siegen, FR Germany. January 1985.

CRITICAL EVALUATION: (continued)

$MgSO_3.3H_2O$

Data on the solubility of $MgSO_3.3H_2O$ in water have also been reported by several authors (8,9,13,14,16). The published data are in relatively good agreement.

RECOMMENDED VALUES

The solubility of $MgSO_3.3H_2O$ in water at 323.2 K (50°C) on the molality scale is 0.082 (±0.003) mol kg^{-1} (8.6 g $MgSO_3$/kg H_2O).

The solubility of $MgSO_3.3H_2O$ decreases with increasing temperature (1,13,16). The following equation is fitted to the data:

$$\log x = -71.873 + 3811.30/T + 22.8142 \log T \qquad (2)$$

This equation is valid for the range 315 - 373 K. The correlation coefficient is 0.997. A graph of this equation is included in Fig. 1.

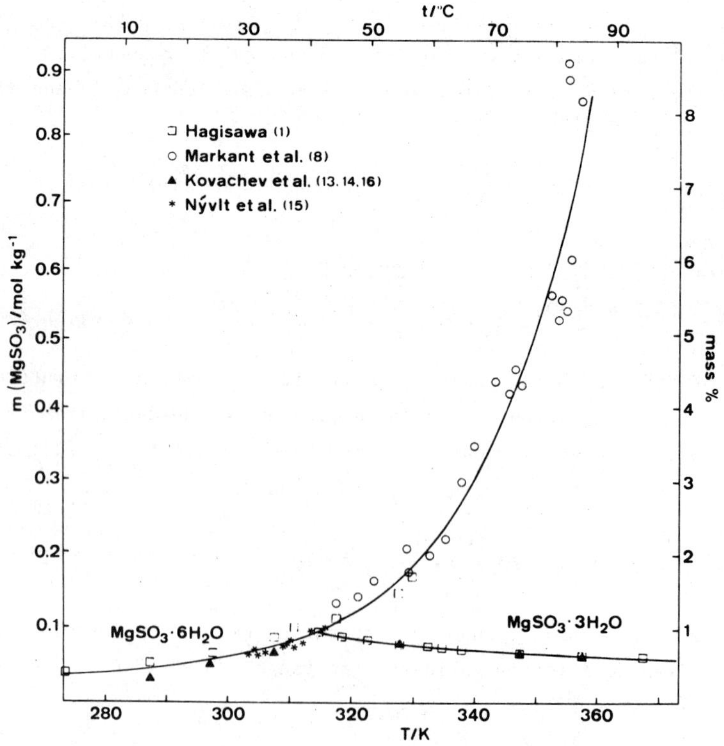

Fig. 1 Solubility of magnesium sulfite in water, as recommended for 273 - 373 K (equations (1) and (2)).

COMPONENTS:	EVALUATOR:
1. Magnesium sulfite; $MgSO_3$; [7757-88-2] 2. Water; H_2O; [7732-18-5]	H.D. Lutz, Dept. of Chemistry, University of Siegen, FR Germany. January 1985.

CRITICAL EVALUATION: (continued)

OTHER HYDRATES

Data on the solubility of other lower hydrates of magnesium sulfite, i.e. $MgSO_3.xH_2O$ and $MgSO_3.xH_2O$, have been reported by Markant *et al.* (8). The temperature coefficient of the solubility of these hydrates has been found to be negative.

SOLUBILITY IN THE SYSTEM $MgSO_3$-SO_2-H_2O

The solubility of magnesium sulfite increases very strongly with increasing SO_2 content of the solution (1,8,10-12,17-20). The reported numerical data are in relatively good agreement. The following equation, fitted (by evaluator) to the values given by Hagisawa (1), Yakimets *et al.* (11), and Conrad *et al.* (17), is recommended for the solubility of $MgSO_3.6H_2O$ in aqueous sulfurous acid solutions at 298.2 K (molality scale)

$$m(MgSO_3) = 0.0347 + 0.4995\, m(SO_2\ \text{tot}). \qquad (3)$$

The correlation coefficient is 0.99988. A graph derived from this equation is shown in Fig. 2.

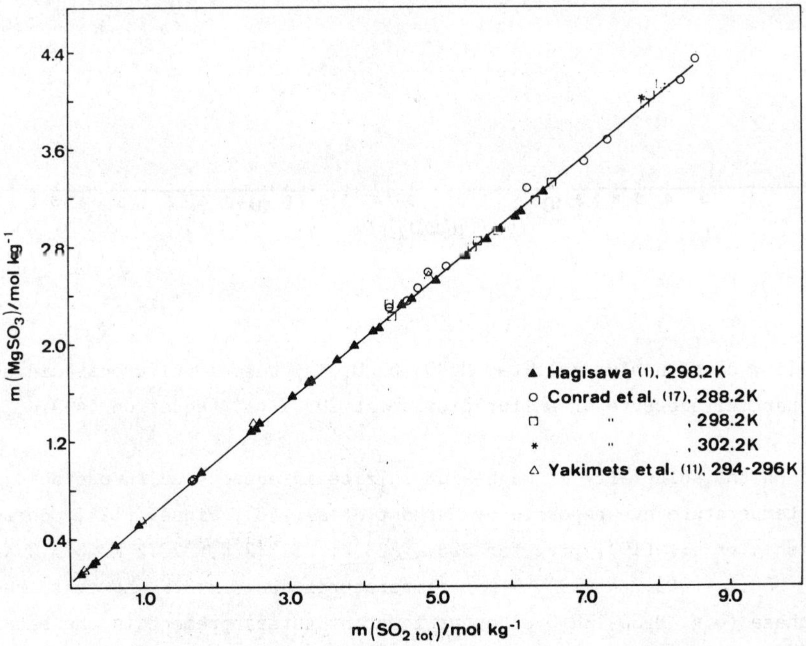

Fig. 2 Solubility of magnesium sulfite, $MgSO_3.6H_2O$, in aqueous sulfurous acid solutions, as recommended for 298 K (equation (3)).

COMPONENTS:	EVALUATOR:
1. Magnesium sulfite; $MgSO_3$; [7757-88-2] 2. Water; H_2O; [7732-18-5]	H.D. Lutz, Dept. of Chemistry, University of Siegen, FR Germany. January 1985.

CRITICAL EVALUATION: (continued)

Studies on the partial pressure of SO_2 over saturated solutions of magnesium sulfite are scarce (1,17,20). The data given by Hagisawa (1) for 298 K are recommended (see Fig. 3). The following equation is fitted (by evaluator) to these data:

$$\log m(MgSO_3) = -0.66763 + 0.26370 \log p(SO_2) \qquad (4)$$

The correlation coefficient is 0.9977.

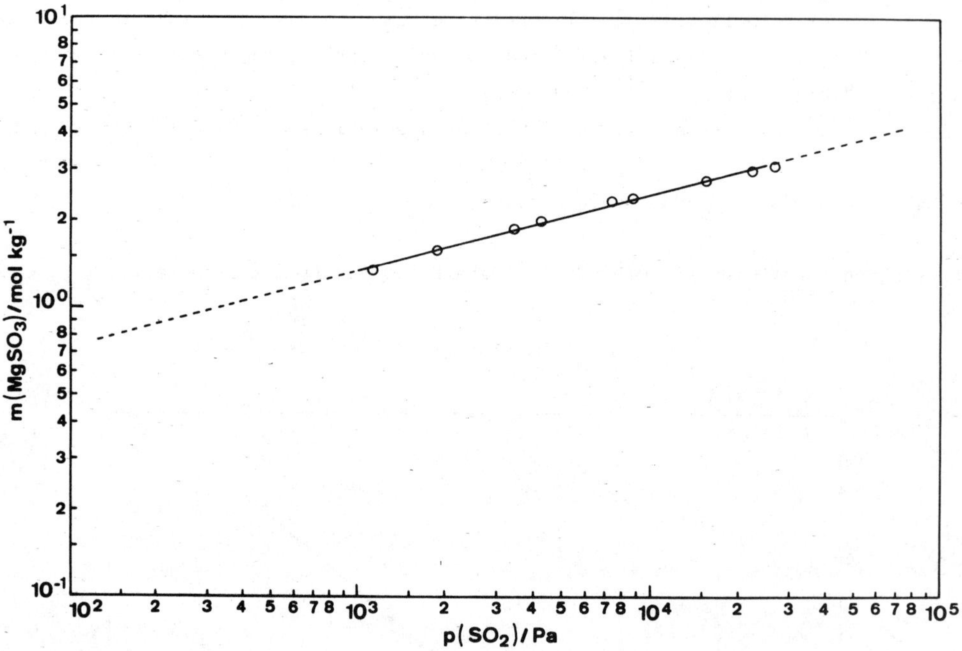

Fig. 3 Solubility of magnesium sulfite, $MgSO_3.6H_2O$, in aqueous sulfurous acid solutions vs. partial pressure of sulfur dioxide at 298 K (1) (equation (4)).

Numerical data on the solubility of magnesium sulfite in aqueous sulfurous acid solutions above ambient temperature are reported by Markant *et al.* (8), Pinaev (12), Conrad *et al.* (17), and Semishin *et al.* (19), *viz.* for 308 - 338 K; 313.2 K, 323.2 K, 333.2 K (12); 308.2 K (17); 308.2 - 343.2 K (19). For temperatures above 313 K, however, the nature of the solid phase (e.g. $MgSO_3.3H_2O$ or other lower hydrates) present in the saturated solution has not been specified by most authors. The data indicate that the temperature dependence of the solubility of magnesium sulfite is relatively small in the presence of large amounts of free SO_2.

COMPONENTS:

1. Magnesium sulfite; $MgSO_3$; [7757-88-2]
2. Water; H_2O; [7732-18-5]

EVALUATOR:

H.D. Lutz,
Dept. of Chemistry,
University of Siegen,
FR Germany.
January 1985.

CRITICAL EVALUATION: (continued)

SOLUBILITY OF MAGNESIUM SULFITE IN THE PRESENCE OF $MgSO_4$, $MgCl_2$, Na_2SO_3, AND SUCROSE

The solubility of magnesium sulfite in water is strongly affected by the presence of a third component. Numerical data are reported for the systems $MgSO_3$-$MgSO_4$-H_2O (12,13,15), $MgSO_3$-$MgSO_4$-SO_2-H_2O (19,20), $MgSO_3$-MgCl-H_2O (16), $MgSO_3$-$MgCl_2$-SO_2-H_2O (21), $MgSO_3$-Na_2SO_3-H_2O (14), and $MgSO_3$-sucrose-H_2O (22).

In the system $MgSO_3$-$MgSO_4$-H_2O the solubilities of both $MgSO_3.6H_2O$ and $MgSO_3.3H_2O$ at first increase with increasing $MgSO_4$ content (12,13,15) of up to $m(MgSO_4)$ = 1.5 - 2.5 mol kg^{-1} (10 - 20 mass %) to approximately twice the amount soluble in pure water and then decrease up to saturation with $MgSO_4$ (12,13). This behaviour has been observed for all temperatures investigated, *viz*. 288 K (13), 306.8 - 314.3 K (15), 308 K (13), 313.2 K (12), 323.2 K (12), 328 K (13), 333.2 K (12), and 348 K (13), irrespective of the solid phase present, and also in the presence of excess of sulfur dioxide (19,20). Graphs drawn (by evaluator) from data reported by Kovachev *et al*. (13) are shown in Fig. 4.

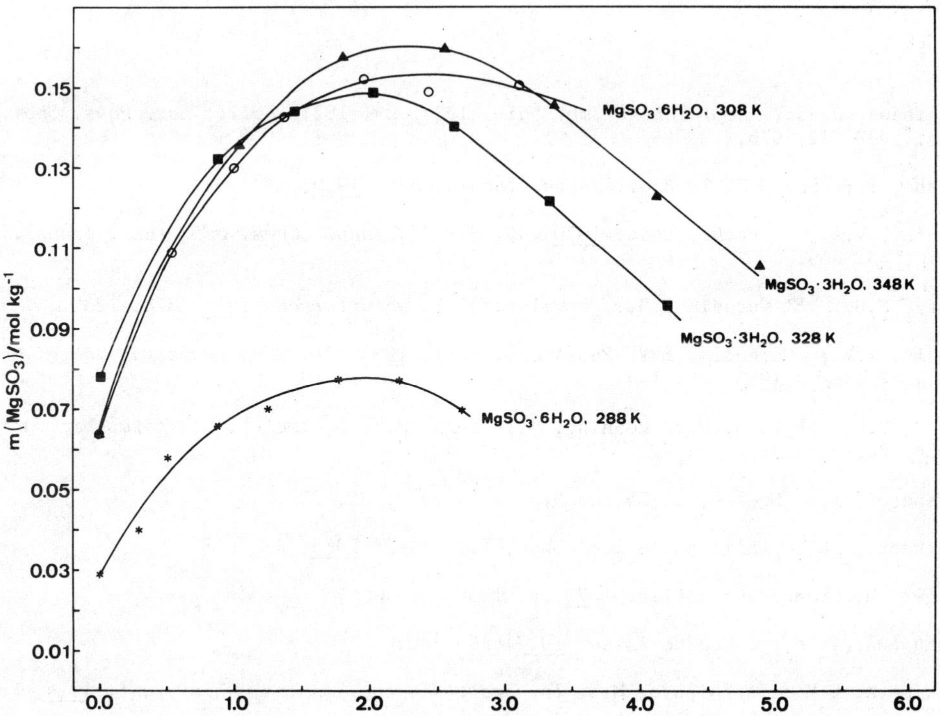

Fig. 4 Solubility of magnesium sulfite in aqueous magnesium sulfate solutions (13).

COMPONENTS:	EVALUATOR:
1. Magnesium sulfite; $MgSO_3$; [7757-88-2] 2. Water; H_2O; [7732-18-5]	H.D. Lutz, Dept. of Chemistry, University of Siegen, FR Germany. January 1985.

CRITICAL EVALUATION: (continued)

Numerical data on the solubility of $MgSO_3.3H_2O$ in the presence of magnesium chloride are given by Bakalov *et al.* (16) and McIlroy (21). Bakalov's data (16) indicate that the solubility of magnesium sulfite, i.e. $MgSO_3.3H_2O$, decreases to 20 and 30% of the amount soluble in pure water at a $MgCl_2$ concentration of approximately 40 mass % at 348 K and 358 K, respectively.

In the system $MgSO_3$-Na_2SO_2-H_2O the solubilities of both $MgSO_3.6H_2O$ and $MgSO_3.3H_2O$ at first increase with increasing Na_2SO_3 content (at 298.2 K to approximately four times the amount soluble in pure water). At higher concentrations of Na_2SO_3, i.e. higher than 10 mass %, the solubility of magnesium sulfite decreases to values less than the solubility in pure water.

The solid phases at high Na_2SO_3 concentrations are hydrates of ternary sodium magnesium sulfites of not exactly known composition.

The solubility of magnesium sulfite is not much affected by the presence of sucrose (22) in the solution.

REFERENCES

1. Hagisawa, H. *Sci. Rep. Tohoku Imp. Univ.* 1934, *23*, 182; *Bull. Inst. Phys. Chem. Res.* 1933, *12*, 976.

2. Okabe, T.; Hori, S. *Technol. Repts. Tohoku Univ.* 1959, *23*, 85.

3. Pinaev, V.A. *Zh. Prikl. Khim.* 1970, *43*, 871; *J. Appl. Chem. USSR (Eng. Transl.)* 1970, *43*, 869.

4. Lutz, H.D.; El-Suradi, S.M.; Engelen, B. *Z. Naturforsch.* 1977, *32b*, 1230.

5. Rodin, I.V.; Margulis, E.V. *Zh. Neorg. Khim.* 1983, *28*, 258; *Russ. J. Inorg. Chem. (Eng. Transl.)* 1983, *28*, 144.

6. Lutz, H.D.; Eckers, W.; Henning, J.; Jung, M.; Buchmeier, W. *Thermochim. Acta* 1984, *74*, 323.

7. Szendrei, T.; Ho-Tun, E. *Thermochim. Acta* 1980, *35*, 239.

8. Markant, H.P.; Phillips, N.D.; Shah, I.S. *Tappi* 1965, *48*, 648.

9. Hager, H. *Chem. Zentralblatt* 1875, p. 135.

10. Marusawa, T. *Kogyo Kagaku Zasshi* 1917, *20*, 280.

11. Yakimets, E.M.; Arkhipova, M.S. *Tr. Ural. Nauchno.-Issled. Khim. Inst.* 1954, No. 1, 112.

12. Pinaev, V.A. *Zh. Prikl. Khim.* 1964, *37*, 1361; *J. Appl. Chem. USSR (Eng. Transl.)* 1964, *37*, 1353; *Vestn. Tekhn. i. Ekon. Inform. Nauch.-Issled. Inst. Tekhn.-Ekon. Issled. Gos. Kom. Khim. i. Neft: Prom. pri Gosplane SSSR* 1963, Na 8, 25.

13. Kovachev, Ts.; Bakalov, V.; Trendafelov, D. *Khim. Ind. (Sofia)* 1970, *42*, 209.

COMPONENTS:	EVALUATOR:
1. Magnesium sulfite; $MgSO_3$; [7757-88-2] 2. Water; H_2O; [7732-18-5]	H.D. Lutz, Dept. of Chemistry, University of Siegen, FR Germany. January 1985.

CRITICAL EVALUATION: (continued)

14. Trendafelov, D.; Kovachev, Ts.; Bakalov, V. *Izv. Otd. Khim. Nauki. Bulg. Akad. Nauk.* 1971, *4*, 643.

15. Nývlt, J.; Rychlý, R.; Kricková, J. *Chem. Prům.* 1977, *27*, 552.

16. Bakalov, V.D.; Kovachev, Ts.B.; Trendafelov, D. *Khim. Ind. (Sofia)* 1971, *43*, 351.

17. Conrad, F.H.; Brice, D.B. *J. Am. Chem. Soc.* 1948, *70*, 2179.

18. Simon, A.; Waldmann, K. *Naturwissenschaften* 1958, *45*, 128.

19. Semishin, V.I.; Abramov, I.I.; Vorotnitskaya, L.T. *Izv. Vyss. Uchebn. Zaved., Khim. Technol.* 1959, 2, 834.

20. Kuz'minykh, I.N.; Babushkina, M.D. *Zh. Prikl. Khim.* 1957, *30*, 466; *J. Appl. Chem. USSR (Eng. Transl.)* 1957, *30*, 495.

21. McIlroy, R.A. *Tappi* 1973, *56*, 79.

22. Saillard, E. *Suppl. Circ. Hebd.* 1931, No. 2188; *Sugar Abstracts* in *Facts About Sugar 26*, 222.

COMPONENTS:	ORIGINAL MEASUREMENTS:
1. Magnesium sulfite; $MgSO_3$; [7757-88-2] 2. Water; H_2O; [7732-18-5]	Marusawa, T. *Kogyo Kagaku Zasshi* 1917, *20*, 280-7.
VARIABLES:	PREPARED BY:
Temperature: 291 K	B. Engelen

EXPERIMENTAL VALUES:

The author reports the solubility of $MgSO_3.6H_2O$ [13446-29-2] in water at 18°C to be

$c(MgSO_3) = 0.0501$ mol dm^{-3} (5.229 g/dm^3, compiler).

AUXILIARY INFORMATION

METHOD APPARATUS/PROCEDURE:

Saturation method. Equilibrium was established after several days. SO_3^{2-} was determined iodometrically.

SOURCE AND PURITY OF MATERIALS:

$MgSO_3.6H_2O$ was precipitated from oxygen-free $MgCl_2$ solutions with Na_2SO_3. The precipitate was checked for Cl^- and SO_4^{2-} content.

ESTIMATED ERROR:

The value given is the mean of 4 experiments which differ by 1.8%.

REFERENCES:

COMPONENTS:	ORIGINAL MEASUREMENTS:
1. Magnesium sulfite; $MgSO_3$; [7757-88-2] 2. Water; H_2O; [7732-18-5]	Hagisawa, H. *Sci. Rep. Tohoku Imp. Univ., Ser. 1* 1934, *23*, 182-92; *Bull. Inst. Phys. Chem. Res., Tokyo* 1933, *12*, 976-83.
VARIABLES:	PREPARED BY:
Temperature: 273 - 368 K	B. Engelen, H.D. Lutz

EXPERIMENTAL VALUES:

The solubilities of $MgSO_3.6H_2O$ [13446-29-2] and $MgSO_3.3H_2O$ [19086-20-5] in water at various temperatures are:

t/°C	$MgSO_3$ mass %[a]	$MgSO_3$ m/mol kg^{-1}[b]
	$MgSO_3.6H_2O$	
0	0.338	0.0324
15	0.497	0.0478
25	0.646	0.0622
35	0.846	0.0817
45	1.116	0.1081
55	1.465	0.1424
57.5	1.688	0.1645
62.5	1.950	0.1905
	$MgSO_3.3H_2O$	
38	1.034	0.1001
42	0.937	0.0906
46	0.897	0.0867
50	0.844	0.0815
55	0.817	0.0789
60	0.758	0.0731
62.5	0.748	0.0722
65	0.720	0.0694
75	0.664	0.0640
85	0.623	0.0600
95	0.615	0.0592

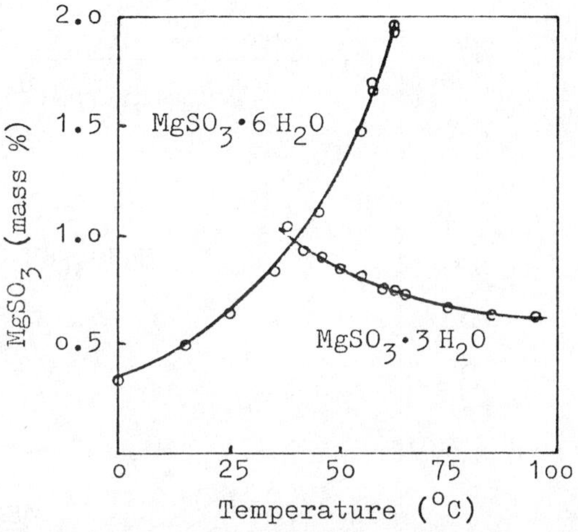

a g/100 ml soln. author
b Calculated by compilers

AUXILIARY INFORMATION

METHOD APPARATUS/PROCEDURE:	SOURCE AND PURITY OF MATERIALS:
Saturation method. Equilibrium was established after several hours. Magnesium was determined as the sulfate, sulfite by iodometric titration.	Magnesium sulfite was precipitated from aqueous $Mg(HSO_3)_2$ solutions obtained from $MgCO_3$ dissolved in oxygen-free water by passing SO_2.
	ESTIMATED ERROR: Deviation in several experiments (2 - 3) is 0.3%.
	REFERENCES:

COMPONENTS:

1. Magnesium sulfite; $MgSO_3$; [7757-88-2]
2. Water; H_2O; [7732-18-5]

ORIGINAL MEASUREMENTS:

Markant, H.P.; Phillips, N.D.; Shah, I.S.

Tappi 1965, *48*, 648-53.

VARIABLES:

Temperature: 318 - 368 K

PREPARED BY:

B. Engelen, H.D. Lutz

EXPERIMENTAL VALUES:

The authors give a solubility diagram for $MgSO_3.6H_2O$ [13446-29-2] and two other magnesium sulfite hydrates ($MgSO_3.xH_2O$ and $MgSO_3.x'H_2O$) of unknown composition. One value of $MgSO_3.3H_2O$ [19086-20-5] at 60°C is also given. The scale is given in mass % of SO_2 as $MgSO_3$ and °F by the authors. A scale in mass % of $MgSO_3$ and °C has been added by the compilers.

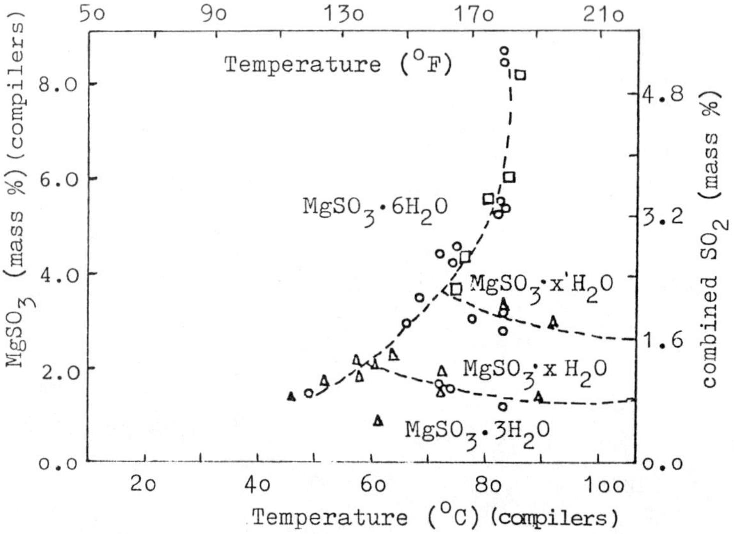

Reprinted by permission

(continued on next page)

AUXILIARY INFORMATION

METHOD APPARATUS/PROCEDURE:

Saturation method. Saturated solutions were prepared by adding:

$MgSO_3.6H_2O$ to water (Δ)

SO_2 gas to a MgO slurry ( )

MgO to a $Mg(HSO_3)_2$ solution (o)

The solutions were analysed for sulfite. Method not given.

SOURCE AND PURITY OF MATERIALS:

$Mg(HSO_3)_2$ solutions and $MgSO_3.6H_2O$ were prepared by adding SO_2 gas to a slurry of MgO in distilled water.

ESTIMATED ERROR:

Not given.

REFERENCES:

COMPONENTS:	ORIGINAL MEASUREMENTS:
1. Magnesium sulfite; $MgSO_3$; [7757-88-2] 2. Water; H_2O; [7732-18-5]	Markant, H.P.; Phillips, N.D.; Shah, I.S. *Tappi* 1965, *48*, 648-53.

EXPERIMENTAL VALUES (continued):

The following numerical data were estimated from the diagram by the compilers.

	$MgSO_3.6H_2O$		$MgSO_3.xH_2O$		$MgSO_3.x'H_2O$	
t/°C	mass %	m/mol kg^{-1}	mass %	m/mol kg^{-1}	mass %	m/mol kg^{-1}
45	1.23	0.119				
50	1.46	0.142				
55	1.81	0.176				
60	2.21	0.216	2.03	0.198	0.75[a]	0.072[a]
65	2.75	0.271	1.85	0.180		
70	3.37	0.334	1.67	0.163		
75	4.21	0.421	1.53	0.149	3.41	0.338
80	5.35	0.541	1.43	0.139	3.14	0.311
83	7.29	0.753	1.38	0.134	3.08	0.304
85			1.34	0.130	2.97	0.293
90			1.27	0.123	2.82	0.278
95			1.26	0.122	2.71	0.267
100			1.24	0.120	2.63	0.259

[a] $MgSO_3.3H_2O$ as solid phase.

COMPONENTS:

1. Magnesium sulfite; $MgSO_3$; [7757-88-2]
2. Water; H_2O; [7732-18-5]

ORIGINAL MEASUREMENTS:

Rodin, I.V.; Margulis, E.V.

Zh. Neorg. Khim. 1983, *28*, 258-9; *Russ. J. Inorg. Chem. (Eng. Transl.)* 1983, *28*, 144.

VARIABLES:

Four temperatures: 293 - 363 K

PREPARED BY:

B. Engelen

EXPERIMENTAL VALUES:

Solubilities of magnesium sulfite in water at various temperatures are reported.

t/°C	$MgSO_3$ 10^4mass %	$MgSO_3$ $10^2 m$/mol kg^{-1} [a]
20	3140	3.018
50	3520	3.384
70	3870	3.722
90	4100	3.944

[a] Calculated by compiler.

AUXILIARY INFORMATION

METHOD APPARATUS/PROCEDURE:

Saturation method. Equilibrium was established by stirring the saturated solutions in thermostatically controlled glass tubes. Equilibrium was tested for analytically. 4 hr are reported to be sufficient. Magnesium was determined gravimetrically.

SOURCE AND PURITY OF MATERIALS:

Magnesium sulfite, claimed to be $MgSO_3.3.5H_2O$ [85017-92-1], was obtained by precipitation from $MgSO_4$ solutions with Na_2SO_3 (1).

ESTIMATED ERROR:

Not given.

REFERENCES:

1. Margulis, E.V.; Grishankina, N.S. *Zh. Neorg. Khim.* 1963, *8*, 2638.

COMPONENTS:

1. Magnesium sulfite; $MgSO_3$; [7757-88-2]
2. Sulfur dioxide; SO_2; [7446-09-5]
3. Water; H_2O; [7732-18-5]

ORIGINAL MEASUREMENTS:

Conrad, F.H.; Brice, D.B.

J. Am. Chem. Soc. 1948, *70*, 2179-82.

VARIABLES:

Three temperatures: 288, 298 and 308 K
Partial pressure of SO_2

PREPARED BY:

B. Engelen, H.D. Lutz

EXPERIMENTAL VALUES:

The authors report the solubility of $MgSO_3 \cdot 6H_2O$ [13446-29-2] in aqueous sulfurous acid solutions for various SO_2 vapour pressures at 15, 25, and 35°C on the basis of equilibrium studies of the system $MgO-SO_2-H_2O$.

t/°C	p_{total} mm Hg	p_{SO_2}[a] 10^{-3} bar	g SO_2 in 100 g H_2O total	g SO_2 in 100 g H_2O combined[b]	$MgSO_3$ g/kg H_2O[c]	$m(MgSO_3)$[c] mol kg^{-1}
15	75.5	83.6	30.4	15.7	256	2.45
	87.5	99.6	28.05	14.75	240.3	2.302
	110.7	130.5	29.65	15.05	245.2	2.349
	156.8	192.0	32.9	16.9	275	2.64
	162.3	199.3	31.4	16.55	269.6	2.583
	245.5	310.3	39.9	21.0	342	3.28
	419.3	541.9	46.9	23.55	383.7	2.676
	448.7	581.1	44.8	22.4	365	3.50
	726.8	951.9	54.6	27.8	453	4.34
	730.3	956.6	53.3	26.75	435.8	4.176

a,b,c See the following page.

(continued on next page)

AUXILIARY INFORMATION

METHOD APPARATUS/PROCEDURE:

Saturation method. Saturation was approached from both the supersaturated and the unsaturated state. Total pressure (SO_2 and H_2O) was measured with a mercury manometer. The amounts of total and combined SO_2 were determined by a combination of acidimetric and iodometric titration (1).

SOURCE AND PURITY OF MATERIALS:

MgO was of p.a. quality. Amounts of sulfate and CaO were negligible. SO_2 used was free from SO_3. The amount of inert or non-absorbable gases was about 0.15%.

ESTIMATED ERROR:

Not given.

REFERENCES:

1. Birchard, W.H. *Pap. Ind.* 1926, *8*, 793.

COMPONENTS:	ORIGINAL MEASUREMENTS:
1. Magnesium sulfite; $MgSO_3$; [7757-88-2]	Conrad, F.H.; Brice, D.B.
2. Sulfur dioxide; SO_2; [7446-09-5]	*J. Am. Chem. Soc.* 1948, *70*, 2179-82.
3. Water; H_2O; [7732-18-5]	

EXPERIMENTAL VALUES (continued):

t/°C	p_{total} mm Hg	p_{SO_2}[a] 10^{-3} bar	g SO_2 in 100 g H_2O total	g SO_2 in 100 g H_2O combined[b]	$MgSO_3$ g/kg H_2O[c]	$m(MgSO_3)$[c] mol kg^{-1}
25	130.2	141.9	28.2	14.2	231	2.22
	154.4	174.2	28.0	14.85	241.9	2.318
	213.2	252.6	35.6	18.2	297	2.84
	264.3	320.7	35.2	17.9	292	2.79
	288.0	352.3	37.5	18.75	305.5	2.927
	326.5	403.6	40.7	20.35	331.6	3.177
	410.9	516.1	42.1	21.3	347	3.32
	560.7	715.9	50.3	25.4	414	3.96
	638.4	819.4	51.6	26.3	428	4.11
	726.8	937.3	50.7	25.85	421.2	4.035
	728.4	939.4	52.0	26.2	427	4.09
35	728.7	915.3	50.1	25.7	419	4.01

[a] Corrected for p_{H_2O} and converted to bar by compilers.

[b] SO_2 required to form the monosulfite.

[c] Calculated from the amount of combined SO_2 by the compilers.

COMPONENTS:	ORIGINAL MEASUREMENTS:
1. Magnesium sulfite; $MgSO_3$; [7757-88-2] 2. Sulfur dioxide; SO_2; [7446-09-5] 3. Water; H_2O; [7732-18-5]	Markant, H.P.; Phillips, N.D. Shah, I.S. *Tappi* 1965, *48*, 648-53.
VARIABLES:	PREPARED BY:
Temperature: 305 - 338 K Concentration of SO_2	B. Engelen, H.D. Lutz

EXPERIMENTAL VALUES:

The authors report the solubility of magnesium sulfite in aqueous sulfurous acid solutions for various values of temperature and total amount of SO_2. Experimental data are given in a graph of amount of combined SO_2 (mass %), i.e. SO_2 required to form the monosulfite *vs.* temperature (in °F) for several amounts of total SO_2 (mass %), shown on the Figure as T. A scale in °C has been added by the compilers.

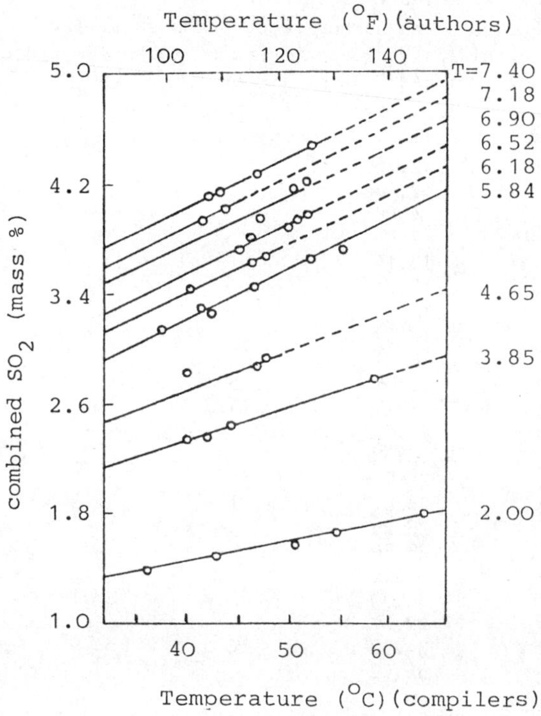

Reprinted by permission

AUXILIARY INFORMATION

METHOD APPARATUS/PROCEDURE:	SOURCE AND PURITY OF MATERIALS:
Saturation limits were determined by turbidimetric measurements during cooling solutions of $Mg(HSO_3)_2$ + $MgSO_3$ of known compositions. The first precipitation temperature was determined as the point of intersection of the two straight line portions of the "cooling curves".	The $Mg(HSO_3)_2$ solutions were prepared in closed containers under nitrogen by adding SO_2 to a slurry of MgO in distilled water.
	ESTIMATED ERROR: Not given.
	REFERENCES:

COMPONENTS:

1. Magnesium sulfite; $MgSO_3$; [7757-88-2]
2. Magnesium hydrogen sulfite; $Mg(HSO_3)_2$ [13774-25-9]
3. Water; H_2O; [7732-18-5]

ORIGINAL MEASUREMENTS:

Hagisawa, H.

Sci. Rep. Tohoku Imp. Univ., Ser. 1 1934, *23*, 182-92; *Bull. Inst. Phys. Chem. Res., Tokyo* 1933, *12*, 976-83.

VARIABLES:

One temperature: 298 K

PREPARED BY:

B. Engelen, H.D. Lutz

EXPERIMENTAL VALUES:

For 25°C the author reports the solubility of $MgSO_3.6H_2O$ [13446-29-2] in solutions containing various amounts of magnesium hydrogen sulfite.

$Mg(HSO_3)_2$	p_{SO_2}	p_{SO_2}[a]	$MgSO_3$[b]	$MgSO_3$[c]	
mass %	mm Hg	10^{-3}bar	mass %[d]	mass %[d]	m/mol kg^{-1}[a]
0.73	-	-	0.70	1.11	0.108
2.26	-	-	0.67	1.94	0.191
2.44	-	-	0.64	2.00	0.198
4.80	-	-	0.68	3.37	0.340
7.45	-	-	0.66	4.84	0.500
13.14	-	-	0.68	8.03	0.880
13.21	-	-	0.65	8.05	0.882
14.00	-	-	0.69	8.53	0.943
18.57	-	-	0.66	11.06	1.283
19.01	-	-	0.67	11.31	1.319
19.30	8.6	11.4	0.65	11.45	1.340
22.15	14.2	18.9	0.60	13.00	1.569
23.54	-	-	0.52	13.69	1.678
23.58	-	-	0.55	13.75	1.686
25.66	25.7	34.3	0.50	14.87	1.866
26.87	31.9	42.5	0.52	15.56	1.982
28.26	-	-	0.36	16.18	2.092
28.76	-	-	0.31	16.41	2.133
30.13	55.7	74.2	0.50	17.37	2.302
30.93	65.4	87.2	0.38	17.70	2.365
32.67	-	-	0.13	18.42	2.508
34.50	116.2	154.9	0.15	19.46	2.715
35.72	-	-	0.14	20.14	2.854

a,b,c,d See the following page. (continued on next page)

AUXILIARY INFORMATION

METHOD APPARATUS/PROCEDURE:

Method as for binary system. Total pressure (SO_2 + H_2O) was measured with a mercury manometer, and vapour pressure of water was determined by the dynamic method.

SOURCE AND PURITY OF MATERIALS:

As for binary system. When partial pressure of SO_2 was determined, $MgCO_3$-free MgO was used instead of $MgCO_3$.

ESTIMATED ERROR:

REFERENCES:

COMPONENTS:	ORIGINAL MEASUREMENTS:
1. Magnesium sulfite; $MgSO_3$; [7757-88-2] 2. Magnesium hydrogen sulfite; $Mg(HSO_3)_2$ [13774-25-9] 3. Water; H_2O; [7732-18-5]	Hagisawa, H. *Sci. Rep. Tohoku Imp. Univ., Ser. 1* 1934, *23*, 182-92; *Bull. Inst. Phys. Chem. Res., Tokyo* 1933, *12*, 976-83.

EXPERIMENTAL VALUES (continued):

$Mg(HSO_3)_2$	p_{SO_2}	p_{SO_2}[a]	$MgSO_3$[b]	$MgSO_3$[c]	
mass %	mm Hg	10^{-3}bar	mass %[d]	mass %[d]	m/mol kg^{-1}[a]
36.52	166.6	222.1	0.10	20.54	2.942
37.45	199.0	265.3	-	20.97	3.036
37.73	-	-	0.05	21.17	3.080
39.04	-	-	-	21.85	3.235

[a] Calculated by the compilers.

[b] $MgSO_3$ in addition to the bisulfate.

[c] Total magnesium sulfite ($MgSO_3$ + $Mg(HSO_3)_2$ as $MgSO_3$).

[d] g/100 g soln, author.

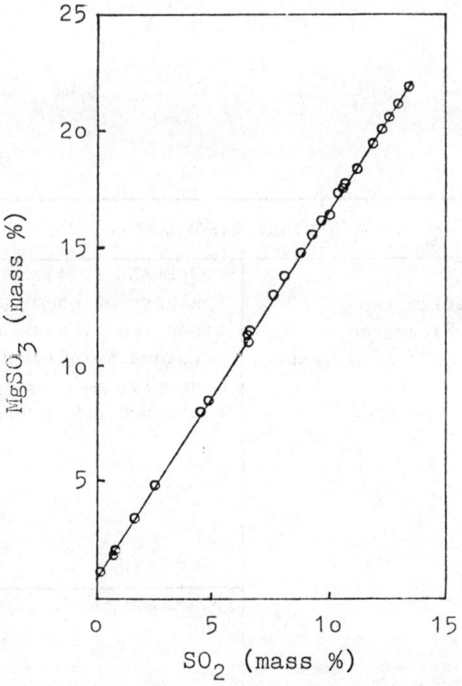

COMPONENTS:

1. Magnesium sulfite; $MgSO_3$; [7757-88-2]
2. Magnesium hydrogen sulfite; $Mg(HSO_3)_2$; [13774-25-9]
3. Water; H_2O; [7732-18-5]

ORIGINAL MEASUREMENTS:

Yakimets, E.M.; Arkhipova, M.S.

Tr. Ural. Nauchno-Issled. Khim. Inst. 1954, No. 1, 112-8.

VARIABLES:

Temperature: 294 - 296 K
Concentration of $Mg(HSO_3)_2$

PREPARED BY:

B. Engelen

EXPERIMENTAL VALUES:

The authors report the solubility of magnesium sulfite in water and in $Mg(HSO_3)_2$ solutions at 21-23°C.

t/°C	$Mg(HSO_3)_2$	$MgSO_3$	$MgSO_3$[b]	
	mass %[a]	mass %	mass %	m/mol kg^{-1}
23	-	0.615	-	0.059
23	1.298	0.630	1.356	0.132
22	7.307	1.117	5.207	0.541
22	18.231	1.308	11.513	1.342
21	30.881	1.539	18.825	2.556

[a] Calculated by the compiler from g/dm^3 and the density of the solutions, both given by the authors.

[b] Total amount of dissolved magnesium sulfite ($MgSO_3$ + $Mg(HSO_3)_2$, given as $MgSO_3$), calculated by the compiler.

AUXILIARY INFORMATION

METHOD APPARATUS/PROCEDURE:

Saturation method. Equilibrium was established after 1-4 hr. Saturated solutions were analysed for Mg^{2+} SO_3^{2-}, and SO_4^{2-} content. Method not given.

SOURCE AND PURITY OF MATERIALS:

A slurry of chemical pure MgO was dissolved by passing in SO_2. The obtained solution then was saturated with magnesium sulfite. The solutions contained 2-6 g/dm^3 SO_4^{-2} (authors).

ESTIMATED ERROR:

REFERENCES:

COMPONENTS:

1. Magnesium sulfite; $MgSO_3$; [7757-88-2]

2. Magnesium hydrogen sulfite; $Mg(HSO_3)_2$; [13774-25-9]

3. Water; H_2O; [7732-18-5]

ORIGINAL MEASUREMENTS:

Semishin, V.I.; Abramov, I.I; Vorotnitskaya, L.T.

Izv. Vyssh. Uchebn. Zaved., Khim. Khim. Technol. 1959, 2, 834-9.

VARIABLES:

Temperature: 308 - 343 K
Concentration of $Mg(HSO_3)_2$
pH: 4 - 9

PREPARED BY:

B. Engelen

EXPERIMENTAL VALUES:

The authors report the solubility of magnesium sulfite in aqueous solutions containing 0.5 - 4.7 mass % of $Mg(HSO_3)_2$, and in solutions of 0.4 - 4.4 mass % of $Mg(HSO_3)_2$ and 10 mass % of $MgSO_4$ at various temperatures from 35 to 70°C. The pH values of the saturated solutions are also given.

t/°C	pH[a]	$Mg(HSO_3)_2$[b] mass %	$MgSO_3$[c] mass %	$MgSO_3$[d] mass %	$m(MgSO_3)$[d] mol kg^{-1}
			without $MgSO_4$		
35	6.00	0.53	0.595	0.89	0.086
	5.38	1.03	0.470	1.05	0.102
	5.20	2.09	0.448	1.62	0.159
	5.18	3.06	0.437	2.15	0.213
	4.80	3.88	0.417	2.59	0.259
	4.67	4.71	0.411	3.05	0.308
40	6.40	0.53	0.752	1.05	0.102
	5.89	1.03	0.659	1.24	0.120
	5.00	2.09	0.561	1.73	0.170
	5.20	3.06	0.538	2.25	0.224
	4.93	3.88	0.504	2.68	0.268
	4.72	4.71	0.463	3.10	0.313

a,b,c,d See the following page. (continued on next page)

AUXILIARY INFORMATION

METHOD APPARATUS/PROCEDURE:

Saturation method. Equilibrium was established after 1.5 hr in a thermostatically controlled flask. Amounts of HSO_3^- and SO_3^{2-} were determined by a combination of iodometric and acidimetric titration. pH was determined potentiometrically.

SOURCE AND PURITY OF MATERIALS:

Magnesium sulfite was prepared by bubbling SO_2 through a suspension of MgO in water. The precipitated $MgSO_3.6H_2O$ was analysed for Mg^{2+}, SO_3^{2-}, and SO_4^{2-} content. $Mg(HSO_3)_2$ solutions were obtained and analysed in the same way as above.

ESTIMATED ERROR:

Deviations in temperature and pH, (authors) are ±0.01 K and ±0.2-0.4, respectively.

REFERENCES:

COMPONENTS:	ORIGINAL MEASUREMENTS:
1. Magnesium sulfite; $MgSO_3$; [7757-88-2] 2. Magnesium hydrogen sulfite; $Mg(HSO_3)_2$; [13774-25-9] 3. Water; H_2O; [7732-18-5]	Semishin, V.I.; Abramov, I.I; Vorotnitskaya, L.T. *Izv. Vyssh. Uchebn. Zaved., Khim. Khim. Technol.* 1959, 2, 834-9.

EXPERIMENTAL VALUES (continued):

t/°C	pH[a]	$Mg(HSO_3)_2$[b] mass %	$MgSO_3$[c] mass % without $MgSO_4$	$MgSO_3$[d] mass %	$m(MgSO_3)$[d] mol kg^{-1}
45	6.60	0.53	0.939	1.24	0.120
	6.00	1.03	0.838	1.41	0.138
	5.37	2.09	0.796	1.97	0.194
	5.30	3.06	0.651	2.36	0.235
	5.00	3.88	0.612	2.78	0.279
	4.80	4.71	0.581	3.22	0.326
50	7.05	0.53	1.060	1.36	0.132
	6.27	1.03	0.976	1.55	0.152
	5.60	2.09	0.940	2.11	0.208
	5.40	3.06	0.800	2.51	0.250
	5.03	3.88	0.736	2.91	0.292
	5.00	4.71	0.724	3.36	0.341
55	7.81	0.53	1.418	1.71	0.168
	6.29	1.03	1.307	1.88	0.185
	6.18	2.09	1.246	2.42	0.239
	5.50	3.06	1.199	2.91	0.291
	5.20	3.88	1.084	3.26	0.328
	5.08	4.71	0.978	3.61	0.367
60	8.20	0.53	1.825	2.12	0.208
	6.40	1.03	1.535	2.11	0.208
	6.20	2.09	1.391	2.56	0.254
	5.54	3.06	1.368	3.08	0.309
	5.26	3.88	1.289	3.46	0.350
	5.20	4.71	1.154	3.79	0.386
70	8.45	0.53	1.968	2.26	0.223
	6.61	1.03	1.891	2.47	0.244
	6.40	2.09	1.844	3.01	0.301
	5.60	3.06	1.549	3.26	0.328
	5.32	3.88	1.538	3.71	0.376
	5.26	4.71	1.512	4.15	0.424

[a] pH determined 20-25°C.

[b] Concentration of the solution in which solid magnesium sulfite has been dissolved.

[c] Determined analytically as the amount of SO_3^{2-} present in addition to HSO_3^- after establishing equilibrium.

[d] Total amount of dissolved magnesium sulfite ($MgSO_3$ + $Mg(HSO_3)_2$ as $MgSO_3$), calculated by the compiler.

COMPONENTS:	ORIGINAL MEASUREMENTS:
1. Magnesium sulfite; $MgSO_3$; [7757-88-2] 2. Magnesium sulfate; $MgSO_4$; [7487-88-9] 3. Water; H_2O; [7732-18-5]	Kovachev, Ts.; Bakalov, V.; Trendafelov, D. *Khim. Ind. (Sofia)* 1970, *42*, 209-11.
VARIABLES:	PREPARED BY:
Four temperatures: 288 - 348 K Amount of $MgSO_4$	B. Engelen

EXPERIMENTAL VALUES:

The authors report the solubility of $MgSO_3.6H_2O$ [13446-29-2] at 15 and 35°C and of $MgSO_3.3H_2O$ [19086-20-5] at 55 and 75°C in water and in solutions with various amounts of $MgSO_4$.

t/°C	$MgSO_4$ mass %	$MgSO_3$ mass %	$m(MgSO_3)$ mol kg^{-1} [a]	Solid phase
15	0.0	0.30	0.029	$MgSO_3.6H_2O$
	3.39	0.40	0.040	"
	5.73	0.57	0.058	"
	9.48	0.62	0.066	"
	13.00	0.63	0.070	"
	17.38	0.63	0.074	"
	20.93	0.61	0.074	"
	24.26	0.55	0.070	"
	24.70	0.55	0.070	$MgSO_3.6H_2O + MgSO_4.7H_2O$
35	0.0	0.66	0.064	$MgSO_3.6H_2O$
	6.16	1.06	0.109	"
	10.50	1.20	0.130	"
	13.90	1.27	0.143	"
	18.80	1.27	0.152	"
	22.36	1.19	0.149	"
	26.95	1.13	0.151	"
	28.80	0.92	0.125	$MgSO_3.6H_2O + MgSO_4.7H_2O$

[a] Calculated by the compiler.

(continued on next page)

AUXILIARY INFORMATION

METHOD APPARATUS/PROCEDURE:

Saturation method. Equilibrium was established by stirring the given solutions with excess of magnesium sulfite under N_2. Time not given, but is assumed to be the same (4 - 8 hr) as in a second paper by the authors (1). Mg^{2+} was determined complexometrically, SO_3^{2-} iodometrically. From these two figures the amounts of $MgSO_3$ and $MgSO_4$ were calculated by the authors.

SOURCE AND PURITY OF MATERIALS:

Not given.

ESTIMATED ERROR:

Not given.

REFERENCES:

1. Trendafelov, D.; Kovachev, Ts.; Bakalov, V. *Izv. Otd. Khim. Nauki, Bulg. Akad. Nauk.* 1971, 4, 643.

COMPONENTS:

1. Magnesium sulfite; $MgSO_3$; [7757-88-2]
2. Magnesium sulfate; $MgSO_4$; [7487-88-9]
3. Water; H_2O; [7732-18-5]

ORIGINAL MEASUREMENTS:

Kovachev, Ts.; Bakalov, V.; Trendafelov, D.

Khim. Ind. (Sofia) 1970, *42*, 209-11.

EXPERIMENTAL VALUES (continued):

t/°C	$MgSO_4$ mass %	$MgSO_3$ mass %	$m(MgSO_3)$ mol kg^{-1} [a]	Solid phase
55	0.0	0.81	0.078	$MgSO_3.3H_2O$
	9.43	1.23	0.132	"
	14.45	1.27	0.144	"
	19.27	1.24	0.149	"
	23.63	1.10	0.140	"
	28.30	0.90	0.122	"
	33.36	0.66	0.096	"
	33.92	0.65	0.095	$MgSO_3.3H_2O + MgSO_4.6H_2O$
75	0.0	0.66	0.064	$MgSO_3.3H_2O$
	10.95	1.25	0.136	"
	17.50	1.34	0.158	"
	23.18	1.26	0.160	"
	28.50	1.07	0.146	"
	32.86	0.85	0.123	"
	36.73	0.69	0.106	"
	36.88	0.67	0.103	$MgSO_3.3H_2O + MgSO_4.H_2O$

[a] Calculated by the compiler.

COMPONENTS:

1. Magnesium sulfite; $MgSO_3$; [7757-88-2]
2. Magnesium sulfate; $MgSO_4$; [7487-88-9]
3. Water; H_2O; [7732-18-5]

ORIGINAL MEASUREMENTS:

Nývlt, J.; Rychlý, R.; Kricková, J.

Chem. Prům. 1977, 27, 552-6.

VARIABLES:

Temperature: 303 - 318 K
Concentration of $MgSO_4$

PREPARED BY:

B. Engelen

EXPERIMENTAL VALUES:

The authors report the solubility of $MgSO_3.6H_2O$ [13446-29-2] in water and in a solution containing 10 mass % of $MgSO_4$. The experimental data are given only in a graph, from which the following figures are estimated by the compiler.

t/°C	$MgSO_3$ mass %	$m(MgSO_3)$ mol kg^{-1}
	pure water	
31.0	0.599	0.058
32.1	0.681	0.065
32.6	0.597	0.058
33.6	0.662	0.064
36.8	0.714	0.069
37.7	0.768	0.074
37.9	0.810	0.078
38.4	0.706	0.068
39.8	0.776	0.075
41.4	0.950	0.092
43.7	0.965	0.093
	with 10 mass % $MgSO_4$	
33.60	1.007	0.108
37.10	1.266	0.137
41.10	1.484	0.161

Reprinted by permission

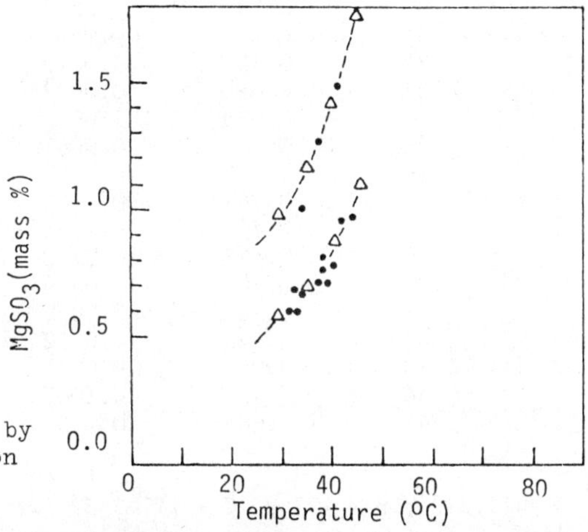

• Experimental values
Δ Values from least squares fit (authors)

(continued on next page)

AUXILIARY INFORMATION

METHOD APPARATUS/PROCEDURE:

Saturation method. Equilibrium was established after 5 hr. Aliquots of the supernatant solutions were analysed for magnesium sulfite complexometrically, iodometrically, and by formaldehyde titration against HCl.

SOURCE AND PURITY OF MATERIALS:

Not given.

ESTIMATED ERROR:

Not given.

REFERENCES:

COMPONENTS:	ORIGINAL MEASUREMENTS:
1. Magnesium sulfite; $MgSO_3$; [7757-88-2]	Nývlt, J.; Rychlý, R.; Kricková, J.
2. Magnesium sulfate; $MgSO_4$; [7487-88-9]	*Chem. Prům.* 1977, 27, 552-6.
3. Water; H_2O; [7732-18-5]	

EXPERIMENTAL VALUES (continued):

The authors report that experimental data can be represented by $\log x = A + B/T + C.\log T$ (x = mole fraction; T = temperature in K)[a,b]. The values calculated from this equation by the authors are:

	without $MgSO_4$[a]		with 10 mass % $MgSO_4$[b]	
t/°C	$MgSO_3$ (mass %)	$m(MgSO_3)$[c] $mol\ kg^{-1}$	$MgSO_3$ (mass %)	$m(MgSO_3)$[c] $mol\ kg^{-1}$
30	0.595	0.0573	0.976	0.105
31	0.614	0.0591	1.008	0.109
32	0.635	0.0612	1.042	0.112
33	0.657	0.0633	1.079	0.116
34	0.680	0.0655	1.118	0.121
35	0.706	0.0681	1.160	0.125
36	0.733	0.0707	1.206	0.130
37	0.762	0.0735	1.255	0.135
38	0.793	0.0765	1.307	0.141
39	0.827	0.0798	1.363	0.147
40	0.863	0.0834	1.423	0.154
41	0.901	0.0871	1.487	0.161
42	0.943	0.0912	1.556	0.169
43	0.987	0.0955	1.629	0.177
44	1.034	0.1001	1.708	0.185
45	1.085	0.1050	1.793	0.195

[a] A = -411.3726, B = 17342, C = 141.3459 (without $MgSO_4$)

[b] A = -411.0565, B = 17342, C = 141.3459 (with 10 mass % of $MgSO_4$)

[c] Calculated by the compiler.

COMPONENTS:

1. Magnesium sulfite; $MgSO_3$; [7757-88-2]
2. Magnesium chloride; $MgCl_2$; [7786-30-3]
3. Water; H_2O; [7732-18-5]

ORIGINAL MEASUREMENTS:

Bakalov, V.D.; Kovachev, Ts.B.; Trendafelov, D.

Khim. Ind. (Sofia) 1971, *43*, 351-3.

VARIABLES:

Two temperatures: 348 and 358 K
Amount of $MgCl_2$

PREPARED BY:

B. Engelen

EXPERIMENTAL VALUES:

The authors report the solubility of $MgSO_3.3H_2O$ [19086-20-5] in water and in solutions of different $MgCl_2$ concentrations at 75 and 85°C.

t/°C	$MgCl_2$ mass %	$MgSO_3$ mass %	$m(MgSO_3)^a$ mol kg^{-1}
75	0.0	0.66	0.064
	10.60	0.52	0.056
	16.53	0.39	0.045
	22.47	0.28	0.035
	28.53	0.17	0.023
	32.07	0.16	0.022
	35.75	0.12	0.018
	37.98	0.10	0.015
85	0.0	0.62	0.060
	12.38	0.55	0.061
	20.62	0.35	0.042
	26.83	0.27	0.035
	33.07	0.17	0.024
	36.56	0.14	0.021
	39.26	0.13	0.021

[a] Calculated by the compiler.

AUXILIARY INFORMATION

METHOD APPARATUS/PROCEDURE:

Saturation method. Equilibrium was established by stirring the given solutions with excess of magnesium sulfite under N_2. Time not given, but is assumed to be the same (4 - 8 hr) as in a second paper by the authors (1). Mg^{2+} was determined complexometrically, SO_3^{2-} iodometrically. From these two figures the amounts of $MgSO_3$ and $MgCl_2$ were calculated by the authors.

SOURCE AND PURITY OF MATERIALS:

Not given.

ESTIMATED ERROR:

Not given.

REFERENCES:

1. Trendafelov, D.; Kovachev, Ts.; Bakalov, V. *Izv. Otd. Khim. Nauki, Bulg. Akad. Nauk.* 1971, 4, 643.

COMPONENTS:

1. Magnesium sulfite; $MgSO_3$; [7757-88-2]
2. Sodium sulfite; Na_2SO_3; [7757-83-7]
3. Water; H_2O; [7732-18-5]

ORIGINAL MEASUREMENTS:

Trendafelov, D.; Kovachev, Ts.; Bakalov, V.

Izv. Otd. Khim. Nauki, Bulg. Akad. Nauk. 1971, *4*, 643-52.

VARIABLES:

Three temperatures: 298, 308 and 328 K
Amount of Na_2SO_3

PREPARED BY:

B. Engelen

EXPERIMENTAL VALUES:

t/°C	Na_2SO_3 mass %	$MgSO_3$ mass %	$m(MgSO_3)$ mol kg^{-1a}	Solid phase
25	pure water	0.48	0.046	$MgSO_3.6H_2O$
	4.29	0.88	0.089	"
	8.15	1.07	0.113	"
	10.08	1.76	0.191	$MgSO_3.6H_2O.Na_2SO_3.7H_2O$
	12.94	0.45	0.050	"
	16.66	0.31	0.036	"
	19.28	0.24	0.029	"
	21.04	0.16	0.019	"
	21.42	0.15	0.018	$Na_2SO_3.7H_2O$+ $MgSO_3.6H_2O.Na_2SO_3.7H_2O$
35	pure water	0.66	0.064	$MgSO_3.6H_2O$
	4.80	1.20	0.122	"
	8.42	1.47	0.156	"
	10.67	1.61	0.176	$MgSO_3.6H_2O$ + $Na_2SO_3.MgSO_3.3H_2O$
	11.33	1.21	0.133	$Na_2SO_3.MgSO_3.3H_2O$
	14.17	0.64	0.072	"
	17.08	0.43	0.050	"
	19.30	0.24	0.029	"
	21.31	0.17	0.021	"
	23.89	0.09	0.011	"
	25.21	0.06	0.008	"
	27.10	0.04	0.005	$Na_2SO_3.MgSO_3.3H_2O$ + Na_2SO_3

(continued on next page)

AUXILIARY INFORMATION

METHOD APPARATUS/PROCEDURE:

Saturation method. Equilibrium was established after 4 - 8 hr by stirring the given solutions with excess of magnesium sulfite under N_2. Mg^{2+} was determined complexometrically, SO_3^{2-} iodometrically. From these two figures the amounts of $MgSO_3$ and Na_2SO_3 were calculated by the authors.

SOURCE AND PURITY OF MATERIALS:

Not given.

ESTIMATED ERROR:

Not given.

REFERENCES:

COMPONENTS:	ORIGINAL MEASUREMENTS:
1. Magnesium sulfite; $MgSO_3$; [7757-88-2] 2. Sodium sulfite; Na_2SO_3; [7757-83-7] 3. Water; H_2O; [7732-18-5]	Trendafelov, D.; Kovachev, Ts.; Bakalov, V. *Izv. Otd. Khim. Nauki, Bulg. Akad. Nauk.* 1971, 4, 643-52.

EXPERIMENTAL VALUES (continued):

t/°C	Na_2SO_3 mass %	$MgSO_3$ mass %	$m(MgSO_3)$ mol kg^{-1} [a]	Solid phase
55	pure water	0.81	0.078	$MgSO_3.3H_2O$
	4.81	1.09	0.111	$MgSO_3.3H_2O$-type[b]
	9.07	0.80	0.085	"
	12.53	0.38	0.042	Na_2SO_3-type[c]
	15.87	0.23	0.026	"
	19.50	0.15	0.018	"
	21.92	0.11	0.014	"
	24.53	0.05	0.006	"
	26.66	0.04	0.005	"
	27.13	0.03	0.004	"
	27.18	0.02	0.003	"

[a] Calculated by the compiler.

[b] Na_2SO_3-$MgSO_3$ mixed crystals of type $MgSO_3.3H_2O$.

[c] Na_2SO_3-$MgSO_3$ mixed crystals of type Na_2SO_3.

COMPONENTS:

1. Magnesium sulfite; $MgSO_3$; [7757-88-2]
2. Magnesium sulfate; $MgSO_4$; [7487-88-9]
3. Sulfur dioxide; SO_2; [7446-09-5]
4. Water; H_2O; [7732-18-5]

ORIGINAL MEASUREMENTS:

Kuz'minykh, I.N.; Babushkina, M.D.

Zh. Prikl. Khim. 1957, *30*, 466-9; *J. Appl. Chem. USSR (Eng. Transl.)* 1957, *30*, 495-8.

VARIABLES:

Four temperatures: 283 - 444 K
Concentration of $MgSO_4$
Concentration of SO_2

PREPARED BY:

B. Engelen, H.D. Lutz

EXPERIMENTAL VALUES:

The authors report the SO_2 partial pressure over aqueous solutions saturated with solid magnesium sulfite at various temperatures, various concentrations of sulfurous acid, and small amounts of magnesium sulfate.

t/°C	p_{SO_2}	p_{SO_2}[a]	pH	SO_2 (mass %)		$MgSO_4$	$MgSO_3$[c]	
	mm Hg	10^{-3}bar	at 20°C	total	free[b]	mass %	mass %	m/mol kg^{-1}[a]
10	2.05	2.73	3.8	9.62	4.48	1.386	8.37	0.950
	4.95	6.60	3.5	13.04	6.24	0.486	11.08	1.319
	7.61	10.14	3.3	14.94	6.84	0.673	13.2	1.634
	17.51	23.34	3.25	18.35	9.13	0.396	15.0	1.975
	25.53	34.03	3.0	19.89	10.13	1.25	15.9	2.181
	90.0	120.0	-	24.56	12.36	0.168	19.89	2.970
23.3	1.29	1.72	4.1	7.05	2.74	0.099	6.97	0.753
	1.57	2.09	4.1	8.48	3.96	0.709	7.29	0.812
	3.11	4.15	3.88	9.22	4.41	1.014	7.89	0.878
	7.29	9.72	3.5	12.46	5.75	0.598	11.24	1.291
	13.56	18.08	3.45	14.11	6.99	0.653	11.63	1.411
	22.35	29.79	3.26	17.2	7.94	0.445	15.09	1.946
	37.14	49.51	-	19.1	9.34	0.881	15.57	2.138
	38.14	50.84	3.0	19.1	9.3	0.414	15.97	2.133
	47.5	63.32	-	20.2	9.92	1.028	16.75	2.308
	72.18	96.22	2.55	21.86	10.4	0.619	18.53	2.655
	81.16	108.19	2.71	22.66	10.92	0.297	19.18	2.752

[a,b,c] See the following page.

(continued on next page)

AUXILIARY INFORMATION

METHOD APPARATUS/PROCEDURE:

Aqueous solutions of sulfurous acid were kept over magnesium sulfite for several days. Equilibrium pressure of SO_2 was determined dynamically in a special apparatus (1) by passing oxygen-free N_2 gas through the solution-precipitate mixture and analysing the moist inert gas for SO_2 with iodine. Gas volume was measured in an aspirator. Further details are given in ref. (1).

SOURCE AND PURITY OF MATERIALS:

Not given.

ESTIMATED ERROR:

Data given are the results of several experiments which are in satisfactory agreement (given by the authors).

REFERENCES:

1. Kuz'minykh, I.N.; Kuznetsova, A.G. *Zh. Prikl. Khim.* 1954, 27, 816.

COMPONENTS:	ORIGINAL MEASUREMENTS:
1. Magnesium sulfite; $MgSO_3$; [7757-88-2] 2. Magnesium sulfate; $MgSO_4$; [7487-88-9] 3. Sulfur dioxide; SO_2; [7446-09-5] 4. Water; H_2O; [7732-18-5]	Kuz'minykh, I.N.; Babushkina, M.D. *Zh. Prikl. Khim.* 1957, *30*, 466-9; *J. Appl. Chem. USSR (Eng. Transl.)* 1957, *30*, 495-8.

EXPERIMENTAL VALUES (continued):

t/°C	p_{SO_2} mm Hg	p_{SO_2}[a] 10^{-3}bar	pH at 20°C	SO_2 (mass %) total	SO_2 (mass %) free[b]	$MgSO_4$ mass %	$MgSO_3$[c] mass %	$MgSO_3$[c] m/mol kg^{-1}[a]
48.3	1.33	1.77	4.25	4.36	2.06	1.18	3.75	0.388
	2.88	3.84	4.1	6.37	2.66	0.545	6.06	0.643
	9.5	12.66	3.8	10.2	4.64	0.283	9.10	1.025
	13.2	17.60	3.6	11.78	5.67	0.566	9.99	1.160
	16.36	21.81	3.5	13.04	6.33	0.452	10.95	1.301
	26.07	34.75	-	15.15	7.45	0.188	12.55	1.546
	28.7	38.26	3.3	15.0	7.64	0.282	12.03	1.474
	35.2	46.92	3.1	16.65	8.2	1.51	13.8	1.777
	93.1	124.10	3	19.8	9.7	0.282	16.38	2.226
70.6	4.12	5.49	4.25	6.37	3.82	0.545	5.76	0.440
	19.86	26.47	3.85	10.19	4.19	0.357	8.61	0.973
	27.4	36.52	3.6	11.77	5.68	0.63	9.95	1.157
	42.8	57.05	3.5	13.1	6.38	0.047	10.98	1.298
	92	122.6	3.2	16.44	8.06	1.81	13.7	1.763

[a] Calculated by the compilers.

[b] Excess over the amount required to form the monosulfite.

[c] Calculated from total and free SO_2 by the authors.

COMPONENTS:	ORIGINAL MEASUREMENTS:
1. Magnesium sulfite; $MgSO_3$; [7757-88-2] 2. Magnesium sulfate; $MgSO_4$; [7487-88-9] 3. Magnesium hydrogen sulfite; $Mg(HSO_3)_2$; [13774-25-9] 4. Water; H_2O; [7732-18-5]	Pinaev, V.A. *Zh. Prikl. Khim.* 1964, *37*, 1361-3; *J. Appl. Chem. USSR (Eng. Transl.)* 1964, *37*, 1353-5.
VARIABLES:	PREPARED BY:
Three temperatures: 313, 323 and 333 K Concentration of $MgSO_4$ Concentration of $Mg(HSO_3)_2$	B. Engelen, H.D. Lutz

EXPERIMENTAL VALUES:

The author gives solubility diagrams for magnesium sulfite in the systems $MgSO_3$-$MgSO_4$-H_2O (Fig. 1) and $MgSO_3$-$Mg(HSO_3)_2$-H_2O (Fig. 2) at 40, 50 and 60°C.

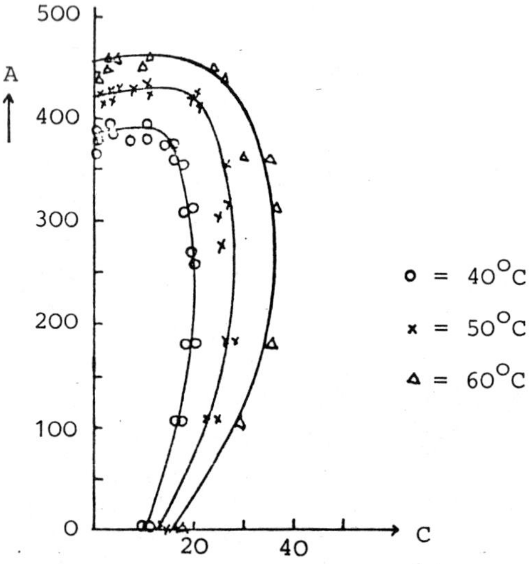

Fig. 1

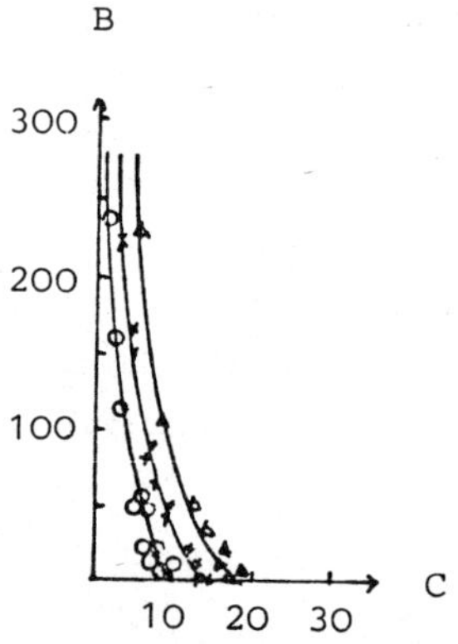

Fig. 2

A: g $MgSO_4/dm^3$ B: g $Mg(HSO_3)_2/dm^3$

C: g $MgSO_3/dm^3$

(continued on next page)

AUXILIARY INFORMATION

METHOD APPARATUS/PROCEDURE:	SOURCE AND PURITY OF MATERIALS:
Saturation method. Equilibrium (constancy of the phase composition) was obtained after several hours.	Magnesium hydrogen sulfite was prepared from MgO and gaseous SO_2. $MgSO_4$, $MgSO_3.6H_2O$ and MgO used were all commercial products.
	ESTIMATED ERROR:
	Solubility: ± 3% Temperature: ±0.2 K (author)
	REFERENCES:

COMPONENTS:	ORIGINAL MEASUREMENTS:
1. Magnesium sulfite; $MgSO_3$; [7757-88-2] 2. Magnesium sulfate; $MgSO_4$; [7487-88-9] 3. Magnesium hydrogen sulfite; $Mg(HSO_3)_2$; [13774-25-9] 4. Water; H_2O; [7732-18-5]	Pinaev, V.A. *Zh. Prikl. Khim.* 1964, *37*, 1361-3; *J. Appl. Chem. USSR (Eng. Transl.)* 1964, *37*, 1353-5.

EXPERIMENTAL VALUES (continued):

For pure water and several concentration ranges of $MgSO_4$ the author also gives numerical data on the solubility of magnesium sulfite (Fig. 1).

$MgSO_4$ g dm^{-3}	$MgSO_3$ g dm^{-3}			$MgSO_3$[a] mol dm^{-3} x 10^3		
	40°C	50°C	60°C	40°C	50°C	60°C
pure water	10.2-10.9	12.5-13.0	16.0-17.4	97.7-104.4	120-125	153-167
175 - 200	18-20	26-28	34-35	172-192	249-268	326-335
200 - 300	19-20	26-28	34-35	182-192	249-268	326-335
eutonic point	15.0	20.5	24.5	144	196.4	234.7
saturated soln.	10	10	10	96	96	96

The following numerical data were estimated, from the diagram given in Fig. 2, by the compilers.

t/°C	$Mg(HSO_3)_2$ g dm^{-3}	$MgSO_3$[b] g dm^{-3}	$MgSO_3$[c] g dm^{-3}	$(MgSO_3)$[c] mol dm^{-3} x 10^3
40	0	10.8	10.8	103.5
	50	5.5	33.5	320.9
	100	3.8	59.8	572.8
	150	2.7	86.7	830.4
	200	1.7	113.7	1089.0
	250	1.2	141.2	1352.4
50	0	13.0	13.0	124.6
	50	8.6	36.6	350.6
	100	6.6	62.6	599.6
	150	4.7	88.7	849.6
	200	3.8	115.8	1109.2
	250	2.7	142.7	1366.8
60	0	16.4	16.4	157.1
	50	11.6	39.6	379.3
	100	8.3	64.3	615.9
	150	6.8	90.8	869.7
	200	5.5	117.5	1125.5
	250	4.5	144.5	1384.1

[a] Calculated by the compilers.

[b] Determined analytically as the amount of SO_3^{2-} present in addition to HSO_3^- after establishing equilibrium.

[c] Total amount of dissolved magnesium sulfite ($MgSO_3$ + $Mg(HSO_3)_2$ as $MgSO_3$, calculated by the compilers.

COMPONENTS:

1. Magnesium sulfite; $MgSO_3$; [7757-88-2]
2. Magnesium hydrogen sulfite; $Mg(HSO_3)_2$; [13774-25-9]
3. Magnesium sulfate; $MgSO_4$; [7487-88-9]
4. Water; H_2O; [7732-18-5]

ORIGINAL MEASUREMENTS:

Semishin, V.I.; Abramov, I.I.; Vorotnitskaya, L.T.

Izv. Vyssh. Uchebn. Zaved., Khim. Khim. Technol. 1959, 2, 834-9.

VARIABLES:

Temperature: 308 - 343 K
Concentration of $Mg(HSO_3)_2$
Amount of $MgSO_4$ pH: 4 - 9

PREPARED BY:

B. Engelen

EXPERIMENTAL VALUES:

t/°C	pH^a	$Mg(HSO_3)_2{}^b$ mass %	$MgSO_3{}^c$ mass %	$MgSO_3{}^d$ mass %	$m(MgSO_3)^d$ mol kg^{-1}
		with 10 mass % $MgSO_4$			
35	6.20	0.40	0.751	0.97	0.105
	5.36	1.14	0.679	1.32	0.143
	4.75	2.21	0.643	1.88	0.207
	4.40	2.91	0.493	2.12	0.235
	3.82	3.86	0.397	2.56	0.286
	4.25	4.42	0.578	3.05	0.344
40	6.40	0.40	0.867	1.09	0.118
	5.59	1.14	0.837	1.48	0.161
	4.80	2.21	0.901	2.14	0.236
	4.75	2.91	0.716	2.34	0.260
	4.21	3.86	0.680	2.84	0.318
	4.70	4.42	0.729	3.21	0.363
45	6.58	0.40	1.076	1.30	0.141
	5.80	1.14	0.927	1.57	0.171
	4.82	2.21	0.964	2.20	0.243
	4.87	2.91	1.015	2.64	0.294
	4.37	3.86	0.932	3.09	0.348
	4.72	4.42	0.903	3.38	0.382

(continued on next page)

AUXILIARY INFORMATION

METHOD APPARATUS/PROCEDURE:

Saturation method. Equilibrium was established after 1.5 hr in a thermostatically controlled flask. Amounts of HSO_3^- and SO_3^{2-} were determined by a combination of iodometric and acidimetric titration. pH was determined potentiometrically.

SOURCE AND PURITY OF MATERIALS:

Magnesium sulfite was prepared by bubbling SO_2 through a suspension of MgO in water. The precipitated $MgSO_3.6H_2O$ was analysed for Mg^{2+}, SO_3^{2-}, and SO_4^{2-} content. $Mg(HSO_3)_2$ solutions were obtained and analysed in the same way as above.

ESTIMATED ERROR:

Deviations in temperature and pH (authors) are ±0.01 K and ±0.2- 0.4, respectively.

REFERENCES:

COMPONENTS:	ORIGINAL MEASUREMENTS:
1. Magnesium sulfite; $MgSO_3$; [7757-88-2] 2. Magnesium hydrogen sulfite; $Mg(HSO_3)_2$; [13774-25-9] 3. Magnesium sulfate; $MgSO_4$; [7487-88-9] 4. Water; H_2O; [7732-18-5]	Semishin, V.I.; Abramov, I.I.; Vorotnitskaya, L.T. *Izv. Vyssh. Uchebn. Zaved., Khim. Khim. Technol.* 1959, 2, 834-9.

EXPERIMENTAL VALUES (continued):

t/°C	pH[a]	$Mg(HSO_3)_2$[b] mass %	$MgSO_3$[c] mass %	$MgSO_3$[d] mass %	$m(MgSO_3)$[d] mol kg^{-1}
			with 10 mass % $MgSO_4$		
50	7.62	0.40	1.205	1.43	0.155
	5.95	1.14	1.340	1.98	0.217
	4.86	2.21	1.073	2.31	0.255
	5.00	2.91	1.072	2.70	0.301
	4.47	3.86	1.091	3.25	0.366
	4.75	4.42	1.198	3.67	0.417
55	8.00	0.40	1.351	1.57	0.171
	6.03	1.14	1.729	2.37	0.260
	5.20	2.21	1.437	2.67	0.297
	5.12	2.91	1.390	3.02	0.338
	4.62	3.86	1.388	3.55	0.401
	4.87	4.42	1.579	4.05	0.462
60	8.12	0.40	1.520	1.74	0.190
	6.36	1.14	2.050	2.69	0.297
	5.45	2.21	1.787	3.02	0.337
	5.12	2.91	1.596	3.22	0.361
	4.75	3.86	1.725	3.89	0.441
	4.93	4.42	2.014	4.49	0.515
70	8.70	0.40	2.078	2.30	0.252
	6.40	1.14	2.676	3.31	0.368
	5.60	2.21	2.661	3.90	0.439
	5.20	2.91	2.599	4.23	0.479
	4.90	3.86	2.627	4.79	0.549
	4.95	4.42	2.764	5.24	0.606

[a] pH determined at 20 - 25°C.

[b] Concentration of the solution in which solid magnesium sulfite has been dissolved.

[c] Determined analytically as the amount of SO_3^{2-} present in addition to HSO_3^- after establishing equilibrium.

[d] Total amount of dissolved magnesium sulfite ($MgSO_3$ + $Mg(HSO_3)_2$ as $MgSO_3$), calculated by the compiler.

COMPONENTS:

1. Magnesium sulfite; $MgSO_3$; [7757-88-2]
2. Magnesium chloride; $MgCl_2$; [7786-30-3]
3. Magnesium hydrogen sulfite; $Mg(HSO_3)_2$; [13774-25-9]
4. Water; H_2O; [7732-18-5]

ORIGINAL MEASUREMENTS:

McIlroy, R.A.

Tappi 1973, *56*, 79-82.

VARIABLES:

One temperature: 344 K
Amount of $MgCl_2$

PREPARED BY:

B. Engelen

EXPERIMENTAL VALUES:

The author reports the solubility of $MgSO_3.3H_2O$ [19086-20-5] in solutions of different $MgCl_2$ concentrations at 160°F (71.1°C). The pH values of the solutions are also given.

$MgCl_2$ mass %	total SO_2 g/100 cm^3	free SO_2[a] g/100 cm^3	pH	$MgSO_3$[b] mass %	$MgSO_3$[b] m/mol kg^{-1}[c]
19.8	0.30	0.00	7.8	0.42	0.050
35.7	0.32	0.02	3.4	0.34	0.050

a Excess over the amount necessary to form the monosulfite.

b Calculated from total and free SO_2 content by the author.

c Calculated by the compiler.

AUXILIARY INFORMATION

METHOD APPARATUS/PROCEDURE:

Saturation method. MgO was added to a solution of $MgCl_2$ containing $Mg(HSO_3)_2$. The solutions were analysed for total and free SO_2 by a combination of iodometric and acidimetric titration. Chloride was determined by precipitation titration as AgCl and magnesium complexometrically.

SOURCE AND PURITY OF MATERIALS:

The $MgCl_2$-$Mg(HSO_3)_2$ solutions were prepared by either adding $MgCl_2.6H_2O$ to a $Mg(HSO_3)_2$ solution or adding MgO to a nearly saturated $MgCl_2$ solution, and then bubbling SO_2 through the solution.

ESTIMATED ERROR:

Not given.

REFERENCES:

COMPONENTS:	EVALUATOR
1. Calcium sulfite; $CaSO_3$; [10257-55-3] 2. Water; H_2O; [7732-18-5]	H.D. Lutz, Dept. of Chemistry, University of Siegen, FR Germany. December 1983.

CRITICAL EVALUATION:

Calcium sulfite crystallizes from aqueous solutions in the form of the hemihydrate, $CaSO_3.\frac{1}{2}H_2O$ [29501-28-8] (1-3). A dihydrate, $CaSO_3.2H_2O$ [10035-03-7], claimed in the older literature (4), could not be confirmed (1-3,5). Very recently a tetrahydrate, $CaSO_3.4H_2O$ [72878-03-6], has been reported (6,7). This crystallizes in the presence of sodium citrate (6) or nitrilotri(methylene phosphonic acid) (7).

Numerical data on the solubility of calcium sulfite, i.e. $CaSO_3.\frac{1}{2}H_2O$, have been given by many authors (7-42). The published figures, however, differ to a great extent. This is probably caused by the very great tendency of calcium sulfite to form supersaturated solutions (7) and by the existence of several different modifications of $CaSO_3.\frac{1}{2}H_2O$ (8,9,43). The difficulty in preparing sulfate-free samples of calcium sulfite and the tendency of $CaSO_3.\frac{1}{2}H_2O$ to form solid solutions with calcium sulfate (44,45) must further be taken into account.

SOLUBILITY OF CALCIUM SULFITE HEMIHYDRATE IN PURE WATER

The values of the solubility of calcium sulfite in pure water at ambient temperature (288 - 303 K) given in the literature (7-20,42) vary from 8.7×10^{-5} (14) to 3.3×10^{-3} (7) mol dm^{-3} (molarity scale). The data reported by Farnell (15), 2.0×10^{-4} mol dm^{-3} at 303 K, Weisberg (10), 3.58×10^{-4} mol dm^{-3} at 291 K, Sano (20), 5.21×10^{-4} mol dm^{-3} at 298 K, Cohen *et al.* (8-9), 7.0×10^{-4} mol dm^{-3} at 298 K, Rengemo *et al.* (17), 7.6×10^{-4} mol dm^{-3} at 298.2 K, Marusawa (13), 7.91×10^{-4} mol dm^{-3} at 291 K, all on the molarity scale, and by Van der Linden (12), 4.10×10^{-4} mol kg^{-1} (molality scale) at 303 K, seem to be nearest to the true value.

The solubility product of calcium sulfite was first determined by Marusawa (13,21) to be 4.4×10^{-7} mol^2 dm^{-6} (molarity scale) at 291.2 K, on the assumption of incomplete dissociation of the dissolved calcium sulfite. From Debye-Hückel theory, a value of $3.66(2) \times 10^{-7}$ mol^2 dm^{-6} can be obtained from the original data given by Marusawa (13). In 1958, the activity solubility product was calculated by Rengemo *et al.* (17) from equilibrium studies of the reaction, $CaSO_4.2H_2O(s)+SO_3^{2-} \rightleftharpoons CaSO_3.\frac{1}{2}H_2O(s)+SO_4^{2-}+1\frac{1}{2}H_2O$, at 298.2 K in aqueous $NaClO_4$ solutions of various ionic strengths to be 3.1 $(\pm 1.5) \times 10^{-7}$ mol^2 dm^{-6} (molarity scale).

RECOMMENDED VALUES

The solubility of $CaSO_3.\frac{1}{2}H_2O$ in water at 298.2 K (25°C), on the molarity scale, is 4.5 $(\pm 1.0) \times 10^{-4}$ mol dm^{-3} (0.054 ± 0.012 g $CaSO_3/dm^3$).

The solubility product, based on the activities, is 3.1 $(\pm 1.5) \times 10^{-7}$ mol^2 dm^{-6} (molarity scale).

COMPONENTS:	EVALUATOR:
1. Calcium sulfite; $CaSO_3$; [10257-55-3] 2. Water; H_2O; [7732-18-5]	H.D. Lutz, Dept. of Chemistry, University of Siegen, FR Germany. December 1983.

CRITICAL EVALUATION: (continued)

The solubility of $CaSO_3.1/2H_2O$ decreases slightly with increasing temperature (12-13,16,19,22). The following equation, fitted (by evaluator) from data given by Van der Linden (12) and Bobrovnik *et al*. (22), who are in relatively good agreement, is recommended.

$$\log S = -15.367 + 1155.67/T + 3.290 \log T \qquad (1)$$

with S = solubility of $CaSO_3.1/2H_2O$ on the molality scale (mol kg^{-1}) and T = temperature (K). The equation is valid for the range 293 - 373 K. A graph derived from this equation is shown in Fig. 1.

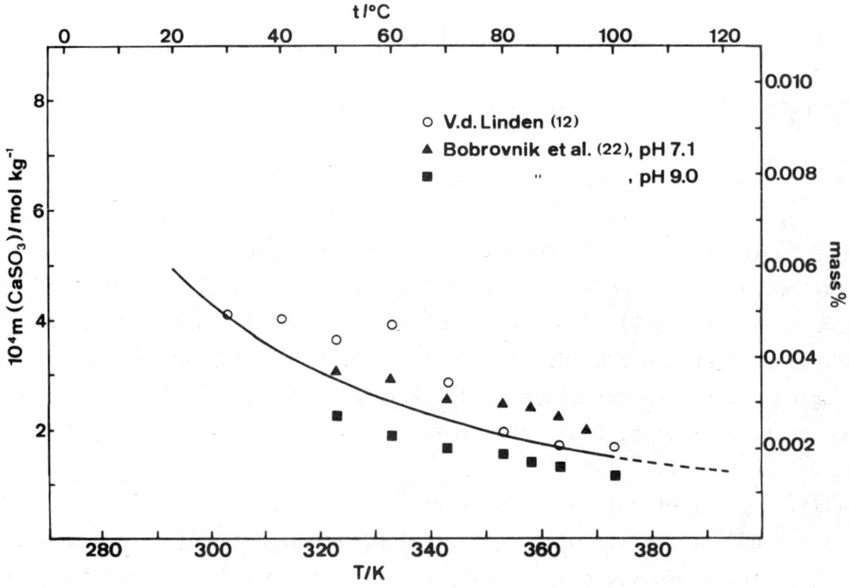

Fig 1. Solubility of calcium sulfite, $CaSO_3.1/2H_2O$, in pure water (equation 1)

SOLUBILITY IN THE SYSTEM $CaSO_3$-SO_2-H_2O

The solubility of calcium sulfite increases very much with increasing SO_2 content of the solution (13,23-26). The numerical data given for ambient temperature and low SO_2 concentrations (up to 2.0 mol kg^{-1}), e.g. the results reported by Mebane *et al*. (25), Conrad *et al*. (27), and Kuz'minykh *et al*. (30), all on the molality scale, and by Marusawa (13) and Engelhardt (34), both on the molarity scale, are in good agreement. The following equation, fitted (by evaluator) with values given by Conrad *et al*. (27) and Kuz'minykh *et al*. (30), is recommended for the solubility of calcium sulfite in aqueous sulfurous acid solutions at 298 K (molality scale, mol kg^{-1}),

$$S = 0.460 \times m(SO_2\ \text{tot}) - 0.026 \times m^2\ (SO_2\ \text{tot}) \qquad (2)$$

COMPONENTS:	EVALUATOR:
1. Calcium sulfite; $CaSO_3$; [10257-55-3] 2. Water; H_2O; [7732-18-5]	H.D. Lutz, Dept. of Chemistry, University of Siegen, FR Germany. December 1983.

CRITICAL EVALUATION: (continued)

This means that the ratio of dissolved calcium sulfite to total SO_2 concentration is nearly constant. A graph derived from this equation is shown in Fig. 2. The equation is valid for SO_2 concentrations up to 2.0 mol kg^{-1} SO_2 at equilibrium pressure.

At higher SO_2 concentrations (up to 5.5 mol kg^{-1}) the data reported in the literature, e.g. by Conrad *et al.* (27), Simon *et al.* (33), and Humm (26), differ, as shown in Fig. 2. Data obtained by extrapolation of the graph (Fig. 2) seem to be more reliable.

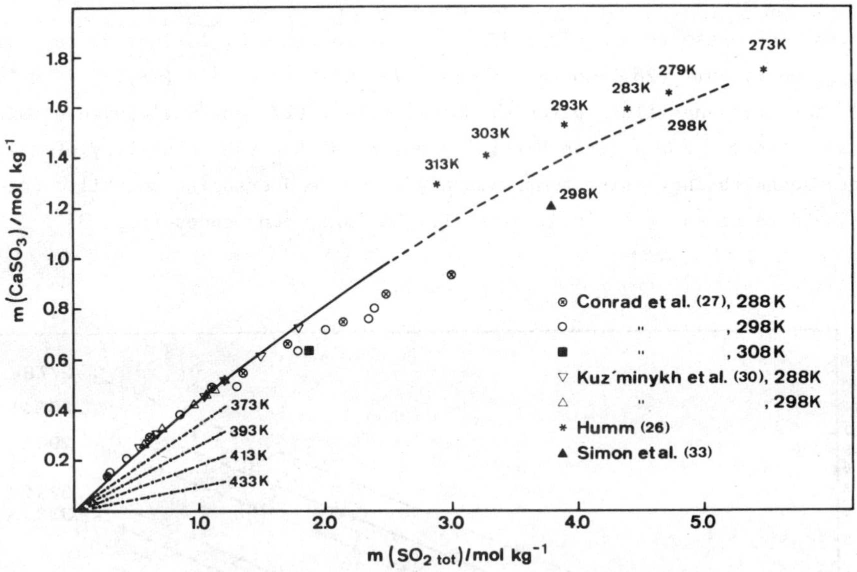

Fig. 2 Solubility of calcium sulfite in aqueous sulfurous acid solutions
----- recommended for 288 - 308 K, see equation (2),
-•-•- from data given by Marusawa (23), Otuka (29), and Engelhardt (34)

The solubility of calcium sulfite at an equilibrium partial pressure of sulfur dioxide equal to 1 bar (= 10^5 Pa) at ambient temperature has been determined by several authors, but the values obtained differ greatly, e.g. 1.45 (26), 1.22 (33), and 0.76 (27) mol kg^{-1} (molality scale) at concentrations of total SO_2 of 3.6, 3.79, and 2.4 mol kg^{-1}, respectively. Extrapolation of the data reported by Kuz'minykh *et al.* (30) to p_{SO_2} = 1 bar leads to the values $m(CaSO_3)$ = 0.98 mol kg^{-1} and $m(SO_2\ tot)$ = 2.46 mol kg^{-1}, which are recommended.

COMPONENTS:

1. Calcium sulfite; $CaSO_3$; [10257-55-3]
2. Water; H_2O; [7732-18-5]

EVALUATOR:

H.D. Lutz,
Dept. of Chemistry,
University of Siegen,
FR Germany.
December 1983.

CRITICAL EVALUATION: (continued)

Numerical data on the solubility of calcium sulfite in aqueous sulfurous acid solutions below and above ambient temperature (23-32) indicate that the solubility of calcium sulfite decreases with increasing temperature at a given SO_2 concentration, but the temperature coefficient is relatively small. Therefore the equation given above is approximately valid, too, for temperatures other than 298 K, especially in the range from 273 to 333 K. Data at temperatures above 373 K have also been reported (23,28,29,34). Isotherms derived from these values (by evaluator) are included in Fig. 2. There are some indications that the ratio $m(CaSO_3)/m(SO_2 \text{ tot})$ of aqueous sulfurous acid solutions saturated with $CaSO_3$ increases from 273 to 313 K and decreases from 313 to 433 K.

Data on the partial pressure of sulfur dioxide over saturated solutions of calcium sulfite are given by Humm (26), Conrad *et al.* (27), Gishler *et al.* (28), and Kuz'minykh *et al.* (30), but only the values given by Conrad *et al.* (27) and Kuz'minykh *et al.* (30) agree to some extent. For a given partial pressure of SO_2 the solubility of calcium sulfite decreases with increasing temperature due to the decreasing solubility of sulfur dioxide. The data given by Kuz'minykh *et al.* (30) are recommended (Fig. 3).

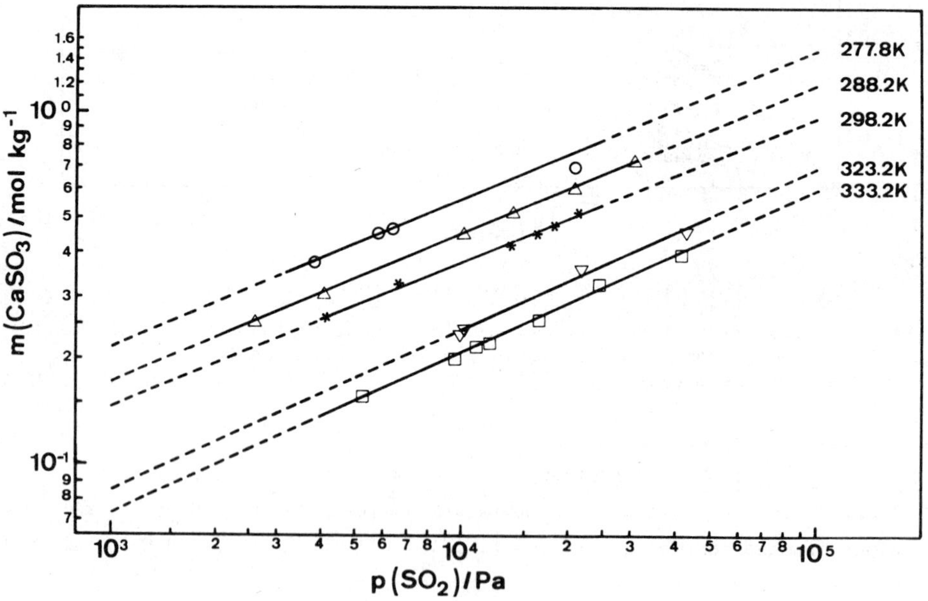

Fig. 3 Solubility of calcium sulfite in aqueous sulfurous acid solutions and partial pressure of sulfur dioxide (30)

The solubility products, $K_1 = [Ca^{2+}] \times [HSO_3^-]^2/p_{SO_2}$ and $K_2 = [Ca^{2+}] \times [HSO_3^-]^2/[H_2SO_3]$, of $CaSO_3.1/2H_2O$ in the presence of gaseous SO_2 at 298.2, 308.2, and 348.2 K have been

COMPONENTS:	EVALUATOR:
1. Calcium sulfite; $CaSO_3$; [10257-55-3] 2. Water; H_2O; [7732-18-5]	H.D. Lutz, Dept. of Chemistry, University of Siegen, FR Germany. December 1983.

CRITICAL EVALUATION: (continued)

determined by Nilsson *et al.* (31,32) from equilibrium studies of the reactions $SO_2(g)$ + H_2O H_2SO_3 and $CaSO_3.1/2H_2O$ + $SO_2(g)$ + $1/2H_2O$ Ca^{2+} + $2HSO_3^-$ at different concentrations of $NaClO_4$ as ionic medium. The data obtained are:

$m(NaClO_4)$/mol kg^{-1}		1			3.5	
T/K	298.2	308.2	348.2	298.2	308.2	348.2
K_1/$mol^3 kg^{-3} bar^{-1}$	1.208	0.543	0.048	0.294	0.121	0.013
K_2/$mol^2 kg^{-2}$	0.871	0.741	0.170	0.263	0.151	0.041

SOLUBILITY OF CALCIUM SULFITE IN THE PRESENCE OF $CaSO_4$, NH_4NO_3, NaCl, SODIUM PHOSPHATE, $NaClO_4$, HCl, H_3PO_4, NaOH, SUCROSE, GLUCOSE, XYLOSE, ACETIC ACID, CITRIC ACID, LIGNOSULFONIC ACID, SODIUM FORMATE, SODIUM ACETATE, SODIUM CITRATE, ALCOHOL, AND SEA-WATER.

In the system $CaSO_3$-$CaSO_4$-H_2O (12-14,16,37) the solubility of $CaSO_3.1/2H_2O$ decreases with increasing $CaSO_4$ content to approximately half the amount soluble in pure water for solutions saturated with gypsum, $CaSO_4.2H_2O$, (12,14,16). This behaviour has been observed for all temperatures studied, *viz.* 293.2 K (16), 298.2 K (14), 303.2 K (12), 313.2 K (12,14), 323.2 K (12), 333.2 K (12,14), 343.2 K (12), 353.2 K (12,14), 363.2 K (12), and 373.2 K (12). The solubility of calcium sulfite in solutions saturated with gypsum decreases with increasing temperature (12), as found for pure water.

Experimental data on the solubility of $CaSO_3.1/2H_2O$ in the presence of ammonium nitrate at 303 K are given by Dubovaya *et al.* (38). The figures indicate that the solubility at first increases and then decreases with increasing NH_4NO_3 content due to ionic strength effects. The same is obviously true in solutions containing sodium chloride (39). A relatively large increase in the solubility of calcium sulfite has been found (by Wurz *et al.* (16)) in the presence of Na_3PO_4, but the reported numerical data given for one concentration, $c(Na_3PO_4)$ = 0.033 mol dm^{-3}, and four temperatures (293, 313, 333, and 353 K) cannot be directly compared with the solubility in pure water because of the altered pH. Data on the solubility in aqueous sulfurous acid solutions containing 1 and 3.5 mol kg^{-1} (molality) sodium perchlorate as ionic medium have been reported by Nilsson *et al.* (32) for 298.2, 308.2, and 348.2 K. The solubility of calcium sulfite in sea-water has been studied by Kurota *et al.* (37), but the figures seem to be too high.

The solubility of calcium sulfite increases greatly in the presence of acids. Numerical data are given for HCl (39), H_3PO_4 (25), acetic acid (16,20), and citric and lignosulfonic acid (16). The sodium salts of the organic acids under consideration also increase the solubility of calcium sulfite (16). The same has been found in the presence of sodium hydroxide (20).

COMPONENTS:	EVALUATOR:
1. Calcium sulfite; $CaSO_3$; [10257-55-3] 2. Water; H_2O; [7732-18-5]	H.D. Lutz, Dept. of Chemistry, University of Siegen, FR Germany. December 1983.

CRITICAL EVALUATION: (continued)

Numerical data on the solubility of calcium sulfite at different pH values have been given by several authors (16,18-20,22,39). The solubility increases with both increasing and decreasing pH (20). The minimum value of the solubility at pH 8.5 is found to be smaller than that of pure water (20,22).

The solubility of calcium sulfite is also affected by the presence of sugar (10-12,15,16,18,19,22,40). Numerical data have been reported on sucrose, glucose (12,16), and xylose (16). The solubility of $CaSO_3.1/2H_2O$ at first increases and then decreases with increasing sucrose content of the solution (10,22). The values obtained by Bobrovnik *et al.* (22) are recommended.

Arnal *et al.* (41) reported on the solubility of calcium sulfite in aqueous ethanol solutions. The value of 8×10^{-4} mol dm^{-3} (molarity scale) in a solution with 67.2 mass % ethanol seems to be too high.

REFERENCES

1. Matthews, F.W.; McIntosh, A.O. *Can. J. Res., Sect. B* 1948, *26*, 747.

2. Setoyama, K.; Takahashi, S.; Sekiya, M. *Sekko to Sekkai* 1976, *141*, 57.

3. Lutz, H.D.; El-Suradi, S. *Z. Anorg. Allg. Chem.* 1976, *425*, 134.

4. *Gmelins Handbuch der Anorganischen Chemie*, 8. Auflage, Springer Verlag, Heidelberg, 1974, Vol. 28, Part B 3, p 664.

5. Tsuyuki, N.; Kasai, J. *Nippon Kagaku Kaishi* 1976, (1), 59.

6. Matsuno, T.; Koishi, M. *Nippon Kagaku Kaishi* 1979, 1687.

7. McCall, M.T.; Tadros, M.E. *Colloids and Surfaces* 1980, *1*, 161.

8. Cohen, A.; Zangen, M.; Goldschmidt, J.M.E. *Inorg. Nucl. Chem. Lett.* 1980, *16*, 165.

9. Cohen, A.; Zangen, M.; Koenisbuch, M.; Goldschmidt, J.M.E. *Desalination* 1982, *41*, 215.

10. Weisberg, J. *Bull. Soc. Chim. Fr.* 1896, 15, 1247.

11. Robart, J. *Bull. Assoc. Chim. Sucr. Distill. Fr. Colon.* 1913, *31*, 108.

12. Van der Linden, T. *Arch. Suikerind. Ned.-Indie* 1916, 24, 1113; *Dtsch. Zuckerind.* 1916, 41, 815; *J. Soc. Chem. Ind., London* 1917, *36*, 96.

13. Marusawa, T. *Kogyo Kagaku Zasshi* 1917, *20*, 287.

14. Bichowsky, F.R. *J. Am. Chem. Soc.* 1923, *45*, 2225.

15. Farnell, R.G.W. *J. Soc. Chem. Ind. London, Trans. Commun.* 1925, *44*, 530.

16. Wurz, O.; Swoboda, O. *Text.-Rundsch.* 1948, *3*, 201.

COMPONENTS:	EVALUATOR:
1. Calcium sulfite; $CaSO_3$; [10257-55-3] 2. Water; H_2O; [7732-18-5]	H.D. Lutz, Dept. of Chemistry, University of Siegen, FR Germany. December 1983.

CRITICAL EVALUATION (continued)

17. Rengemo, T.; Brune, U.; Sillén, L.G. *Acta Chem. Scand.* 1958, *12*, 873.

18. Ramaiah, N.A.; Sharma, J.K. *Proc. Ann. Conv. Sugar Technol. Assoc. India* 1960, *28*, 64.

19. Gupta, S.C.; Ramaiah, N.A.; Kumar, K. *Proc. Ann. Conv. Sugar Technol. Assoc. India* 1965, *33*, 175.

20. Sano, H. *Osaka Kogyo Gijutsu Shikensho Kiho* 1974, *25*, 179.

21. Marusawa, T.; Naito, D.-I.; Uchida, J.I. *Mem. Ryojun Coll. Eng.* 1929, *1*, 351.

22. Bobrovnik, L.D.; Kotel'nikova, L.P. *Izv. Vyssh. Uchebn. Zaved., Pishch. Tekhnol.* 1974, (4), 155.

23. Marusawa, T. *Kogyo Kagaku Zasshi* 1917, *20*, 737.

24. Schwarz, R.; Müller-Clemm, H. *Z. Anorg. Allg. Chem.* 1921, *34*, 272.

25. Mebane, W.M.; Dobbins, J.T.; Cameron, F.K. *J. Phys. Chem.* 1929, *33*, 961.

26. Humm, W. *Untersuchungen an Sulfitlaugenturmen, Diss.*, ETH Zurich, 1929; Güntter-Staib Verlag, Biberach-Riss, 1929.

27. Conrad, F.H.; Beuschlein, W.L. *J. Am. Chem. Soc.* 1934, *56*, 2554.

28. Gishler, P.E.; Maass, O. *Can. J. Res., Sect. B* 1935, *13*, 370.

29. Otuka, Y. *J. Soc. Chem. Ind. Jpn.* 1939, 42, 205.

30. Kuz'minykh, I.N.; Babushkina, M.D. *Zh. Prikl. Khim.* 1956, 29, 1488; *J. Appl. Chem. USSR (Eng. Transl.)* 1956, 29, 1607.

31. Frydman, M.; Nilsson, G.; Rengemo, T.; Sillén, L.G. *Acta Chem. Scand.* 1958, *12*, 878.

32. Nilsson, G.; Rengemo, T.; Sillén, L.G. *Acta Chem. Scand.* 1958, *12*, 868.

33. Simon, A.; Waldmann, K. *Naturwissenschaften* 1958, *45*, 128.

34. Engelhardt, G. *Zellst. Pap.* 1962, 43.

35. Dickens, W.A.; Plummer, A.W. *Tappi* 1957, *40*, 895.

36. Porubszky, I.; Jeszenovics, A.; Koltai, L. *Magy. Kem. Lapja* 1974, *29*, 89.

37. Kurota, O.; Takahashi, S.; Nakaoka, A. *Japan Kokai* 1977, 52-89561, 287.

38. Dubovaya, V.K.; Nabiev, M.N. *Uzb. Khim. Zh.* 1959, *5*, 6.

39. Templeton, C.C.; Rushing, S.S.; Rodgers, J.C. *Mater. Prot.* 1963, *2*, 42.

40. Geese, J.W. *Z. Ver. Dtsch. Zucker-Ind., Allg. Teil* 1898, *48*, 99.

41. Arnal, T.G.; Mesorama, J.M.P. *An. Fis. Quim.* 1947, *43*, 439.

42. Rodin, I.V.; Margulis, E.V. *Zh. Neorg. Khim.* 1983, *28*, 258; *Russ. J. Inorg. Chem. (Eng. Transl.)* 1983, *28*, 144.

43. Yasue, T.; Shiino, H.; Arai, Y. *Yogyo Kyokai Shi* 1980, *88*, 197.

44. Setoyama, K.; Takahashi, S. *Yogyo Kyokai Shi* 1978, *86*, 244.

45. Setoyama, K.; Takahashi, S. *Sekko to Sekkai* 1979, *161*, 133.

COMPONENTS:

1. Calcium sulfite; $CaSO_3$; [10257-55-3]
2. Water; H_2O; [7732-18-5]

ORIGINAL MEASUREMENTS:

Farnell, R.G.W.

J. Soc. Chem. Ind. London, Trans. Commun. 1925, *44*, 530.

VARIABLES:

One temperature: 303 K

PREPARED BY:

B. Engelen, H.D. Lutz

EXPERIMENTAL VALUES:

The author reports the solubility of $CaSO_3.2H_2O$ [10035-03-7] in water at 30°C. The pH value of the solution was 10.

Composition of saturated solution

24 mg $CaSO_3/dm^3$

$= 2.0 \times 10^{-4}$ mol dm^{-3} (compilers)

AUXILIARY INFORMATION

METHOD APPARATUS/PROCEDURE:

A solution of calcium sulfite was stirred over solid calcium sulfite for 20 hr in a thermostatically controlled vessel. The solution was analysed for CaO, SO_2, and pH. CaO was determined manganometrically after precipitation as calcium oxalate. Total SO_2 was determined iodometrically. The determination was performed in duplicate.

SOURCE AND PURITY OF MATERIALS:

Calcium sulfite was prepared by adding calcium hydroxide to a sulfurous acid solution.

ESTIMATED ERROR:

REFERENCES:

COMPONENTS:

1. Calcium sulfite; $CaSO_3$; [10257-55-3]
2. Water; H_2O; [7732-18-5]

ORIGINAL MEASUREMENTS:

Rodin, I.V.; Margulis, E.V.

Zh. Neorg. Khim. 1983, *28*, 258; *Russ. J. Inorg. Chem. (Eng. Transl.)* 1983, *28*, 144.

VARIABLES:

Four temperatures: 293 - 363 K

PREPARED BY:

B. Engelen

EXPERIMENTAL VALUES:

Solubilities of calcium sulfite in water at different temperatures are reported.

t/°C	$CaSO_3$ 10^4 mass %	$10^3 m$/mol kg^{-1} [a]
20	185.5	1.544
50	246.3	2.051
70	297.8	2.479
90	355.4	2.959

[a] Calculated by the compiler.

AUXILIARY INFORMATION

METHOD APPARATUS/PROCEDURE:

Saturation method. Equilibrium was established by stirring the saturated solutions in thermostatically controlled glass tubes. Equilibrium was tested for analytically - 4 hr was reported to be sufficient. Calcium was determined gravimetrically.

SOURCE AND PURITY OF MATERIALS:

Calcium sulfite, claimed to be $CaSO_3.1.5H_2O$ [96247 22-2], was obtained by precipitation from $CaSO_4$ solutions with Na_2SO_3 (1).

ESTIMATED ERROR:

Not given.

REFERENCES:

1. Margulis, E.V.; Grishankina, N.S. *Zh. Neorg. Khim.* 1963, *8*, 2638.

COMPONENTS:

1. Calcium sulfite; $CaSO_3$; [10257-55-3]
2. Sulfur dioxide; SO_2; [7446-09-5]
3. Water; H_2O; [7732-18-5]

ORIGINAL MEASUREMENTS:

Kuz'minykh, I.N.; Babushkina, M.D.

Zh. Prikl. Khim. 1956, *29*, 1488-93; **J. Appl. Chem. USSR (Eng. Transl.)* 1956, *29*, 1607-11.

VARIABLES:

Five temperatures: 278 - 333 K
Partial pressure of SO_2

PREPARED BY:

H.D. Lutz, B. Engelen

EXPERIMENTAL VALUES:

Solubility data of calcium sulfite in aqueous sulfurous acid solutions at different pressures of sulfur dioxide given by the authors are:

t/°C	p_{SO_2}[a] mm Hg	p_{SO_2}[a,b] 10^{-3}bar	g SO_2/100 g H_2O total	g SO_2/100 g H_2O combined[c]	$CaSO_3$[d] g/kg H_2O	$m(CaSO_3)$[d] mol kg^{-1}
	28.5	38.0	5.49	2.4	45.0	0.375
	43.2	57.6	6.76	2.92	54.7	0.456
4.5	47.2	62.9	7.0	3.0	56.3	0.468
	155.5	207.3	11.78	4.46	83.6	0.696
	19.2	25.6	3.51	1.6	30.0	0.250
	30.3	40.4	4.36	1.95	36.6	0.304
	75.7	100.9	6.8	2.92	54.8	0.456
15	103.75	138.3	7.81	3.3	61.9	0.515
	155.5	207.3	9.61	3.92	73.5	0.612
	227	302.6	11.55	4.67	87.6	0.729

[a] Corrected for p_{H_2O} by the authors.
[b] Calculated by the compilers.
[c] SO_2 required to form the monosulfite.
[d] Calculated from the amount of combined SO_2 by the compilers.

(continued on next page)

AUXILIARY INFORMATION

METHOD APPARATUS/PROCEDURE:

Aqueous solutions of sulfurous acid were kept over a calcium sulfite precipitate for several days. The equilibrium pressure of SO_2 was determined dynamically in a special apparatus (1) by passing oxygen-free N_2 gas through the solution-precipitate mixture, and analysing the moist inert gas for SO_2 with iodine. Gas volume was measured in an aspirator. The solution was analysed for total SO_2 and calcium content (1). The results are the means of several measurements. The authors report good agreement.

SOURCE AND PURITY OF MATERIALS:

Not given.

ESTIMATED ERROR:

Not given.

REFERENCES:

1. Kuz'minykh, I.N.; Kuznetsova, A.G. *Zh. Prikl. Khim.* 1954, *27*, 816.

COMPONENTS:	ORIGINAL MEASUREMENTS:
1. Calcium sulfite; $CaSO_3$; [10257-55-3] 2. Sulfur dioxide; SO_2; [7446-09-5] 3. Water; H_2O; [7732-18-5]	Kuz'minykh, I.N.; Babushkina, M.D. *Zh. Prikl. Khim.* 1956, *29*, 1488-93; **J. Appl. Chem. USSR (Eng. Transl.)* 1956, *29*, 1607-11.

EXPERIMENTAL VALUES (continued):

t/°C	p_{SO_2}[a] mm Hg	p_{SO_2}[a,b] 10^{-3}bar	g SO_2/100 g H_2O total	g SO_2/100 g H_2O combined[c]	$CaSO_3$[d] g/kg H_2O	$m(CaSO_3)$[d] mol kg^{-1}
	31	41.3	3.64	1.68	31.5	0.262
	49.3	65.7	4.55	2.08	39.0	0.325
	102.5	136.6	6.19	2.68	50.3	0.418
25	121	161.3	6.74	2.88	54.0	0.450
	136.5	182.0	7.18	3.05	57.2	0.476
	159	212.0	7.81	3.29	61.7	0.514
	73.4	97.9	3.3	1.5	28.1	0.234
	74.9	99.9	3.41	1.55	29.1	0.242
50	165.2	220.2	5.15	2.3	43.1	0.359
	319	425.3	6.6	2.91	54.6	0.454
	39.3	52.4	2.15	1.01	18.9	0.158
	71.1	94.8	2.75	1.28	24.0	0.200
60	81.8	109.1	3.04	1.4	26.3	0.219
	88.9	118.5	3.11	1.43	26.8	0.223
	123	164.0	3.6	1.66	31.1	0.259
	182	242.6	4.6	2.13	39.9	0.332
	308	410.6	5.47	2.54	47.6	0.396

[a] Corrected for p_{H_2O} by the authors.
[b] Calculated by the compilers.
[c] SO_2 required to form the monosulfite.
[d] Calculated from the amount of combined SO_2 by the compilers.

COMPONENTS:

1. Calcium sulfite; $CaSO_3$; [10257-55-3]
2. Sulfur dioxide; SO_2; [7446-09-5]
3. Water; H_2O; [7732-18-5]

ORIGINAL MEASUREMENTS:

Marusawa, T.

Kogyo Kagaku Zasshi 1917, *20*, 287-301.

VARIABLES:

One temperature: 291 K
Concentration of sulfur dioxide

PREPARED BY:

B. Engelen

EXPERIMENTAL VALUES:

The author reports the composition of aqueous sulfurous acid solutions saturated with calcium sulfite at 18°C.

Composition of the saturated solution (mol dm^{-3})

10^2x $c(Ca(HSO_3)_2)$	10^2x $c(H_2SO_3)$	10^2x $c(Ca^{2+})$	10^2x $c(HSO_3^-)$	10^5x $c(SO_3^{2-})$	10^3x $c(H^+)$	$c(\text{total } SO_2{}^a)$	$c(CaSO_3)^b$
2.07	1.07	1.525	2.41	3.16	3.56	0.0762	0.036
15.25	5.06	8.39	17.3	10.9	5.25	0.5286	0.2364
18.23	5.74	9.70	19.9	13.1	5.22	0.6210	0.2793
28.43	8.58	13.50	27.5	15.7	5.95	0.9294	0.4193
42.15	11.9	17.5	35.6	18.4	6.40	1.3180	0.5965

a Calculated as $2[Ca(HSO_3)_2] + [H_2SO_3] + [HSO_3^-] + [SO_3^{2-}]$ by the compiler.

b Calculated as $[Ca^{2+}] + [Ca(HSO_3)_2]$ by the compiler.

AUXILIARY INFORMATION

METHOD APPARATUS/PROCEDURE:

Saturation method. Time for establishing equilibrium not given. Calcium was determined manganometrically after precipitation as calcium oxalate. Sulfite was determined iodometrically. The amount of sulfate formed by oxidation of the sulfite was calculated from the difference between the calcium and sulfite concentrations.

SOURCE AND PURITY OF MATERIALS:

Calcium sulfite was precipitated by passing SO_2 through a suspension of $CaCO_3$ in water, and analysed for calcium and sulfate.

ESTIMATED ERROR:

REFERENCES:

1. Melcher, A.C. *J. Am. Chem. Soc.* 1910, *32*, 50.

COMPONENTS:

1. Calcium sulfite; $CaSO_3$; [10257-55-3]
2. Sulfur dioxide; SO_2; [7446-09-5]
3. Water; H_2O; [7732-18-5]

ORIGINAL MEASUREMENTS:

Schwarz, R.; Müller-Clemm, H.

Z. Anorg. Allg. Chem. 1921, *34*, 272-5.

VARIABLES:

Temperature: 275 - 306 K
Concentration of sulfur dioxide

PREPARED BY:

B. Engelen

EXPERIMENTAL VALUES:

The authors report the solubility of calcium sulfite in aqueous sulfurous acid solutions at various temperatures.

t/°C	SO_2/mass %		$m(SO_2 total)^a$	$m(CaSO_3)^a$
	total	combined[b]	mol kg^{-1}	mol kg^{-1}
2	7.90	5.4	1.374	0.47
11	6.75	4.56	1.155	0.390
22	6.47	4.33	1.102	0.369
23	6.48	4.34	1.104	0.370
24	6.49	4.35	1.106	0.371
24	6.58	4.40	1.123	0.375
24	7.24	4.80	1.247	0.413
24	7.96	5.17	1.384	0.449
24	10.23	6.56	1.838	0.589
24	10.74	6.86	1.944	0.621
25	6.70	4.51	1.145	0.385
26	6.75	4.55	1.155	0.389
26	9.0	6.6	1.594	0.585

[a] Calculated from combined[b] SO_2 by the compiler.

[b] Amount required to form $Ca(HSO_3)_2$.

(continued on next page)

AUXILIARY INFORMATION

METHOD APPARATUS/PROCEDURE:

Aqueous sulfurous acid solutions were saturated with calcium sulfite within 30 - 48 hr, with exclusion of oxygen. After stirring the solution-precipitate mixture, the solid phase was allowed to settle under the SO_2 equilibrium pressure of the solution for 12 hr. The supernatant solution was analysed for total and combined SO_2 by iodometric and acidimetric titration, for calcium manganometrically after precipitation of calcium as oxalate, and for sulfate gravimetrically.

SOURCE AND PURITY OF MATERIALS:

Calcium sulfite was Merck "pure" grade. Sulfur dioxide was taken from a gas cylinder.

ESTIMATED ERROR:

REFERENCES:

COMPONENTS:	ORIGINAL MEASUREMENTS:
1. Calcium sulfite; $CaSO_3$; [10257-55-3] 2. Sulfur dioxide; SO_2; [7446-09-5] 3. Water; H_2O; [7732-18-5]	Schwarz, R.; Müller-Clemm, H. *Z. Anorg. Allg. Chem.* 1921, *34*, 272-5.

EXPERIMENTAL VALUES (continued):

t/°C	SO_2/mass %		$m(SO_2 total)$[a]	$m(CaSO_3)$[a]
	total	combined[b]	mol kg^{-1}	mol kg^{-1}
27	6.69	4.50	1.143	0.385
29	6.32	4.22	1.074	0.359
29	6.65	4.50	1.136	0.384
29	10.32	6.76	1.858	0.608
30	6.65	4.44	1.136	0.379
33	6.53	4.42	1.114	0.377
33	8.30	5.52	1.451	0.483
33	9.25	6.12	1.639	0.542
33	9.90	6.49	1.771	0.580

a Calculated from combined[b] SO_2 by the compiler.

b Amount required to form $Ca(HSO_3)_2$.

COMPONENTS:

1. Calcium sulfite; $CaSO_3$; [10257-55-3]
2. Sulfur dioxide; SO_2; [7446-09-5]
3. Water; H_2O; [7732-18-5]

ORIGINAL MEASUREMENTS:

Humm, W.

Untersuchungen an Sulfitlaugenturmen. *Diss.*, ETH Zürich, 1929 Güntter-Staib Verlag, Biberach-Riß, 1929.

VARIABLES:

Temperature: 273 - 313 K

PREPARED BY:

H.D. Lutz, B. Engelen

EXPERIMENTAL VALUES:

The author reports the composition of aqueous CaO solutions saturated with SO_2 at 1 and 0.1 Atm (1.013 and 0.1013 bar, $p_{SO_2} + p_{H_2O}$, compilers) pressure.

p/bar	t/°C	total SO_2 mass %	free SO_2[a] mass %	CaO mass %	m(SO_2 total)[b] mol kg^{-1}	m($CaSO_3$)[b] mol kg^{-1}
0.1013	0	14.94	1.52	5.87	2.945	1.322
0.1013	5	14.34	1.27	5.72	2.800	1.276
0.1013	10	13.87	1.08	5.59	2.688	1.238
0.1013	15	12.91	0.90	5.25	2.462	1.144
0.1013	20	12.06	0.78	4.92	2.268	1.057
0.1013	22	11.73	0.74	4.81	2.194	1.028
0.1013	24	11.51	0.72	4.72	2.145	1.005
0.1013	26	11.32	0.70	4.64	2.103	0.985
0.1013	28	11.10	0.68	4.56	2.054	0.964
0.1013	30	10.87	0.66	4.47	2.004	0.942
1.013	0	24.21	8.47	6.89	5.485	1.783
1.013	6	21.73	6.56	6.64	4.736	1.653
1.013	10	20.58	5.67	6.52	4.407	1.595
1.013	20	18.76	4.11	6.41	3.913	1.528
1.013	30	16.31	2.37	6.10	3.281	1.402
1.013	40	14.69	1.54	5.76	2.883	1.291

[a] Excess over the amount necessary to form $Ca(HSO_3)_2$.

[b] Calculated from total SO_2 and CaO content, respectively, by the compilers.

AUXILIARY INFORMATION

METHOD APPARATUS/PROCEDURE:

Saturated solutions were prepared in a special tube (1) by bubbling SO_2 through a suspension of $CaCO_3$ in water at 0°C. The solutions were slowly heated under a continuous stream of SO_2 and analysed at the temperatures given. Total SO_2 was determined iodometrically, free SO_2 acidimetrically and Ca gravimetrically.

SOURCE AND PURITY OF MATERIALS:

SO_2 was obtained by decomposition of $Na_2S_2O_5$ with HCl. The quality of the materials is said to be "pure". No other details are given.

ESTIMATED ERROR:

REFERENCES:

1. Schwarz, R.; Müller-Clemm, H. *Z. Angew. Chem.* 1921, *34*, 272.

COMPONENTS:

1. Calcium sulfite; $CaSO_3$; [10257-55-3]
2. Sulfur dioxide; SO_2; [7446-09-5]
3. Water; H_2O; [7732-18-5]

ORIGINAL MEASUREMENTS:

Conrad, F.H.; Beuschlein, W.L.

J. Am. Chem. Soc. 1934, *56*, 2554-62.

VARIABLES:

Three temperatures: 288, 298 and 308 K
Pressure of SO_2

PREPARED BY:

H.D. Lutz, B. Engelen

EXPERIMENTAL VALUES:

The authors report the solubility of calcium sulfite in aqueous sulfurous acid solutions at 15, 25, and 35°C for various SO_2 vapour pressures, from equilibrium studies of the system $CaO-SO_2-H_2O$.

t/°C	p_{total} mm Hg	p_{SO_2}[a] 10^{-3}bar	SO_2 g/100 g H_2O total	SO_2 g/100 g H_2O combined[b]	CaO g/100 g H_2O total[c]	CaO g/100 g H_2O calc.[d]	$CaSO_3$ g/kg H_2O (compilers)	m($CaSO_3$) mol kg^{-1} (compilers)
	22	12	1.83	0.89	0.78	0.78	16.7	0.139
	49	49	3.99	1.85	1.65	1.62	34.7	0.289
	115	137	7.47	3.12	2.79	2.73	58.5	0.487
	154	198	8.72	3.51	3.15	3.07	65.8	0.547
15	263	334	11.00	4.23	3.82	3.70	79.3	0.660
	408	528	13.76	4.78	4.22	4.19	89.8	0.747
	517	673	15.92	5.47	4.81	4.79	102.6	0.854
	761	999	19.25	5.95	5.23	5.21	111.6	0.929
	40	22	1.91	0.91	0.84	0.80	17.1	0.143
	52	38	2.81	1.33	1.12	1.16	24.9	0.207
	136	150	5.55	2.43	2.04	2.13	45.6	0.380
	254	308	8.37	3.14	2.79	2.75	58.9	0.490
25	461	584	11.52	4.06	3.59	3.55	76.1	0.633
	594	762	13.28	4.59	4.10	4.02	86.1	0.717
	756	978	15.38	5.12	4.46	4.48	96.0	0.799
	763	987	15.16	4.88	4.26	4.27	91.5	0.761
35	761	961	12.02	4.03	3.53	3.53	75.6	0.629

[a] Corrected for p_{H_2O} and converted to bar by the compilers.
[b] SO_2 required to form the monosulfite.
[c] Gravimetric determination of calcium.
[d] CaO equivalent to combined SO_2 in $CaSO_3$, calculated by the authors.

AUXILIARY INFORMATION

METHOD APPARATUS/PROCEDURE:

Equilibrium between CaO, SO_2 and H_2O, which was established after more than 12 hr, was studied in a special flask with connections to a weighing pipette for analysing the solutions saturated with calcium sulfite, and to a mercury manometer to measure the pressure of the gas (sulfur dioxide and water vapour) over the solution. Calcium was determined by precipitation as oxalate and ignition of the precipitate to calcium oxide. Total, free and combined SO_2 were determined by acidimetric and iodometric titration, respectively (1).

By extrapolation of the experimental results, equilibrium data are ascertained for the temperatures 5, 35, 50, and 60°C (2).

SOURCE AND PURITY OF MATERIALS:

CaO was prepared by heating calcium oxalate monohydrate of p.a. quality to constant weight. The sulfate and MgO content was negligible. The SO_2 used was SO_3-free. The amount of inert or non-absorbable gases was about 0.15%.

ESTIMATED ERROR:

Agreement between duplicate determinations: 0.1%

REFERENCES:

1. Birchard, W.H. *Pap. Ind.* 1926, *8*, 793.
2. Conrad, F.H.; Beuschlein, W.L. *Pap. Trade J.* 1937, *4*, 105.

COMPONENTS:	ORIGINAL MEASUREMENTS:
1. Calcium sulfite; $CaSO_3$; [10257-55-3] 2. Sulfur dioxide; SO_2; [7446-09-5] 3. Water; H_2O; [7732-18-5]	Gishler, P.E.; Maass, O. *Can. J. Res., Sect. B* 1935, *13*, 370-9.
VARIABLES:	PREPARED BY:
Concentration of calcium sulfite Concentration of sulfur dioxide	B. Engelen

EXPERIMENTAL VALUES:

The authors report the precipitation temperatures, i.e. the temperatures at which the first precipitation appeared or disappeared, of aqueous calcium sulfite/sulfurous acid solutions of known composition.

Precipit. temp./°C	H_2O cm^3	CaO g	SO_2 g	m(total SO_2)[a] mol kg^{-1}	m($CaSO_3$)[a] mol kg^{-1}
45.0	25	0.500	0.965	0.603	0.357
50.5	50	0.600	0.966	0.302	0.214
57[b]	88.29	2.06	5.30	0.937	0.416
60.4[c]	88.29	2.061	6.845	1.210	0.4162
65[b]	88.29	2.06	6.80	1.202	0.416
75.7	25	0.500	1.091	0.681	0.357
93.0	25	0.500	1.411	0.881	0.357
95[b]	84.26	0.99	2.70	0.500	0.210
95.5[b]	25	0.498	1.192	0.744	0.355
97.0	25	0.300	0.814	0.508	0.214

a,b,c,d See the following page.

(continued on next page)

AUXILIARY INFORMATION

METHOD APPARATUS/PROCEDURE:

a) Sealed pyrex glass bombs of about 95 cm^3, filled with known amounts of calcium oxide, water, and sulfur dioxide, were heated up or cooled down several times at a rate of about 2°C/hr with shaking until a precipitate formed or disappeared.

b) In one experiment precipitation temperature was determined by extrapolation from vapour pressure measurements over a solution first in the supersaturated and then in the saturated state at various temperatures.

c) Specific conductivity measurements were made of solutions with known concentration of calcium oxide, water and sulfur dioxide in the unsaturated and saturated state at various temperatures. Precipitation temperature was then determined by extrapolation. Further details are given in ref. (1).

SOURCE AND PURITY OF MATERIALS:

CaO was prepared by subjecting ground crystals of Iceland spar ($CaCO_3$) to a temperature of 1000°C for a week under air, free from carbon dioxide and water vapour. SO_2 was distilled for purification (1).

ESTIMATED ERROR:

Not given.

REFERENCES:

1. Gurd, G.W.; Gishler, P.E.; Maass, O. *Can. J. Res., Sect. B* 1935, *13*, 209.

COMPONENTS:

1. Calcium sulfite; $CaSO_3$; [10257-55-3]
2. Sulfur dioxide; SO_2; [7446-09-5]
3. Water; H_2O; [7732-18-5]

ORIGINAL MEASUREMENTS:

Gishler, P.E.; Maass, O.

Can. J. Res., Sect. B 1935, *13*, 370-9.

EXPERIMENTAL VALUES (continued):

Precipit. temp./°C	H_2O cm^3	CaO g	SO_2 g	m(total SO_2)[a] mol kg^{-1}	m($CaSO_3$)[a] mol kg^{-1}
100.0	25	0.151	0.311	0.194	0.108
103.5	25	0.499	1.644	1.027	0.356
107.5	25	0.495	1.802	1.125	0.353
116.0	25	0.150	0.352	0.220	0.107
117.0	50	0.600	2.282	0.712	0.214
117.0	50	0.600	2.531	0.790	0.214
117.8[d]	25	0.501	1.330	0.830	0.357
130.0	25	0.150	0.377	0.235	0.107
135.0	25	0.300	1.558	0.973	0.214
137.0	25	0.300	1.822	1.138	0.214

[a] Calculated on the assumption that 1 cm^3 of H_2O = 1 g of H_2O, by the compiler.

[b] Results from conductivity measurements.

[c] Results from SO_2 vapour pressure measurements, p_{SO_2} = 93.2 mm Hg.

[d] Solutions contain 0.5 mass % of a peptizing reagent.

COMPONENTS:	ORIGINAL MEASUREMENTS:
1. Calcium sulfite; $CaSO_3$; [10257-55-3] 2. Sulfur dioxide; SO_2; [7446-09-5] 3. Water; H_2O; [7732-18-5]	Otuka, Y. *J. Soc. Chem. Ind. Jpn.* 1939, *42*, 205-9.
VARIABLES: Concentration of calcium sulfite Concentration of sulfur dioxide Temperature: 370 - 433 K	PREPARED BY: B. Engelen

EXPERIMENTAL VALUES:

The author reports precipitation temperatures of aqueous solutions with known concentrations of calcium sulfite and sulfurous acid.

Precipitation temperature/°C	total SO_2 mol dm^{-3}	$CaSO_3$ mol dm^{-3}
96.7	0.5002	0.1513
112.0	0.6784	0.1454
115.0	0.3393	0.0746
117.8	0.6649	0.2054
119.4	0.6294	0.1513
121.8	0.5096	0.1502
124.3	0.7248	0.1454
125.6	0.6922	0.1518
128.3	0.8066	0.2054
129.5	0.3798	0.0727
133.0	0.6638	0.1518
134.6	0.9091	0.2054
136.0	0.7835	0.1508

(continued on next page)

AUXILIARY INFORMATION

METHOD APPARATUS/PROCEDURE:

Solutions of known concentration of H_2SO_3 and $Ca(HSO_3)_2$ were heated slowly in a sealed glass bulb (heating rate not given) and the temperature at which the solid phase (calcium sulfite) appeared or disappeared from the solution was determined. The volume ratio of the liquid to free space in the bulb was about 1:1, 3:1, and 6:1, respectively. No further details are given.

SOURCE AND PURITY OF MATERIALS:

Not given.

ESTIMATED ERROR:

Not given.

REFERENCES:

COMPONENTS:

1. Calcium sulfite; $CaSO_3$; [10257-55-3]
2. Sulfur dioxide; SO_2; [7446-09-5]
3. Water; H_2O; [7732-18-5]

ORIGINAL MEASUREMENTS:

Otuka, Y.

J. Soc. Chem. Ind. Jpn. 1939, 42, 205-9.

EXPERIMENTAL VALUES (continued):

Precipitation temperature/°C	total SO_2 mol dm^{-3}	$CaSO_3$ mol dm^{-3}
137.0	0.8277	0.1454
139.4	1.0000	0.2054
141.2	0.7503	0.1498
142.8	0.9196	0.1515
143.5	1.0966	0.2054
145.2	0.9431	0.1472
145.3	0.8785	0.1471
146.5	0.4785	0.0746
147.7	1.1905	0.2054
149.5	0.5212	0.0746
151.2	0.9273	0.1471
160.0	0.6099	0.0723

Precipitation temperatures of solutions containing wood powder and wood chips in addition to the calcium sulfite-sulfurous acid mixtures are also given by the author.

COMPONENTS:

1. Calcium sulfite; $CaSO_3$; [10257-55-3]
2. Sodium perchlorate; $NaClO_4$; [7601-89-0]
3. Sulfur dioxide; SO_2; [7446-09-5]
4. Water; H_2O; [7732-18-5]

ORIGINAL MEASUREMENTS:

Nilsson, G.; Rengemo, T.; Sillén, L.G.

Acta Chem. Scand. 1958, *12*, 868-72.

VARIABLES:

Three temperatures: 298, 308 and 348 K
Concentration of $NaClO_4$ (ionic strength)

PREPARED BY:

H.D. Lutz, B. Engelen

EXPERIMENTAL VALUES:

The authors determined the solubility of calcium sulfite in aqueous sulfurous acid solutions from equilibrium studies of the reaction

$$CaSO_3(s) + SO_2(g) + H_2O \rightleftharpoons Ca^{2+} + 2HSO_3^-$$

at 25, 35, and 75°C, and a SO_2 vapour pressure of 1 atm (1.013 bar, compilers). $NaClO_4$ solutions were used as ionic medium, with ionic strengths of 1 and 3.5 mole Na^+/kg H_2O. From the obtained data the equilibrium constants

$$K = [Ca^{2+}] \times [HSO_3^-]^2/P_{SO_2} = 4\ m^3_{Ca^{2+}}/P_{SO_2}\ [mol^3kg^{-3}Atm^{-1}],$$

given as log K, were calculated by the authors.

(continued on next page)

AUXILIARY INFORMATION

METHOD APPARATUS/PROCEDURE:

Solutions were prepared isothermally by bubbling SO_2 through solutions of $NaClO_4$ with an excess of solid calcium sulfite. Time for establishing equilibrium is not given, but seems to be the same as given in a second paper by these authors (1), namely 5 - 38 days. Calcium was precipitated as the oxalate and titrated with $KMnO_4$ after dissolving in sulfuric acid. Total content of SO_2 was determined iodometrically.

SOURCE AND PURITY OF MATERIALS:

$CaCO_3$ was precipitated from a solution of calcium chloride with Na_2SO_3, washed, then sucked dry, all under a nitrogen atmosphere. SO_2 was taken from a gas cylinder, washed with water and then with a $NaClO_4$ solution of desired concentration and temperature. $NaClO_4$ was prepared from Na_2CO_3 and $HClO_4$. $CaCl_2$, Na_2SO_3, Na_2CO_3, and $HClO_4$ were all of p.a. purity.

ESTIMATED ERROR:

Temperature: ±0.1K for 25 and 35°C, ±1 K for 75°C. Log K: ±0.01 (authors).

REFERENCES:

1. Rengemo, T.; Brune, U.; Sillén, L.G. *Acta Chem. Scand.* 1958, *12*, 873.
2. Jones, J.H. *J. Phys. Colloid. Chem.* 1947, *51*, 516.

COMPONENTS:	ORIGINAL MEASUREMENTS:
1. Calcium sulfite; $CaSO_3$; [10257-55-3] 2. Sodium perchlorate; $NaClO_4$; [7601-89-0] 3. Sulfur dioxide; SO_2; [7446-09-5] 4. Water; H_2O; [7732-18-5]	Nilsson, G.; Rengemo, T.; Sillén, L.G. *Acta Chem. Scand.* 1958, *12*, 868-72.

EXPERIMENTAL VALUES (continued):

$m(NaClO_4)/mol\ kg^{-1}$		1			3.5	
t/°C	25	35	75	25	35	75
partial pressure of SO_2/Atm[a]	0.970	0.947	0.632	0.973	0.951	0.664
mol Ca^{2+}/kg soln.	0.497	0.395	0.168	0.260	0.198	0.0877
$m(Ca^{2+})/mol\ kg^{-1}$[b]	0.667	0.507	0.198	0.417	0.308	0.129
$10^3\ w(Ca^{2+})$[c] (compilers)	19.92	15.83	6.73	10.42	7.94	3.515
log *K*	0.09	-0.26	-1.30	-0.52	-0.91	-1.89
$K/mol^3\ kg^{-3}\ Atm^{-1}$ (compilers)	1.224	0.550	0.049	0.298	0.123	0.013

[a] Calculated (by compilers) from total pressure (1.013 bar) and some additional data given by the authors. The data are the equilibrium vapour pressure of water and the activity of water in the solutions, calculated (by the authors) from Jones's (2) osmotic coefficients.

[b] Calculated from the original analytical data by the authors.

[c] w(mass fraction).

COMPONENTS:

1. Calcium sulfite; $CaSO_3$; [10257-55-3]
2. Sodium perchlorate; $NaClO_4$; [7601-98-0]
3. Sulfur dioxide; SO_2; [7446-09-5]
4. Water; H_2O; [7732-18-5]

ORIGINAL MEASUREMENTS:

Frydman, M.; Nilsson, G.; Rengemo, T.; Sillén, L.G.

Acta Chem. Scand. 1958, *12*, 878-84.

VARIABLES:

Three temperatures: 298, 308 and 348 K

PREPARED BY:

B. Engelen, H.D. Lutz

EXPERIMENTAL VALUES:

The authors report the solubility products

$$K_{s0_a} = [Ca^{2+}] \times [SO_3^{2-}] \text{ and } K_{s0_b} = [Ca^{2+}] \times [HSO_3^-]^2/[H_2SO_3]$$

in aqueous solutions of $NaClO_4$ with ionic strengths of 1 and 3.5 mol/kg of Na^+ in H_2O at 25, 35, and 75°C. The data were calculated by combining the equilibrium constants of the following reactions

$$CaSO_4(s) \rightleftharpoons Ca^{2+} + SO_4^{2-} \quad (1)$$

$$CaSO_4(s) + SO_3^{2-} \rightleftharpoons CaSO_3(s) + SO_4^{2-} \quad (2)$$

$$SO_2(g) + H_2O \rightleftharpoons H_2SO_3 \quad (3)$$

$$CaSO_3(s) + SO_2(g) + H_2O \rightleftharpoons Ca^{2+} + 2HSO_3^- \quad (4)$$

The equilibrium studies are fully described in two other papers by the authors (1,2).

(continued on next page)

AUXILIARY INFORMATION

METHOD APPARATUS/PROCEDURE:

The equilibrium was studied in $NaClO_4$ solutions of the given ionic strengths. The time to establish equilibrium is not given, but seems to be 5 - 38 days as described in ref. (2). Equilibrium 4 was studied under nitrogen with solutions containing 0.3 cm^3 of benzyl alcohol per dm^3 to avoid oxidation of the sulfite. The composition of the saturated solutions was determined as in ref. (1). Calcium was determined oxidimetrically with $KMnO_4$ after precipitating as oxalate and dissolving in sulfuric acid, total amount of SO_2 iodometrically, and sulfate gravimetrically as $BaSO_4$.

SOURCE AND PURITY OF MATERIALS:

Calcium sulfite was precipitated from a solution of calcium chloride with Na_2SO_3, washed, and sucked dry, all under a N_2 atmosphere. SO_2 was taken from a gas cylinder, washed first with water, and then with an $NaClO_4$ solution of desired concentrations and temperature. $NaClO_4$ was prepared from Na_2CO_3 and $HClO_4$. $CaCl_2$, $CaSO_4$, Na_2SO_3, and $HClO_4$ were all of p.a. quality. N_2 was purified by Meyer and Ronge's method (3).

ESTIMATED ERROR:

Temperature: ±0.1 K for 25 and 35°C, ±1 K for 75°C. Log *K*: 0.01 for equilibrium 1 and 3, 0.02 for equilibrium 4 and 0.1 for equilibrium 2 (authors).

REFERENCES:

1. Nilsson, G.; Rengemo, T.; Sillén, L.G. *Acta Chem. Scand.* 1958, *12*, 868.
2. Rengemo, T.; Brune, U.; Sillén, L.G. *Acta Chem. Scand.* 1958, *12*, 873.
3. Meyer, F.R.; Ronge, G. *Angew. Chem.* 1939, 52, 637.

COMPONENTS:	ORIGINAL MEASUREMENTS:
1. Calcium sulfite; $CaSO_3$; [10257-55-3] 2. Sodium perchlorate; $NaClO_4$; [7601-98-0] 3. Sulfur dioxide; SO_2; [7446-09-5] 4. Water; H_2O; [7732-18-5]	Frydman, M.; Nilsson, G.; Rengemo, T.; Sillén, L.G. *Acta Chem. Scand.* 1958, *12*, 878-84.

EXPERIMENTAL VALUES (continued):

$m(NaClO_4)$ mol kg^{-1}			1			3.5	
t/°C		25	35	75	25	35	75
No.	Equilibrium		log *K* (measured)				
1	$[Ca^{2+}] \times [SO_4^{2-}]$[a]	-2.92	-2.91	-2.94	-3.16	-3.09	-3.34
2	$[SO_4^{2-}]/[SO_3^{2-}]$[b]	1.88	1.96	2.23	1.88	1.96	2.23
3	$[H_2SO_3]/p_{SO_2}$[c]	0.03	-0.13	-0.53	0.06	-0.09	-0.50
4	$[Ca^{2+}] \times [HSO_3^-]^2/p_{SO_2}$[d]	0.09	-0.26	-1.30	-0.52	-0.91	-1.89
			log *K* (derived)				
1-2	$[Ca^{2+}] \times [SO_3^{2-}]$[e]	-4.80	-4.87	-5.17	-5.04	-5.05	-5.57
3-4	$[Ca^{2+}] \times [HSO_3^-]^2/[H_2SO_3]$[f]	-0.06	-0.13	-0.77	-0.58	-0.82	-1.39

[a] $K_1 = m_{Ca^{2+}} \times m_{SO_4^{2-}}$ [$mol^2\ kg^{-2}$]

[b] $K_2 = m_{SO_4^{2-}}/m_{SO_3^{2-}}$

[c] $K_3 = m_{H_2SO_3}/p_{SO_2}$ [$mol\ kg^{-1}\ Atm^{-1}$]

[d] $K_4 = m_{Ca^{2+}} \times m^2_{HSO_3^-}/p_{SO_2}$ [$mol^3\ kg^{-3}\ Atm^{-1}$]

[e] $K_{1-2} = K_{s0_a} = m_{Ca^{2+}} \times m_{SO_3^{2-}}$ [$mol^2\ kg^{-2}$]

[f] $K_{3-4} = K_{s0_b} = m_{Ca^{2+}} \times {}^2m_{HSO_3^-}/m_{H_2SO_3}$ [$mol^2\ kg^{-2}$]

COMPONENTS:

1. Calcium sulfite; $CaSO_3$; [10257-55-3]
2. Sulfur dioxide; SO_2; [7446-09-5]
3. Water; H_2O; [7732-18-5]

ORIGINAL MEASUREMENTS:

Engelhardt, G.

Zellst. Pap. 1962, 43-50.

VARIABLES:

Temperature: 291 - 403 K
Concentration of SO_2

PREPARED BY:

B. Engelen, H.D. Lutz

EXPERIMENTAL VALUES:

The author reports the solubility of calcium sulfite in aqueous sulfurous acid solutions at various temperatures.

t/°C	SO_2 g/100 cm^3 total	SO_2 g/100 cm^3 combined[b]	CaO g/100 cm^3	$CaSO_3$[a] c/mol dm^{-3}
18	1.024	0.480	0.420	0.0749
18	1.523	0.777	0.682	0.1213
18	2.402	1.265	1.110	0.1975
18	2.930	1.450	1.270	0.2264
18	4.627	2.307	2.019	0.3601
40	7.96	3.660	3.206	0.5713
55	7.30	3.17	2.814	0.495
60	2.80	1.260	1.106	0.1967
75	4.0	1.6		0.250
93	4.0	1.33		0.208
108	6.0	1.7		0.265
115	6.0	1.5		0.234
115	3.0	0.75		0.117
130	5.0	1.0		0.156

[a] Calculated by the compilers from combined SO_2.

[b] SO_2 required to form the monosulfite.

AUXILIARY INFORMATION

METHOD APPARATUS/PROCEDURE:

a) Temperature range 18 - 60°C: solutions of sulfur dioxide were saturated with calcium sulfite.

b) Temperature range 75 - 130°C: calcium hydrogen sulfite solutions of known composition were slowly heated under their equilibrium pressures in a thermostatically controlled flask until the first precipitation of calcium sulfite was observed. No further details are given.

SOURCE AND PURITY OF MATERIALS:

Not given.

ESTIMATED ERROR:

REFERENCES:

COMPONENTS:

1. Calcium sulfite; $CaSO_3$; [10257-55-3]
2. Calcium sulfate; $CaSO_4$; [7778-18-9]
3. Water; H_2O; [7732-18-5]

ORIGINAL MEASUREMENTS:

Bichowsky, F.R.

J. Am. Chem. Soc. 1923, *45*, 2225-35.

VARIABLES:

One temperature: 298 K

PREPARED BY:

H.D. Lutz, B. Engelen

EXPERIMENTAL VALUES:

The author reports the solubility of calcium sulfite in water and in saturated solutions of calcium sulfite at 25°C for samples of calcium sulfite prepared by different methods.

Sample[a]	Composition of saturated solutions $10^5c(CaSO_3)/mol\ dm^{-3}$	
	in water	in sat. $CaSO_4$ soln.
1	20.2	3.71
2	10.7	-
3	8.7	3.67
4	-	3.72

[a] $CaSO_3$ samples 1, 2, 3, and 4 were prepared as given under source.

AUXILIARY INFORMATION

METHOD APPARATUS/PROCEDURE:

Saturation method. Sulfite was determined iodometrically.

SOURCE AND PURITY OF MATERIALS:

Calcium sulfite was obtained:
1) by precipitation from $CaCl_2$ solutions with SO_2 and NaOH,
2) by dissolving commercial sulfite in sulfurous acid and precipitating by boiling off the SO_2 in a vacuum,
3) by washing the commercial salt with air-free distilled water,
4) commercial salt.

ESTIMATED ERROR:

Not given.

REFERENCES:

COMPONENTS:

1. Calcium sulfite; $CaSO_3$; [10257-55-3]
2. Calcium sulfate; $CaSO_4$; [7778-18-9]
3. Water; H_2O; [7732-18-5]

ORIGINAL MEASUREMENTS:

Marusawa, T.

Kogyo Kagaku Zasshi 1917, *20*, 287-301.

Marusawa, T.; Naito, D.-I.; Uchida, J.I.

Mem. Ryojun Coll. Eng. 1929, *1*, 351-93.

VARIABLES:

One temperature: 291 K
Concentration of sulfate

PREPARED BY:

B. Engelen, H.D.Lutz

EXPERIMENTAL VALUES:

The author determined both the solubility of $CaSO_3.2H_2O$ [10035-03-7] in solutions containing different amounts of calcium sulfate, and the solubility product $K_{s0}(CaSO_3) = c_{Ca^{2+}} \times c_{SO_3^{2-}}$, of this compound at 18°C.

Concentration of the saturating solutes			Degree of dissociation[a]	Solubility product[b]
SO_4^{2-}	Ca^{2+}	SO_3^{2-}	γ	
$10^4 c_1$/mol dm^{-3}	$10^4 c_2$/mol dm^{-3}	$10^4 c_3$/mol dm^{-3}		$10^7 K_{s0}$/mol^2 dm^{-6}
1.29	8.59	7.30	0.835	4.37
1.62	8.84	7.22	0.833	4.43
2.22	9.20	6.98	0.830	4.42

From the values obtained for the solubility product, the solubility of calcium sulfite in pure water was calculated (by author) to be 7.91 x 10^{-4} mol dm^{-3}.[c] The solubility at 94°C is said to be 6.6 x 10^{-4} mol dm^{-3}.

[a] Determined by the authors from conductance measurements of calcium sulfate solutions made by Melcher (1) on the assumption that the degree of dissociation (activity, compilers) of $CaSO_4$ and $CaSO_3$ solutions is the same.

[b] Calculated from $c_{Ca^{2+}}$, $c_{CO_3^{2-}}$, and γ by the author.

[c] On the same assumptions as stated under[a].

AUXILIARY INFORMATION

METHOD APPARATUS/PROCEDURE:

Saturation method. Time for establishing equilibrium not given. Calcium was determined manganometrically after precipitation as calcium oxalate. Sulfite was determined iodometrically. The amount of sulfate formed by oxidation of the sulfite was calculated from the difference between calcium and sulfite concentrations.

SOURCE AND PURITY OF MATERIALS:

Calcium sulfite was precipitated by passing SO_2 through a suspension of $CaCO_3$ in water, and analysed for calcium and sulfate.

ESTIMATED ERROR:

REFERENCES:

1. Melcher, A.C. *J. Am. Chem. Soc.* 1910, *32*, 50.

COMPONENTS:

1. Calcium sulfite; $CaSO_3$; [10257-55-3]
2. Calcium sulfate; $CaSO_4$; [7778-18-9]
3. Water; H_2O; [7732-18-5]

ORIGINAL MEASUREMENTS:

Wurz, O.; Swoboda, O.

Text. -Rundsch. 1948, *3*, 201-6.

VARIABLES:

Four temperatures: 293 - 353 K

PREPARED BY:

B. Engelen, H.D. Lutz

EXPERIMENTAL VALUES:

The authors report the solubility of calcium sulfite in water and in solutions of various non-saturating solutes at different temperatures.

Solvent	Composition of saturated solutions $CaSO_3$			
t/°C	20	40	60	80
	mg/100 g soln.			
pure water	9.95	7.27	8.31	5.90
$CaSO_4$ sat. soln.	2.90	3.28	3.66	4.30
	10^{-4} mole/kg soln. (compilers)			
pure water	8.28	7.72	6.92	4.91
$CaSO_4$ sat. soln.	2.41	2.73	3.05	3.58

AUXILIARY INFORMATION

METHOD APPARATUS/PROCEDURE:

Oxygen-free water or the given solutions were saturated with calcium sulfite. The solutions were stirred for 0.5 hr and kept for 12 hr at the given temperature. An aliquot of the solutions was then analysed for sulfite with iodine or chloramine. The mean of 4 measurements is given.

SOURCE AND PURITY OF MATERIALS:

An oxygen-free slurry of calcium oxide prepared by adding ignited calcium carbonate to oxygen-free water was saturated with sulfur dioxide under nitrogen. After removing excess SO_2 by passing nitrogen through the solution the obtained calcium sulfite slurry was evaporated to dryness. The product contained 2% sulfate.

ESTIMATED ERROR:

REFERENCES:

COMPONENTS:

1. Calcium sulfite; $CaSO_3$; [10257-55-3]
2. Ammonium nitrate; NH_4NO_3; [6484-52-2]
3. Water; H_2O; [7732-18-5]

ORIGINAL MEASUREMENTS:

Dubovaya, V.K.; Nabiev, M.N.

Uzb. Khim. Zh. 1959, 5, 6-12.

VARIABLES:

One temperature: 303 K
Concentration of NH_4NO_3

PREPARED BY:

B. Engelen

EXPERIMENTAL VALUES:

The authors report the solubility of calcium sulfite in solutions of various NH_4NO_3 concentration at 30°C.

NH_4NO_3	CaO	$CaSO_3$ (compiler)	
mass %	g/dm^3	g/dm^3	10^2c/mol dm^{-3}
20	1.136	2.433	2.026
40	1.314	2.815	2.343
60	1.058	2.267	1.887

AUXILIARY INFORMATION

METHOD APPARATUS/PROCEDURE:

Saturation method. Equilibrium was tested for analytically. Time and method not given. To avoid oxidation of the sulfite, phenylenediamine was added during the analysis procedure. The type of ions determined and analytical method are not given.

SOURCE AND PURITY OF MATERIALS:

Calcium sulfite was freshly prepared, method not given. NH_4NO_3 of commercial quality was used. Calcium sulfite was contaminated with a small amount of sulfate.

ESTIMATED ERROR:

REFERENCES:

COMPONENTS:

1. Calcium sulfite; $CaSO_3$; [10257-55-3]
2. Sucrose; $C_{12}H_{22}O_{11}$; [471-34-1]
3. Water; H_2O; [7732-18-5]

ORIGINAL MEASUREMENTS:

Weisberg, J.

Bull. Soc. Chim. Fr. 1896, *15*, 1247-50.

VARIABLES:

One temperature: 291 K
Concentration of sugar

PREPARED BY:

B. Engelen

EXPERIMENTAL VALUES:

The author reports the solubility of calcium sulfite in water and in solutions of sugar at 18°C.

Sucrose mass %	g $CaSO_3/dm^3$	$10^4c(CaSO_3)/mol\ dm^{-3}$ (compiler)
0	0.043	3.58
10	0.0825	6.87
30	0.0800	6.66

AUXILIARY INFORMATION

METHOD APPARATUS/PROCEDURE:

Saturation method. Sulfite was determined iodometrically.

SOURCE AND PURITY OF MATERIALS:

$CaSO_3$ was precipitated by passing SO_2 through a suspension of $Ca(OH)_2$ in water. The precipitate was washed and dried over sulfuric acid in a desiccator.

ESTIMATED ERROR:

REFERENCES:

COMPONENTS:

1. Calcium sulfite; $CaSO_3$; [10257-55-3]
2. Sucrose; $C_{12}H_{22}O_{11}$; [471-34-1]
3. Water; H_2O; [7732-18-5]

ORIGINAL MEASUREMENTS:

Gupta, S.C.; Ramaiah, N.A.; Kumar, K.

Proc. Ann. Conv. Sugar Technol. Assoc. India, 1965, *33*, 175-9.

VARIABLES:
Temperature: 273 - 333 K
Concentration of sucrose
pH

PREPARED BY:

B. Engelen

EXPERIMENTAL VALUES:
The following data have been estimated by the compiler from a graph given by the authors.

Sucrose mass %[b]	CaO mg/dm^3					$10^3c(CaSO_3)$[a] mol dm^{-3}					
	5	10	30	40	60	5	10	30	40	60	t/°C
-	131	121	103[c]	100	97[c]	2.34	2.14	1.84	1.78	1.73	
10	120	108	100	93	88	2.14	1.93	1.78	1.66	1.57	
20	117	102	97	92	85[c]	2.09	1.82	1.73	1.64	1.52	
30	113	98	90	85	77	2.02	1.75	1.60	1.52	1.37	
40	104	89	83	75	70[c]	1.85	1.59	1.48	1.34	1.25	
50	-	-	80	72	66	-	-	1.43	1.28	1.18	

Sucrose mass %[b]	pH	CaO/mg/dm^{3}[d]	$10^3c(CaSO_3)$[a]/mol dm^{-3}
10	7	74	1.32
10	7.5	215	3.83
10	8	266	4.74
10	9	306	5.46
10	10	305	5.44
10	11	286	5.10

[a] Calculated by the compiler.

[b] Given as °Brix by the authors.

[c] Numerical data given by the authors.

[d] Temperature not given, but it seems to be room temperature.

AUXILIARY INFORMATION

METHOD APPARATUS/PROCEDURE:

Sugar solutions of the given concentrations were saturated with calcium sulfite in a thermostatically controlled vessel. Equilibrium was reached after 6 hr. Calcium was determined complexometrically.

SOURCE AND PURITY OF MATERIALS:

Calcium sulfite of p.a. quality was recrystallized in doubly distilled water and dried in a vacuum at 150°C. Sucrose of a purity of 99.9% was used.

ESTIMATED ERROR:

Error in temperature: ±0.2 K for 5 - 10°C: ±0.1 K for 30 - 60°C.

REFERENCES:

COMPONENTS:

1. Calcium sulfite; $CaSO_3$; [10257-55-3]

2. Sucrose; $C_{12}H_{22}O_{11}$; [471-34-1]

3. Water; H_2O; [7732-18-5]

ORIGINAL MEASUREMENTS:

Bobrovnik, L.D.; Kotel'nikova, L.P.

Izv. Vyssh. Uchebn. Zaved., Pishch. Tekhnol. 1974, (4), 155-6.

VARIABLES:

Temperature: 323 - 368 K
Concentration of sucrose
Two pH values: 7.1 and 9.0

PREPARED BY:

B. Engelen

EXPERIMENTAL VALUES:

The authors report the solubility of $CaSO_3$ in water and in solutions of various sucrose concentration for two pH values (7.1 and 9.0) at various temperatures.

Composition of the saturated solutions

Sucrose	$CaSO_3$						
	50	60	70	80	85	90	95 t/°C
mass %	10^6w (mass fraction, compiler)[a]						
	pH = 7.1						
0	37	35	31	30	29	27	24
10	48	44	38	36	33	31	29
15	41	36	32	31	28	25	21
25	40	37	30	28	27	26	25
30	34	33	28	27	26	24	23
40	31	29	22	21	21	21	21
50	24	20	17	18	18	19	20
60	19	18	16	18	18	20	20

[a] Converted from g/100 g soln, given by the authors.

(continued on next page)

AUXILIARY INFORMATION

METHOD APPARATUS/PROCEDURE:

Saturation method. Equilibrium was established by stirring the saturated solution in a thermostatically controlled vessel at the given temperatures, time not given. Calcium was determined complexometrically, sulfite iodometrically. Method of adjusting the pH not given.

SOURCE AND PURITY OF MATERIALS:

Not given.

ESTIMATED ERROR:

REFERENCES:

COMPONENTS:	ORIGINAL MEASUREMENTS:
1. Calcium sulfite; $CaSO_3$; [10257-55-3] 2. Sucrose; $C_{12}H_{22}O_{11}$; [471-34-1] 3. Water; H_2O; [7732-18-5]	Bobrovnik, L.D.; Kotel'nikova, L.P. *Izv. Vyssh. Uchebn. Zaved., Pishch. Tekhnol.* 1974, (4), 155-6.

EXPERIMENTAL VALUES (continued):

Sucrose	$CaSO_3$						
	50	60	70	80	85	90	95 t/°C
mass %	10^6w(mass fraction, compiler)[a]						
	pH = 9.0						
0	27	23	20	19	17	16	14
10	34	29	26	22	20	18	15
15	32	28	26	23	21	18	17
25	32	28	21	16	14	11	10
30	28	23	17	12	9	7	4
40	23	21	16	8	5	3	1
50	17	9	5	1.6	1.6	1.2	1.2
60	7	5	3	1.5	1.5	1.5	0.8
	10^{-5}mole/kg soln (compiler)						
	pH = 7.1						
0	30.8	29.1	25.8	25.0	24.1	22.5	20.0
10	40.0	36.6	31.6	30.0	27.5	25.8	24.1
15	34.1	30.0	26.6	25.8	23.3	20.8	17.5
25	33.3	30.8	25.0	23.3	22.5	21.6	20.8
30	28.3	27.5	23.3	22.5	21.6	20.0	19.1
40	25.8	24.1	18.3	17.5	17.5	17.5	17.5
50	20.0	16.6	14.2	15.0	15.0	15.8	16.6
60	15.8	15.0	13.3	15.0	15.0	16.6	16.6
	pH = 9.0						
0	22.5	19.1	16.6	15.8	14.2	13.3	11.7
10	28.3	24.1	21.6	18.3	16.6	15.0	12.5
15	26.6	23.3	21.6	19.1	17.5	15.0	14.2
25	26.6	23.3	17.5	13.3	11.7	9.2	8.3
30	23.3	19.1	14.2	10.0	7.5	5.8	3.3
40	19.1	17.5	13.3	6.7	4.2	2.5	0.8
50	14.2	7.5	4.2	1.33	1.33	1.00	1.00
60	5.83	4.16	2.50	1.25	1.25	1.25	0.67

[a] Converted from g/100 g soln, given by the authors.

COMPONENTS:

1. Calcium sulfite; $CaSO_3$; [10257-55-3]
2. Calcium sulfate; $CaSO_4$; [7778-18-9]
3. Glucose; $C_6H_{12}O_6$; [50-99-7]
4. Sucrose; $C_{12}H_{22}O_{11}$; [50-99-7]
5. Water; H_2O; [7732-18-5]

ORIGINAL MEASUREMENTS:

Van der Linden, T.

Arch. Suikerind.Ned.-Indie 1916, *24*, 1113-28; *Dtsch. Zuckerind.* 1916, *41*, 815; *J. Soc. Chem. Ind., London* 1917, *36*, 96.

VARIABLES:

Temperature: 303 - 373 K

PREPARED BY:

H.D. Lutz, B. Engelen

EXPERIMENTAL VALUES:

The author reports the solubility of $CaSO_3.2H_2O$ [10035-03-7] in water, various aqueous solutions of glucose and sucrose, and in all these solvents saturated with gypsum.

Content of saccharose (suc) and glucose (glu) (mass %) and of $CaSO_4$ (saturated solution)	30	40	50	60	70	80	90	100 °C
				$CaSO_3.2H_2O$ mg/kg soln				
pure water	64	63	57	61	45	31	27	11
15% suc	103	83	73	80	59	41	36	41
15% suc and 1.5% glu	104	81	85	71	60	47	40	29
$CaSO_4$ (sat.)	31	29	25	19	12	9	8	6
$CaSO_4$ (sat.) and 15% suc	35	32	22	19	21	17	20	21
$CaSO_4$ (sat.), 15% suc, and 1.5% glu	32	27	22	20	19	19	19	23

(continued on next page)

AUXILIARY INFORMATION

METHOD APPARATUS/PROCEDURE:

Equilibrium was established by stirring the solvents with excess of solid calcium sulfite under nitrogen in a thermostatically controlled vessel for 10 - 36 hr. After filtering at the given temperature sulfite was determined iodometrically. Calcium was determined as CaO after precipitation as $CaCO_3$ and sulfate as $BaSO_4$. The values given are means of 2 - 13 measurements which differ by up to 10%.

SOURCE AND PURITY OF MATERIALS:

Calcium sulfite was precipitated from calcium chloride or calcium acetate solutions with a freshly prepared solution of $NaHSO_3$.
Calcium sulfate was the commercial pure salt.
Sucrose was precipitated from aqueous solutions with ethanol and washed with ether.
Glucose was the chemically pure product of Merck.

ESTIMATED ERROR:

REFERENCES:

COMPONENTS:

1. Calcium sulfite; $CaSO_3$; [10257-55-3]
2. Calcium sulfate; $CaSO_4$; [7778-18-9]
3. Glucose; $C_6H_{12}O_6$; [50-99-7]
4. Sucrose; $C_{12}H_{22}O_{11}$; [50-99-7]
5. Water; H_2O; [7732-18-5]

ORIGINAL MEASUREMENTS:

Van der Linden, T.

Arch. Suikerind.Ned.-Indie 1916, *24*, 1113-28; *Dtsch. Zuckerind.* 1916, *41*, 815; *J. Soc. Chem. Ind., London* 1917, *36*, 96.

EXPERIMENTAL VALUES (continued):

Content of saccharose (suc) and glucose (glu) (mass %) and of $CaSO_4$ (saturated solution)	30	40	50	60	70	80	90	100 °C
				$CaSO_3$[a] mg/kg soln				
pure water	49	48	44	47	35	24	21	8
15% suc	79	64	56	62	45	32	28	32
15% suc and 1.5% glu	80	62	65	55	46	36	31	22
$CaSO_4$ (sat.)	24	22	19	15	9	7	6	5
$CaSO_4$ (sat.) and 15% suc	27	25	17	15	16	13	15	16
$CaSO_4$ (sat.), 15% suc, and 1.5% glu	25	21	17	15	15	15	15	18
				$CaSO_3$[a] 10^{-4}mole/kg soln				
pure water	4.10	4.03	3.65	3.91	2.88	1.99	1.73	0.70
15% suc	6.60	5.31	4.67	5.12	3.78	2.63	2.31	2.63
15% suc and 1.5% glu	6.66	5.19	5.44	4.55	3.84	3.01	2.56	1.86
$CaSO_4$ (sat.)	1.99	1.86	1.60	1.22	0.77	0.58	0.51	0.38
$CaSO_4$ (sat.) and 15% suc	2.24	2.05	1.41	1.22	1.34	1.09	1.28	1.34
$CaSO_4$ (sat.), 15% suc, and 1.5% glu	2.05	1.73	1.41	1.28	1.22	1.22	1.22	1.47

[a] Calculated by the compilers.

COMPONENTS:

1. Calcium sulfite; $CaSO_3$; [10257-55-3]
2. Calcium sulfate; $CaSO_4$; [7778-18-9]
3. Sodium perchlorate; $NaClO_4$; [7601-89-0]
4. Water; H_2O; [7732-18-5]

ORIGINAL MEASUREMENTS:

Rengemo, T.; Brune, U.; Sillén, L.G.

Acta Chem. Scand. 1958, *12*, 873.

VARIABLES:

Temperature: 298, 308 and 348 K
Concentration of $NaClO_4$

PREPARED BY:

H.D. Lutz, B. Engelen

EXPERIMENTAL VALUES:

The authors studied the equilibrium

$$CaSO_4.2H_2O(s) + SO_3^{2-} \rightleftharpoons CaSO_3.1/2H_2O(s) + SO_4^{2-} + 3/2H_2O$$

in aqueous solutions of $NaClO_4$. From the equilibrium constants obtained $K = [SO_4^{2-}]/[SO_3^{2-}]$ = 76, 91, and 170 at 25, 35, and 75°C, respectively, and the solubility product of gypsum [10101-41-4], given by Latimer *et al.* (1).

$$K_{s0} = (CaSO_4.2H_2O) = 10^{-4.63}\ [mol^2\ dm^{-6}]$$

the activity solubility product of calcium sulfite at 25°C.

$$K_{s0}(CaSO_3.1/2H_2O) = a_{Ca^{2+}} \times a_{SO3^{2-}}$$

was determined by the authors[a] to be

$$10^{-6.51}\ (= 3.1 \times 10^{-7})\ mol^2\ dm^{-6}.$$

From this value the solubility of $CaSO_3.1/2H_2O$ in pure water at 25°C is estimated by the authors to be

$$91\ mg\ CaSO_3/dm^3\ \text{or}\ 7.6 \times 10^{-4}\ mol\ dm^{-3}.$$

[a] On the assumption that the activity coefficients of SO_4^{2-} and SO_3^{2-} are equal.

AUXILIARY INFORMATION

METHOD APPARATUS/PROCEDURE:

Equilibrium was studied under nitrogen in solutions which contained $NaClO_4$ as ionic medium, m_{Na^+} = 1 and 3.5 mol kg^{-1}, respectively. The solutions were made by adding an excess of $CaSO_4.2H_2O$ to a solution of $Na_2SO_3 + NaClO_4$. Equilibrium was established after 5 - 38 days. Sulfite was determined iodometrically and sulfate acidimetrically after precipitation as $BaSO_4$ and conversion into $BaCO_3$.

SOURCE AND PURITY OF MATERIALS:

$NaClO_4$ was prepared from Na_2CO_3 and $HClO_4$. $HClO_4$, Na_2CO_3, Na_2SO_3, and $CaSO_4.2H_2O$ were all commercial products of p.a. quality. N_2 was purified by Meyer and Ronge's method (2).

ESTIMATED ERROR:

Temperature: ±0.1 K. Log K_s: ±0.2 (authors)

REFERENCES:

1. Latimer, W.M.; Hicks, J.F.G.; Schutz, P.W. *J. Chem. Phys.* 1933, *1*, 620.
2. Meyer, F.R.; Ronge, G. *Angew. Chem.* 1939, 52, 637.

COMPONENTS:

1. Calcium sulfite; $CaSO_3$; [10257-55-3]
2. Calcium sulfate; $CaSO_4$; [7778-18-9]
3. Water; H_2O; [7732-18-5]
4. Sea-water

ORIGINAL MEASUREMENTS:

Kurota, O.; Takahashi, S.; Nakaoka, A.

Japan. Kokai 1977, 52-89561, 287-8.

VARIABLES:

pH value; composition

PREPARED BY:

B. Engelen, H.D. Lutz

EXPERIMENTAL VALUES:

The authors report the calcium content of saturated calcium sulfite solutions containing 0.125 mass % (compilers) of sulfate (saturated gypsum) or sea-water of various pH values, at room temperature.

Solvent	pH	Composition of saturated solutions	
		Ca^{2+a}	$CaSO_3$
		ppm	10^{-3}mole/kg soln
1 (sat. gypsum)	6.5	120	2.99
1 (sat. gypsum)	8	90	2.25
1 (sat. gypsum)	9	80	2.00
2 (sea-water)	8.2	110	2.74
3 (sea-water)	9	85	2.12

[a] Calculated by the compiler.

AUXILIARY INFORMATION

METHOD APPARATUS/PROCEDURE:

1. 1.6 g Na_2SO_3 was added to 1 dm^3 of a saturated solution of gypsum containing 520 ppm Ca^{2+} after adjusting the pH with NaOH to 6.5, 8, or 9, respectively.

2. 2.6 g Na_2SO_3 was added to 1 dm^3 of concentrated sea-water containing 820 ppm Ca^{2+}.

3. 4.6 dm^3 gaseous SO_2 was added to 10 dm^3 of concentrated sea-water containing 820 ppm Ca^+ after adjusting the pH value with NaOH to 9.

SOURCE AND PURITY OF MATERIALS:

Not given.

ESTIMATED ERROR:

REFERENCES:

COMPONENTS:	ORIGINAL MEASUREMENTS:
1. Calcium sulfite; $CaSO_3$; [10257-55-3] 2. Hydrochloric acid; HCl; [7647-01-0] 3. Sodium chloride; NaCl; [7647-14-5] 4. Water; H_2O; [7732-18-5]	Templeton, C.C.; Rushing, S.S.; Rodgers, J.C. *Mater. Prot.* 1963, 2, 42-7.
VARIABLES:	**PREPARED BY:**
Concentration of NaCl at 298 K pH: range 5 to 8	H.D. Lutz, B. Engelen

EXPERIMENTAL VALUES:

The authors report the concentration solubility product of calcium sulfite for various concentrations of sodium chloride (0.2 - 20 mass %) over a pH range from 5.0 to 8.0 at 25°C. The "analytical solubility product" has been defined by the authors as K'_{s0} = (mg analytical Ca^{2+}/dm^3 soln) x (mg analytical sulfite/dm^3 soln)[a]. These K'_{s0}-values were given as a graph of mass % of sodium chloride vs. pH.

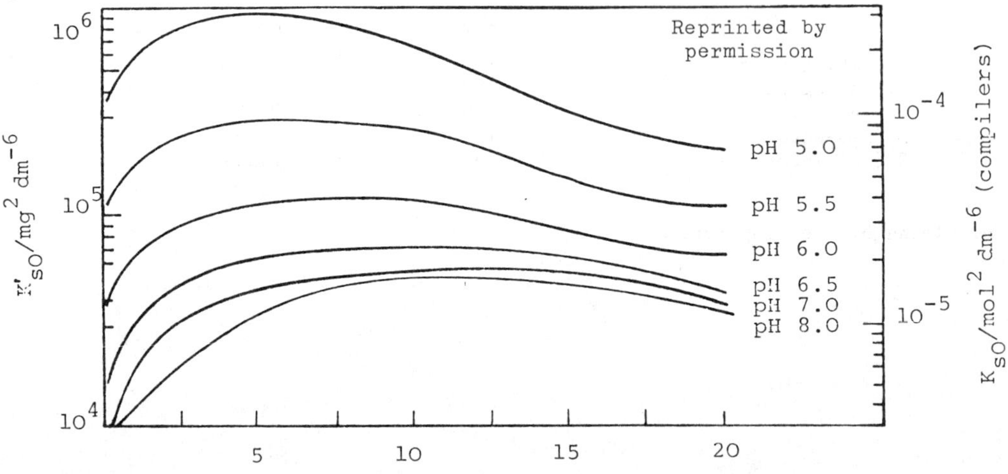

[a] See the following page.

(continued on next page)

AUXILIARY INFORMATION

METHOD APPARATUS/PROCEDURE:

Equilibrium was established by stirring the precipitate-brine mixture, acidified with HCl, for 48 hr. Calcium was determined by titration with EDTA, total dissolved sulfite (SO_3^{2-}, HSO_3^-, "H_2SO_3") iodometrically. The final pH was measured with glass and calomel electrodes.

SOURCE AND PURITY OF MATERIALS:

Purity of $CaSO_3.1/2H_2O$ is said to be 95 - 99%. No further details are reported.

ESTIMATED ERROR:

Not given.

REFERENCES:

COMPONENTS:	ORIGINAL MEASUREMENTS:
1. Calcium sulfite; $CaSO_3$; [10257-55-3] 2. Hydrochloric acid; HCl; [7647-01-0] 3. Sodium chloride; NaCl; [7647-14-5] 4. Water; H_2O; [7732-18-5]	Templeton, C.C.; Rushing, S.S.; Rodgers, J.C. *Mater. Prot.* 1963, 2, 42-7.

EXPERIMENTAL VALUES (continued):

The following data have been estimated from the graph (by compilers).

NaCl	$K'_{s0}/10^4$ mg^2 dm^{-6}						$K_{s0}/10^{-5}$ mol^2 dm^{-6}						
mass %	5.0	5.5	6.0	6.5	7.0	8.0	5.0	5.5	6.0	6.5	7.0	8.0	pH
2.5	81	25	9.2	5.1	3.1	2.0	25	7.8	2.9	1.6	1.0	0.6	
5	90	30	11	6.2	4.3	3.3	38	9.3	3.4	1.9	1.3	1.0	
7.5	87	28	12	6.9	5.1	4.5	27	8.7	3.7	2.2	1.6	1.4	
10	65	26	12	7.0	5.3	5.0	20	8.1	3.7	2.2	1.7	1.6	
12.5	45	20	11	6.9	5.4	5.0	14	6.2	3.4	2.2	1.7	1.6	
15	31	14	8.8	6.6	5.3	4.8	9.7	4.4	2.7	2.1	1.7	1.5	
17.5	24	12	7.2	5.4	5.0	4.2	7.5	3.7	2.2	1.7	1.6	1.3	
20	21	11	7.0	4.3	3.9	3.7	6.5	3.4	2.2	1.3	1.2	1.2	

a $[SO_3^{2-}] + [HSO_3^-] + ["H_2SO_3"]$, calculated as SO_3^{2-} by the authors.

COMPONENTS:

1. Calcium sulfite; $CaSO_3$; [10257-55-3]
2. Sodium chloride; NaCl; [7647-14-5]
3. Water; H_2O; [7732-18-5]
4. Sea-water

ORIGINAL MEASUREMENTS:

Cohen, A.; Zangen, M.; Koenigsbuch, M.; Goldschmidt, J.M.E.

Desalination 1982, *41*, 215-32.

VARIABLES:

Four temperatures: 298 - 348 K
pH values: 7.5 and 8.5
Time of stirring

PREPARED BY:

H.D. Lutz

EXPERIMENTAL VALUES:

The authors report the solubility of different samples of hydrated calcium sulfite, probably $CaSO_3.1/2H_2O$ [29501-28-8], in pure water and in sea-water at various pH values, temperatures, and stirring times.

Solubility[a] of hydrated calcium sulfite, in a large excess, in pure water at 25°C and pH 7.5.

Stirring time	Sample I	Sample II	Sample III	Sample IV
	10^{-3} mol kg^{-1}[a]	10^{-3} mol kg^{-1}[a]	10^{-3} mol kg^{-1}[a]	10^{-3} mol kg^{-1}[a]
5 min	0.81 ± 0.01	1.08 ± 0.02	1.03 ± 0.02	0.80 ± 0.01
1 hr	-	1.05 ± 0.02	1.06 ± 0.02	-
24 hr	0.69 ± 0.01	1.02 ± 0.02	1.02 ± 0.02	0.70 ± 0.01
3 days	0.68 ± 0.01	1.00 ± 0.02	0.94 ± 0.02	-
2 months	-	0.90 ± 0.02	-	0.58 ± 0.01

[a] mmolal total sulfite ($10^3m(CaSO_3)$/mol kg^{-1}), determined by iodometry.

(continued on next page)

AUXILIARY INFORMATION

METHOD APPARATUS/PROCEDURE:

500 cm^3 of solvent (pure water or sea-water) of preadjusted pH were introduced into a thermostatically controlled glass vessel. Sulfite oxidation was prevented by addition of 1 cm^3 benzyl alcohol/dm^3 solvent. The solvent was brought to the required temperature, with stirring and under a continuous stream of nitrogen, and then a large excess (100 times the solubility in pure water) or a small excess (10 times the solubility) of the calcium sulfite sample was added. Stirring and passage of nitrogen were continued and, at fixed intervals, samples of solution were pumped out through a sintered-glass filter, weighed, analysed for sulfite and calcium, and the pH was determined. Sulfite was determined iodometrically, calcium complexometrically.

SOURCE AND PURITY OF MATERIALS:

Different samples of calcium sulfite were prepared at ambient temperature by the following methods:
1) By mixing equivalent amounts of aqueous Na_2SO_3 and $CaCl_2$ solutions at concentrations of 0.05 (samples I), 0.4 (samples II), and 2.5 mol dm^{-3} (samples III).
2) By passing SO_2 through suspensions of $CaCO_3$ in water, until CO_2 was completely removed (samples IV). Commercial analytical purity reagents were used.
The calcium sulfite samples were washed with oxygen-free distilled water and characterized, both before and after the solubility determinations, by chemical, thermal, and X-ray powder diffraction analyses. The sea-water was taken from the Mediterranean Sea near Yavne, Israel. The sea-water was first filtered to remove solids, and then refluxed with HCl in a nitrogen atmosphere to decompose carbonates so as to avoid precipitation of $CaSO_3$ on heating.

ESTIMATED ERROR:

Temperature: ± 0.1 K
pH-value ± 0.05 (authors)

COMPONENTS:	ORIGINAL MEASUREMENTS:
1. Calcium sulfite; $CaSO_3$; [10257-55-3]	Cohen, A.; Zangen, M.; Koenigsbuch, M.; Goldschmidt, J.M.E.
2. Sodium chloride; NaCl; [7647-14-5]	*Desalination* 1982, *41*, 215-32.
3. Water; H_2O; [7732-18-5]	
4. Sea-water	

EXPERIMENTAL VALUES (continued):

Solubility[a] of hydrated calcium sulfite, sample II in large excess, in natural sea-water at various temperatures and pH 7.5.

Stirring	35°C		45°C		75°C
time	total sulfite[b]	total calcium[c]	total sulfite[b]	total calcium[c]	total sulfite[b]
	10^{-3} mol kg^{-1a}		10^{-3} mol kg^{-1a}		10^{-3} mol kg^{-1a}
20 min	2.39	12.6	2.63	12.3	-
1 hr	2.31	-	2.10	11.8	1.21
2 hr	2.17	-	2.05	11.7	-
3 hr	2.14	12.3	1.94	-	1.00
5 hr	2.09	-	1.90	-	-
24 hr	2.00	12.1	1.81	11.4	-
3 days	-	-	-	-	0.91

a mmolal total sulfite ($10^3 m(CaSO_3)/mol\ kg^{-1}$) or total calcium.

b Includes SO_3^{2-}, HSO_3^- and associated sulfite species. Values accurate to ±0.02.

c Includes the amount originally present in sea-water. Values accurate to ±0.1.

Sulfite concentration[a] in natural sea-water[b], after addition of 5 x 10^{-3} mole Na_2SO_3/dm^3 sea-water at 25°C and pH 8.1.

Stirring time	Without addition of solid calcium sulfite	With addition of solid calcium sulfite
	10^{-3} mol kg^{-1a}	10^{-3} mol kg^{-1a}
2 hr	4.72	4.49
3 hr	4.74	4.22
5 hr	4.72	4.09
7 hr	4.70	4.02
9 hr	4.65	3.92
24 hr	4.43	3.08
4 days	-	2.52
11 days	-	2.04

a mmolal total sulfite ($=10^3 m(CaSO_3)/mol\ kg^{-1}$) (error ±0.02, authors).

b Containing 11.7 x 10^{-3} mol dm^{-3} calcium (molarity).

(continued on next page)

COMPONENTS:	ORIGINAL MEASUREMENTS:
1. Calcium sulfite; $CaSO_3$; [10257-55-3] 2. Sodium chloride; NaCl; [7647-14-5] 3. Water; H_2O; [7732-18-5] 4. Sea-water	Cohen, A.; Zangen, M.; Koenigsbuch, M.; Goldschmidt, J.M.E. *Desalination* 1982, *41*, 215-32.

EXPERIMENTAL VALUES (continued):

Solubility[a] of hydrated calcium sulfite, sample II, in sea-water at 25°C.

Stirring time	Sea-water natural			Sea-water synthetic[d]	Sea-water concentrated[e]
	pH 7.5[b]	pH 7.5[c]	pH 8.5[b]	pH 7.5[b]	pH 7.5[b]
	10^{-3} mol kg^{-1}[a]	10^{-3} mol kg^{-1}[a]	10^{-3} mol kg^{-1}[a]	10^{-3} mol kg^{-1}[a]	10^{-3} mol kg^{-1}[a]
20 min	2.93[f]	2.22	2.93	3.90	2.64
1 hr	2.72	2.12	2.67	3.71	2.91
2 hr	2.57	2.10	2.54	-	2.78
3 hr	2.51	2.05	2.47	-	2.69
5 hr	2.44	2.00	2.33	-	2.51
24 hr	2.24	1.83	2.18	3.46	2.17
2 days	2.17	1.68	-	3.36	-
6 days	-	1.48	-	-	-
3 weeks	-	-	-	2.94	-
4 months	1.52	-	-	2.74	-

Solubility[a] of hydrated calcium sulfite, sample II after equilibration with sea-water at room temperature, in pure water at 25°C.

Stirring time	Equilibration with natural sea-water	Equilibration with synthetic sea-water[d]	
	for 3 weeks	for 48 hr	for 3 weeks
	10^{-3} mol kg^{-1}[a]	10^{-3} mol kg^{-1}[a]	10^{-3} mol kg^{-1}[a]
5 min	0.78[g]	0.82	-
1 hr	0.81	0.82	-
2 hr	0.76	-	0.64
24 hr	0.67	-	-
5 days	0.65	0.65	-
12 days	-	0.60	-

a mmolal total sulfite (= $10^3 m(CaSO_3)/mol\ kg^{-1}$), determined by iodometry.

b Solid calcium sulfite in large excess.

c Solid calcium sulfite in small excess.

d Synthetic sea-water, i.e. solution containing 0.7 mol dm^{-3} NaCl.

e Double-concentration natural sea-water, prepared by evaporating the decarbonated natural sea-water to half its volume.

f Error in the data ±0.02 (authors).

g Error in the data ±0.01 (authors).

COMPONENTS:	ORIGINAL MEASUREMENTS:
1. Calcium sulfite; $CaSO_3$; [10257-55-3] 2. Calcium hydroxide phosphate 3. Phosphorus pentoxide; P_2O_5; [1314-56-3] 4. Sulfur dioxide; SO_2; [7446-09-5] 5. Water; H_2O; [7732-18-5]	Mebane, W.M.; Dobbins, J.T.; Cameron, F.K. *J. Phys. Chem.* 1929, *33*, 961-9.
VARIABLES:	PREPARED BY:
Two temperatures: 273 and 298 K Composition	H.D. Lutz

EXPERIMENTAL VALUES:

The authors report the solubility of calcium sulfite and of mixtures of calcium sulfite and basic calcium phosphate solid solutions as solid phases in solutions containing various amounts of phospherous pentoxide and saturated with gaseous sulfur dioxide at 0 and 25°C.

Composition of saturated solutions					Solid phase
P_2O_5	CaO		SO_2		
mass %	mass %	mole/kg soln[a]	mass %	mole/kg soln[a]	
Temperature = 0°C					
-	0.03	0.0053	4.03	0.629	calcium sulfite?
0.67	1.42	0.253	0.43	0.067	sulfite phosphate mixture[b]
1.23	1.72	0.307	1.22	0.190	"
1.58	2.31	0.412	2.09	0.326	"
1.63	2.40	0.428	2.36	0.368	"
1.91	2.50	0.446	2.75	0.429	"
3.29	4.85	0.865	7.00	1.093	"

[a] Calculated by the compiler.

[b] Mixture of calcium sulfite and calcium hydroxide phosphate solid solution.

(continued on next page)

AUXILIARY INFORMATION

METHOD APPARATUS/PROCEDURE:	SOURCE AND PURITY OF MATERIALS:
Aqueous solutions of phosphoric acid were placed in wide-mouthed bottles fitted with rubber stoppers carrying inlet and outlet tubes of glass. Solid mixtures of calcium phosphate and calcium carbonate were added to the solutions until a solid persisted after saturation with gaseous sulfur dioxide. Equilibrium was established after 14 days. Samples of the supernatant solutions were analysed for calcium, sulfur dioxide, and phosphoric acid. Calcium was determined manganometrically after precipitation as the oxalate, sulfur dioxide iodometricaly, and phosphoric acid acidimetrically after precipitation as ammonium phosphomolybdate.	Not given. ESTIMATED ERROR: REFERENCES:

COMPONENTS:	ORIGINAL MEASUREMENTS:
1. Calcium sulfite; $CaSO_3$; [10257-55-3] 2. Calcium hydroxide phosphate 3. Phosphorus pentoxide; P_2O_5; [1314-56-3] 4. Sulfur dioxide; SO_2; [7446-09-5] 5. Water; H_2O; [7732-18-5]	Mebane, W.M.; Dobbins, J.T.; Cameron, F.K. *J. Phys. Chem.* 1929, *33*, 961-9.

EXPERIMENTAL VALUES (continued):

Composition of saturated solutions

P_2O_5 mass %	CaO mass %	CaO mole/kg soln[a]	SO_2 mass %	SO_2 mole/kg soln[a]	Solid phase
Temperature = 25°C					
-	1.92	0.342	4.57	0.713	calcium sulfite
1.30	1.27	0.226	1.60	0.250	sulfite phosphate mixture[b]
1.80	1.54	0.275	2.36	0.368	"
1.63	1.57	0.280	2.83	0.442	"
1.53	1.63	0.291	2.65	0.414	"
3.28	2.11	0.376	1.66	0.259	"
3.45	2.30	0.410	1.92	0.300	"
4.10	2.62	0.467	2.18	0.340	"
4.38	2.73	0.487	2.57	0.401	"
3.80	3.00	0.535	4.45	0.695	"
4.33	3.52	0.628	6.18	0.965	"
5.94	4.34	0.774	7.18	1.121	"
6.11	4.80	0.856	4.32	0.674	"
6.11	5.40	0.963	2.75	0.429	"
5.86	5.57	0.993	4.00	0.624	"

[a] Calculated by the compiler.

[b] Mixture of calcium sulfite and calcium hydroxide phosphate solid solution.

COMPONENTS:

1. Calcium sulfite; $CaSO_3$; [10257-55-3]
2. Calcium hydrogen phosphate; $CaHPO_4$; [7757-93-9]
3. Sucrose; $C_{12}H_{22}O_{11}$; [471-34-1]
4. Water; H_2O; [7732-18-5]

ORIGINAL MEASUREMENTS:

Ramaiah, N.A.; Sharma, J.K.

Proc. Ann. Conv. Sugar Technol. Assoc. India 1960, 28, 64-8.

VARIABLES:

One temperature: 323 K
Concentration of sucrose and $CaHPO_4$

PREPARED BY:

B. Engelen and H.D. Lutz

EXPERIMENTAL VALUES:

The authors report the solubility of calcium sulfite in water and solutions of various sucrose and calcium hydrogen phosphate concentrations at 50°C. The pH-values of the saturated solutions are also given.

Solvent	pH	mg CaO per dm^3 of solution	mg $CaSO_3$[a] per dm^3 of solution	$10^3c(CaSO_3)$[a] mol dm^{-3}
pure water	7.5	142	304.2	2.533
15 mass % sucrose	7.6	87	186.4	1.552
50 cm^3 suc[b] + 5 cm^3 phos[c]	6.0	100	214.2	1.783
50 cm^3 suc[b] + 10 cm^3 phos[c]	5.8	123	263.5	2.194
50 cm^3 suc[b] + 15 cm^3 phos[c]	5.7	144.6	309.8	2.579

[a] Calculated by the compilers.

[b] 15 mass % sucrose.

[c] $CaHPO_4$ content was 24 mg Ca per dm^3 of solution.

AUXILIARY INFORMATION

METHOD APPARATUS/PROCEDURE:

Water, sucrose solutions containing 15 mass % of sucrose, and the same sucrose solutions with the specified quantities of a solution of $CaHPO_4$ (triple superphosphate) saturated with lime and neutralized to pH 6.3 were saturated with excess of calcium sulfite. Calcium was determined complexometrically, and in solutions containing $CaHPO_4$, polarographically by using sodium zincate as the amperometric indicator and EDTA as the titrant.

SOURCE AND PURITY OF MATERIALS:

Reagents of commercial quality were used.

ESTIMATED ERROR:

REFERENCES:

COMPONENTS:	ORIGINAL MEASUREMENTS:
1. Calcium sulfite; $CaSO_3$; [10257-55-3] 2. Acetic acid; $C_2H_4O_2$; [64-19-7] 3. Sodium hydroxide; NaOH; [1310-73-2] 4. Water; H_2O; [7732-18-5]	Sano, H. *Osaka Kogyo Gijutsu Shikensho Kiho* 1974, *25*, 179-82.
VARIABLES:	**PREPARED BY:**
One temperature: 298 K pH: range 4 to 13	B. Engelen

EXPERIMENTAL VALUES:

The author reports the solubility of calcium sulfite in water at 25°C as a function of the pH of the solution. Experimental values are given in the form of a graph of total dissolved sulfite *vs*. pH. From this graph numerical data were estimated by the compiler.

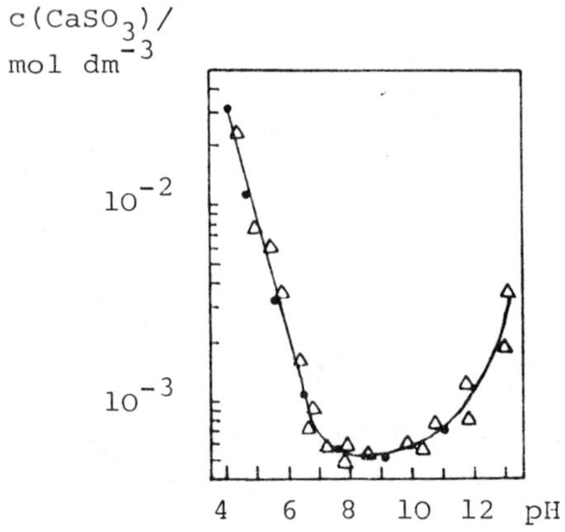

Reprinted by permission

pH	$CaSO_3$ g/dm^3	$10^3c(CaSO_3)$ mol dm^{-3}
4	4.00	33.3
5	1.01	8.4
6	0.256	2.13
7	0.082	0.685
8	0.063	0.521
9	0.060	0.50
10	0.065	0.545
11	0.084	0.70
12	0.135	1.12
13	0.266	2.21

AUXILIARY INFORMATION

METHOD APPARATUS/PROCEDURE:

1.0 g of $CaSO_3.1/2H_2O$ was added to 100 ml of water (pH = 7.9). CH_3COOH or NaOH was used to adjust the pH. After stirring for 1 hr the pH was measured (method not given), then the undissolved calcium sulfite was strained off and the content of SO_3^{2-} and HSO_3^- determined iodometrically.

SOURCE AND PURITY OF MATERIALS:

1) Commercial $CaSO_3.1/2H_2O$ (purity 90%) was dissolved in dilute HCl.NaOH was added to this solution to precipitate $CaSO_3.1/2H_2O$ (pH = 6). The precipitate was dispersed and filtered to give a product of 99% purity (triangles in the figure).
2) In a second procedure a mixture of $CaSO_3.1/2H_2O$ and $CaCO_3$ (10:1) was dissolved and precipitated in the same manner (black dots in the figure).

ESTIMATED ERROR:

REFERENCES:

COMPONENTS:

1. Calcium sulfite; $CaSO_3$; [10257-55-3]
2. Acetic acid; CH_3COOH; [64-19-7]
3. Citric acid; $C_6H_8O_7$; [77-92-9]
4. Water; H_2O; [7732-18-5]

ORIGINAL MEASUREMENTS:

Wurz, O.; Swoboda, O.

Text.-Rundsch. 1948, *3*, 201-6.

VARIABLES:

Four temperatures: 293 - 353 K

PREPARED BY:

B. Engelen, H.D. Lutz

EXPERIMENTAL VALUES:

Non-saturating solutes[a] Concentration/mol dm^{-3}	pH[b]	pH[c]	Composition of saturated solutions $CaSO_3$				
			20	40	60	80	t/°C
			mass % (= g/100 g soln, authors)				
Acetic acid 0.1	3	4.1	0.3	-	-	-	
	5	5.0	0.127	0.142	0.129	0.116	
	7	7.5	0.024	0.021	0.015	0.014	
	9	8.3	0.022	0.019	0.014	0.013	
Citric acid 0.1 ?	3	3.4	0.3	-	-	-	
	4	5.7	0.165	0.205	0.178	0.167	
	7	7.2	0.139	0.133	0.127	0.100	
	9	8.4	0.136	0.127	0.118	0.110	
			10^{-3} mole/kg soln (compilers)				
Acetic acid 0.1	3	4.1	25	-	-	-	
	5	5.0	10.57	11.82	10.74	9.66	
	7	7.5	2.00	1.75	1.25	1.17	
	9	8.3	1.83	1.58	1.17	1.08	
Citric acid 0.1 ?	3	3.4	25	-	-	-	
	4	5.7	13.73	17.06	14.82	13.90	
	7	7.2	11.57	11.07	10.57	9.16	
	9	8.4	11.32	10.57	9.82	9.16	

[a] Adjusted with NaOH to the pH given.

[b] pH of the solvent, adjusted with NaOH.

[c] pH of the equilibrated solution.

AUXILIARY INFORMATION

METHOD APPARATUS/PROCEDURE:

Oxygen-free water or the given solutions were saturated with calcium sulfite. The solutions were stirred for 0.5 hr and kept for 12 hr at the given temperature. An aliquot of the solutions was then analysed for sulfite with iodine or chloramine. The mean of 4 measurements is given.

SOURCE AND PURITY OF MATERIALS:

An oxygen-free slurry of calcium oxide preparation by adding ignited calcium carbonate to oxygen-free water was saturated with sulfur dioxide under nitrogen. After removing excess SO_2 by passing nitrogen through the solution the obtained calcium sulfite slurry was evaporated to dryness. The product contained 2% sulfate.

ESTIMATED ERROR:

REFERENCES:

COMPONENTS:

1. Calcium sulfite; $CaSO_3$; [10257-55-3]
2. Sodium formate; $NaHCO_2$; [141-53-7]
3. Sodium acetate; $NaCH_3CO_2$; [127-09-3]
4. Sodium phosphate; Na_3PO_4; [7601-54-9]
5. Water; H_2O; [7732-18-5]

ORIGINAL MEASUREMENTS:

Wurz, O.; Swoboda, O.

Text.-Rundsch. 1948, *3*, 201-6.

VARIABLES:

Four temperatures: 293 - 353 K

PREPARED BY:

B. Engelen, H.D. Lutz

EXPERIMENTAL VALUES:

Non-saturating solutes Concentration/mol dm^{-3}	pH^a	Composition of saturated solutions $CaSO_3$ mg/100 cm^3				
		20	40	60	80	t/°C
Na_3PO_4 0.033	6.3	47.20	48.75	47.38	46.50	
$NaCH_3CO_2$ 0.1	6.3	63.80	67.50	65.20	64.70	
$NaHCO_2$ 0.01, $NaCH_3CO_2$ 0.01	5.3	14.40	16.49	15.49	15.00	
$NaHCO_2$ 0.01, $NaCH_3CO_2$ 0.01	7.2	9.17	10.61	10.57	10.27	
$NaHCO_2$ 0.01, $NaCH_3CO_2$ 0.01	8.5	6.40	7.10	6.78	6.36	
		10^4c(mol dm^{-3}) (compilers)				
Na_3PO_4 0.033	6.3	39.29	40.58	39.44	38.70	
$NaCH_3CO_2$ 0.1	6.3	53.10	56.18	54.27	53.85	
$NaHCO_2$ 0.01, $NaCH_3CO_2$ 0.01	5.3	11.99	13.73	12.89	12.49	
$NaHCO_2$ 0.01, $NaCH_3CO_2$ 0.01	7.2	7.63	8.83	8.80	8.55	
$NaHCO_2$ 0.01, $NaCH_3CO_2$ 0.01	8.5	5.33	5.91	5.64	5.29	

[a] pH was adjusted with NaOH.

AUXILIARY INFORMATION

METHOD APPARATUS/PROCEDURE:

Oxygen-free water or the given solutions were saturated with calcium sulfite. The solutions were stirred for 0.5 hr and kept for 12 hr at the given temperature. An aliquot of the solutions was then analysed for sulfite with iodine or chloramine. The mean of 4 measusurements is given.

SOURCE AND PURITY OF MATERIALS:

An oxygen-free slurry of calcium oxide prepared by adding ignited calcium carbonate to oxygen-free water was saturated with sulfur dioxide under nitrogen. After removing excess SO_2 by passing nitrogen through the solution the obtained calcium sulfite slurry was evaporated to dryness. The product contained 2% sulfate.

ESTIMATED ERROR:

REFERENCES:

COMPONENTS:
1. Calcium sulfite; $CaSO_3$; [10257-55-3]
2. Xylose; $C_5H_{10}O_5$; [58-86-6]
3. Glucose; $C_6H_{12}O_6$; [50-99-7]
4. Lignosulfonic acid; [8062-15-5]
5. Water; H_2O; [7732-18-5]

ORIGINAL MEASUREMENTS:
Wurz, O.; Swoboda, O.
Text.-Rundsch. 1948, *3*, 201-6.

VARIABLES:
Four temperatures: 293 - 353 K

PREPARED BY:
B. Engelen, H.D. Lutz

EXPERIMENTAL VALUES:

Non-saturating solutes	Concentration/mass %	pH	Composition of saturated solutions $CaSO_3$			
			20	40	60	80 t/°C
			mg/100 g soln			
Glucose/Xylose	1	7	9.97	8.81	7.67	6.49
Lignosulfonic acid	10	7	31	46	39	36[a]
Lignosulfonic acid	10[a]	7.2	42	57	58	60[a]
			10^{-4} mole/kg soln (compilers)			
Glucose/Xylose	1	7	8.30	7.33	6.38	5.40
Lignosulfonic acid	10	7	25.8	38.3	32.5	30.0
Lignosulfonic acid	10[a]	7.2	35.0	47.4	48.3	49.9

[a] Adjusted with $NaHCO_2$ and $NaCH_3CO_2$.

AUXILIARY INFORMATION

METHOD APPARATUS/PROCEDURE:
Oxygen-free water or the given solutions were saturated with calcium sulfite. The solutions were stirred for 0.5 hr and kept for 12 hr at the given temperature. An aliquot of the solution was then analysed for sulfite with iodine or chloramine. The mean of 4 measurements is given.

SOURCE AND PURITY OF MATERIALS:
An oxygen-free slurry of calcium oxide prepared by adding ignited calcium carbonate to oxygen-free water was saturated with sulfur dioxide under nitrogen. After removing excess SO_2 by passing nitrogen through the solution the obtained calcium sulfite slurry was evaporated to dryness. The product contained 2% sulfate.

ESTIMATED ERROR:

REFERENCES:

COMPONENTS:	EVALUATOR:
1. Strontium sulfite; $SrSO_3$; [13451-02-0] 2. Water; H_2O; [7732-18-5]	H.D. Lutz, Dept. of Chemistry, University of Siegen, FR Germany. August 1983.

CRITICAL EVALUATION:

Strontium sulfite crystallizes from aqueous solutions as the anhydrous salt, $SrSO_3$. Numerical data on the solubility of strontium sulfite in water have been given by Autenrieth *et al.* (1), Rodin *et al.* (2), and Cohen *et al.* (3). Autenrieth *et al.* (1) report a value of 3.3×10^{-2} g $SrSO_3$/kg H_2O ($m(SrSO_3) = 2.0 \times 10^{-4}$ mol kg^{-1}) at 289 - 291 K. Rodin *et al.* (2) report that the solubility of $SrSO_3$ in water increases from 1.38×10^{-3} g/dm^3 ($c(SrSO_3) = 8.2 \times 10^{-6}$ mol dm^{-3}) at 293 K to 3.22×10^{-3}g/dm^3 (1.91×10^{-5} mol dm^{-3}) at 363 K. Cohen *et al.* (3) studied the change of the amount of dissolved strontium sulfite in solutions containing 0.7 mol dm^{-3} NaCl and rapidly precipitated $SrSO_3$. They report that the $SrSO_3$ concentration at 298 K decreases from 7.0×10^{-4} to 3.5×10^{-4} mol kg^{-1} (molality) within 24 hr. The change in the solubility is attributed to a phase transition of the precipitated strontium sulfite.

TENTATIVE VALUES

The solubility of $SrSO_3$ in water at room temperature is approximately 1×10^{-4} mol kg^{-1} (molality scale) (0.015 g/kg H_2O). The temperature coefficient of solubility is positive.

Fragmentary investigations on the ternary systems $SrSO_3$-SO_2-H_2O (4) and $SrSO_3$-ethanol-H_2O (5) indicate that the solubility of strontium sulfite increases to 17.3 g (0.103 mole)/kg soln with increasing SO_2 content of the solution (4) and decreases with increasing ethanol content (5). $SrSO_3$ is described to be practically insoluble in dilute acetic acid (1,6) and readily soluble in other acids, undergoing decomposition.

REFERENCES

1. Autenrieth, W.; Windaus, A. *Z. Anal. Chem.* 1898, *37*, 290.

2. Cohen, A.; Zangen, M.; Goldschmidt, J.M.E. *Rev. Chim. Miner.* 1981, *18*, 142.

3. Rodin, I.V.; Margulis, E.V. *Zh. Neorg. Khim.* 1983, *28*, 255; *Russ. J. Inorg. Chem. (Eng. Transl.)* 1983, *28*, 142.

4. Simon, A.; Waldmann, K. *Naturwissenschaften* 1958, *45*, 128.

5. Arnal, T.G.; Mesorana, J.M.P. *An. Fis. Quim* 1947, *43*, 439.

6. Hinds, J.I.D. *J. Am. Chem. Soc.* 1911, *33*, 510; *Chem. News* 1911, *103*, 157.

COMPONENTS:

1. Strontium sulfite; $SrSO_3$; [13451-02-0]

2. Water; H_2O; [7732-18-5]

ORIGINAL MEASUREMENTS:

Autenrieth, W.; Windaus, A.

Z. Anal. Chem. 1898, *37*, 290-300.

VARIABLES:

One temperature: 290 K

PREPARED BY:

H.D. Lutz

EXPERIMENTAL VALUES:

The authors report the solubility of strontium sulfite in water at 16-18°C to be

1 part of $SrSO_3$ per 30 000 parts of H_2O.

This value is converted by the compiler to

3.3×10^{-2}g $SrSO_3$/kg H_2O or $m(SrSO_3) = 2.0 \times 10^{-4}$ mol kg^{-1}.

AUXILIARY INFORMATION

METHOD APPARATUS/PROCEDURE:

Saturation method. Equilibrium was established with frequent shaking after several days. Sulfite was determined iodometrically and strontium gravimetrically as the sulfate.

The content of strontium in the saturated solution was somewhat higher than the sulfite content due to a little amount of $SrSO_4$. Numerical data are not given by the authors.

SOURCE AND PURITY OF MATERIALS:

Not given.

ESTIMATED ERROR:

REFERENCES:

COMPONENTS:

1. Strontium sulfite; $SrSO_3$; [13451-02-0]

2. Water; H_2O; [7732-18-5]

ORIGINAL MEASUREMENTS:

Rodin, I.V.; Margulis, E.V.

Zh. Neorg. Khim. 1983, *28*, 255-6; **Russ. J. Inorg. Chem. (Eng. Transl.)* 1983, *28*, 142.

VARIABLES:

Four temperatures: 293 - 363 K

PREPARED BY:

B. Engelen

EXPERIMENTAL VALUES:

The solubilities of strontium sulfite in water at various temperatures are reported. The solubility products are defined as $K_{s0}(SrSO_3) = [Sr^{2+}]^2$.

t/°C	Sr	$SrSO_3$		$10^{11}K_{s0}$
	mg/dm^3	mg/dm^{3a}	$10^5c/mol\ dm^{-3}$	$mol^2\ dm^{-6}$
20	0.72	1.38	0.82	6.72
50	1.13	2.16	1.28	16.4
70	1.49	2.85	1.69	28.6
90	1.68	3.22	1.91	36.5

[a] Calculated by the compiler.

AUXILIARY INFORMATION

METHOD APPARATUS/PROCEDURE:

Saturation method. Equilibrium was established by stirring the saturated solution in thermostatically controlled glass tubes. Equilibrium was tested for analytically - 3 hr was reported to be sufficient. Strontium was determined gravimetrically.

SOURCE AND PURITY OF MATERIALS:

Strontium sulfite was precipitated from $SrCl_2$ solutions with Na_2SO_3.

ESTIMATED ERROR:

Temperature: ±0.5 K

REFERENCES:

COMPONENTS:	ORIGINAL MEASUREMENTS:
1. Strontium sulfite; $SrSO_3$; [13451-02-0] 2. Sodium chloride; NaCl; [7647-14-5] 3. Water; H_2O; [7732-18-5]	Cohen, A.; Zangen, M.; Goldschmidt, J.M.E. *Rev. Chim. Miner.* 1981, *18*, 142-7.
VARIABLES: One temperature: 298 K One concentration of NaCl: 0.7 mol dm^{-3} Time of stirring	PREPARED BY: H.D. Lutz

EXPERIMENTAL VALUES:

The solubility of rapidly precipitated $SrSO_3$ at 25°C decreases from 7.0 x 10^{-4} to 3.5 x 10^{-4} mol kg^{-1} (molality) within 24 hr. The change of the solubility is attributed to a phase transition of the solid strontium sulfite.

AUXILIARY INFORMATION

METHOD APPARATUS/PROCEDURE:	SOURCE AND PURITY OF MATERIALS:
The solubility of $SrSO_3$ was determined in a stirred solution of aqueous NaCl (concentration 0.7 mol dm^{-3}). Method of analysis not given.	Strontium sulfite was precipitated by mixing equivalent amounts of $SrCl_2$ and Na_2SO_3.
	ESTIMATED ERROR:
	REFERENCES:

COMPONENTS:	EVALUATOR:
1. Barium sulfite; $BaSO_3$; [7787-39-5] 2. Water; H_2O; [7732-18-5]	H.D. Lutz, Dept. of Chemistry, University of Siegen, FR Germany. August 1983.

CRITICAL EVALUATION:

Barium sulfite crystallizes from aqueous solutions in the form of the anhydrous salt, $BaSO_3$. The solubility of $BaSO_3$ in water has not yet been thoroughly investigated. Numerical data have been reported by Autenrieth *et al.* (1), Rogowicz (2), Parsons (3), Marećek *et al.* (4), Rodin *et al.* (5), and Cohen *et al.* (6), yet not under directly comparable experimental conditions. In 1898, Autenrieth *et al.* (1) reported the value of 0.022 g/kg H_2O ($m(BaSO_3) = 1.0 \times 10^{-4}$ mol kg^{-1}) at 289 K. Rogowicz (2) gave the values 0.1974 and 0.0177 g/dm^3 ($c(BaSO_3)$ = 9.080 x 10^{-4} and 8.14 x 10^{-5} mol dm^{-3}) at 293 and 353 K, respectively. Parsons (3) reported the solubility product of $BaSO_3$ (K_{s0}) as 9.5 x 10^{-10} mol^2 dm^{-6} at 298.2 K from a source which is not given nor could be traced. This value is equivalent to a solubility of 6.7 mg/dm^3 ($c(BaSO_3)$ = 3.08 x 10^{-5} mol dm^{-3}). In 1983 Rodin *et al.* (5) determined some values from 1.23 x 10^{-3} g/dm^3 ($c(BaSO_3)$ = 5.7 x 10^{-6} mol dm^{-3}) at 293 K to 3.5 x 10^{-3} g/dm^3 (1.61 x 10^{-5} mol dm^{-3}) at 353 K. Cohen *et al.* (6) reported that the amount of dissolved barium sulfite in solutions containing 0.7 mol dm^{-3} NaCl and rapidly precipitated $BaSO_3$ at 298 K decreases from 3.0 x 10^{-4} to 2.3 x 10^{-4} mol kg^{-1} (molality) within 24 hr. The change in the solubility is attributed to a phase transition of the precipitated barium sulfite. In 1970, Marećek *et al.* (4) determined the activity solubility product from equilibrium studies of the reaction $BaSO_3(s) + CO_3^{2-} \rightleftharpoons BaCO_3(s) + SO_3^{2-}$ as 4.9 x 10^{-10} mol^2 kg^{-2} (molality scale) at 298.2 K.

The value of Marećek *et al.* (4) appears to be the most reliable solubility value with respect to the activity of Ba^{2+} (aq) and SO_3^{2-} (aq) at the present time. The total amount of dissolved barium sulfite including the undissociated $BaSO_3$ seems to be greater than that of Marećek *et al.* (4) and is probably approximately the same as the values given by Autenrieth *et al.* (1) and Parsons (3). The scarce information on the temperature shift in the solubility of $BaSO_3$ is contradictory. A negative temperature coefficient of the solubility is reported by Rogowicz (2), a positive temperature coefficient by Rodin *et al.* (5).

TENTATIVE VALUES

The solubility of $BaSO_3$ in water at room temperature is approximately 5 x 10^{-5} mol kg^{-1} (0.01 g/kg H_2O). The solubility product, based on the activities, is 5 x 10^{-10} mol^2 kg^{-2} (molality scale). The temperature coefficient of solubility is positive.

The solubility of barium sulfite in water is affected by the presence of a third component. This becomes evident by fragmentary experimental data on the ternary systems $BaSO_3$-SO_2-H_2O (7,8), $BaSO_3$-sucrose-H_2O (2,9), and $BaSO_3$-ethanol-H_2O (10). The solubility of $BaSO_3$ increases by about 10^3 with increasing SO_2 content (7,8) and decreases with increasing sucrose (2,9) and ethanol (10) content. In acids $BaSO_3$ is readily dissolved, but undergoes decomposition.

COMPONENTS:	EVALUATOR:
1. Barium sulfite; $BaSO_3$; [7787-39-5] 2. Water; H_2O; [7732-18-5]	H.D. Lutz, Dept. of Chemistry, University of Siegen, FR Germany. August 1983.

CRITICAL EVALUATION: (continued)

REFERENCES

1. Autenrieth, W.; Windaus, A. *Z. Anal. Chem.* 1898, *37*, 290.
2. Rogowicz, J. *Z. Ver. Dtsch. Zucker-Ind., Allg. Teil* 1905, 938.
3. Parsons, R. *Handbook of Electrochemical Constants*, Academic Press, London, 1959, p. 56.
4. Marećek, J.; Erdös, E. *Collect. Czech. Chem. Commun.* 1970, *35*, 1017.
5. Rodin, I.V.; Margulis, E.V. *Zh. Neorg. Khim.* 1983, *28*, 255; *Russ. J. Inorg. Chem. (Eng. Transl.)* 1983, *28*, 142.
6. Cohen, A.; Zangen, M.; Goldschmidt, J.M.E. *Rev. Chim. Miner.* 1981, *18*, 142.
7. Simon, A.; Waldmann, K. *Naturwissenschaften* 1958, *45*, 128.
8. Conrad, F.H.; Beuschlein, W.L. *J. Am. Chem. Soc.* 1934, *56*, 2554.
9. Geese, J.W. *Z. Ver. Dtsch. Zucker-Ind., Allg. Teil* 1898, *48*, 99.
10. Arnal, T.G.; Mesorana, J.M.P. *An. Fis. Quim.* 1947, *43*, 439.

COMPONENTS:	ORIGINAL MEASUREMENTS:
1. Barium sulfite; $BaSO_3$; [7787-39-5] 2. Water; H_2O; [7732-18-5]	Autenrieth, W.; Windaus, A. *Z. Anal. Chem.* 1898, *37*, 290-300.
VARIABLES: One temperature: 289 K	PREPARED BY: H.D. Lutz

EXPERIMENTAL VALUES:

The solubility of barium sulfite in water at 16°C is

1 part of $BaSO_3$ per 46 000 parts of water.

This value is equal to (compiler)

2.2×10^{-2} g $BaSO_3$/kg H_2O or $m(BaSO_3) = 1.0 \times 10^{-4}$ mol kg^{-1}.

AUXILIARY INFORMATION

METHOD APPARATUS/PROCEDURE:	SOURCE AND PURITY OF MATERIALS:
Saturation method. Equilibrium was established, with frequent shaking, after several days. The sulfite content of the saturated solution was determined iodometrically.	$BaSO_3$ was precipitated from a solution of barium chloride with Na_2SO_3 and washed carefully with distilled water. ESTIMATED ERROR: REFERENCES:

COMPONENTS:

1. Barium sulfite; $BaSO_3$; [7787-39-5]
2. Water; H_2O; [7732-18-5]

ORIGINAL MEASUREMENTS:

Rodin, I.V.; Margulis, E.V.

Zh. Neorg. Khim. 1983, *28*, 255; *Russ. J. Inorg. Chem. (Eng. Transl.)* 1983, *28*, 142.

VARIABLES:

Four temperatures: 293 - 363 K

PREPARED BY:

B. Engelen

EXPERIMENTAL VALUES:

The solubilities of barium sulfite in water at various temperatures are reported. The solubility products are defined as $K_{s0}(BaSO_3) = [Ba^{2+}]^2$.

t/°C	Ba	$BaSO_3$		$10^{11}K_{s0}$
	mg/dm^3	mg/dm^{3a}	$10^5 c/mol\ dm^{-3}$	$mol^2\ dm^{-6}$
20	0.78	1.23	0.57	3.25
50	1.19	1.88	0.87	7.57
70	1.62	2.56	1.18	13.9
80	2.21	3.50	1.61	25.9

[a] Calculated by the compiler.

AUXILIARY INFORMATION

METHOD APPARATUS/PROCEDURE:

Saturation method. Equilibrium was established by stirring the saturated solution in thermostatically controlled glass tubes. Equilibrium was tested for analytically - 3 hr is reported to be sufficient. Barium was determined gravimetrically.

SOURCE AND PURITY OF MATERIALS:

Barium sulfite was precipitated from $BaCl_2$ solutions with Na_2SO_3.

ESTIMATED ERROR:

Temperature: ±0.5 K.

REFERENCES:

COMPONENTS:	ORIGINAL MEASUREMENTS:
1. Barium sulfite; $BaSO_3$; [7787-39-5] 2. Sucrose; $C_6H_{12}O_6$; [50-99-7] 3. Water; H_2O; [7732-18-5]	Rogowicz, J. *Z. Ver. Dtsch. Zucker-Ind., Allg. Teil* 1905, 938.
VARIABLES:	PREPARED BY:
Two temperatures: 293 and 353 K	H.D. Lutz

EXPERIMENTAL VALUES:

Composition of saturated solution

Sucrose	g $BaSO_3$/100 cm^3		$10^3c(BaSO_3)$/mol dm^{-3} [a]	
mass %	20°C	80°C	20°C	80°C
0	0.01974	0.00177	0.9080	0.0814
10	0.01040	0.00335	0.4784	0.1541
20	0.00968	0.00289	0.4453	0.1329
30	0.00782	0.00223	0.3597	0.1026
40	0.00484	0.00158	0.2226	0.0727
50	0.00298	0.00149	0.1371	0.0685
66 (saturated soln.)	0.00223	0.00112	0.1026	0.0515

[a] Calculated by the compiler.

AUXILIARY INFORMATION

METHOD APPARATUS/PROCEDURE:

Saturation method. Sulfite was determined iodometrically, barium gravimetrically as $BaSO_4$.

SOURCE AND PURITY OF MATERIALS:

Barium sulfite was precipitated by mixing freshly prepared solutions of barium hydroxide and sulfurous acid. Sucrose was recrystallized from alcohol.

ESTIMATED ERROR:

REFERENCES:

COMPONENTS:

1. Barium sulfite; $BaSO_3$; [7787-39-5]
2. Sulfur dioxide; SO_2; [7446-09-5]
3. Water; H_2O; [7732-18-5]

ORIGINAL MEASUREMENTS:

Conrad, F.H.; Beuschlein, W.L.

J. Am. Chem. Soc. 1934, *56*, 2554-62.

VARIABLES:

One temperature: 298 K
Pressure of SO_2

PREPARED BY:

B. Engelen

EXPERIMENTAL VALUES:

t/°C	p_{total}	p_{SO_2}[a]	g SO_2/100 g H_2O		$BaSO_3$[c]	$m(BaSO_3)$[c]
	mm Hg	10^{-3} bar	total	combined[b]	g/kg H_2O	mol kg^{-1}
25	659	847	7.23	0.25	8.48	0.039

[a] Corrected for p_{H_2O} and converted to bar by the compiler.

[b] SO_2 required to form the monosulfite.

[c] Calculated from combined SO_2 by the compiler.

AUXILIARY INFORMATION

METHOD APPARATUS/PROCEDURE:

Equilibrium, which was established after more than 12 hr, was studied in a special flask with connections to a weighing pipette for analysing the saturated solution, and to a mercury monometer to measure the pressure of the gas over the solution.

Total, free and combined SO_2 were determined by acidimetric and iodometric titration, respectively (1).

SOURCE AND PURITY OF MATERIALS:

Preparation of $BaSO_3$ was not given. The SO_2 used was SO_3-free. The amount of inert or non-absorbable gases was about 0.15%.

ESTIMATED ERROR:

REFERENCES:

1. Birchard, W.H. *Pap. Ind.* 1926, *8*, 793.

COMPONENTS:	ORIGINAL MEASUREMENTS:
1. Barium sulfite; $BaSO_3$; [7787-39-5] 2. Sodium chloride; NaCl; [7647-14-5] 3. Water; H_2O; [7732-18-5]	Cohen, A.; Zangen, M.; Goldschmidt, J.M.E. *Rev. Chim. Miner.* 1981, *18*, 142-7.
VARIABLES: One temperature: 298 K One concentration of NaCl: 0.7 mol dm^{-3} Time of stirring	PREPARED BY: H.D. Lutz

EXPERIMENTAL VALUES:

The solubility of rapidly precipitated $BaSO_3$ at 25°C decreases from 3.0 x 10^{-4} to 2.3 x 10^{-4} mol kg^{-1} (molality) within 24 hr. The change in the solubility is attributed to phase transition of the obtained barium sulfite.

AUXILIARY INFORMATION

METHOD APPARATUS/PROCEDURE:	SOURCE AND PURITY OF MATERIALS:
The solubility of $BaSO_3$ was determined in a stirred solution of aqueous NaCl (concentration 0.7 mol dm^{-3}). Method of analysis not given.	Barium sulfite was precipitated by mixing equivalent amounts of aqueous solutions of $BaCl_2$ and Na_2SO_3.
	ESTIMATED ERROR:
	REFERENCES:

COMPONENTS:

1. Barium sulfite; $BaSO_3$; [7787-39-5]
2. Sodium carbonate; Na_2CO_3; [497-19-8]
3. Sodium sulfite; Na_2SO_3; [7757-83-7]
4. Water; H_2O; [7732-18-5]

ORIGINAL MEASUREMENTS:

Mareček, J.; Erdös, E.

Collect. Czech. Chem. Commun. 1970, *35*, 1017-29.

VARIABLES:

One temperature: 298 K
Concentration of Na_2CO_3 and Na_2SO_3

PREPARED BY:

H.D. Lutz

EXPERIMENTAL VALUES:

The authors studied the equilibrium of the reaction

$$BaSO_3(s) + CO_3^{2-} \quad BaCO_3(s) + SO_3^{2-}$$

in aqueous solutions of Na_2CO_3 and Na_2SO_3. From the equilibrium constant found at 25°C

$$K = [SO_3^{2-}]/[CO_3^{2-}] = 0.243$$

and the activity solubility product of $BaCO_3$ [513-77-9] reported by Townley *et al.* (1) and Hogge *et al.* (2)

$$K_{s0}(BaCO_3) = 2.03 \times 10^{-9} \text{ [mol}^2 \text{ kg}^{-2}]$$

the activity solubility product of barium sulfite at 25°C

$$K_{s0}(BaSO_3) = 4.9 \times 10^{-10} \text{ [mol}^2 \text{ kg}^{-2}]^a$$

was calculated by the authors.

[a] Molality scale.

AUXILIARY INFORMATION

METHOD APPARATUS/PROCEDURE:

Equilibrium was studied under nitrogen in a special apparatus with connections to a weighing pipette, to analyse the solutions saturated with barium sulfite and barium carbonate. The solutions were made by adding an excess of solid $BaSO_3$, $BaCO_3$, or mixtures of $BaSO_3$ and $BaCO_3$ to solutions containing various amounts of Na_2SO_3 and Na_2CO_3 (m_{Na^+} = 0.1 to 1 mol kg^{-1}). Equilibrium was established after 20 hr. Sulfite was determined iodometrically and carbonate acidimetrically after oxidizing the sulfite with H_2O_2. Furthermore, the content of sodium was determined gravimetrically as sulfate to check the amount of sulfite oxidized to sulfate in the saturated solution.

SOURCE AND PURITY OF MATERIALS:

$BaSO_3$ was prepared from $BaCO_3$ and gaseous SO_2. $BaCO_3$ was precipitated from saturated $Ba(OH)_2$ solution with gaseous CO_2. Sodium carbonate was prepared by thermal decomposition of $NaHCO_3$. Water was redistilled, boiled and saturated with N_2. The other chemicals used were commercial products of p.a. purity.

ESTIMATED ERROR:

Temperature: ±0.05 K;
K (equilibrium): ±0.7% (estimated by authors).

REFERENCES:

1. Townley, R.W.; Whitney, W.B.; Felsing, W.A. *J. Am. Chem. Soc.* 1937, *59*, 631.
2. Hogge, E.; Johnston, H.L. *J. Am. Chem. Soc.* 1939, *61*, 2154.

COMPONENTS:	EVALUATOR:
1. Manganese (II) sulfite; $MnSO_3$; [13568-71-3] 2. Water; H_2O; [7732-18-5]	H.D. Lutz, Dept. of Chemistry, University of Siegen, FR Germany. April 1983.

CRITICAL EVALUATION:

Manganese sulfite crystallizes from aqueous solutions in the form of several hydrates. The formation of the various hydrates depends on temperature, amount of sulfurous acid, and preparation technique.

Two different trihydrates, monoclinic α-$MnSO_3.3H_2O$ (1-3) [60365-38-0] and orthorhombic β-$MnSO_3.3H_2O$ (4) [60365-38-0], and $MnSO_3.5/2H_2O$ (2,5-6) [75042-11-4] are reported to exist at room temperature. At higher temperatures $MnSO_3.2H_2O$ (2) [75042-12-5] and $MnSO_3.H_2O$ (1,7-9) [65410-83-5] are formed. The existence of a $MnSO_3.\frac{1}{2}H_2O$ [60365-38-0] (10) has not been fully verified.

The solubility of these hydrates has not yet been throroughly investigated. Several authors report that manganese sulfite is insoluble in water (7,11-12), readily soluble in sulfurous acid (7) or in other acids, undergoing decomposition (11,12). Only Gorgeu (7) and Rodin *et al.* (13) give some numerical data, namely for $MnSO_3.3H_2O$, $MnSO_3.2.5H_2O$, and $MnSO_3.1H_2O$. It is not known which of the two different trihydrates Gorgeu investigated, but it seems to be the monoclinic α-$MnSO_3.3H_2O$ (2). The data given by Gorgeu (7) are 0.1 g $MnSO_3$/kg H_2O ($m(MnSO_3) = 7.4 \times 10^{-4}$ mol kg^{-1}) at room temperature (cold water) and 0.2 g $MnSO_3$/kg H_2O ($m(MnSO_3) = 1.5 \times 10^{-3}$ mol kg^{-1}) in hot water. The solubility given by Gorgeu seems to be rather too high, because the samples used were probably not fully free of sulfate. Rodin *et al.* (13) report that the solubility of $MnSO_3.2.5H_2O$ in water increases from 6.93×10^{-3} mass % ($m(MnSO_3) = 5.13 \times 10^{-4}$ mol kg^{-1}) at 293 K to 1.23×10^{-2} mass % ($m(MnSO_3) = 9.13 \times 10^{-4}$ mol kg^{-1}) at 343 K. The solubility of $MnSO_3.1H_2O$ is said to be 1.48×10^{-2} mass % ($m(MnSO_3) = 1.10 \times 10^{-3}$ mol kg^{-1}) (13) at 363 K.

TENTATIVE VALUE

The solubility of $MnSO_3$ (i.e. of the hydrates present) in water at room temperature is approximately 5×10^{-4} mol kg^{-1} (molality).

REFERENCES

1. Lutz, H.D.; El-Suradi, S.M.; Engelen, B. *Z. Naturforsch., Teil B* 1977, *32b*, 1230.
2. Lutz, H.D.; El-Suradi, S.M.; Mertins, Ch.; Engelen, B. *Z. Naturforsch., Teil B* 1980, *35*, 808.
3. Engelen, B.; Freiburg, C. *Z. Naturforsch., Teil B* 1979, *34*, 1495.
4. Baggio, R.F.; Baggio, S. *Acta Crystallogr. Sect. B* 1976, *32*, 1959.
5. Rammelsberg, C. *Ann. Phys. Chem.* 1846, *67*, 245.
6. Rohrig, A.J. *J. Prakt. Chem.* 1888, *37*, 217.
7. Gorgeu, A. *C.R. Hebd. Seances Acad. Sci.* 1883, *96*, 341.
8. Denigès, M.G. *Bull. Soc. Chim. Fr.* 1892, *7*, 569.

COMPONENTS:	EVALUATOR:
1. Manganese (II) sulfite; $MnSO_3$; [13568-71-3] 2. Water; H_2O; [7732-18-5]	H.D. Lutz, Dept. of Chemistry, University of Siegen, FR Germany. April 1983.

CRITICAL EVALUATION: (continued)

9. Cola, M.; Tarantino, S. *Gazz. Chim. Ital.* 1962, *92*, 174.

10. Buttler, F.G.; Mitchell, A.J. *J. Therm. Anal.* 1976, *10*, 257.

11. Matroff, G. *Diss.* Braunschweig T.H. 1930, 15/16.

12. Muspratt, J.S. *Justus Liebigs Ann. Chem.* 1844, *50*, 259.

13. Rodin, I.V.; Margulis, E.V. *Zh. Neorg. Khim.* 1983, *28*, 258; *Russ. J. Inorg. Chem. (Eng. Transl.)* 1983, *28*, 144.

COMPONENTS:

1. Manganese sulfite; $MnSO_3$; [13568-71-3]
2. Water; H_2O; [7732-18-5]

ORIGINAL MEASUREMENTS:

Rodin, I.V.; Margulis, E.V.

Zh. Neorg. Khim. 1983, *28*, 258-9; *Russ. J. Inorg. Chem. (Eng. Transl.)* 1983, *28*, 144.

VARIABLES:

Four temperatures: 293 - 363 K

PREPARED BY:

B. Engelen

EXPERIMENTAL VALUES:

t/°C	$MnSO_3$ 10^4 mass %	$MnSO_3$ $10^4 m$/mol kg^{-1} [a]
20[b]	69.3	5.134
50[b]	95.4	7.067
70[b]	123.2	9.127
90[c]	148.4	10.994

[a] Calculated by the compiler.

[b] The solid phase is claimed to be $MnSO_3.2.5H_2O$ [75042-11-4].

[c] The solid phase is claimed to be $MnSO_3.1H_2O$ [65410-83-5].

AUXILIARY INFORMATION

METHOD APPARATUS/PROCEDURE:

Saturation method. Equilibrium was established by stirring the saturated solutions in thermostatically controlled glass tubes. Equilibrium was tested for analytically - 3 hr was reported to be sufficient. Manganese was determined titrimetrically using peroxodisulfate/Ag^+.

SOURCE AND PURITY OF MATERIALS:

Not well defined.

ESTIMATED ERROR:

Not given.

REFERENCES:

COMPONENTS:
1. Manganese sulfite; $MnSO_3$; [13568-71-3]
2. Carbon dioxide; CO_2; [124-38-9]
3. Sulfur dioxide; SO_2; [7746-09-5]
4. Water; H_2O; [7732-18-5]

ORIGINAL MEASUREMENTS:
Gorgeu, A.
C. R. Hebd. Seances Acad. Sci. 1883, *96*, 341-3.

VARIABLES:
Two approximate temperatures
CO_2 and SO_2 content

PREPARED BY:
B. Engelen

EXPERIMENTAL VALUES:
Solubility of manganese sulfite in cold water and hot water, and in cold water saturated with CO_2 or SO_2 is reported.

Temperature	Solvent	g $MnSO_3$/ kg solvent	$m(MnSO_3)$/mol kg^{-1}[a]
cold	pure water	0.1	7.4×10^{-4}
hot	pure water	0.2	1.5×10^{-3}
cold	water saturated with CO_2	1[b]	7.4×10^{-3}[c]
cold	water saturated with SO_2	150-170	1.11 - 1.26

[a] Calculated by the compiler.
[b] Units are in g/dm^3 solvent.
[c] Units are in mol/dm^3 solvent.

AUXILIARY INFORMATION

METHOD APPARATUS/PROCEDURE:
Not given.

SOURCE AND PURITY OF MATERIALS:
$MnSO_3 \cdot 3H_2O$ was prepared by precipitation from an aqueous solution of $MnCl_2$ with alkaline sulfite, added in small portions. The sulfite contained 2-3% of sulfate (author).

ESTIMATED ERROR:

REFERENCES:

COMPONENTS:	EVALUATOR:
1. Iron(II) sulfite; $FeSO_3$; [51092-74-1] 2. Water; H_2O; [7732-18-5]	H.D. Lutz, Dept. of Chemistry, University of Siegen, FR Germany. February 1983.

CRITICAL EVALUATION:

Iron(II) sulfite forms several well established hydrates, three polymorphic forms of $FeSO_3.3H_2O$ [21006-12-1] (α (1-3), β (4), and γ (3), $FeSO_3.5/2H_2O$ (3,5,6) [13450-81-2], and $FeSO_3.2H_2O$ (3) [21006-12-1]. The trihydrates crystallize from aqueous solutions at or below room temperature, $FeSO_3.5/2H_2O$ and $FeSO_3.2H_2O$ above 330 and 355 K, respectively (3). Iron(III) sulfite or hydrates of it have not been established.

Detailed data on the solubility of iron(II) sulfite are not available. In the older literature (7) it is claimed that iron(II) sulfite is nearly insoluble in water, insoluble in alcohol, and readily soluble in excess of sulfurous acid. Numerical data have been reported only by Terres *et al.* (8) and by Margulis *et al.* (9). Margulis *et al.* (9) report on the solubility of a sesquihydrate, $FeSO_3.3/2H_2O$ [50820-24-1]. The solubility of the hydrate studied increases from 0.0276 mass % of $FeSO_3$ (2.03×10^{-3} mol kg^{-1}, molality) at 293 K to 0.0475 mass % (3.50×10^{-3} mol kg^{-1}) at 353 K. However it is not clear whether this hydrate was really $FeSO_3.3/2H_2O$, not found elsewhere, or another hydrate of $FeSO_3$. Terres *et al.* (8) report the solubility of iron(II) sulfite identified as $FeSO_3.5H_2O$? [96247-21-1] increases with increasing SO_2 content from <<0.3 mol kg^{-1} (molality) to 2.0 and 1.4 mol kg^{-1} at 288.2 and 298.2 K, respectively.

The data given are doubtful.

REFERENCES

1. Bugli, G.; Pannetier, G. *Bull. Soc. Chim. Fr.* 1968, 2355.
2. Johansson, L.G.; Lindqvist, O. *Acta Crystallogr., Sect. B* 1979, *35*, 1017.
3. Lutz, H.D.; Eckers, W.; Engelen, B. *Z. Anorg. Allg. Chem.* 1981, *475*, 165.
4. Johansson, L.G.; Ljungstroem, E. *Acta Crystallogr., Sect. B* 1979, *35*, 2683.
5. Bugli, G. *Bull. Soc. Chim. Fr.* 1977, 639.
6. Johansson, L.G.; Ljungstroem, E. *Acta Crystallogr., Sect. B* 1980, *36*, 1184.
7. Koene *Pogg. Ann.* 1844, *63*, 440.
8. Terres, E.; Ruhl, G. *Angew. Chem.* 1934, *47*, 332.
9. Margulis, E.V.; Rodin, I.V.; Gubieva, D.N. *Zh. Neorg. Khim.* 1981, *26*, 2267; *Russ. J. Inorg. Chem. (Eng. Transl.)* 1981, *26*, 1220.

COMPONENTS:	ORIGINAL MEASUREMENTS:
1. Iron(II) sulfite; $FeSO_3$; [51092-74-1] 2. Water; H_2O; [7732-18-5]	Margulis, E.V.; Rodin, I.V.; Gubieva, D.N. *Zh. Neorg. Khim.* 1981, *26*, 2267-9; *Russ. J. Inorg. Chem. (Eng. Transl.)* 1981, *26*, 1220-1.
VARIABLES:	PREPARED BY:
Four temperatures: 293 - 353 K	H.D. Lutz

EXPERIMENTAL VALUES:

The authors report the solubility of $FeSO_3.3/2H_2O$ [50820-24-1] in pure water at 20, 50, 70, and 80°C.

t/°C	$FeSO_3$ mass %	$m(FeSO_3)$[a] mol kg^{-1}
20	0.0276	0.002031
50	0.0355	0.002613
70	0.0432	0.003180
80	0.0475	0.003497

[a] Calculated by the compiler.

AUXILIARY INFORMATION

METHOD APPARATUS/PROCEDURE:

The solubility of iron sulfite was determined from the concentration of Fe^{2+} in a saturated solution.

Iron sulfite was dissolved in deoxygenated distilled water (solid/liquid ratio 1:4) in closed flasks placed in a water thermostat, with mechanical stirring. Saturation was assumed when $c_{Fe^{2+}}$ stopped increasing with time. In all cases, 1 hr was sufficient for equilibrium. Iron sulfite is hydrolysed at 90°C, with the liberation of SO_2 and the formation of a pale-yellow hydroxide sulfite. Iron was determined colorimetrically.

SOURCE AND PURITY OF MATERIALS:

Iron sulfite was synthesized by precipitation from a concentrated solution of the sulfite by adding Na_2SO_3 (105% of the stoichiometric quantity) at room temperature with mechanical stirring. The sulfite precipitate was washed with distilled water which had been deoxygenated by boiling, to avoid oxidation of the sulfite.

ESTIMATED ERROR:

Temperature: ±0.5 K (authors).

REFERENCES:

COMPONENTS:	ORIGINAL MEASUREMENTS:
1. Iron(II) sulfite; $FeSO_3$; [51092-74-1] 2. Sulfur dioxide; SO_2; [7446-09-5] 3. Water; H_2O; [7732-18-5]	1. Terres, E.; Ruhl, G. *Angew. Chem.* 1934, *47*, 332-4. 2. Terres, E.; Ruhl, G. *Beitrage zur Chemie der schwefligen Saure, Beiheft zu den Zeitschriften des Vereins deutscher Chemiker No 8* 1934.
VARIABLES:	PREPARED BY:
Two temperatures: 288 and 298 K Concentration of SO_2	H.D. Lutz, B. Engelen

EXPERIMENTAL VALUES:

The authors report the solubility of $FeSO_3.5(?)H_2O$ in aqueous sulfurous acid solutions at 15 and 25°C. The first paper reports the experimental data only in a graph. Numerical data are reported in the second paper.

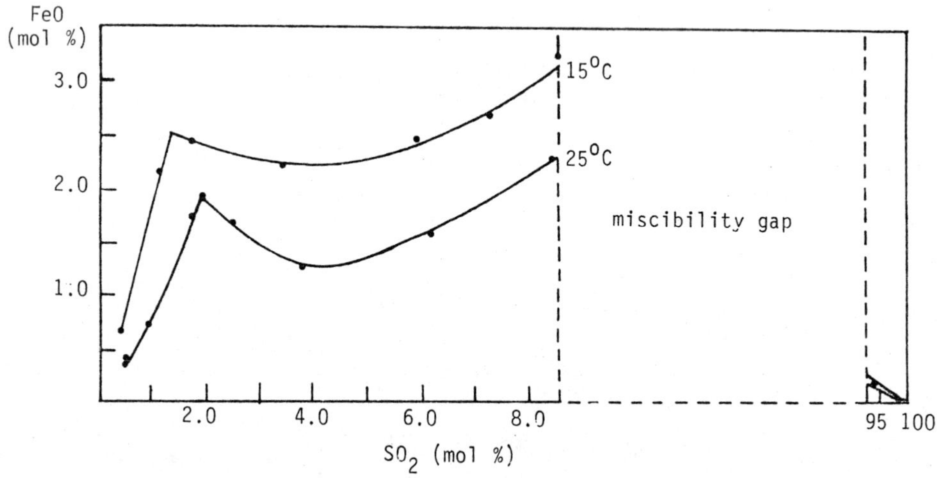

Reprinted by permission

(continued on next page)

AUXILIARY INFORMATION

METHOD APPARATUS/PROCEDURE:	SOURCE AND PURITY OF MATERIALS:
SO_2-H_2O mixtures were treated together with solid iron sulfite in closed glass ampoules at the stated temperatures. The solutions obtained were filtered through a fine glass frit and after oxidation of sulfite to sulfate were analysed for sulfate and iron.	Iron(II) sulfite was precipitated from a solution of iron(II) chloride with sodium sulfite.
	ESTIMATED ERROR:
	REFERENCES:

COMPONENTS:	ORIGINAL MEASUREMENTS:
1. Iron(II) sulfite; $FeSO_3$; [51092-74-1] 2. Sulfur dioxide; SO_2; [7446-09-5] 3. Water; H_2O; [7732-18-5]	1. Terres, E.; Ruhl, G. *Angew. Chem.* 1934, *47*, 332-4. 2. Terres, E.; Ruhl, G. *Beitrage zur Chemie der schwefligen Saure, Beiheft zu den Zeitschriften des Vereins deutscher Chemiker No 8* 1934.

EXPERIMENTAL VALUES (continued):

Composition of the saturated solutions[a]

SO_2 mol %	FeO mol %	m(FeO)[b] mol kg^{-1}	Solid phase
Temperature = 15°C			
0.35	0.65	0.36	$FeSO_3.5(?)H_2O$
1.16	2.15	1.23	"
1.75	2.42	1.40	$Fe(HSO_3)_2$?
3.44	2.18	1.28	"
5.92	2.44	1.48	"
7.32	2.63	1.62	"
8.53	3.21	2.02	"
Temperature = 25°C			
0.45	0.32	0.18	$FeSO_3.5(?)H_2O$
0.96	0.71	0.40	"
1.78	1.72	0.99	"
2.51	1.65	0.96	$Fe(HSO_3)_2(?)$
3.85	1.24	0.73	"
6.20	1.52	0.91	"
8.43	2.26	1.40	"

[a] The mixtures separate into two liquid layers at concentrations of SO_2 between 8.6 and 93 mol %.

[b] Calculated by the compilers.

COMPONENTS:	EVALUATOR:
1. Cobalt(II) sulfite; $CoSO_3$; [32702-66-2] 2. Water; H_2O; [7732-18-5]	H.D. Lutz, Dept. of Chemistry, University of Siegen, FR Germany. February 1983.

CRITICAL EVALUATION:

Cobalt sulfite crystallizes from aqueous solutions in the form of several hydrates. The formation of the various hydrates depends on temperature, composition of the solution, especially the amount of sulfurous acid, and preparation technique. Furthermore numerous sulfite complexes of tervalent cobalt are known. The existence of the following cobalt sulfite hydrates has been established: $CoSO_3.6H_2O$ (1,2) [60936-55-2], $CoSO_3.3H_2O$ (o.-rh.) (1,3,4) [20911-44-8], $CoSO_3.3H_2O$ (mon.) (1,3) [20911-44-8], $CoSO_3.5/2H_2O$ (1,3,4) [20911-45-9] and $CoSO_3.2H_2O$ (1,4) [65410-84-6]. The solubility of these hydrates has not yet been thoroughly investigated. Several authors report that cobalt sulfite is nearly insoluble in water (5,6), insoluble in alcohol (6,7), readily soluble in sulfurous acid (5-7) and in other acids, with decomposition. The solid phases studied in the older literature were $CoSO_3.6H_2O$ (5-7) and probably $CoSO_3.3H_2O$ (o.-rh.) (6,7). Numerical data on the solubility of cobalt sulfite were given by Margulis *et al.* (8), who report that the solubility of $CoSO_3.3H_2O$ (modification not given, but probably the o.-rh. form) in water increases from 0.209 mass % of $CoSO_3$ ($m(CoSO_3)$ = 0.0151 mol kg^{-1}) at 293 K to 0.316 mass % (0.0228 mol kg^{-1}) at 363 K. Preliminary studies in our laboratory showed a solubility of about 10^{-3} mol dm^{-3} at ambient temperature. A tentative value cannot be given.

REFERENCES

1. Lutz, H.D.; Eckers, W.; Engelen, B. *Z. Anorg. Allg. Chem.* 1981, *475*, 165.
2. Klasens, H.A.; Perdok, W.G.; Terpstra, P. *Z. Kristallogr.* 1936, *94*, 1.
3. Bugli, G.; Pannetier, G. *C.R. Hebd. Seances Acad. Sci., Paris, Ser. C* 1968, *267*, 234.
4. Lutz, H.D.; El-Suradi, S.M.; Engelen, B. *Z. Naturforsch., Teil B* 1977, *32*, 1230.
5. Shelton, F.K. *U.S. Bur. Mines, Rep. Invest.* 1946, No. 3836, 14.
6. Rammelsberg, C. *Pogg. Ann.* 1846, *67*, 245.
7. Rohrig, A. *J. Prakt. Chem.* 1888, *37*, 217.
8. Margulis, E.V.; Rodin, I.V.; Gubieva, D.N. *Zh. Neorg. Khim.* 1981, *26*, 2267; *Russ. J. Inorg. Chem. (Eng. Transl.)* 1981, *26*, 1220.

COMPONENTS:

1. Cobalt(II) sulfite; $CoSO_3$; [32702-66-2]
2. Water; H_2O; [7732-18-5]

ORIGINAL MEASUREMENTS:

Margulis, E.V.; Rodin, I.V.; Gubieva, D.N.

Zh. Neorg. Khim. 1981, *26*, 2267-9; *Russ. J. Inorg. Chem. (Eng. Transl.)* 1981, *26*, 1220-1.

VARIABLES:

Four temperatures: 293 - 353 K

PREPARED BY:

H.D. Lutz

EXPERIMENTAL VALUES:

The authors report the solubility of $CoSO_3.3H_2O$ [20911-44-8] in pure water at 20, 50, 70, and 90°C.

t/°C	$CoSO_3$[a] mass %	$m(CoSO_3)$[a,b] mol kg^{-1}	$CoSO_3$[c] mass %	$m(CoSO_3)$[b,c] mol kg^{-1}
20	0.217	0.01565	0.209	0.01507
50	0.248	0.01789	-	-
70	0.297	0.02143	-	-
90	0.326	0.02353	0.316	0.02281

[a] From concentration of Co^{2+}.

[b] Calculated by the compiler.

[c] From the SO_3^{2-} concentration.

AUXILIARY INFORMATION

METHOD APPARATUS/PROCEDURE:

The solubility of cobalt sulfite was determined from the concentration of Co^{2+} in the saturated solution, and in some experiments also from the SO_3^{2-} concentration. Cobalt sulfite was dissolved in deoxygenated distilled water (solid/liquid ratio 1:4) in closed flasks placed in a water thermostat, with mechanical stirring. Saturation was assumed when $c_{Co^{2+}}$ stopped increasing with time. In all cases, 3 hr was sufficient for equilibrium. Cobalt was determined colorimetrically, sulfite iodometrically.

SOURCE AND PURITY OF MATERIALS:

Cobalt sulfite was synthesized by precipitation from a concentrated solution of the sulfite by adding Na_2SO_3 (105% of the stoichiometric quantity) at room temperature with mechanical stirring. The sulfite precipitate was washed with distilled water which had been deoxygenated by boiling, to avoid oxidation of the sulfite.

ESTIMATED ERROR:

Temperature: ±0.5 K (authors).

REFERENCES:

COMPONENTS:	EVALUATOR:
1. Nickel(II) sulfite; $NiSO_3$; [7757-95-1] 2. Water; H_2O; [7732-18-5]	H.D. Lutz, Dept. of Chemistry, University of Siegen, FR Germany. February 1983.

CRITICAL EVALUATION:

Nickel sulfite crystallizes from aqueous solutions in the form of the hydrates $NiSO_3.6H_2O$ [1344-81-0] at room temperature (1,2) and $NiSO_3.3H_2O$ [77902-26-2], $NiSO_3.5/2H_2O$ [77902-27-3], and $NiSO_3.2H_2O$ [77902-28-4] above 40, 55, and 85°C, respectively (2). Besides the crystalline hydrates, amorphous nickel sulfite hydrate is formed very easily by precipitation of nickel salts with sulfites (2). Nickel sulfite, i.e. $NiSO_3.6H_2O$, is claimed to be nearly insoluble in water (3,4), readily soluble in sulfurous acid (3,4) and in other acids, with decomposition. Numerical data on the solubility of nickel sulfite were given by Margulis *et al.* (5), who report that the solubility of $NiSO_3.5/2H_2O$ in water increases from 0.190 mass % of $NiSO_3$ ($m(NiSO_3)$ = 0.0137 mol kg^{-1}) at 293 K to 0.286 mass % (0.0207 mol kg^{-1}) at 363 K. The data available may be around the correct order of magnitude, but a tentative value cannot be given.

REFERENCES

1. Klasens, H.A.; Perdok, W.G.; Terpstra, P. *Z. Kristallogr.* 1936, *94*, 1.
2. Lutz, H.D.; Eckers, W.; Engelen, B. *Z. Anorg. Allg. Chem.* 1981, *475*, 165.
3. Muspratt, J.S. *Justus Liebigs Ann. Chem.* 1844, *50*, 259.
4. Rammelsberg, C. *Ann. Phys. Chem.* 1846, *67*, 391.
5. Margulis, E.V.; Rodin, I.V.; Gubieva, D.N. *Zh. Neorg. Khim.* 1981, *26*, 2267; *Russ. J. Inorg. Chem. (Eng. Transl.)* 1981, *26*, 1220.

COMPONENTS:

1. Nickel(II) sulfite; $NiSO_3$; [7757-95-1]
2. Water; H_2O; [7732-18-5]

ORIGINAL MEASUREMENTS:

Margulis, E.V.; Rodin, I.V.; Gubieva, D.N.

Zh. Neorg. Khim. 1981, *26*, 2267-9; **Russ. J. Inorg. Chem.* 1981, *26*, 1220-1.

VARIABLES:

Four temperatures: 293 - 363 K

PREPARED BY:

H.D. Lutz

EXPERIMENTAL VALUES:

The authors report the solubility of $NiSO_3.5/2H_2O$ [77902-27-3] in pure water at 20, 50, 70, and 90°C.

t/°C	$NiSO_3$[a] mass %	$m(NiSO_3)$[a,b] mol kg^{-1}	$NiSO_3$[c] mass %	$m(NiSO_3)$[b,c] mol kg^{-1}
20	0.198	0.01430	0.190	0.01372
50	0.215	0.01553	-	-
70	0.254	0.01835	-	-
90	0.292	0.02110	0.286	0.02067

a From concentration of Ni^{2+}.

b Calculated by the compiler.

c From the SO_3^{2-} concentration.

AUXILIARY INFORMATION

METHOD APPARATUS/PROCEDURE:

The solubility of nickel sulfite was determined from the concentration of Ni^{2+} in the saturated solution, and in some experiments also from the SO_3^{2-} concentration. The solution of nickel sulfite was carried out in deoxygenated distilled water (solid/liquid ratio 1:4) in closed flasks placed in a water thermostat, with mechanical stirring. Saturation was assumed when c_{Ni} stopped increasing with time. In all cases, 3 hr were sufficient for equilibrium. Nickel was determined colorimetrically, sulfite iodometrically.

SOURCE AND PURITY OF MATERIALS:

Nickel sulfite was synthesized by precipitation from a concentrated solution of the sulfite by adding Na_2SO_3 (105% of the stoichiometric quantity) at room temperature with mechanical stirring. The obtained sulfite precipitate was washed, using distilled water which had been deoxygenated by boiling to avoid oxidation of the sulfite.

ESTIMATED ERROR:

Temperature: ±0.5 K (authors).

REFERENCES:

COMPONENTS:	EVALUATOR:
1. Copper(I) sulfite; Cu_2SO_3; [35788-00-2] 2. Water; H_2O; [7732-18-5]	H.D. Lutz, Dept. of Chemistry, University of Siegen, FR Germany. May 1983.

CRITICAL EVALUATION:

Copper(I) sulfite crystallizes from aqueous solution in the form of the hemihydrate $Cu_2SO_3 . 1/2H_2O$ [35788-00-2] (Etard's salt) (1,2). The monohydrate, $Cu_2SO_3 . 1H_2O$ [10294-49-2] (Rogojski's salt) (3) has not been confirmed (2). $Cu_2SO_3 . 1/2H_2O$ is claimed to be insoluble in water, alcohol and ether (1). Numerical data on the solubility of copper(I)sulfite were given by Margulis *et al*. (4), who reported the existence of the hydrate $Cu_2SO_3 . 9/2H_2O$ [35788-00-2]. The solubility of this hydrate increases from 2.2 mg Cu/dm^3 ($c(Cu_2SO_3) = 1.73 \times 10^{-5}$ mol dm^{-3}) at 293 K to 26 mg Cu/dm^3 (2.04×10^{-4} mol dm^{-3}) at 363 K.

REFERENCES

1. Etard, A. *C.R. Hebd. Seances Acad. Sci.* 1882, *95*, 36 and 137.

2. Dasent, W.E.; Morrison, D. *J. Inorg. Nucl. Chem.* 1964, *26*, 1122.

3. Rogojski, J.-B. *J. Prakt. Chem.* 1851, *53*, 409; *C.R. Trav. Chim.* 1851, *7*, 156; *Justus Liebigs Ann. Chem.* 1851, *79*, 255.

4. Margulis, E.V.; Rodin, I.V. *Zh. Neorg. Khim.* 1982, *27*, 374; *Russ. J. Inorg. Chem.* 1982, *27*, 211.

COMPONENTS:

1. Copper(I) sulfite; Cu_2SO_3; [35788-00-2]
2. Water; H_2O; [7732-18-5]

ORIGINAL MEASUREMENTS:

Margulis, E.V.; Rodin, I.V.

Zh. Neorg. Khim. 1982, *27*, 374-7; *Russ. J. Inorg. Chem.* 1982, *27*, 211-3.

VARIABLES:

Six temperatures: 293 - 363 K

PREPARED BY:

H.D. Lutz

EXPERIMENTAL VALUES:

The authors report the solubility of $Cu_2SO_3.9/2H_2O$ [35788-00-2] in water at various temperatures, and the solubility product of this compound defined as $K_{s0}(Cu_2SO_3.9/2H_2O) = [Cu^+]^2 [SO_3^{2-}]$.

Composition of the saturated solutions

t/°C	mg Cu/dm^3	$10^5c(Cu)$ mol dm^{-3}	$K_{s0}(Cu_2SO_3)$ mol^3 dm^{-9}	$10^5c(Cu_2SO_3)^a$ mol dm^{-3}
20	2.2	3.47	4.18×10^{-14}	1.73
30	3.2	5.04	1.28×10^{-13}	2.52
40	9.0	14.2	2.86×10^{-12}	7.08
50	12.4	19.5	7.42×10^{-12}	9.76
70	18.5	29.1	2.46×10^{-11}	14.6
90	25.9	40.8	6.79×10^{-11}	20.4

[a] Calculated by the compiler from mg Cu/dm^3.

AUXILIARY INFORMATION

METHOD APPARATUS/PROCEDURE:

The solubility of copper(I)sulfite was studied by the isothermal method. The experiments were carried out in a water thermostat with mechanical stirring in closed flasks. The time required for saturation was 2.5 hr. The solutions were analysed for sulfite (method not given).

SOURCE AND PURITY OF MATERIALS:

Copper(I)sulfite was precipitated from $CuSO_4$ solutions with Na_2SO_3 (molar ratio $Na_2SO_3/CuSO_4$ = 1:1) at 20°C. After stirring for 2 hr, the precipitate was filtered off, washed with water and acetone, and dried in air at room temperature.

ESTIMATED ERROR:

Temperature: ±0.5 K (authors).

REFERENCES:

COMPONENTS:	EVALUATOR:
1. Copper(I,II) sulfite; $Cu_2SO_3.CuSO_3$; [15293-86-4] 2. Water; H_2O; [7732-18-5]	H.D. Lutz, Dept. of Chemistry, University of Siegen, FR Germany. June 1983.

CRITICAL EVALUATION:

Copper(I,II) sulfite crystallizes from aqueous solution in the form of the hydrate $Cu_2SO_3.CuSO_3.2H_2O$ (Chevreul's salt) (1) [13814-81-8]. The formation of other hydrates, e.g. Pean's salt (2), could not be confirmed (3). Copper(I,II) sulfite has been claimed to be insoluble in water (4,5), slightly soluble in sulfurous acid (4,5) and in other acids (4,6), soluble in aqueous NH_3 solutions (4,7), and insoluble in $CuSO_4$ solutions (8). Numerical data have been given by Terres *et al.* (9), Pesin *et al.* (10), and Margulis *et al.* (11). Pesin *et al.* (10) report that the solubility of $Cu_2SO_3.CuSO_3.2H_2O$ in pure water is 1.09×10^{-3} mol kg^{-1} (molality) (0.042 mass %) at 298.2 K and 3.62×10^{-3} mol kg^{-1} (0.140 mass %) at 333.2 K. Margulis *et al.* (11) determined $c(Cu_2SO_3.CuSO_3) = 8.71 \times 10^{-5}$ mol dm^{-3} at 293 K and 1.9×10^{-6} mol dm^{-3} at 363K. The data given by Terres *et al.* (9) may be extrapolated to pure water to be in the range 1×10^{-3} to 1×10^{-2} mol kg^{-1}. Thus the solubility given by Pesin *et al.* (10) may be of the right order of magnitude. The same seems to be true for the positive temperature coefficient of solubility reported by Pesin *et al.* (10). Pesin *et al.* (10) also report that the solubility of copper(I,II) sulfite increases to 3.20×10^{-3} and 1.22×10^{-2} mol kg^{-1} at 298.2 K and 333.2 K, respectively, with concentration of $CuSO_4.5H_2O$ increasing to 30 mass %. The solubility of copper(I,II) sulfite increases to m(CuO) = 0.170 and 0.043 mol kg^{-1} with increasing SO_2 content at 288.2 and 298.2 K, respectively, as reported by Terres *et al.* (9). Normal copper(II) sulfite or hydrates of it are not known.

REFERENCES

1. Chevreul *Ann. Chim. (Paris)* 1812, *83*, 181.
2. Péan de St. Gilles, L. *Ann. Chim. Phys.* 1854, *42*, 23.
3. Brauer, G.; Eichner, M. *Z. Anorg. Allg. Chem.* 1956, *287*, 95.
4. Berthier, P. *Ann. Chim. Phys.* 1843, *7*, 74.
5. Omori, K.; Okuwaki, A.; Suzuki, T.; Ito, H.; Okabe, T. *Bull. Chem. Soc. Japan* 1966, *39*, 78.
6. Doepping, O. *Bull. Acad. Imp. Sci. St.-Petersbourg* 1851, *9*, 179.
7. Rogojski, J.-B. *C.R. Trav. Chim.* 1851, *7*, 156.
8. Gin, G. *Chem. News J. Phys. Sci.* 1903, *88*, 5.
9. Terres, E.; Ruhl, G. *Angew. Chem.* 1934, *47*, 332.
10. Pesin, Ya.M.; Shabashova, M.L. *Zh. Prikl. Khim.* 1950, *23*, 350; *J. Appl. Chem. USSR (Eng. Transl.)* 1950, *23*, 365.
11. Margulis, E.V.; Rodin, I.V. *Zh. Neorg. Khim.* 1982, *27*, 374; *Russ. J. Inorg. Chem. (Eng. Transl.)* 1982, *27*, 211.

COMPONENTS:	ORIGINAL MEASUREMENTS:
1. Copper(I,II) sulfite; $Cu_2SO_3.CuSO_3$; [15293-86-4] 2. Sulfur dioxide; SO_2; [7446-09-5] 3. Water; H_2O; [7732-18-5]	1. Terres, E.; Ruhl, G. *Angew. Chem.* 1934, 47, 332-4. 2. Terres, E.; Ruhl, G. *Beitrage zur Chemie der schwefligen Saure, Beiheft zu den Zeitschriften des Vereinsdeutscher Chemiker No 8*, 1934.
VARIABLES:	PREPARED BY:
Two temperatures: 288 and 298 K Concentration of SO_2	H.D. Lutz, B. Engelen

EXPERIMENTAL VALUES:

The authors report the solubility of $Cu_2SO_3.CuSO_3.2H_2O$ (Chevreul's salt) [13814-81-8] in aqueous sulfurous acid solutions at 15 and 25°C. The experimental data are given in a graph, in the first paper. Numerical data are reported in the second paper.

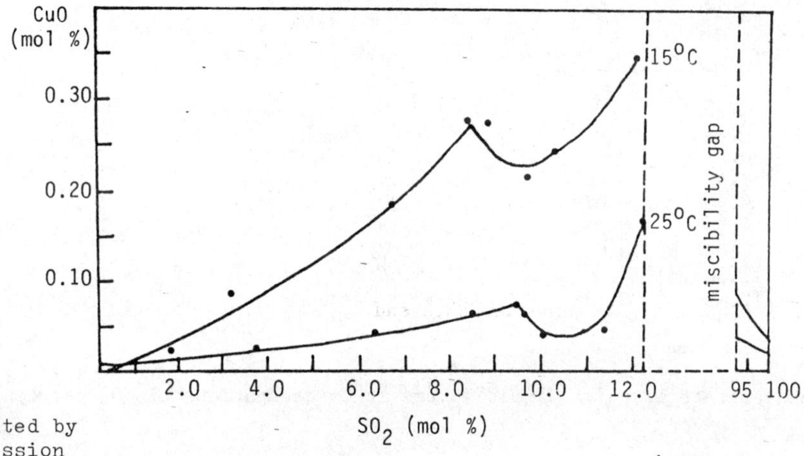

Reprinted by permission

(continued on next page)

AUXILIARY INFORMATION

METHOD APPARATUS/PROCEDURE:	SOURCE AND PURITY OF MATERIALS:
SO_2-H_2O mixtures were treated together with solid copper(I,II) sulfite in closed glass ampoules at the stated temperatures. The solutions obtained were filtered through a fine glass frit, and after oxidation of the sulfite were analysed for sulfate and copper content.	Chevreul's salt was obtained from copper sulfate and sodium sulfite.
	ESTIMATED ERROR:
	REFERENCES:

COMPONENTS:	ORIGINAL MEASUREMENTS:
1. Copper(I,II) sulfite; $Cu_2SO_3.CuSO_3$; [15293-86-4] 2. Sulfur dioxide; SO_2; [7446-09-5] 3. Water; H_2O; [7732-18-5]	1. Terres, E.; Ruhl, G. *Angew. Chem.* 1934, *47*, 332-4. 2. Terres, E.; Ruhl, G. *Beitrage zur Chemie der schwefligen Saure, Beiheft zu den Zeitschriften des Vereinsdeutscher Chemiker No 8*, 1934.

EXPERIMENTAL VALUES (continued):

Composition of the saturated solutions[a]

SO_2 mol %	CuO mol %	m(CuO)[b] 10^{-2} mol kg^{-1}	Solid phase
Temperature = 15°C			
1.84	0.02	1.13	$Cu_2SO_3.CuSO_3.2H_2O$
3.18	0.08	4.59	"
6.72	0.18	10.7	"
8.26	0.28	17.0	"
8.73	0.27	16.5	$Cu(HSO_3)_2$?
9.65	0.21	12.9	"
10.25	0.24	14.9	"
Temperature = 25°C			
0.05	0.01	0.56	$Cu_2SO_3.CuSO_3.2H_2O$
3.72	0.02	1.15	"
6.32	0.04	2.37	"
8.47	0.06	3.64	"
9.45	0.07	4.29	"
9.54	0.06	3.68	$Cu(HSO_3)_2$?
10.05	0.04	2.47	"
11.32	0.09	5.64	"
12.18	0.17	10.8	"

a The mixtures separate into two liquid layers at concentrations of SO_2 between 12.2 and 93 mol %.

b Compilers.

COMPONENTS:

1. Copper(I,II) sulfite; $Cu_2SO_3.CuSO_3$; [15293-86-4]

2. Copper(II) sulfate; $CuSO_4$; [7758-98-7]

3. Water; H_2O; [7732-18-5]

ORIGINAL MEASUREMENTS:

Pesin, Ya.M.; Shabashova, M.L.

Zh. Prikl. Khim. 1950, *23*, 350-6; **J. Appl. Chem. USSR (Eng. Transl.)* 1950, *23*, 365-72.

VARIABLES:

Two temperatures: 298 and 333 K
Concentration of $CuSO_4$

PREPARED BY:

H.D. Lutz

EXPERIMENTAL VALUES:

The authors report the solubility of $Cu_2SO_3.CuSO_3.2H_2O$ (Chevreul's salt) [13814-81-8] in water and in solutions of various $CuSO_4$ concentration at 25 and 60°C.

Concentration of $CuSO_4.5H_2O$	Composition of saturated solutions 25°C				60°C			
mass %	mass % Determination Copper	Sulfur	Mean value	$10^3 m$ mol kg^{-1} (compiler)	mass % Determination Copper	Sulfur	Mean value	$10^3 m$ mol kg^{-1} (compiler)
0	0.042	0.042	0.042	1.09	0.141	0.139	0.140	3.62
10	0.086	0.090	0.088	2.43	0.309	0.315	0.312	8.64
20	0.090	0.102	0.096	2.85	0.331	0.337	0.334	9.94
30	0.098	0.102	0.100	3.20	0.380	0.378	0.379	12.18

AUXILIARY INFORMATION

METHOD APPARATUS/PROCEDURE:

Saturation method. The water and the solutions of $CuSO_4$ were boiled to drive off the oxygen. Equilibrium was established after 24 hr. The solutions were analysed for copper and sulfur (methods not given, but are assumed to be the same as in a previous paper by the authors (1), i.e. copper was determined iodometrically and sulfur as $BaSO_4$).

SOURCE AND PURITY OF MATERIALS:

$Cu_2SO_3.CuSO_3.2H_2O$ was prepared (as described in a previous paper by the authors (1)) by adding a 10% solution of $CuSO_4$ to a solution of sodium sulfite at 70-80°C. The crystalline precipitate that formed (after standing for 2-3 hr) was washed with water until the reaction for the SO_4^{2-} ion disappeared, and then dried at 90-100°C.

ESTIMATED ERROR:

REFERENCES:

1. Pesin, Ya.M.; Shabashova, M.L. *Zh. Prikl. Khim.* 1950, *23*, 278.

COMPONENTS:

1. Copper(I,II) sulfite; $Cu_2SO_3.CuSO_3$; [15293-86-4]

2. Water; H_2O; [7732-18-5]

ORIGINAL MEASUREMENTS:

Margulis, E.V.; Rodin, I.V.

Zh. Neorg. Khim. 1982, *27*, 374-7; **Russ. J. Inorg. Chem. (Eng. Transl.)* 1982, *27*, 211-3.

VARIABLES:

Six temperatures: 293 - 363 K

PREPARED BY:

H.D. Lutz

EXPERIMENTAL VALUES:

The authors report the solubility of $Cu_2SO_3.CuSO_3.2H_2O$ (Chevreul's salt) [13814-81-8] in water at various temperatures, and the solubility product of this compound, defined as $K_{s0}(Cu_2SO_3.CuSO_3.2H_2O) = [Cu^{2+}][Cu^{+}]^2[SO_3^{2-}]^2$.

Composition of the saturated solutions

t/°C	Cu mg/dm³	$10^5c(Cu)$ mol dm⁻³	$Cu_2SO_3.CuSO_3$ K_{s0}/mol⁵ dm⁻¹⁵	$Cu_2SO_3.CuSO_3$ 10^5c^a/mol dm⁻³
20	16.6	26.1	1.21×10^{-18}	8.71
30	12.4[b]	22.1[b]	5.27×10^{-19}[b]	6.50
40	2.7	4.25	1.39×10^{-22}	1.42
50	1.8	2.84	1.84×10^{-23}	0.94
70	0.94	1.48	7.10×10^{-25}	0.49
90	0.37	0.58	6.56×10^{-27}	0.19

[a] Calculated by the compiler from mg Cu/dm³.

[b] Inconsistent data (12.4 mg Cu/dm³ $\neq 22.1 \times 10^{-5}$ mol dm⁻³).

AUXILIARY INFORMATION

METHOD APPARATUS/PROCEDURE:

The solubility of copper(I,II)sulfite was studied by the isothermal method. The experiments were carried out in a water thermostat with mechanical stirring in closed flasks. The time required for saturation was 2.5 hr. The solutions were analysed for sulfite (method not given).

SOURCE AND PURITY OF MATERIALS:

Copper(I,II)sulfite was precipitated from $CuSO_4$ solutions with Na_2SO_3 (molar ratio $Na_2SO_3/CuSO_4$ = 1:1) at temperatures > 40°C. After stirring for 2 hr, the precipitate was filtered off, washed with water and acetone, and dried in air at room temperature.

ESTIMATED ERROR:

Temperature: ±0.5 K.

REFERENCES:

COMPONENTS:

1. Silver sulfite; Ag_2SO_3; [13465-98-0]

2. Water; H_2O; [7732-18-5]

EVALUATOR:

H.D. Lutz,
Dept. of Chemistry,
University of Siegen,
FR Germany.

July 1984.

CRITICAL EVALUATION:

Silver sulfite crystallizes from aqueous solutions as the anhydrous salt Ag_2SO_3 (1). Numerical data on solubility of Ag_2SO_3 in water are scarce (2-4). In 1910, Baubigny (2) reported that the upper limit of the solubility is 1.60 x 10^{-4} mol dm^{-3}, molarity, at 290 K. Rodin *et al*. (4) give a value of 8.1 x 10^{-6} mol dm^{-3} at 293 K. They also found a positive temperature coefficient of solubility. Data on the solubility product of Ag_2SO_3, defined as K_{s0} $(Ag_2SO_3) = [Ag^+]^2[SO_3^{2-}]$, are given by Chateau *et al*. (3) and Rodin *et al*. (4), *viz*. 1.5 x 10^{-14} mol^3 dm^{-6} at 298.2 K and 4.17 x 10^{-15} mol^3 dm^{-9}, both molarity scale, at 293 K, respectively.

TENTATIVE VALUES

The solubility of Ag_2SO_3 in water at room temperature is approximately 1 x 10^{-5} mol dm^{-3} (molarity scale) (3 x 10^{-3} g/dm^3). The temperature coefficient of solubility is probably positive.

Silver sulfite is described to be soluble in excess of sulfurous acid or alkaline metal sulfites, forming complex ions, and readily soluble in acids, with decomposition (1).

REFERENCES

1. *Gmelins Handbuch der Anorganischen Chemie*, 8. Aufl., Band 61 Silber, Teil B, Lieferung 3, Springer-Verlag, Berlin 1973, p.62.

2. Baubigny, H. *Ann. Chim.* 1910, Ser. 8, *20*, 5.

3. Chateau, H.; Duranté, M.; Hervier, B. *Sci. Ind. Photogr.* 1956, *27*, 257.

4. Rodin, I.V.; Margulis, E.V.; Zhigur'yanova, S.A. *Zh. Neorg. Khim.* 1983, *28*, 1619; *Russ. J. Inorg. Chem. (Eng. Transl.)* 1983, *28*, 916.

COMPONENTS:	ORIGINAL MEASUREMENTS:
1. Silver sulfite; Ag_2SO_3; [13465-98-0] 2. Water; H_2O; [7732-18-5]	Baubigny, H. *Ann. Chim.* 1910, Ser. 8, *20*, 5-57.
VARIABLES:	PREPARED BY:
One temperature: 291 K	B. Engelen

EXPERIMENTAL VALUES:

The author reports the solubility of silver sulfite in water at 16-19°C to be 4.74×10^{-3} g in 100 cm^3 of soln ($c(Ag_2SO_3) = 1.60 \times 10^{-4}$ mol dm^{-3}, compiler).

This value is said by the author to be a maximum value for the solubility of silver sulfite because of a small impurity of Ag_2SO_4.

AUXILIARY INFORMATION

METHOD APPARATUS/PROCEDURE:	SOURCE AND PURITY OF MATERIALS:
Saturation method. Equilibrium was established after 24 hr. An aliquot of the supernatant solution was analysed gravimetrically for silver (as the chloride).	Silver sulfite was precipitated by bubbling sulfur dioxide through an oxygen-free solution of silver nitrate. The precipitate is said by the author to be nearly free from silver sulfate.
	ESTIMATED ERROR:
	Temperature: ± 1.5 K.
	REFERENCES:

COMPONENTS:

1. Silver sulfite; Ag_2SO_3; [13465-98-0]
2. Water; H_2O; [7732-18-5]

ORIGINAL MEASUREMENTS:

Chateau, H.; Duranté, M.; Hervier, B.

Sci. Ind. Photogr. 1956, 27, 257-62.

VARIABLES:

One temperature: 298 K

PREPARED BY:

B. Engelen

EXPERIMENTAL VALUES:

The solubility product of Ag_2SO_3, defined as $K_{s0}(Ag_2SO_3) = [Ag^+]^2[SO_3^{2-}]$, is reported to be $K_{s0} = 1.5 \times 10^{-14}$ $mol^3\ dm^{-9}$ at 25°C.

AUXILIARY INFORMATION

METHOD APPARATUS/PROCEDURE:

The solubility product was determined by potentiometric measurements in the system $Hg/Hg_2Cl_2/KCl_{sat}/KNO_{3sat}/NaHSO_3+AgNO_3/Ag$ from the point of first precipitation of Ag_2SO_3 for various pH values. The pH was adjusted with HNO_3 and HCl, respectively, and measured with a glass electrode.

SOURCE AND PURITY OF MATERIALS:

Commercial Na_2SO_3, HNO_3, HCl, and $AgNO_3$ were used.

ESTIMATED ERROR:

REFERENCES:

COMPONENTS:

1. Silver sulfite; Ag_2SO_3; [13465-98-0]
2. Water; H_2O; [7732-18-5]

ORIGINAL MEASUREMENTS:

Rodin, I.V.; Margulis, E.V.; Zhigur'yanova, S.A.

Zh. Neorg. Khim. 1983, *28*, 1619-20; **Russ. J. Inorg. Chem. (Eng. Transl.)* 1983, *28*, 916-7.

VARIABLES:

Four temperatures: 293 - 363 K

PREPARED BY:

B. Engelen

EXPERIMENTAL VALUES:

The authors report the solubility of $Ag_2SO_3.1H_2O$ in water at various temperatures, and its solubility product, defined as $K_{s0}(Ag_2SO_3.1H_2O) = [Ag^+]^2[SO_3^{2-}] = [Ag^+]^3$.

Composition of the saturated solutions

t/°C	mg Ag/dm^3	10^5(Ag) mol dm^{-3}	Ag_2SO_3 10^6 mass %	$K_{s0}(Ag_2SO_3)$ mol^3 dm^{-9}	$10^5c(Ag_2SO_3)^a$ mol dm^{-3}
20	1.74	1.61	2.38	4.17×10^{-15}	0.81
50	2.94	2.72	4.03	2.01×10^{-14}	1.36
70	4.18	3.87	5.73	5.80×10^{-14}	1.94
80	5.32	4.93	7.29	1.19×10^{-13}	2.47

[a] Calculated by the compiler from mg Ag/dm^3.

AUXILIARY INFORMATION

METHOD APPARATUS/PROCEDURE:

The solubility of silver sulfite was studied by the isothermal saturation method. The experiments were done in closed flasks kept in a water thermostat, with mechanical stirring. The time required for saturation was 3 hr. Silver was determined colorimetrically.

SOURCE AND PURITY OF MATERIALS:

Silver sulfite was precipitated from Ag_2SO_4 solutions with Na_2SO_3. The mixture was stirred for 2 hr, then the precipitate was filtered off, washed with water and acetone, and dried in air at room temperature.

ESTIMATED ERROR:

Temperture: ±0.5 K.

REFERENCES:

COMPONENTS:	EVALUATOR:
1. Zinc sulfite; $ZnSO_3$; [13597-44-9] 2. Water; H_2O; [7732-18-5]	H.D. Lutz, Dept. of Chemistry, University of Siegen, FR Germany. August 1984.

CRITICAL EVALUATION:

Zinc sulfite crystallizes from aqueous solutions in the form of various hydrates, $ZnSO_3.nH_2O$, with n = 3, 5/2, 2, and 1 (1-4). The formation of these hydrates depends on temperature, composition of the solution, especially the amount of sulfurous acid, and preparation technique. Furthermore several basic zinc sulfite hydrates (5-7) and double salts with alkali sulfites (1,6,8) exist. The existence of the following zinc sulfite hydrates has been confirmed: $ZnSO_3.3H_2O$ [75042-13-6] (2), three polymorphic forms of $ZnSO_3.5/2H_2O$ [14460-28-7] (α (3,9,10), β (2,9) and γ (2,9)), $ZnSO_3.2H_2O$ [7488-52-0] (3), and two polymorphic forms of $ZnSO_3.1H_2O$ [66516-57-2] (α (1,4) and β (1,4)). The trihydrate crystallizes from aqueous solutions below 275 K, the three forms of $ZnSO_3.5/H_2O$ at ambient temperature (2,5,9,10), the dihydrate at temperatures above 338 K (2,5), and the α-monohydrate at temperatures above 363 K in the presence of excess of sulfur dioxide (2).

The solubility of zinc sulfite in water has not been thoroughly investigated. Numerical data are scarce (8,11-15). In 1890, Heuston *et al.* (11) reported that the solubility of $ZnSO_3.2H_2O$ in water is 1.1×10^{-2} mol kg^{-1} (molality) (0.16 mass % $ZnSO_3$). Murooka *et al.* (8) report the solubility of $ZnSO_3.2\ 1/2H_2O$ in water to be 1.733×10^{-2} mole/kg soln at 298.2 K. Because of the finding that α-$ZnSO_3.5/2H_2O$ is the stable hydrate at ambient temperature (2), it is assumed that both authors (8,11) have had α-$ZnSO_3.5/2H_2O$ as solid phase. In 1983, Margulis *et al.* (15) determined 5 values of the solubility of $ZnSO_3.5/2H_2O$ (presumably the α-form) from 0.1786 mass % $ZnSO_3$ ($m(ZnSO_3)$ = 0.0123 mol kg^{-1}) at 293 K to 0.1939 mass % (0.01336 mol kg^{-1}) at 368 K.

TENTATIVE VALUE

The solubility of α-$ZnSO_3.5/2H_2O$ in water at ambient temperature is approximately 1×10^{-2} mol kg^{-1} (molality scale) (1.5 g $ZnSO_3$/kg H_2O). The temperature coefficient of solubility is positive.

The solubility of zinc sulfite in water is affected by the presence of a third compound. This is shown by some experimental data on the systems $ZnSO_3$-SO_2-H_2O (12-14), $ZnSO_3$-Na_2SO_3-H_2O (8,15), and $ZnSO_3$-$ZnSO_4$-SO_2-H_2O (13-15). The solubility of zinc sulfite increases with increasing concentration of sulfurous acid (12-14) or increasing partial pressure of sulfur dioxide (13,14), and slightly with increasing sulfate content (13-15). In solutions containing Na_2SO_3, the solubility decreases with up to 1.1 mass % of Na_2SO_3 and increases at higher concentrations (15). It has been further claimed that zinc sulfite is nearly insoluble in alcohol and ether (16).

The numerical data on the solubility of zinc sulfite (presumably α-$ZnSO_3.2\ 1/2H_2O$) in the presence of excess of sulfurous acid given by Terres *et al.* (12), Peisakhov *et al.* (13),

COMPONENTS:	EVALUATOR:
1. Zinc sulfite; $ZnSO_3$; [13597-44-9] 2. Water; H_2O; [7732-18-5]	H.D. Lutz, Dept. of Chemistry, University of Siegen, FR Germany. August 1984.

CRITICAL EVALUATION: (continued)

and Kuz'minykh *et al.* (14) are not directly comparable, but seem to be in satisfactory agreement. Figures 1 and 2 show graphs derived (by evaluator) from data reported by Kuz'minykh *et al.* (14) and Terres *et al.* (12). Figure 1 indicates that there is a linear increase of solubility with increasing SO_2 content. The following equation is fitted to the data obtained by Kuz'minykh *et al.* (14) and Terres *et al.* (12) at 293.5 K and 298 K, respectively,

$$m(ZnSO_3) = 0.035 + 0.436 \times m(SO_2\ \text{tot}) \qquad (1)$$

The correlation coefficient is 0.997. This equation and the graph in Figure 1 is approximately valid (deviation < 2.2 %) for other temperatures from 280 to 360 K.

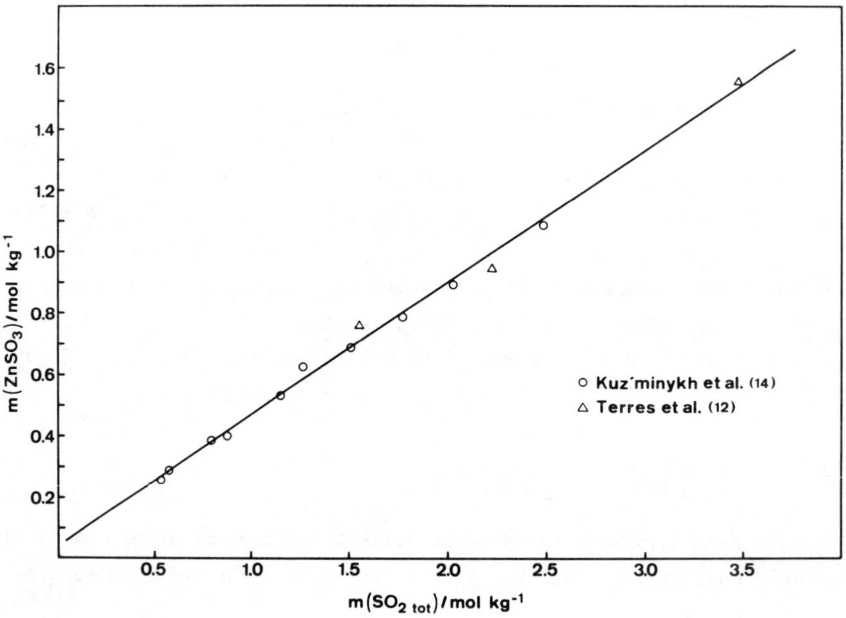

Fig. 1 Solubility of zinc sulfite (presumably α-$ZnSO_3 \cdot 2\ 1/2H_2O$) in aqueous sulfurous acid solutions, as recommended for 293 - 298 K (equation 1).

COMPONENTS:	EVALUATOR:
1. Zinc sulfite; $ZnSO_3$; [13597-44-9] 2. Water; H_2O; [7732-18-5]	H.D. Lutz, Dept. of Chemistry, University of Siegen, FR Germany. August 1984.

CRITICAL EVALUATION: (continued)

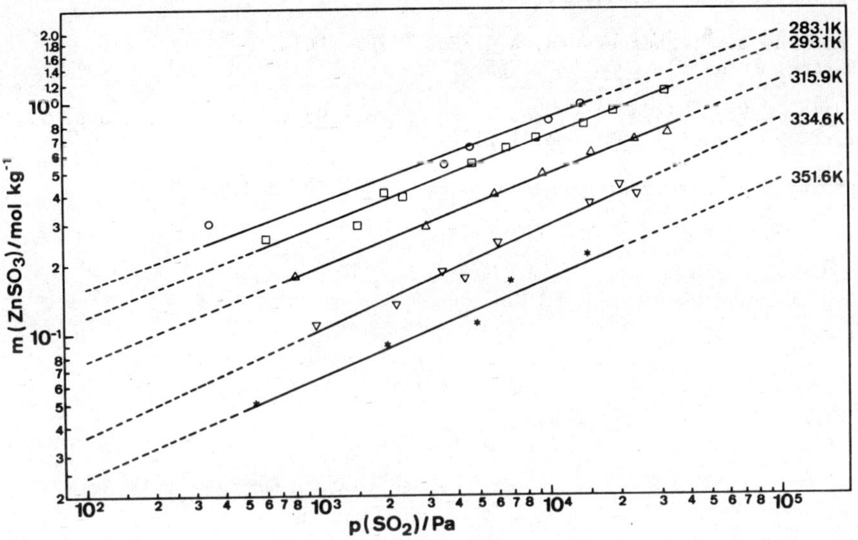

Fig. 2 Solubility of zinc sulfite (presumably α-$ZnSO_3 . 2\ 1/2H_2O$) in aqueous sulfurous acid solutions, *vs.* partial pressure of sulfur dioxide (14).

REFERENCES

1. Lutz, H.D.; Eckers, W.; Engelen, B. *Z. Anorg. Allg. Chem.* 1981, *475*, 165.
2. Lutz, H.D.; El-Suradi, S.M.; Mertins, Ch.; Engelen, B. *Z. Naturforsch. Teil B* 1980, *35*, 808.
3. Engelen, B. *Habilitationsschrift*. Univ. Siegen 1983.
4. Lutz, H.D.; Eckers, W.; Henning, J.; Jung, M.; Buchmeier, W. *Thermochim. Acta* 1984, *74*, 323.
5. Lutz, H.D.; El-Suradi, S.M.; Engelen, B. *Z. Naturforsch., Teil B* 1977, *32*, 1230.
6. Johnstone, H.F.; Singh, A.D. *Ind. Eng. Chem.* 1940, *32*, 1037.
7. Jager, L. *Chem. Prům.* 1957, *7*, 544.
8. Murooka, T.; Sato, H. *Bull. Inst. Phys. Chem. Res. (Tokyo)* 1937, *16*, 636.
9. Okabe, T.; Kamisawa, K.; Hori, S. *Nippon Kagaku Zasshi* 1960, *81*, 529.
10. Pannetier, G.; Djega-Mariadassou, G.; Bregeault, J.-M. *Bull. Soc. Chim. Fr.* 1964, 1749.
11. Heuston, F.T.; Tichborne, C.R. *Br. Med. J.* 1890, 1063.
12. Terres, E.; Ruhl, G. *Angew. Chem.* 1934, *47*, 332.

COMPONENTS:	EVALUATOR:
1. Zinc sulfite; $ZnSO_3$; [13597-44-9] 2. Water; H_2O; [7732-18-5]	H.D. Lutz, Dept. of Chemistry, University of Siegen, FR Germany. August 1984.

CRITICAL EVALUATION: (continued)

13. Peisakhov, I.L.; Karmazina, V.D. *Zh. Prikl. Khim.* 1959, *32*, 70; **J. Appl. Chem. USSR (Eng. Transl.)* 1959, *32*, 71.

14. Kuz'minykh, I.N.; Kuznetsova, A.G. *Zh. Prikl. Khim.* 1954, *27*, 816; **J. Appl. Chem. USSR (Eng. Transl.)* 1954, *27*, 765.

15. Margulis, E.V.; Rodin, I.V. *Zh. Neorg. Khim.* 1981, *26*, 2269; *Russ. J. Inorg. Chem. (Eng. Transl.)* 1981, *26*, 1221.

16. Muspratt, J.S. *Justus Liebigs Ann. Chem.* 1844, *50*, 283.

COMPONENTS:

1. Zinc sulfite; $ZnSO_3$; [13597-44-9]
2. Water; H_2O; [7732-18-5]

ORIGINAL MEASUREMENTS:

Heuston, F.T.; Tichborne, C.R.

Br. Med. J. 1890, 1063.

VARIABLES:

Room temperature

PREPARED BY:

B. Engelen

EXPERIMENTAL VALUES:

The solubility of zinc sulfite in water at room temperature is reported to be

0.16 mass %

= 0.011 mol kg^{-1} (molality, compiler).

AUXILIARY INFORMATION

METHOD APPARATUS/PROCEDURE:

Not given, but probably saturation method. Sulfite was determined iodometrically.

SOURCE AND PURITY OF MATERIALS:

Zinc sulfite was obtained by precipitation from zinc sulfate solutions with sodium sulfite.

ESTIMATED ERROR:

REFERENCES:

COMPONENTS:

1. Zinc sulfite; $ZnSO_3$; [13597-44-9]
2. Sulfur dioxide; SO_2; [7446-09-5]
3. Water; H_2O; [7732-18-5]

ORIGINAL MEASUREMENTS:

1. Terres, E.; Ruhl, G. *Angew. Chem.* 1934, *47*, 332-4.
2. Terres, E.; Ruhl, G. *Beitrage zur Chemie der schwefligen Saure, Beiheft zu den Zeitschriften des Vereins deutscher Chemiker No 8*, 1934.

VARIABLES:

Two temperatures: 288 and 298 K
Concentration of SO_2

PREPARED BY:

H.D. Lutz, B. Engelen

EXPERIMENTAL VALUES:

The authors report the solubility of $ZnSO_3.5/2H_2O$ [14460-28-7] in aqueous sulfurous acid solutions at 15 and 25°C. In the first paper, the experimental data are given in a graph. Numerical data are reported in the second paper.

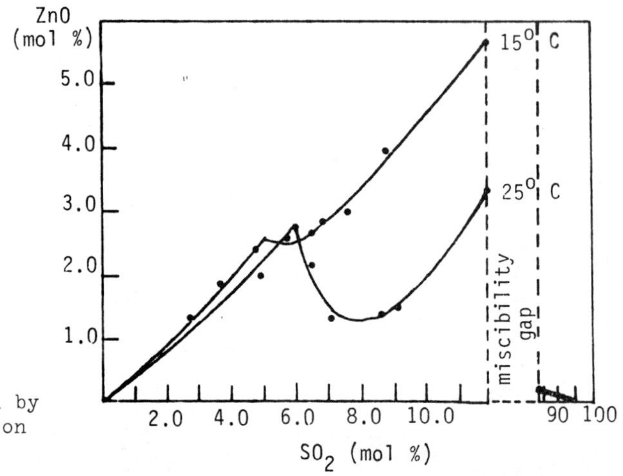

Reprinted by permission

(continued on next page)

AUXILIARY INFORMATION

METHOD APPARATUS/PROCEDURE:

SO_2-H_2O mixtures were treated together with solid zinc sulfite in closed glass ampoules at the stated temperatures. The solutions obtained were filtered through a fine glass frit and after oxidation of sulfite were analysed for sulfate and zinc content.

SOURCE AND PURITY OF MATERIALS:

Zinc sulfite was precipitated from a solution of zinc sulfite with Na_2SO_3.

ESTIMATED ERROR:

REFERENCES:

COMPONENTS:	ORIGINAL MEASUREMENTS:
1. Zinc sulfite; $ZnSO_3$; [13597-44-9] 2. Sulfur dioxide; SO_2; [7446-09-5] 3. Water; H_2O; [7732-18-5]	1. Terres, E.; Ruhl, G. *Angew. Chem.* 1934, 47, 332-4. 2. Terres, E.; Ruhl, G. *Beitrage zur Chemie der schwefligen Saure, Beiheft zu den Zeitschriften des Vereins deutscher Chemiker No 8*, 1934.

EXPERIMENTAL VALUES (continued):

Composition of saturated solutions

SO_2 mol %	ZnO mol %	m(ZnO)[a] mol kg^{-1}	Solid phase
Temperature = 15°C			
2.68	1.27	0.73	$ZnSO_3.5/2H_2O$
3.70	1.84	1.08	"
4.70	2.37	1.42	"
5.43	2.49	1.50	$Zn(HSO_3)_2$?
6.48	2.64	1.61	"
6.80	2.80	1.72	"
7.56	2.89	1.79	"
8.75	3.92	2.49	"
11.90[b]	5.62	3.78	"
Temperature = 25°C			
2.68	1.32	0.76	$ZnSO_3.5/2H_2O$
3.78	1.62	0.95	"
4.86	1.98	1.18	"
5.72	2.58	1.56	"
6.45	2.16	1.31	$Zn(HSO_3)_2$?
7.05	1.32	0.80	"
8.66	1.39	0.86	"
9.10	1.48	0.92	"
11.82[b]	3.32	2.17	"

[a] Compilers.

[b] Between 11.9 and 84 mol % SO_2 the mixtures separate into two liquid layers.

COMPONENTS:

1. Zinc sulfite; $ZnSO_3$; [13597-44-9]
2. Zinc sulfate; $ZnSO_4$; [7733-02-0]
3. Sulfur dioxide; SO_2; [7446-09-5]
4. Water; H_2O; [7732-18-5]

ORIGINAL MEASUREMENTS:

Kuz'minykh, I.N.; Kuznetsova, A.G.

Zh. Prikl. Khim. 1954, 27, 816-21;
**J. Appl. Chem. USSR (Eng. Transl.)* 1954, 27, 765-70.

VARIABLES:

Five temperatures: 283 - 351 K
Partial pressure and concentration of sulfur dioxide
Concentration of zinc sulfate

PREPARED BY:

B. Engelen

EXPERIMENTAL VALUES:

The authors report the composition of saturated solutions of zinc sulfite containing various amounts of zinc sulfate and excess of sulfur dioxide. The partial pressure of sulfur dioxide over the solutions is also given.

t/°C	p_{SO_2}	p_{SO_2} [a]	SO_4^{2-}	total SO_2	free SO_2 [b]	Zn	$m(ZnSO_3)$ [c]
	mm Hg	10^2Pa	mass %	mass %	mass %	mass %	mol kg^{-1}
9.9	2.55	3.39	0.381	3.822	2.022	2.099	0.2998
	26.36	35.14	0.323	6.628	3.588	3.322	0.5288
	34.06	45.40	0.76	7.703	4.161	4.142	0.6326
	74.54	99.37	0.99	9.95	5.532	5.204	0.8224
	101.43	135.22	0.465	11.78	6.345	5.875	1.0361
20.3	4.50	5.99	0.48	3.275	1.719	1.59	0.2566
	11.04	14.71	0.16	3.5	1.75	1.907	0.2892
	14.48	19.30	0.357	5.153	2.787	2.66	0.4021
	17.44	23.25	0.246	4.72	2.418	2.52	0.3884
	34.84	46.44	0.46	6.62	3.547	3.45	0.5360
	48.88	65.16	0.3	7.227	3.655	3.65	0.6277
	65.96	87.93	0.5	8.397	4.545	4.21	0.6919
	104.67	139.54	0.72	9.621	5.321	4.88	0.7919

a,b,c See the following page.

(continued on next page)

AUXILIARY INFORMATION

METHOD APPARATUS/PROCEDURE:

Aqueous solutions of sulfurous acid were kept over zinc sulfite precipitate in several thermostatically controlled glass vessels, joined in series, as part of a special apparatus for determination of the equilibrium pressure of SO_2. The supernatant solution in the last vessel was analysed for total SO_2 content, for sulfate (from oxidation of the sulfite), and for zinc content. The equilibrium pressure of SO_2 was determined dynamically by passing oxygen-free N_2 gas through the vessels and analysing the moist inert gas for SO_2 with iodine and thiosulfate. The same was done with solution-precipitate mixtures containing different amounts of sulfate. Also the density of the solutions was measured at 20°C.

SOURCE AND PURITY OF MATERIALS:

Not given.

ESTIMATED ERROR:

Not given.
The figures are mean values of several measurements. The authors report good agreement.

REFERENCES:

COMPONENTS:

1. Zinc sulfite; $ZnSO_3$; [13597-44-9]
2. Zinc sulfate; $ZnSO_4$; [7733-02-0]
3. Sulfur dioxide; SO_2; [7446-09-5]
4. Water; H_2O; [7732-18-5]

ORIGINAL MEASUREMENTS:

Kuz'minykh, I.N.; Kuznetsova, A.G.

Zh. Prikl. Khim. 1954, *27*, 816-21;
**J. Appl. Chem. USSR (Eng. Transl.)* 1954, *27*, 765-70.

EXPERIMENTAL VALUES (continued):

t/°C	p_{SO_2} mm Hg	p_{SO_2}[a] 10^2Pa	SO_4^{2-} mass %	total SO_2 mass %	free SO_2[b] mass %	Zn mass %	$m(ZnSO_3)$[c] mol kg^{-1}
20.3	142.2	189.58	0.53	10.81	6.015	5.27	0.8975
	232.8	310.37	0.48	12.51	6.82	5.813	1.0938
42.7	5.95	7.93	0.214	2.14	1.07	1.27	0.1733
	21.89	29.18	0.35	3.456	1.728	2.025	0.2864
	43.24	57.64	0.27	4.863	2.512	2.586	0.3976
	70.0	93.32	0.07	5.797	2.996	2.91	0.4793
	113.17	150.88	0.424	7.193	3.84	3.713	0.5902
	174.1	232.11	0.342	8.23	4.445	4.098	0.6765
	238.55	318.04	0.483	9.219	5.239	4.396	0.7232
61.4	7.28	9.70	0.1845	1.359	0.679	0.833	0.1087
	16.17	21.55	0.303	1.795	0.971	1.049	0.1328
	25.63	34.17	0.342	2.645	1.516	1.387	0.1832
	32	42.66	0.76	3.001	1.956	1.59	0.1723
	44.76	59.67	0.471	3.142	1.654	1.84	0.2456
	84.42	112.55	0.877	4.1	2.28	2.46	0.3069
	110.7	147.58	0.418	4.652	2.528	2.455	0.3585
	148.63	198.15	0.733	5.55	3.03	3.082	0.4340
	175.4	233.84	0.75	5.48	3.204	2.835	0.3906
78.4	4.03	5.37	0.267	0.631	0.315	0.505	0.0500
	14.7	16.59	0.168	1.112	0.556	0.683	0.0885
	36.17	48.22	0.33	1.646	0.971	0.914	0.1085
	50.69	67.58	0.27	2.208	1.206	1.208	0.1623
	107.7	143.58	0.1343	2.979	1.646	1.453	0.2180
20.3	2.81	3.75	8.95	1.665	1.019	6.756	0.1220
	14.79	19.71	6.233	4.23	2.413	6.101	0.3399
	19.24	25.65	7.52	4.175	2.172	7.17	0.3853
	19.3	25.73	2.637	4.547	2.61	3.776	0.3395
	19.88	26.50	4.45	4.38	2.394	5.062	0.3600
	23.05	30.73	8.409	4.418	2.23	7.96	0.4311
	25.98	34.63	9.26	4.075	2.103	8.33	0.3929
	30.48	40.63	0.935	6.57	3.43	4.018	0.5539
	41.34	55.11	12.38	2.761	1.88	10.38	0.3995

a,b,c See the following page.

(continued on next page)

COMPONENTS:	ORIGINAL MEASUREMENTS:
1. Zinc sulfite; $ZnSO_3$; [13597-44-9] 2. Zinc sulfate; $ZnSO_4$; [7733-02-0] 3. Sulfur dioxide; SO_2; [7446-09-5] 4. Water; H_2O; [7732-18-5]	Kuz'minykh, I.N.; Kuznetsova, A.G. *Zh. Prikl. Khim.* 1954, *27*, 816-21; **J. Appl. Chem. USSR (Eng. Transl.)* 1954, *27*, 765-70.

EXPERIMENTAL VALUES (continued):

t/°C	p_{SO_2} mm Hg	p_{SO_2} [a] 10^2Pa	SO_4^{2-} mass %	total SO_2 mass %	free SO_2 [b] mass %	Zn mass %	$m(ZnSO_3)$ [c] mol kg^{-1}
20.3	37.33	49.76	11.36	3.968	2.034	9.7	0.4026
	40.24	53.64	3.578	6.15	3.445	5.203	0.4963
	49.41	65.87	4.233	6.063	3.18	5.83	0.5365
	56.47	75.28	5.01	5.79	3.341	6.498	0.4622
	60.19	80.24	9.464	5.105	2.713	8.89	0.4878
	66.76	89.00	8.33	5.56	3.301	7.98	0.4513
	70.08	93.43	6.4	6.217	3.338	7.307	0.5612
	76.5	101.99	11.08	4.962	2.808	9.75	0.4531
	79.62	106.15	8.034	6.155	3.27	8.42	0.5819
	83.56	111.40	8.47	5.522	3.04	8.306	0.4986
	85.54	114.04	17.2	3.476	2.025	13.2	0.3425
	92.52	123.34	12.98	4.969	2.714	11.15	0.4964
	108.0	143.98	18.00	3.56	2.082	13.5	0.3552
	134.7	179.58	16.33	4.431	2.464	13.14	0.4645
	136.8	182.38	8.375	7.076	3.99	8.86	0.6364
	176.2	234.91	8.064	7.54	4.077	9.099	0.7179

[a] Calculated by the compiler.

[b] Excess over the amount necessary to form the monosulfite.

[c] Calculated from the concentration of combined SO_2 (amount necessary to form the monosulfite = SO_2 - free SO_2) by the compiler.

COMPONENTS:

1. Zinc sulfite; $ZnSO_3$; [13597-44-9]
2. Zinc sulfate; $ZnSO_4$; [7733-02-0]
3. Sulfur dioxide; SO_2; [7446-09-5]
4. Water; H_2O; [7732-18-5]

ORIGINAL MEASUREMENTS:

Peisakhov, I.L.; Karamazina, V.D.

Zh. Prikl. Khim. 1959, *32*, 70-8;
**J. Appl. Chem. USSR (Eng. Transl.)* 1959, *32*, 71-7;

VARIABLES:

Concentration of sulfur dioxide and sulfate

PREPARED BY:

H.D. Lutz

EXPERIMENTAL VALUES:

The authors report the composition and the partial pressure of sulfur dioxide of saturated solutions of zinc sulfite containing various amounts of zinc sulfate and excess sulfur dioxide.

Composition of saturated solutions[a] g/dm^3					10^2c/mol dm^{-3} (compiler)			pH[b]	Partial pressure of SO_2[c]	
$ZnSO_4$	$Zn(HSO_3)_2$	Zn^{2+}	SO_2 total	SO_2 free[d]	$ZnSO_4$	$Zn(HSO_3)_2$	SO_2 total		mm Hg	10^2Pa (compiler)
4.52	5.32	3.38	2.98	0	2.80	2.34	4.65	4.8	0.057	0.076
3.52	10.2	4.44	5.84	0	2.18	4.5	9.12	4.2	0.149	0.199
8.1	13.9	7.25	7.97	0.19	5.0	6.1	12.44	4.2	0.465	0.620
4.44	13.9	5.79	7.83	0	2.75	6.1	12.22	3.9	0.302	0.403
3.71	24.2	8.47	14.0	0.26	2.30	10.6	21.9	3.6	1.68	2.24
6.6	25.6	10.05	15.0	0.58	4.1	11.3	23.4	3.6	1.67	2.23
11.4	39.3	15.9	23.65	1.54	7.1	17.3	36.92	3.15	3.8	5.1
9.35	41.4	15.6	25.0	1.73	5.79	18.2	39.0	3.4	3.6	4.8
24.8	5.56	11.62	3.36	0.23	15.4	2.44	5.24	4.3	0.206	0.275
26.8	41.3	22.7	23.7	0.45	16.6	18.2	37.0	3.0	6.70	8.93
55	3.86	23.4	2.21	0.32	34	1.70	3.45	4.05	0.07	0.09

[a,b,c,d] See the following page.

(continued on next page)

AUXILIARY INFORMATION

METHOD APPARATUS/PROCEDURE:

Nitrogen was passed through thermostatically controlled absorption flasks containing a pulp of solid zinc sulfite and dissolved $Zn(HSO_3)_2$ and $ZnSO_4$. The partial pressure of SO_2 over the pulp was determined by analysing the gas emerging from the absorption flasks for its SO_2 content, by passing the gas through an absorption flask containing iodine solution. The saturated solutions in the pulp were analysed for SO_2 and Zn after the end of each experiment (method not given).

SOURCE AND PURITY OF MATERIALS:

Not given.

ESTIMATED ERROR:

REFERENCES:

COMPONENTS:

1. Zinc sulfite; $ZnSO_3$; [13597-44-9]
2. Zinc sulfate; $ZnSO_4$; [7733-02-0]
3. Sulfur dioxide; SO_2; [7446-09-5]
4. Water; H_2O; [7732-18-5]

ORIGINAL MEASUREMENTS:

Peisakhov, I.L.; Karamazina, V.D.

Zh. Prikl. Khim. 1959, *32*, 70-8;
**J. Appl. Chem. USSR (Eng. Transl.)* 1959, *32*, 71-7;

EXPERIMENTAL VALUES (continued):

Composition of saturated solutions[a] g/dm^3					10^2c/mol dm^{-3} (compiler)			pH[b]	Partial pressure of SO_2[c]	
$ZnSO_4$	$Zn(HSO_3)_2$	Zn^{2+}	SO_2 total	SO_2 free[d]	$ZnSO_4$	$Zn(HSO_3)_2$	SO_2 total		mm Hg	10^2Pa (compiler)
53.2	6.14	23.3	3.5	0.05	33.0	2.70	5.5	3.45	0.363	0.484
49.6	42.3	32.2	25.0	1.15	30.7	18.6	39.0	2.8	8.44	11.25
88.2	3.64	36.7	2.63	0.58	54.6	1.60	4.11	4.1	0.222	0.296
82.6	10.9	36.8	6.9	1.21	51.7	4.8	10.8	3.9	0.948	1.264
82.0	22.9	39.6	13.7	1.02	50.8	10.1	21.4	3.8	2.68	3.57
77.5	35.4	41.5	22.2	2.3	48.0	15.6	34.7	3.2	9.00	12.00
104	3.64	43.5	2.06	0	64.6	1.60	3.22	4.1	1.187	1.583
103	5.67	43.6	3.44	0.24	63.8	2.49	5.37	3.4	0.373	0.497
101.4	42.7	53.1	24.5	0.45	62.8	18.8	38.24	2.8	7.83	10.44
194	3.28	77.8	3.01	1.65	120	1.44	4.70	3.5	0.274	0.365
198	38.1	91.6	21.5	0	123	16.7	33.56	3.4	5.93	7.91

[a] Temperature of equilibration not given, but probably room temperature (compiler).

[b] Before experiment, i.e. before passing nitrogen through the solution.

[c] 20°C, thermostatically controlled.

[d] Excess over the amount necessary to form $Zn(HSO_3)_2$.

The authors also report some graphs and fitting equations of variation of the SO_2 vapour pressure over the zinc salt solutions studied.

COMPONENTS:

1. Zinc sulfite; $ZnSO_3$; [13597-44-9]
2. Sodium sulfite; Na_2SO_3; [7757-83-7]
3. Water; H_2O; [7732-18-5]

ORIGINAL MEASUREMENTS:

Murooka, T.; Sato, H.

Bull. Inst. Phys. Chem. Res. (Tokyo) 1937, *16*, 636-42.

VARIABLES:

Two temperatures: 288 and 298 K
Concentration of Na_2SO_3

PREPARED BY:

B. Engelen, H.D. Lutz

EXPERIMENTAL VALUES:

t/°C	Solvent		Composition of saturated solution				Solid phase
	Na_2SO_3	$NaHSO_3$	$Zn(HSO_3)_2$	$ZnSO_3$	$Zn(HSO_3)_2$[a]	$ZnSO_3$[a]	
	mole/kg soln		g/kg soln		mole/kg soln		
25	0.0		1.460	0.653	0.00642	0.00449	$ZnSO_3.2.5H_2O$
15	0.343[b]					0.0162	"
15	0.641[b]					0.0316	$3ZnSO_3.Na_2SO_3.Zn(OH)_2$?
15	0.499	0.356				0.0641	"
15	0.923	0.346				0.125	"
15	0.977	0.476				0.168	"

a Total concentration of dissolved zinc sulfite is $[ZnSO_3] + 2[Zn(HSO_3)_2]$.

b Only the sum of Na_2SO_3 and $NaHSO_3$ is given by the authors.

AUXILIARY INFORMATION

METHOD APPARATUS/PROCEDURE:

Saturation method. The water or solution of Na_2SO_3 of known composition was stirred with solid $ZnSO_3.2.5H_2O$ in a thermostatically controlled vessel at the given temperatures until equilibrium was reached (time not given). The solutions were analysed for SO_2 iodometrically, and for HSO_3^- acidimetrically. Mean values of 4 measurements are reported.

SOURCE AND PURITY OF MATERIALS:

$ZnSO_3.2.5H_2O$ was prepared by bubbling SO_2 through a suspension of ZnO in water. After dissolving the ZnO, the sulfite was precipitated by heating.

ESTIMATED ERROR:

REFERENCES:

COMPONENTS:

1. Zinc sulfite; $ZnSO_3$; [13597-44-9]
2. Zinc sulfate; $ZnSO_4$; [7733-02-0]
3. Sodium sulfite; Na_2SO_3; [7757-83-7]
4. Sodium sulfate; Na_2SO_4; [7757-82-6]
5. Water; H_2O; [7732-18-5]

ORIGINAL MEASUREMENTS:

Margulis, E.V.; Rodin, I.V.

Zh. Neorg. Khim. 1981, *26*, 2269-70; *Russ. J. Inorg. Chem. (Eng. Transl.)* 1981, *26*, 1221-2.

VARIABLES:

Five temperatures: 293 - 368 K
Concentration of $ZnSO_4$, Na_2SO_3, and Na_2SO_4

PREPARED BY:

H.D. Lutz

EXPERIMENTAL VALUES:

The authors report the solubility of $ZnSO_3.5/2H_2O$ [14460-28-7] in water and in solutions with various amounts of $ZnSO_4$, Na_2SO_3, and Na_2SO_4, at different temperatures.

Pure water

t/°C	$ZnSO_3$ mass %	$m(ZnSO_3)^a$ mol kg^{-1}	Zn kg/m^3	$c(ZnSO_3)^a$ mol dm^{-3}
20	0.1786	0.01230	1.042	0.01594
50	0.1811	0.01248	1.056	0.01615
70	0.1873	0.01290	1.092	0.01671
85	0.1908	0.01314	1.112	0.01701
95	0.1939	0.01336	1.130	0.01728

[a] Calculated by the compiler.

Remark: The data given by the authors for mass % of $ZnSO_3$ and kg Zn/m^3 seem to be inconsistent.

(continued on next page)

AUXILIARY INFORMATION

METHOD APPARATUS/PROCEDURE:

The experiments were carried out in a water thermostat with mechanical stirring in closed flasks. To convert the concentrations in kg m^{-3} to mass %, the density of the solutions were determined pyknometrically at ambient temperature. It has been assumed that the equilibrium had been reached in the systems when the measured concentrations of Zn^{2+} or SO_3^{2-} remained constant. In all cases, 1 hr was sufficient for equilibration. The solubility of zinc sulfite was determined by analysing the solutions for zinc polarographically, for SO_3^{2-} iodometrically, and for sodium by flame photometry.

SOURCE AND PURITY OF MATERIALS:

Zinc sulfite was precipitated from a solution of zinc sulfate with Na_2SO_3 (1).

ESTIMATED ERROR:

Solubility: $\pm 0.5 \times 10^{-3}$ mass %
Temperature: ± 0.5 K (authors).

REFERENCES:

1. Margulis, E.V.; Grishankina, N.S. *Zh. Neorg. Khim.* 1963, *8*, 2638.

COMPONENTS:	ORIGINAL MEASUREMENTS:
1. Zinc sulfite; $ZnSO_3$; [13597-44-9] 2. Zinc sulfate; $ZnSO_4$; [7733-02-0] 3. Sodium sulfite; Na_2SO_3; [7757-83-7] 4. Sodium sulfate; Na_2SO_4; [7757-82-6] 5. Water; H_2O; [7732-18-5]	Margulis, E.V.; Rodin, I.V. *Zh. Neorg. Khim.* 1981, *26*, 2269-70; *Russ. J. Inorg. Chem. (Eng. Transl.)* 1981, *26*, 1221-2.

EXPERIMENTAL VALUES (continued):

t/°C	$ZnSO_4$ mass %	$ZnSO_3$ mass %	$m(ZnSO_3)$[a] mol kg^{-1}	Na_2SO_3 mass %	$ZnSO_3$ mass %	$m(ZnSO_3)$[a] mol kg^{-1}	Na_2SO_4 mass %	$ZnSO_3$ mass %	$m(ZnSO_3)$[a] mol kg^{-1}
20	1.038	0.182	0.01267	0.006	0.179	0.01233	1.053	0.189	0.01316
	5.11	0.196	0.01423	0.048	0.177	0.01220	4.31	0.210	0.01512
	14.94	0.211	0.01710	0.615	0.173	0.01199	7.21	0.239	0.01776
	25.22	0.222	0.02047	0.890	0.140	0.00973	9.83	0.251	0.01919
	32.14	0.228	0.02318	1.050	0.135	0.00939	14.12	0.264	0.02120
	34.81	0.230	0.02434	1.113	0.132	0.00919	15.45	0.268	0.02186
				3.09	0.239	0.01700			
				6.05	0.369	0.02711			
				12.21	0.653	0.05153			
				20.13	1.241	0.10852			
				21.42	1.249	0.11105			
90	1.221	0.207	0.01444	1.242	0.141	0.00983	1.223	0.206	0.01437
	5.43	0.217	0.01581	3.41	0.176	0.01255	5.11	0.241	0.01751
	15.21	0.232	0.01886	6.21	0.182	0.01337	10.01	0.277	0.02123
	25.03	0.249	0.02291	12.41	0.194	0.01526	19.84	0.287	0.02471
	32.38	0.271	0.02767	20.22	0.207	0.01789	27.32	0.291	0.02764
	37.73	0.276	0.03061	21.42	0.209	0.01834	28.23	0.297	0.02857

Solutions containing Na_2SO_3 and Na_2SO_4 at 20°C.

Na_2SO_3 mass %	$ZnSO_3$ mass %	$m(ZnSO_3)$[a] mol kg^{-1}	$ZnSO_3$ mass %	$m(ZnSO_3)$[a] mol kg^{-1}	$ZnSO_3$ mass %	$m(ZnSO_3)$[a] mol kg^{-1}	$ZnSO_3$ mass %	$m(ZnSO_3)$[a] mol kg^{-1}
	1.7 % Na_2SO_4[b]		3.4 % Na_2SO_4[b]		4.2 % Na_2SO_4[b]		5.0 % Na_2SO_4[b]	
0.006	0.207	0.01451	0.216	0.01541	0.234	0.01684	0.247	0.01792
0.048	0.201	0.01409	0.205	0.01463	0.209	0.01504	0.224	0.01626
0.615	0.175	0.01234	0.178	0.01277	0.186	0.01346	0.198	0.01445
0.890	0.145	0.01025	0.154	0.01108	0.164	0.01190	0.172	0.01259
1.050	0.139	0.00984	0.151	0.01110	0.160	0.01163	0.169	0.01239

[a] Calculated by the compiler.

[b] Mass %.

COMPONENTS:	EVALUATOR:
1. Cadmium sulfite; $CdSO_3$; [13477-23-1] 2. Water; H_2O; [7732-18-5]	H.D. Lutz, Dept. of Chemistry, University of Siegen, FR Germany. October 1983

CRITICAL EVALUATION:

Cadmium sulfite crystallizes from aqueous solutions as hydrates and the anhydrous salt. Some are well established, such as $CdSO_3.3/2H_2O$ (1,2) [60943-67-1] and several polymorphic forms of the anhydrous salt $CdSO_3$, I(α) (1), II, and III (3). Numerical data on the solubility of cadmium sulfite in water and in the presence of $CdSO_4$, Na_2SO_3, and Na_2SO_4, respectively, have been given by Margulis and Rodin (4), but from the data it is not clear what kind of solid phase was present in the solutions studied.

REFERENCES

1. Lutz, H.D.; El-Suradi, S. *Z. Anorg. Allg. Chem.* 1976, *425*, 134.
2. Kiers, C.T.; Vos, A. *Cryst. Struct. Commun.* 1978, *7*, 399.
3. Lutz, H.D.; Buchmeier, W.; Eckers, W.; Engelen, B. *Z. Anorg. Allg. Chem.* 1983, *496*, 21.
4. Margulis, E.V.; Rodin, I.V. *Zh. Neorg. Khim.* 1981, *26*, 1428; **Russ. J. Inorg. Chem. (Eng. Transl.)* 1981, *26*, 767.

COMPONENTS:

1. Cadmium sulfite; $CdSO_3$; [13477-23-1]
2. Water; H_2O; [7732-18-5]

ORIGINAL MEASUREMENTS:

Margulis, E.V.; Rodin, I.V.

Zh. Neorg. Khim. 1981, *26*, 1428-30; **Russ. J. Inorg. Chem. (Eng. Transl.)* 1981, *26*, 767-8.

VARIABLES:

Two temperatures: 293 and 363 K

PREPARED BY:

B. Engelen

EXPERIMENTAL VALUES:

t/°C	Cd^{2+} g/dm^3	$CdSO_3$ mass %	$c(CdSO_3)^a$ 10^{-3} mol dm^{-3}	$m(CdSO_3)^a$ 10^{-3} mol kg^{-1}
20	0.248	0.0426	2.21	2.21
90	0.232	0.0398	2.06	2.07

[a] Calculated by the compiler.

AUXILIARY INFORMATION

METHOD APPARATUS/PROCEDURE:

Saturation method. Equilibrium was established by stirring the saturated solutions in closed thermostatically controlled glass tubes. Equilibrium was tested for analytically - 1 hr was reported to be sufficient. Cadmium was determined polarographically, sulfite iodometrically. For conversion into molality units, the density of the solutions was measured pyknometrically at room temperature.

SOURCE AND PURITY OF MATERIALS:

Cadmium sulfite, claimed to be $CdSO_3.2H_2O$, was obtained by precipitation from $CdSO_4$ solutions with Na_2SO_3. The precipitate was washed with water and dry acetone. $CdSO_4.8/3H_2O$ and Na_2SO_3 of p.a. quality were used.

ESTIMATED ERROR:

Temperature: ± 0.5 K
Solubility: 2.7×10^{-5} and 3.1×10^{-5} for molarity and molality units, respectively.

REFERENCES:

COMPONENTS:

1. Cadmium sulfite; $CdSO_3$; [13477-23-1]
2. Cadmium sulfate; $CdSO_4$; [10124-36-4]
3. Water; H_2O; [7732-18-5]

ORIGINAL MEASUREMENTS:

Margulis, E.V.; Rodin, I.V.

Zh. Neorg. Khim. 1981, *26*, 1428-30;
**Russ. J. Inorg. Chem. (Eng. Transl.)* 1981, *26*, 767-8.

VARIABLES:

Two temperatures: 293 and 363 K
Concentration of cadmium sulfate

PREPARED BY:

B. Engelen

EXPERIMENTAL VALUES:

t/°C	$CdSO_4$ mass %	$CdSO_3$ mass %	$m(CdSO_4)$[a] mol kg^{-1}	$m(CdSO_3)$[a] 10^{-3} mol kg^{-1}
20	1.415	0.052	0.6889	2.742
20	5.31	0.056	0.2692	3.075
20	14.89	0.065	0.8399	3.971
20	31.61	0.087	2.220	6.618
20	43.51	0.106	3.702	9.768
90	36.83	0.083	2.801	6.836

[a] Calculated by the compiler.

AUXILIARY INFORMATION

METHOD APPARATUS/PROCEDURE:

$CdSO_4$ solutions were saturated with solid cadmium sulfite and stirred in a thermostatically controlled closed glass tube. Equilibrium was tested for analytically - 1 hr was reported to be sufficient. Cadmium was determined polarographically, sulfite iodometrically. Sulfate determination method is not given. For conversion from g/dm^3 into mass % the density of the solutions was measured at room temperature.

SOURCE AND PURITY OF MATERIALS:

Cadmium sulfite, claimed to be $CdSO_3.2H_2O$, was obtained by precipitation from $CdSO_4$ solutions with Na_2SO_3. The precipitate was washed with water and dry acetone. $CdSO_4.8/3H_2O$ and Na_2SO_3 of p.a. quality were used.

ESTIMATED ERROR:

Temperature: 0.5 K

REFERENCES:

COMPONENTS:

1. Cadmium sulfite; $CdSO_3$; [13477-23-1]
2. Sodium sulfite; Na_2SO_3; [7757-83-7]
3. Sodium sulfate; Na_2SO_4; [7757-82-6]
4. Water; H_2O; [7732-18-5]

ORIGINAL MEASUREMENTS:

Margulis, E.V.; Rodin, I.V.

Zh. Neorg. Khim. 1981, *26*, 1428-30; **Russ. J. Inorg. Chem. (Eng. Transl.)* 1981, *26*, 767-8.

VARIABLES:

Temperature: 293 K (363 K)
Concentration of Na_2SO_3 and Na_2SO_4

PREPARED BY:

B. Engelen

EXPERIMENTAL VALUES:

Composition of the solutions (mass %)			$m(Na_2SO_3)$[a]	$m(Na_2SO_4)$[a]	$m(CdSO_3)$[a]
Na_2SO_3	Na_2SO_4	$CdSO_3$	mol kg^{-1}	mol kg^{-1}	10^{-3} mol kg^{-1}
0.005	-	0.043	0.0003	-	2.236
0.300	-	0.040	0.0239	-	2.085
0.440	-	0.030	0.0351	-	1.566
0.700	-	0.029	0.0559	-	1.518
0.800	-	0.028	0.0640	-	1.467
1.643	-	0.036	0.1326	-	1.902
5.54	-	0.078	0.466	-	4.294
7.99	-	0.106	0.690	-	5.993
12.60	-	0.181	1.146	-	10.78
21.89	-	0.220	2.230	-	14.68
21.89[b]	-	0.217[b]	2.230[b]	-	14.47

[a] Calculated by the compiler.

[b] Experiment done at 90°C.

(continued on next page)

AUXILIARY INFORMATION

METHOD APPARATUS/PROCEDURE:

$CdSO_4$ solutions were saturated with solid cadmium sulfite and stirred in a thermostatically controlled closed glass tube. Equilibrium was tested for analytically - 1 hr was reported to be sufficient. Cadmium was determined polarographically, sulfite iodometrically, and sodium photometrically. Sulfate determination method not given. For conversion from g x dm^{-3} into mass %, the density of the solutions was measured at room temperature.

SOURCE AND PURITY OF MATERIALS:

Cadmium sulfite, claimed to be $CdSO_3.2H_2O$, was obtained by precipitation from $CdSO_4$ solutions with Na_2SO_3. The precipitate was washed with water and dry acetone. $CdSO_4.8/3H_2O$, $Na_2SO_4.10H_2O$, and Na_2SO_3 of p.a. quality were used.

ESTIMATED ERROR:

Temperature: 0.5 K

REFERENCES:

COMPONENTS:	ORIGINAL MEASUREMENTS:
1. Cadmium sulfite; $CdSO_3$; [13477-23-1] 2. Sodium sulfite; Na_2SO_3; [7757-83-7] 3. Sodium sulfate; Na_2SO_4; [7757-82-6] 4. Water; H_2O; [7732-18-5]	Margulis, E.V.; Rodin, I.V. *Zh. Neorg. Khim.* 1981, *26*, 1428-30; **Russ. J. Inorg. Chem. (Eng. Transl.)* 1981, *26*, 767-8.

EXPERIMENTAL VALUES (continued):

Composition of the solutions (mass %)			$m(Na_2SO_3)^a$	$m(Na_2SO_4)^a$	$m(CdSO_3)^a$
Na_2SO_3	Na_2SO_4	$CdSO_3$	mol kg^{-1}	mol kg^{-1}	10^{-3} mol kg^{-1}
-	1.011	0.045	-	0.0719	2.363
-	4.11	0.048	-	0.302	2.602
-	7.30	0.055	-	0.555	3.085
-	10.21	0.062	-	0.8011	3.590
-	15.35	0.078	-	1.278	4.792
-	28.62[b]	0.098[b]	-	2.827[b]	7.143
0.005	1.5	0.048	0.0403	0.107	2.533
0.005	4.5	0.050	0.0416	0.332	2.722
0.005	5.8	0.051	0.0421	0.434	2.815
0.005	7.2	0.057	0.0428	0.547	3.194
0.300	1.5	0.042	2.425	0.108	2.223
0.300	4.5	0.044	2.501	0.333	2.403
0.300	5.8	0.047	2.536	0.435	2.602
0.300	7.2	0.049	2.575	0.548	2.754
0.440	1.5	0.033	3.561	0.108	1.749
0.440	4.5	0.038	3.674	0.333	2.078
0.440	5.8	0.042	3.725	0.436	2.329
0.440	7.2	0.045	3.782	0.549	2.533
0.700	1.5	0.031	5.680	0.108	1.647
0.700	4.5	0.035	5.861	0.334	1.919
0.700	5.8	0.037	5.942	0.437	2.060
0.700	7.2	0.041	6.033	0.551	2.314
0.800	1.5	0.030	6.499	0.108	1.596
0.800	4.5	0.034	6.705	0.335	1.866
0.800	5.8	0.037	6.798	0.437	2.059
0.800	7.2	0.040	6.902	0.551	2.260

[a] Calculated by the compiler.

[b] Experiment done at 90°C.

COMPONENTS:	EVALUATOR:
1. Mercury(I) sulfite; Hg_2SO_3; [89146-33-8] 2. Water; H_2O; [7732-18-5]	H.D. Lutz, Dept. of Chemistry, University of Siegen, FR Germany. January 1984.

CRITICAL EVALUATION:

Information on mercury sulfites is very scarce (1-3). Numerical data on the solubility of Hg_2SO_3 have been reported only by Kryukova (4), who gave a value of 1.5×10^{-11} mol dm^{-3} (molarity) at 298 K. The existence of mercury(I) sulfite, however, is not fully confirmed (1,3). Compounds identified as $Hg_4(SO_3)_2 \cdot H_2O$ and $Hg_2SO_3 \cdot HgSO_3 \cdot 4H_2O$ were claimed as soluble and insoluble in cold water, respectively (1,2).

The data given are doubtful, because the nature of the solid phase is not defined.

REFERENCES

1. *Gmelins Handbuch der Anorganischen Chemie*, 8. Aufl. Quecksilber, Teil B, Lieferung 3, Springer-Verlag, Berlin 1974, p. 1002.
2. Divers, E.; Shimidzu, T. *J. Chem. Soc.* 1886, *49*, 533.
3. Seubert, K.; Elten, M. *Z. Anorg. Allg. Chem.* 1893, *4*, 44.
4. Kryukova, T.A. *Zh. Fiz. Khim.* 1939, *13*, 693.

COMPONENTS:	ORIGINAL MEASUREMENTS:
1. Mercury(I) sulfite; Hg_2SO_3; [89146-33-8] 2. Water; H_2O; [7732-18-5]	Kryukova, T.A *Zh. Fiz. Khim.* 1939, *13*, 693-700.
VARIABLES:	**PREPARED BY:**
One temperature: 298 K	B. Engelen

EXPERIMENTAL VALUES:

The solubility of Hg_2SO_3 is reported to be 1.5×10^{-11} mol dm^{-3}.

AUXILIARY INFORMATION

METHOD APPARATUS/PROCEDURE:

Solubility was determined polarographically from the polarization potentials of a mercury electrode in a solution containing 0.1 mol dm^{-3} KNO_3 and 0.02 mol dm^{-3} sulfite.

SOURCE AND PURITY OF MATERIALS:

Not given.

ESTIMATED ERROR:

REFERENCES:

COMPONENTS:

1. Lead sulfite; $PbSO_3$; [7446-10-8]
2. Water; H_2O; [7732-18-5]

EVALUATOR:

H.D. Lutz,
Dept. of Chemistry,
University of Siegen,
FR Germany.

May 1983.

CRITICAL EVALUATION:

Lead sulfite crystallizes from aqueous solutions as the anhydrous salt, $PbSO_3$. The solubility of $PbSO_3$ in water has not been thoroughly investigated. Numerical data have been given by Terres *et al.* (1), Hanus *et al.* (2), Mangan (3), Ermolaev *et al.* (4), and Rodin *et al.* (5) (precipitate identified as $PbSO_3.1/2H_2O$ (5) ? [7446-10-8]), but not under comparable experimental conditions. Thus only Hanus *et al.* (2) and Rodin *et al.* (5) reported on the solubility of lead sulfite in pure water, namely m($PbSO_3$) = 8 x 10^{-6} mol kg^{-1} in hot water (2) and c($PbSO_3$) = 2.5 x 10^{-6} mol dm^{-3} at 293 K and 1.62 x 10^{-5} mol dm^{-3} at 363 K, respectively (5). The data given by Terres *et al.* (1), Mangan (3), and Ermolaev *et al.* (4), which were all obtained in the presence of a third component, may be extrapolated to pure water to be $\ll$ 2.8 x 10^{-3} mol kg^{-1} (molality scale), < 3.0 x 10^{-4} mol dm^{-3}, and < 1.1 x 10^{-6} mol dm^{-3} (both molarity scale), respectively, at room temperature. The scarce information on the temperature shift of the solubility of $PbSO_3$ is contradictory. A negative temperature coefficient of the solubility is reported by Terres *et al.* (1) and a positive temperature shift by Ermolaev *et al.* (4) and Rodin *et al.* (5). Mangan (3) has found a positive temperature coefficient of solubility below 315 K and a negative one at higher temperatures.

TENTATIVE VALUES

The solubility of $PbSO_3$ in water at room temperature is approximately 1 x 10^{-6} mol dm^{-3} (molarity scale) (3 x 10^{-4} g/dm^3). The temperature coefficient of solubility is probably positive.

The solubility of lead sulfite in water is affected by the presence of a third compound. This is shown by the fragmentary experimental data on the systems $PbSO_3$-SO_4-H_2O (1), $PbSO_3$-$NH_4CH_3CO_2$-H_2O (3), and $PbSO_3$-$NaNO_3$-H_2O (4). The solubility of $PbSO_3$ increases to 4 x 10^{-2} mole (10 g)/kg H_2O with increasing SO_2 content (1,6) at 298 K, to 1.3 x 10^{-5} mol dm^{-3} (molarity scale) (3.8 x 10^{-3} g/dm^3) with $NaNO_3$ concentration increasing to 400 g/dm^3 (4), and to 3.6 x 10^{-4} mol dm^3 (0.10 g/dm^3) with pH value decreasing to 2.8 (4) and decreases with increasing concentration of ammonium acetate (3,7), all at 293.2 K. It has been further claimed that $PbSO_3$ is insoluble in aqueous alcohol (3) and soluble in solutions of sodium hydroxide (3) and that the solubility of $PbSO_3$ increases in the presence of citrate and tartrate (3).

REFERENCES

1. Terres, E.; Ruhl, G. *Angew. Chem.* 1934, *47*, 332.

2. Hanus, J.; Hovorka, V. *Chem. Listy* 1937, *31*, 489.

3. Mangan, J.L. *N. Z. J. Sci. Technol., Sect. B* 1949, *30*, 323.

4. Ermolaev, M.I.; Kudrina, L. *Tr. Voronezh. Tekhnol. Inst.* 1968, *17*, 201.

5. Rodin, I.V.; Margulis, E.V. *Zh. Neorg. Khim.* 1983, *28*, 532; **Russ. J. Inorg. Chem. (Eng. Transl.)* 1983, *28*, 298.

COMPONENTS:	EVALUATOR:
1. Lead sulfite; $PbSO_3$; [7446-10-8] 2. Water; H_2O; [7732-18-5]	H.D. Lutz, Dept. of Chemistry, University of Siegen, FR Germany. May 1983.

CRITICAL EVALUATION: (continued)

6. Rohrig, A. *J. Prakt. Chem.* 1889, *37*, 217

7. Lewis, J.B. *Ind. Eng. Chem., Anal. Ed.* 1940, *12*, 535.

COMPONENTS:	ORIGINAL MEASUREMENTS:
1. Lead sulfite; $PbSO_3$; [7446-10-8] 2. Water; H_2O; [7732-18-5]	Hanus, J.; Hovorka, V. *Chem. Listy* 1937, *31*, 489-500.
VARIABLES:	PREPARED BY:
One temperature: "hot"	H.D. Lutz

EXPERIMENTAL VALUES:

The authors report the solubility of $PbSO_3$ in hot water to be

$$2.2 \times 10^{-3}\ g/dm^3\ H_2O$$

This value is equal to (compiler)

$$m(PbSO_3) = 8.0 \times 10^{-6}\ mol\ kg^{-1}$$

AUXILIARY INFORMATION

METHOD APPARATUS/PROCEDURE:	SOURCE AND PURITY OF MATERIALS:
Tho loss of weight of lead sulfite was determined by washing with different amounts of hot water.	$PbSO_3$ was precipitated with $Na_2S_2O_5$.
	ESTIMATED ERROR:
	Data given are the results of several experiments with different amounts of hot water.
	REFERENCES:

COMPONENTS:	ORIGINAL MEASUREMENTS:
1. Lead sulfite; $PbSO_3$; [7446-10-8] 2. Water; H_2O; [7732-18-5]	Rodin, I.V.; Margulis, E.V. *Zh. Neorg. Khim.* 1983, *28*, 532-3; *Russ. J. Inorg. Chem. (Eng. Transl.)* 1983, *28*, 298-9.
VARIABLES:	PREPARED BY:
Three temperatures: 293, 323 and 363 K	B. Engelen

EXPERIMENTAL VALUES:

The solubility of $PbSO_3.0.5H_2O$ [7446-10-8] in water at various temperatures are reported. The solubility products reported are defined as $K_{s0}(PbSO_3.0.5H_2O) = [Pb^{2+}]^2$.

t/°C	Pb	$PbSO_3$		$10^{12}K_{s0}$
	mg/dm^3	mg/dm^{3a}	$10^5c/mol\ dm^{-3}$	$mol^2\ dm^{-6}$
20	0.52	0.72	0.25	6.25
50	1.41	1.95	0.68	46.2
90	3.35	4.64	1.62	262

[a] Calculated by the compiler.

AUXILIARY INFORMATION

METHOD APPARATUS/PROCEDURE:	SOURCE AND PURITY OF MATERIALS:
Saturation method. Equilibrium was established by stirring the saturated solutions in thermostatically controlled glass tubes. Equilibrium was tested for analytically - 2 hr was reported to be sufficient. Lead was determined polarographically.	Lead sulfite was precipitated from $Pb(CH_3COO)_2$ solutions with Na_2SO_3. ESTIMATED ERROR: Temperature: ±0.5 K REFERENCES:

COMPONENTS:	ORIGINAL MEASUREMENTS:
1. Lead sulfite; $PbSO_3$; [7446-10-8] 2. Sulfur dioxide; SO_2; [7446-09-5] 3. Water; H_2O; [7732-18-5]	1. Terres, E.; Rühl, G. *Angew. Chem.* 1934, 47, 322-4. 2. Terres, E.; Rühl, G. *Beitrage zur Chemie der schwefligen Saure, Beiheft zu den Zeitschriften des Vereins deutscher Chemiker No 8*, 1934.
VARIABLES:	PREPARED BY:
Two temperatures: 288 and 298 K Concentration of SO_2	H.D. Lutz, B. Engelen

EXPERIMENTAL VALUES:

The authors report the solubility of lead sulfite in aqueous sulfurous acid solutions at 15 and 25°C. In the first paper, the experimental data are given in a graph. Numerical data are reported in the second paper.

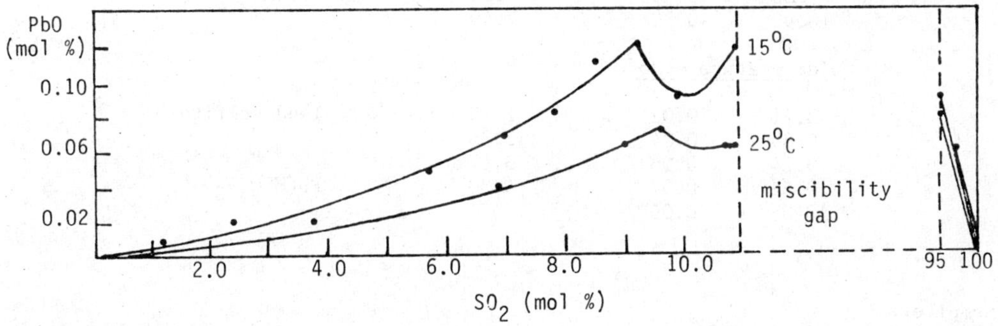

Reprinted by permission

(continued on next page)

AUXILIARY INFORMATION

METHOD APPARATUS/PROCEDURE:

SO_2-H_2O mixtures were treated together with solid $PbSO_3$ in closed glass ampoules at the stated temperatures. The solutions obtained were filtered through a fine glass frit and, after oxidation of the sulfite, analysed for sulfate and lead.

SOURCE AND PURITY OF MATERIALS:

$PbSO_3$ was precipitated from a solution of lead acetate with Na_2SO_3.

ESTIMATED ERROR:

REFERENCES:

COMPONENTS:	ORIGINAL MEASUREMENTS:
1. Lead sulfite; $PbSO_3$; [7446-10-8] 2. Sulfur dioxide; SO_2; [7446-09-5] 3. Water; H_2O; [7732-18-5]	1. Terres, E.; Ruhl, G. *Angew. Chem.* 1934, *47*, 322-4. 2. Terres, E.; Ruhl, G. *Beitrage zur Chemie der schwefligen Saure, Beiheft zu den Zeitschriften des Vereins deutscher Chemiker No 8*, 1934.

EXPERIMENTAL VALUES (continued):

Composition of saturated solutions

SO_2 mol %	PbO mol %	m(PbO)[a] 10^{-2} mol kg^{-1}	Solid phase
Temperature = 15°C			
1.15	0.01	0.56	lead sulfite
2.35	0.02	1.14	"
5.71	0.03	1.77	"
7.82	0.08	4.82	"
8.50	0.11	6.68	"
9.18	0.12	7.34	"
9.80	0.09	5.54	$Pb(HSO_3)_2$?
10.92[b]	0.12	7.49	"
Temperature = 25°C			
3.71	0.02	1.15	lead sulfite
6.85	0.04	2.38	"
9.14	0.06	3.67	"
9.62	0.07	4.30	$Pb(HSO_3)_2$?
10.08[b]	0.06	3.71	"

[a] Compilers.

[b] Between 11 and 95 mol % SO_2 the mixtures separate into two liquid layers.

COMPONENTS:

1. Lead sulfite; $PbSO_3$; [7446-10-8]
2. Ammonium acetate; $NH_4CH_3CO_2$; [631-61-8]
3. Water; H_2O; [7732-18-5]

ORIGINAL MEASUREMENTS:

Mangan, J.L.

N.Z.J. Sci. Technol., Sect. B 1949, *30*, 323-33.

VARIABLES:

Concentration of $NH_4CH_3CO_2$
pH: 4 - 10
Temperature: 273 - 329 K

PREPARED BY:

H.D. Lutz

EXPERIMENTAL VALUES:

The author reports the solubility of $PbSO_3$ in aqueous solutions containing various amounts of ammonium acetate at different temperatures and pH values. The author also reports that lead sulfite is insoluble in aqueous alcohol but soluble in sodium hydroxide solutions, and that the solubility of lead sulfite is considerably increased in the presence of citrate and tartrate.

Solubility of $PbSO_3$[a]			Solubility of $PbSO_3$[b]		
$NH_4CH_3CO_2$ mol/dm^3 soln	10^3c mol dm^{-3}	g/dm^3 (compiler)	pH	10^3c mol dm^{-3}	g/dm^3 (compiler)
0.5	1.80	0.517	4.8	1.04	0.299
1.0	1.20	0.345	5.1	0.86	0.247
1.5	1.40	0.402	5.6	0.72	0.207
2.0	1.40	0.402	5.9	0.56	0.161
3.0	1.40	0.402	6.1	0.44	0.126
4.0	1.00	0.287	7.9	0.36	0.103
5.0	1.60	0.460	9.0	0.74	0.213

a 20°C, pH 6, measured before adding $PbSO_3$.

b 20°C, non-saturating solute $NH_4CH_3CO_2$, 1 mol/dm^3 soln.

(continued on next page)

AUXILIARY INFORMATION

METHOD APPARATUS/PROCEDURE:

$PbSO_3$ was suspended in ammonium acetate solutions of the given concentration and maintained by shaking in a thermostatically controlled flask at the given temperatures for 2 - 3 hr. To prevent oxidation, 5 mass % of sucrose was added to the ammonium acetate solution. The suspended solid was allowed to settle and the saturated solution was filtered rapidly. Sulfite was determined iodometrically. The pH was measured with a glass electrode.

SOURCE AND PURITY OF MATERIALS:

Lead sulfite was precipitated from a solution of lead acetate, washed thoroughly with a 5 mass % sucrose soln, and finally washed with the appropriate ammonium acetate solutions before the solubility was determined.

ESTIMATED ERROR:

REFERENCES:

COMPONENTS:

1. Lead sulfite; $PbSO_3$; [7446-10-8]
2. Ammonium acetate; $NH_4CH_3CO_2$; [631-61-8]
3. Water; H_2O; [7732-18-5]

ORIGINAL MEASUREMENTS:

Mangan, J.L.

N.Z. J. Sci. Technol., Sect. B 1949, *30*, 323-33.

EXPERIMENTAL VALUES (continued):

Solubility of $PbSO_3$[c]

t/°C	10^3c mol dm^{-3}	g/dm^3 (compiler)
0	1.20	0.345
15	1.35	0.388
42	1.70	0.488
56	1.40	0.402

[c] pH 6, measured before adding $PbSO_3$. Non-saturating solute - $NH_4CH_3CO_2$ 1 mol/dm^3 soln.

COMPONENTS:

1. Lead sulfite; $PbSO_3$; [7446-10-8]
2. Sodium nitrate; $NaNO_3$; [7631-99-4]
3. Acetate buffer; CH_3CO_2H-$NaCH_3CO_2$ [64-19-7]; [127-09-3]
4. Water; H_2O; [7732-18-5]

ORIGINAL MEASUREMENTS:

Ermolaev, M.I.; Kudrina, L.T.

Tr. Voronezh. Tekhnol. Inst. 1968, *17*, 201-3.

VARIABLES:

Temperature: 293 - 353 K
pH value
Concentration of $NaNO_3$

PREPARED BY:

H.D. Lutz, B. Engelen

EXPERIMENTAL VALUES:

The authors report the solubility of lead sulfite in solutions of various $NaNO_3$ concentrations for various pH values and temperatures. Experimental data are given in three graphs and as some numerical figures. The solubility increases from 1.1×10^{-6} to 1.32×10^{-5} mol/dm^3 soln with $NaNO_3$ concentration increasing from 100 g to 400 g/dm^3 soln at 20°C (Fig. 1), decreases from 3.64×10^{-4} to 1.76×10^{-5} mol/dm^3 soln with pH value increasing from 2.8 to 7.0 at 20°C and a $NaNO_3$ concentration of 200 g/dm^3 soln (Fig. 2) and increases with increasing temperature; data in Fig. 3 are for pH 4.49 and a $NaNO_3$ concentration of 200 g/dm^3 soln.

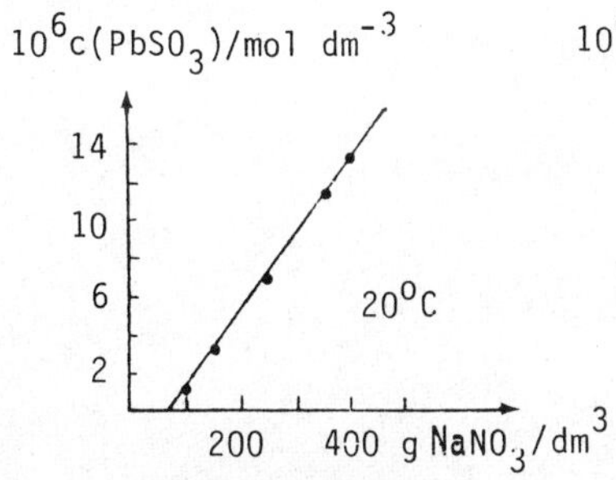

Fig. 1

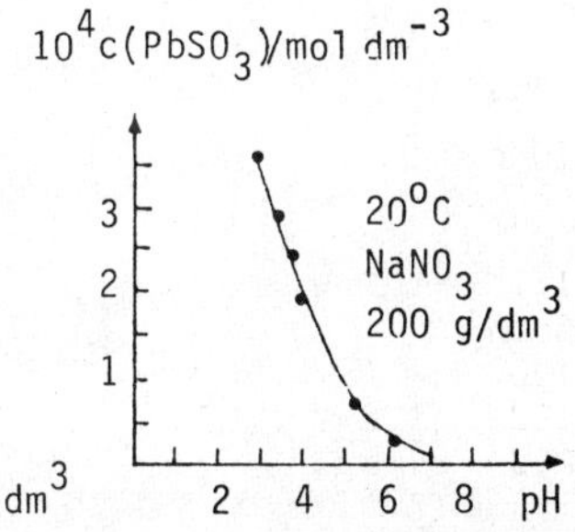

Fig. 2

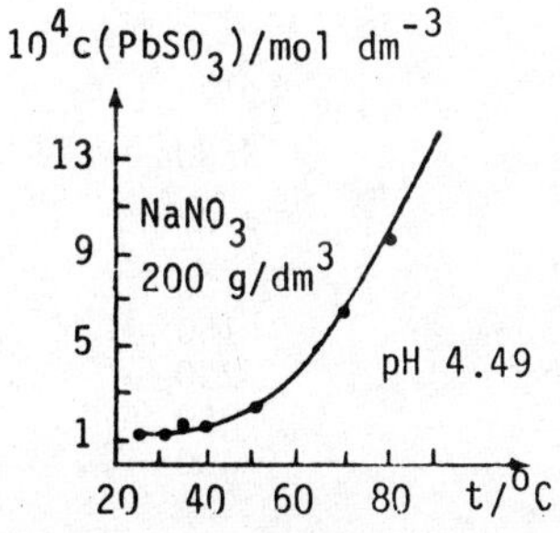

Fig. 3

(continued on next page)

AUXILIARY INFORMATION

METHOD APPARATUS/PROCEDURE:

Saturation method. Equilibrium was established after 8 hr. To avoid oxidation of the sulfite, 0.005% phenylenediamine was added to the solutions. Lead was determined colorimetrically.

SOURCE AND PURITY OF MATERIALS:

Not given.

ESTIMATED ERROR:

REFERENCES:

COMPONENTS:	ORIGINAL MEASUREMENTS:
1. Lead sulfite; $PbSO_3$; [7446-10-8] 2. Sodium nitrate; $NaNO_3$; [7631-99-4] 3. Acetate buffer; $CH_3CO_2H-NaCH_3CO_2$ [64-19-7]; [127-09-3] 4. Water; H_2O; [7732-18-5]	Ermolaev, M.I.; Kudrina, L.T. *Tr. Voronezh. Tekhnol. Inst.* 1968, *17*, 201-3.

EXPERIMENTAL VALUES (continued):

The following figures are estimated by the compilers from the graphs given by the authors.

Solubility of $PbSO_3$[a]

t/°C	10^4c / mol dm^{-3}
25	1.3
30	1.3
35	1.5
40	1.6
50	2.3
60	3.7
70	6.3
80	9.0

Solubility of $PbSO_3$[b]

pH	10^4c / mol dm^{-3}
2.8	3.6
3.3	2.9
3.7	2.4
4.0	2.0
5.0	0.87
5.4	0.63
6.2	0.28
7.0	0.18

Solubility of $PbSO_3$[c]

$NaNO_3$ g/dm^3	10^6c / mol dm^{-3}
100	1.1
150	3.1
250	7.0
350	11.2
400	13.2

a Non-saturating solute 200 g $NaNO_3/dm^3$, pH 4.49.

b Non-saturating solute 200 g $NaNO_3/dm^3$, temperature 20°C.

c 20°C.

COMPONENTS:

1. Sodium selenite; Na_2SeO_3; [10102-18-8]
2. Water; H_2O; [7732-18-5]

EVALUATOR:
Mary R. Masson,
Dept. of Chemistry,
University of Aberdeen,
Meston Walk, Old Aberdeen, AB9 2UE,
Scotland, UK.

June 1984.

CRITICAL EVALUATION:

The binary system sodium selenite - water was studied by Janickis (1,2), and data are also available from studies of ternary systems (3 - 7). The data are all in reasonable agreement, apart from one or two points, which were rejected before final regression equations were derived. There are four equations, corresponding to the equilibria with the four possible solid phases, ice, $Na_2SeO_3.8H_2O$ [41292-05-1], $Na_2SeO_3.5H_2O$ [26970-82-1], and Na_2SeO_3.

The equations are (1) for 253 - 273 K, solid phase ice:

$$(T - 273.15) = 0.0659 - 0.343y + 0.0115y^2 - 0.000419y^3 \qquad s = 0.116 \text{ (8 pts)}$$

or

$$y = -0.199 - 4.33(T - 273.2) - 0.165(T - 273.2)^2 - 0.00227(T - 273.2)^3 \qquad s = 0.140 \text{ (9 pts)}$$

(2) for 253 - 264 K, solid phase $Na_2SeO_3.8H_2O$:

$$y = 48.2 + 0.821(T - 273.2) + 0.0176(T - 173.2)^2 \qquad s = 0.188 \text{ (5 pts)}$$

(3) for 253 - 310 K, solid phase $Na_2SeO_3.5H_2O$:

$$y = 43.2 + 0.108(T - 273.2) + 0.00205(T - 273.2)^2 \qquad s = 0.148 \text{ (20 pts)}$$

and (4) for 310 - 373 K, solid phase Na_2SeO_3:

$$y = 54.0 - 0.133(T - 273.2) + 0.000483(T - 273.2)^2 \qquad s = 0.207 \text{ (14 pts)}$$

where $y = 100w$ is the solubility expressed in mass % of Na_2SeO_3, T is the temperature in K, and s is the standard deviation of the dependent variable about the regression line.

TENTATIVE SOLUBILITIES

The following tentative solubility values for Na_2SeO_3 in water were calculated from equations (3) and (4).

COMPONENTS:	EVALUATOR:
1. Sodium selenite; Na_2SeO_3; [10102-18-8] 2. Water; H_2O; [7732-18-5]	Mary R. Masson, Dept. of Chemistry, University of Aberdeen, Meston Walk, Old Aberdeen, AB9 2UE, Scotland, UK. June 1984.

CRITICAL EVALUATION: (continued)

T/K	Solubility mass %	molality, mol/kg
253.2	41.8	4.15
263.2	42.3	4.24
273.2	43.2	4.40
283.2	44.5	4.64
293.2	46.1	4.95
298.2	47.1	5.15
303.2	48.3	5.40
308.2	49.5	5.67
313.2	49.5	5.67
323.2	48.6	5.47
333.2	47.8	5.29
343.2	47.1	5.15
353.2	46.5	5.03
363.2	45.9	4.91
373.2	45.5	4.83

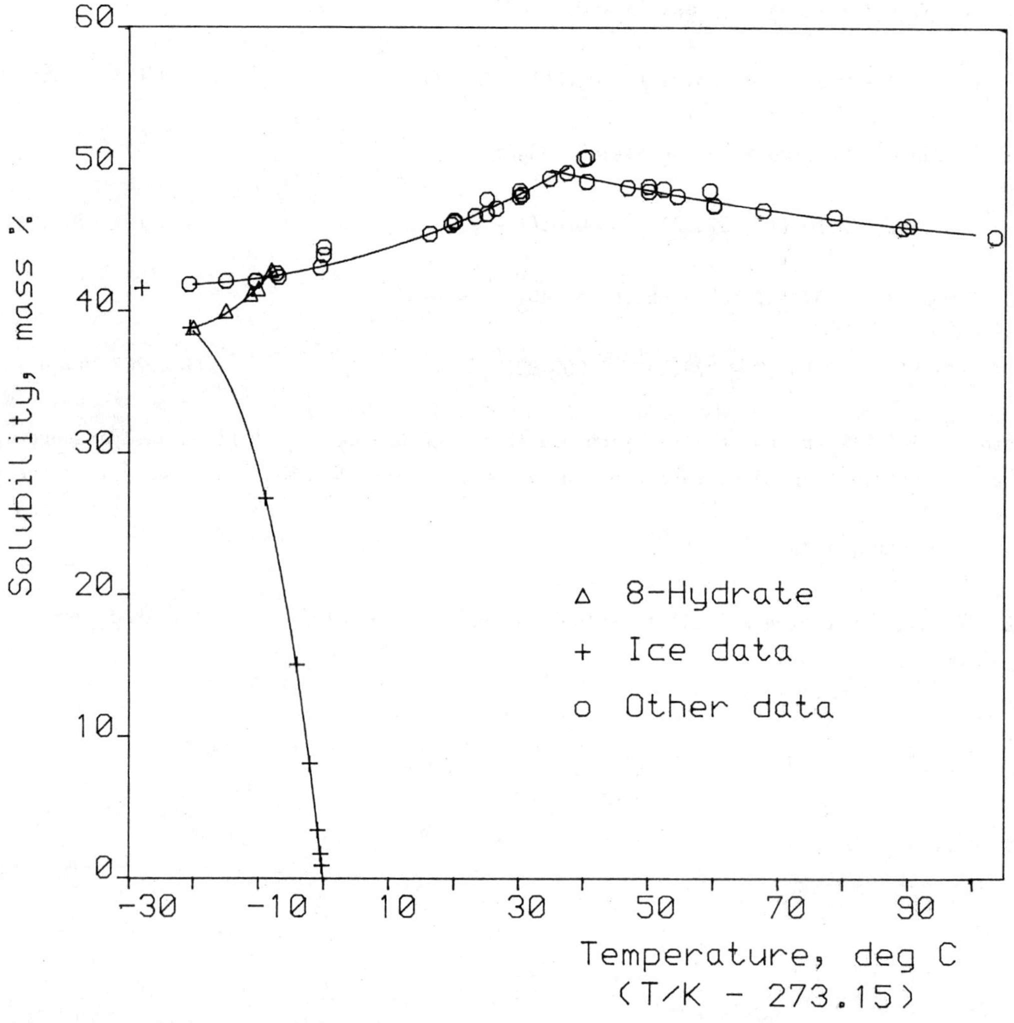

COMPONENTS:	EVALUATOR:
1. Sodium selenite; Na_2SeO_3; [10102-18-8] 2. Water; H_2O; [7732-18-5]	Mary R. Masson, Dept. of Chemistry, University of Aberdeen, Meston Walk, Old Aberdeen, AB9 2UE, Scotland, UK. June 1984.

CRITICAL EVALUATION: (continued)

TERNARY SYSTEMS

The ternary systems involving solium sulfite (3), ethanol (4), sodium selenate (5), sodium carbonate (6) and sodium tellurite (7) are all straightforward systems, and the data appear to be reasonably reliable. In the system involving selenious acid (8) three double salts were observed:

$Na_2SeO_3.7H_2SeO_3$	[15855-80-8]	$(NaH_7(SeO_3)_4)$
$Na_2SeO_3.3H_2SeO_3$	[14013-56-0]	$(NaH_3(SeO_3)_2)$
$Na_2SeO_3.H_2SeO_3$	[7782-82-3]	$(NaHSeO_3)$

The data for the system SeO_2-NaOH-H_2O (9) do not appear to be very reliable, as can be seen from the random nature of the tie-lines in the computer diagram. Also, the data do not appear to support the conclusions reached by the authors.

REFERENCES

1. Janitzki, J. *Z. Anorg. Allgem. Chem.* 1932, *205*, 49.

2. Janickis, J.; Gutmanaite, H. *Z. Anorg. Allgem. Chem.* 1936, *225*, 1.

3. Klebanov, G.S.; Ostapkevich, N.A. *Zh. Priklad. Khim.* 1966, *39*, 2467; **J. Appl. Chem. USSR (Eng. Transl.)* 1966, *39*, 2315.

4. Klebanov, G.S.; Ostapkevich, N.A. *Zh. Priklad. Khim.* 1966, *39*, 1435; **J. Appl. Chem. USSR (Eng. Transl.)* 1966, *39*, 1342.

5. Yanitskii, I.V.; Patkausas, R.M. *Zh. Priklad. Khim.* 1970, *43*, 522; **J. Appl. Chem. USSR (Eng. Transl.)* 1970, *43*, 530.

6. Chimboulev, M.T.; Vassilev, H. *Dokl. Bolg. Akad. Nauk* 1973, *26*, 1509.

7. Chimbulev, M.; Vasilev, Kh. *God. Vissh. Khim.-Tekhnol. Inst. Sofia,* 1977, *22*, 247.

8. Sabbah, R.; Périnet, G. *J. Chim. Phys.* 1965, *62*, 929.

9. Rustamov, P.G.; Mardakhaev, B.N. *Azerb. Khim. Zh.* 1963, 131.

COMPONENTS:

1. Sodium selenite; Na_2SeO_3; [10102-18-8]

2. Water; H_2O; [7732-18-5]

ORIGINAL MEASUREMENTS:

Janitzki, J.

Z. Anorg. Allgem. Chem. 1932, *205*, 49-75.

VARIABLES:

Temperature: 253 - 376 K

PREPARED BY:

Mary R. Masson

EXPERIMENTAL VALUES:

Composition of saturated solutions

t/°C	Na_2SeO_3 mass %	Na_2SeO_3[a] mol/kg	Solid[b] phase
-20.0	38.81	3.667	C
-15.0	40.00	3.855	C
-11.2	41.20	4.052	C
-10.0	41.58	4.116	C
- 8.0	42.90	4.344	C
-20.6	41.88	4.167	B
-15.0	42.11	4.206	B
-10.4	42.15	4.213	B
-10.5	42.13	4.210	B
- 7.2	42.65	4.300	B
- 6.8	42.40	4.256	B
- 0.5	43.09	4.378	B
+16.3	45.48	4.824	B
+19.5	46.12	4.950	B
+23.2	46.73	5.072	B
+26.4	47.28	5.186	B
+30.4	48.27	5.396	B
+34.7	49.40	5.645	B
+40.5	50.90	5.994	B
+37.4	49.79	5.734	A
+40.4	49.18	5.596	A
+46.8	48.76	5.502	A
+50.0	48.85*	5.522	A
+52.3	48.66*	5.481	A
+54.4	48.14	5.368	A
+59.4	48.54*	5.454	A

(continued on next page)

AUXILIARY INFORMATION

METHOD APPARATUS/PROCEDURE:

For each temperature, a saturated solution was prepared by stirring the salt in water inside a stoppered 4-cm diameter test-tube. Small samples of solution were removed at intervals for analysis, in order to test for attainment of equilibrium. The time required varied between 2½ and 19 hr. The solutions were analysed for SeO_2 by the method of Norris and Fay (1).

The solid phases were identified by analysis.

SOURCE AND PURITY OF MATERIALS:

ESTIMATED ERROR:

Temperature:	-20 - 0°C	±0.2°C
	0 - 60°C	±0.1°C
	60 - 110°C	±0.3°C

REFERENCES:

1. Norris, J.F.; Fay, H. *Amer. Chem. J.* 1896, *18*, 703; 1900, *23*, 119.

COMPONENTS:	ORIGINAL MEASUREMENTS:
1. Sodium selenite; Na_2SeO_3; [10102-18-8]	Janitzki, J.
2. Water; H_2O; [7732-18-5]	*Z. Anorg. Allgem. Chem.* 1932, *205*, 49-75.

EXPERIMENTAL VALUES (continued):

t/°C	Na_2SeO_3 mass %	Na_2SeO_3[a] mol/kg	Solid[b] phase
+67.6	47.15	5.159	A
+78.5	46.67	5.060	A
+89.1	45.93	4.912	A
+103.3	45.30	4.789	A

[a] Molalities calculated by the compiler.

[b] Solids: A - Na_2SeO_3, B - $Na_2SeO_3.5H_2O$, C - $Na_2SeO_3.8H_2O$

COMPONENTS:

1. Sodium selenite; Na_2SeO_3; [10102-18-8]
2. Water; H_2O; [7732-18-5]

ORIGINAL MEASUREMENTS:

Janickis, J.; Gutmanaite, H.

Z. Anorg. Allgem. Chem. 1936, *227*, 1 - 16.

VARIABLES:

Temperature: 245 - 273 K

PREPARED BY:

Mary R. Masson

EXPERIMENTAL VALUES:

Composition of equilibrium solutions

t/°C	Na_2SeO_3 mol/dm^3	Na_2SeO_3 mass %	Na_2SeO_3[a] mol/kg	Solid phase
-0.250	0.05	0.860	0.0501	ice
-0.468	0.1	1.706	0.1003	"
-0.890	0.2	3.361	0.2011	"
-2.075	0.5	8.05	0.506	"
-4.095	1	15.06	1.025	"
-8.885	2	26.79	2.115	"
-20.5	satd.	38.8	3.665	ice + $Na_2SeO_3.8H_2O$
-28.1	satd.	41.6	4.12	ice + $Na_2SeO_3.5H_2O$

[a] Molalities calculated by the compiler.

AUXILIARY INFORMATION

METHOD APPARATUS/PROCEDURE:

Freezing points of prepared solutions were measured by use of a Beckman-type apparatus (1). Determinations were repeated until the desired reproducibility was attained. Each reported value is the mean of at least three determinations.

SOURCE AND PURITY OF MATERIALS:

Sodium selenite was prepared by neutralization of selenious acid with sodium hydroxide.

ESTIMATED ERROR:

Temperature reproducibility 0.5%

REFERENCES:

1. Ostwald, W.; Luther, R. *Hand- und Hilfsbuch zur Ausfuhrung physikochemischer Messungen* 5th Ed., Akademische Verlag., Leipzig, 1931.

COMPONENTS:

1. Sodium selenite; Na_2SeO_3; [10102-18-8]
2. Sodium sulfite; Na_2SO_3; [7757-83-7]
3. Water; H_2O; [7732-18-5]

ORIGINAL MEASUREMENTS:

Klebanov, G.S.; Ostapkevich, N.A.

Zh. Priklad. Khim. 1966, *39*, 2467-2470;
**J. Appl. Chem. USSR (Eng. Transl.)* 1966, *39*, 2315-8.

VARIABLES:

Composition
Temperature: 273 and 323 K

PREPARED BY:

Mary R. Masson

EXPERIMENTAL VALUES:

Composition of equilibrium solutions

Na_2SeO_3 mass %	Na_2SO_3 mass %	Na_2SeO_3[a] mol/kg	Na_2SO_3[a] mol/kg	Solid[b] phase
Temperature = 0°C				
43.95	-	4.534	0.	B
43.15	0.78	4.450	0.110	B
42.40	1.40	4.362	0.198	B
42.20	1.94	4.368	0.276	B + D
29.26	3.13	2.502	0.367	D
12.85	7.48	0.933	0.745	D
-	12.30	0.	1.113	D
Temperature = 50°C				
48.48	-	5.441	0.	A
47.20	0.73	5.242	0.111	A
45.13	1.65	4.903	0.246	A
44.41	2.28	4.817	0.339	A + C
37.73	3.58	3.717	0.484	C
28.10	7.45	2.521	0.917	C
22.90	10.01	1.974	1.184	C
17.00	13.21	1.409	1.502	C
9.02	18.50	0.720	2.025	C
-	25.91	0.	2.775	C

[a] Molalities calculated by the compiler.

[b] Solid phases: A - Na_2SeO_3, B - $Na_2SeO_3.5H_2O$, C - Na_2SO_3, D - $Na_2SO_3.7H_2O$

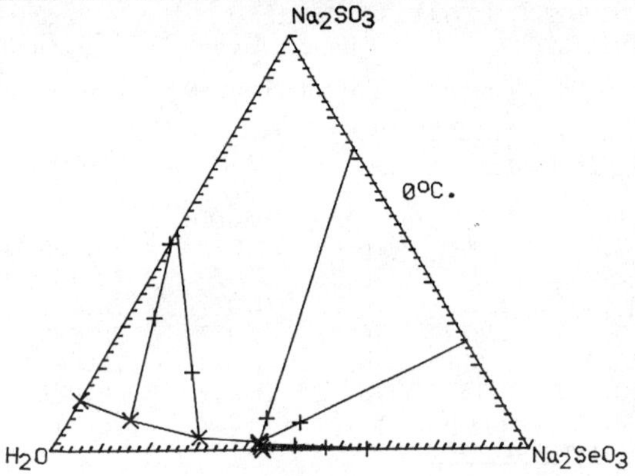

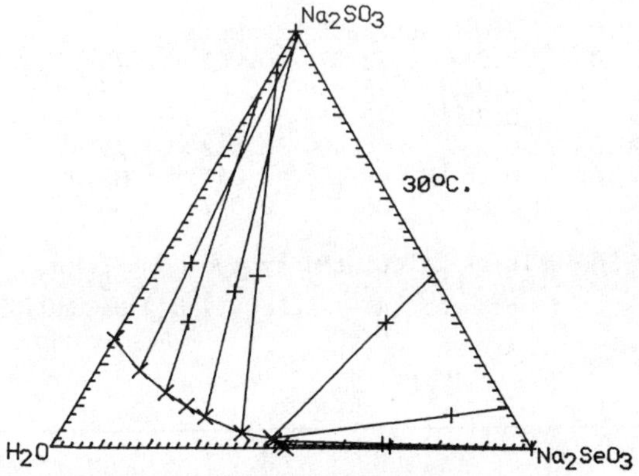

METHOD APPARATUS/PROCEDURE:

Mixtures of Na_2SeO_3 and Na_2SO_3 were dissolved isothermally in water contained in glass vessels with stirrers fitted with hydraulic seals. Trace amounts of p-phenylenediamine were added to the solutions to prevent aerial oxidation of sulfite. 0°C was maintained by melting ice, and 50°C with the aid of a contact thermometer and electromagnetic relay. Equilibrium was reached after 2 days at 0°C and 30 - 35 hr at 50°C.

The solutions were analysed for selenite and sulfite as follows. For sulfite, sodium bicarbonate and excess of 0.1*N* iodine solution were added, then the solution was acidified with acetic acid, and the excess of iodine was titrated with thiosulfate. For selenite, sulfite was bound with formaldehyde, the solution was acidified with HCl, and KI was added. The iodine liberated was titrated with thiosulfate.

The solid residues were also analysed. The compositions of the solid phases were determined by Schreinemakers' remainder method.

SOURCE AND PURITY OF MATERIALS:

Sodium sulfite was of analytical grade, and sodium selenite was of reagent grade.

ESTIMATED ERROR:

Temperature: ±0.1 K
Solubility: no estimate possible.

REFERENCES:

COMPONENTS:

1. Sodium selenite; Na_2SeO_3; [10102-18-8]
2. Ethanol; C_2H_5OH; [64-17-5]
3. Water; H_2O; [7732-18-5]

ORIGINAL MEASUREMENTS:

Klebanov, G.S.; Ostapkevich, N.A.

Zh. Priklad. Khim. 1966, *39*, 1435-7; *J. Appl. Chem. USSR (Eng. Transl.)* 1966, *39*, 1342-4.

VARIABLES:

Ethanol concentration
Temperatures: 293 and 323 K

PREPARED BY:

Mary R. Masson

EXPERIMENTAL VALUES:

Initial	Upper layer				Lower layer				Solid[b]
C_2H_5OH mass %	Na_2SeO_3 mass %	C_2H_5OH mass %	Na_2SeO_3[a] mol/kg	C_2H_5OH[a] mol/kg	Na_2SeO_3 mass %	C_2H_5OH mass %	Na_2SeO_3 mol/kg	C_2H_5OH[a] mol/kg	phase
Temperature = 20°C									
0.0	-	-	-	-	46.30	-	4.986	-	B
1.5	not enough sample				44.28	1.36	4.710	0.543	B
7.0	0.30	76.33	0.074	70.895	43.80	1.51	4.631	0.599	B
10.0	0.32	76.28	0.079	70.758	43.71	1.52	4.615	0.602	B
30.0	0.30	76.23	0.074	70.501	43.53	1.50	4.579	0.592	B
50.0	0.31	76.32	0.077	70.886	43.62	1.51	4.597	0.597	B
70.0	0.32	76.25	0.079	70.640	43.51	1.53	4.578	0.604	B
80.0	0.18	78.52	0.049	80.017	-	-	-	-	A
90.0	0.055	85.63	0.022	129.842	-	-	-	-	A
Temperature = 50°C									
0.0	-	-	-	-	48.48	-	5.441	-	A
3.0	not enough sample				48.10	0.44	5.405	0.186	A
10.0	0.24	78.37	0.065	79.528	46.36	1.23	5.115	0.509	A
30.0	0.23	78.39	0.062	79.586	46.37	1.24	5.118	0.514	A
50.0	0.23	78.36	0.062	79.444	46.35	1.22	5.112	0.505	A
70.0	0.23	78.40	0.062	79.633	46.37	1.21	5.115	0.501	A
80.0	0.16	79.15	0.045	83.037	-	-	-	-	A
90.0	0.04	86.42	0.017	138.541	-	-	-	-	A

[a] Molalities calculated by the compiler.

[b] Solid phases: A - Na_2SeO_3, B - $Na_2SeO_3.5H_2O$

AUXILIARY INFORMATION

METHOD APPARATUS/PROCEDURE:

Aqueous ethanolic solutions were saturated with sodium selenite at 20 and 50°C in glass vessels (100 - 150-ml), fitted with hydraulic seals at 20°C and with reflux condensers at 50°C. Equilibrium was reached after 16 - 28 hr.

Selenite in the solutions and moist solids was determined iodometrically after distillation from the samples.

SOURCE AND PURITY OF MATERIALS:

Reagent-grade sodium selenite was used. Ethanol and water were distilled twice.

ESTIMATED ERROR:

Temperature: ±0.1 K
Analyses: no estimate possible.

REFERENCES:

COMPONENTS:

1. Sodium selenite; Na_2SeO_3; [10102-18-8]

2. Sodium selenate; Na_2SeO_4; [13410-01-0]

3. Water; H_2O; [7732-18-5]

ORIGINAL MEASUREMENTS:

Yanitskii, I.V.; Patkauskas, R.M.

Zh. Priklad. Khim. 1970, *43*, 522-27; **J. Appl. Chem. USSR (Eng. Transl.)* 1970, *43*, 530-5.

VARIABLES:

Concentrations of components
Six temperatures: 273 - 333 K

PREPARED BY:

Mary R. Masson

EXPERIMENTAL VALUES:

Composition of equilibrium solutions

Na_2SeO_3 mass %	Na_2SeO_4 mass %	Na_2SeO_3[a] mol/kg	Na_2SeO_4[a] mol/kg	Solid[b] phase
Temperature = 0°C				
-	11.7	0.	0.701	B
7.63	9.21	0.531	0.586	B
12.9	7.06	0.932	0.467	B
16.9	6.21	1.271	0.427	B
20.9	5.43	1.640	0.390	B
25.1	4.34	2.057	0.326	B
31.3	3.89	2.793	0.318	B
35.2	3.40	3.315	0.293	B
40.1	3.20	4.089	0.299	B
41.8	2.80	4.363	0.268	B + D
42.9	1.78	4.484	0.170	D
44.5	-			D
Temperature = 20°C				
-	30.3	0.	2.301	B
3.39	27.4	0.283	2.095	B
10.39	23.5	0.909	1.881	B
15.2	21.0	1.378	1.742	B
17.3	20.3	1.603	1.722	B
21.0	18.8	2.017	1.653	B
24.3	17.6	2.418	1.603	B
27.1	17.1	2.808	1.622	B
30.5	16.5	3.328	1.648	B
32.9	16.2	3.738	1.685	B + D

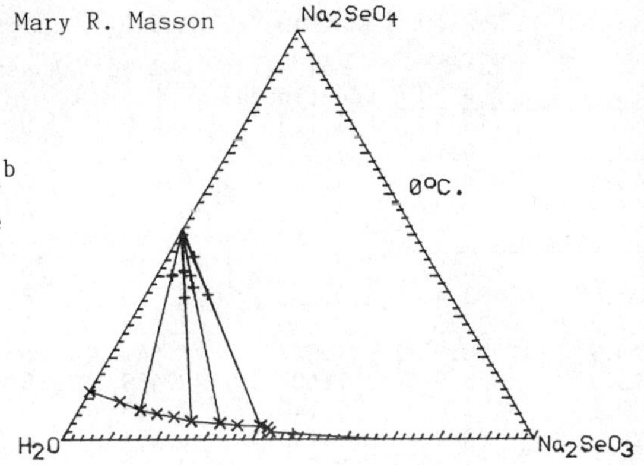

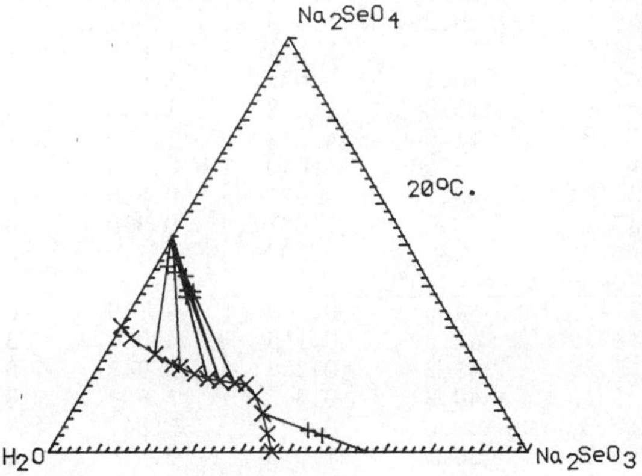

(continued on next page)

AUXILIARY INFORMATION

METHOD APPARATUS/PROCEDURE:

Saturated solutions were prepared in three-necked glass flasks fitted with thermometers and stirrers, which were kept immersed in a water thermostat. Equilibrium was reached after about 36 hr at 0°C, 6-10 hr at 20-40°C, and 10-15 hr at 60°C.

Samples for analysis were removed through the third neck by suction through a cotton plug into tubes. Selenite was determined iodometrically in the solution (1). Total selenium was determined after reduction of the solution as follows: the test solution was boiled gently under reflux for 20 min with 10 ml of satd. KBr and 10ml of 23% HCl solution. A few drops of satd. alcoholic acetanilide were added to remove residual free bromine. All the selenium was then present as selenite, and was determined as above. Selenate was then obtained by difference.

The solid phases were identified by Schreinemakers' method of residuals.

SOURCE AND PURITY OF MATERIALS:

Sodium selenite was prepared by neutralization of analytical grade selenious acid with carbonate-free analytical grade sodium hydroxide. Sodium selenate was prepared by electrolytic oxidation of concentrated selenite solution at a lead anode, at 25 mA/cm^2 current density.

ESTIMATED ERROR:

Temperature: ±0.1 K

REFERENCES:

1. Yanitskii, I.V.; Zelionkaite, V.I.; Patsauskas, E.I. *Zh. Neorg. Khim.* 1957, *2*, 1341.

COMPONENTS:	ORIGINAL MEASUREMENTS:
1. Sodium selenite; Na_2SeO_3; [10102-18-8]	Yanitskii, I.V.; Patkauskas, R.M.
2. Sodium selenate; Na_2SeO_4; [13410-01-0]	*Zh. Priklad. Khim.* 1970, *43*, 522-27; **J. Appl. Chem. USSR (Eng. Transl.)* 1970, *43*, 530-5.
3. Water; H_2O; [7732-18-5]	

EXPERIMENTAL VALUES (continued):

Na_2SeO_3 mass %	Na_2SeO_4 mass %	Na_2SeO_3[a] mol/kg	Na_2SeO_4[a] mol/kg	Solid[b] phase
Temperature = 20°C (continued)				
36.3	13.4	4.173	1.410	D
39.9	9.14	4.527	0.949	D
43.4	4.32	4.800	0.437	D
46.4	-	5.006	0.	D
Temperature = 25°C				
-	38.5	0.	3.313	B
3.88	33.4	0.358	2.818	B
7.28	30.8	0.680	2.633	B
10.9	28.9	1.047	2.541	B
15.9	26.9	1.607	2.489	B
20.1	25.6	2.140	2.495	B
24.5	23.9	2.745	2.451	A + B
27.2	21.5	3.066	2.218	A
29.5	19.2	3.325	1.981	A
31.2	17.9	3.544	1.861	A
32.6	16.1	3.675	1.661	A
35.4	13.8	4.068	1.443	A
37.8	11.8	4.337	1.239	A + D
40.7	8.58	4.640	0.895	D
42.9	5.73	4.829	0.590	D
45.1	3.02	5.027	0.308	D
46.9	-	5.107	0.	D
Temperature = 30°C				
-	44.0	0.	4.159	B
1.12	42.6	0.115	4.006	B
3.52	41.2	0.368	3.945	B
5.15	40.2	0.545	3.893	B
7.34	39.0	0.791	3.847	A + B
9.96	36.5	1.076	3.608	A
12.2	34.2	1.316	3.377	A
14.0	32.2	1.505	3.168	A
16.4	30.3	1.779	3.009	A
19.2	27.8	2.095	2.776	A
22.0	25.3	2.414	2.541	A
25.1	22.4	2.765	2.258	A
30.9	17.8	3.483	1.836	A
33.8	15.7	3.870	1.645	A
37.6	12.2	4.331	1.286	A
40.9	9.74	4.791	1.044	A + D
43.3	6.82	5.020	0.724	D
44.8	4.95	5.155	0.521	D
46.8	2.21	5.307	0.229	D
48.1	-	5.359	0.	D

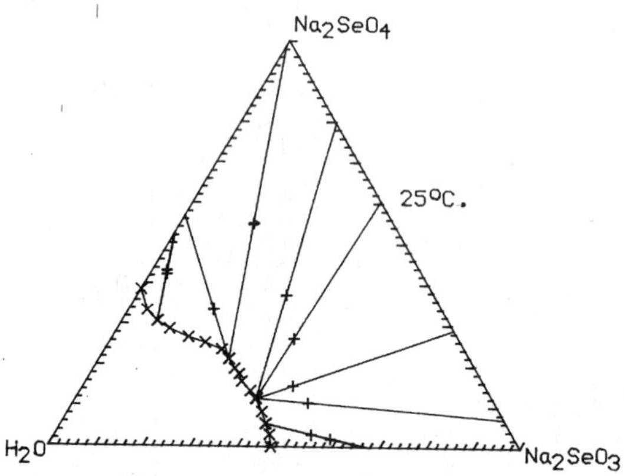

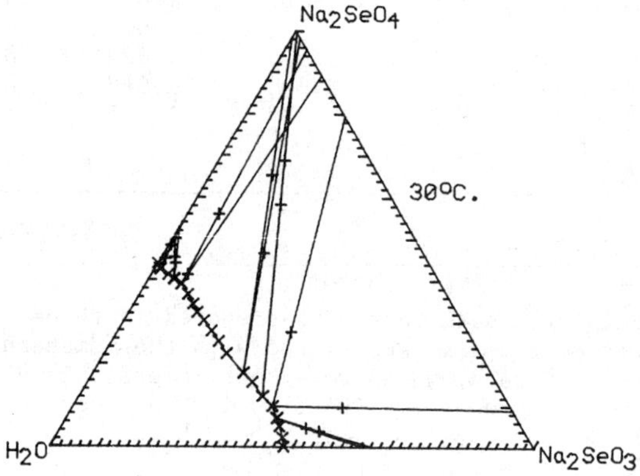

(continued on next page)

COMPONENTS:	ORIGINAL MEASUREMENTS:
1. Sodium selenite; Na_2SeO_3; [10102-18-8]	Yanitskii, I.V.; Patkauskas, R.M.
2. Sodium selenate; Na_2SeO_4; [13410-01-0]	*Zh. Priklad. Khim.* 1970, *43*, 522-27; **J. Appl. Chem. USSR (Eng. Transl.)* 1970, *43*, 530-5.
3. Water; H_2O; [7732-18-5]	

EXPERIMENTAL VALUES (continued):

Na_2SeO_3 mass %	Na_2SeO_4 mass %	Na_2SeO_3[a] mol/kg	Na_2SeO_4[a] mol/kg	Solid[b] phase
Temperature = 40°C				
-	44.9	0.	4.313	A
5.0	41.3	0.538	4.071	A
7.59	38.5	0.814	3.780	A
10.8	35.7	1.167	3.532	A
15.1	31.2	1.626	3.075	A
18.9	27.9	2.054	2.776	A
21.3	25.7	2.324	2.566	A
25.7	22.3	2.858	2.270	A
31.9	16.9	3.603	1.747	A
34.5	15.6	3.998	1.655	A
37.4	12.7	4.334	1.347	A
44.5	7.17	5.324	0.785	A
45.9	6.15	5.535	0.679	A + C
47.5	4.36	5.705	0.479	C
48.5	2.77	5.755	0.301	C
49.7	1.23	5.857	0.133	C
50.8	-	5.970	0.	C
Temperature = 60°C				
-	43.8	0.	4.125	A
1.7	41.9	0.174	3.932	A
6.51	37.4	0.671	3.529	A
11.2	34.1	1.184	3.299	A
16.5	29.0	1.751	2.816	A
20.2	26.3	2.183	2.602	A
27.4	19.9	3.006	1.999	A
31.7	15.8	3.491	1.593	A
35.8	12.9	4.035	1.331	A
37.8	11.3	4.294	1.175	A
39.1	10.5	4.486	1.103	A + C
41.6	7.2	4.698	0.744	C
43.2	5.6	4.879	0.579	C
44.7	3.94	5.033	0.406	C
46.3	1.95	5.173	0.199	C
47.5	-	5.232	0.	

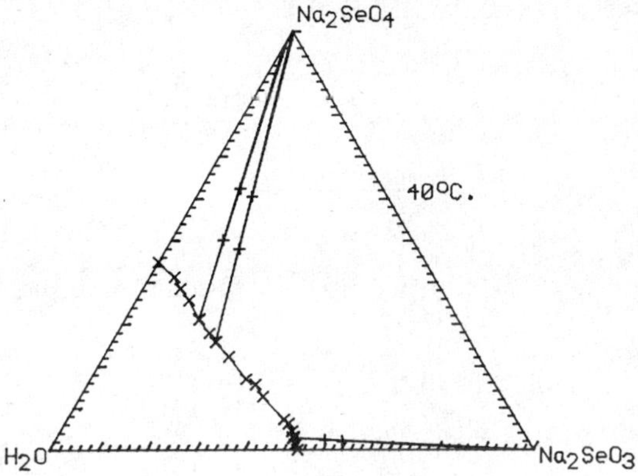

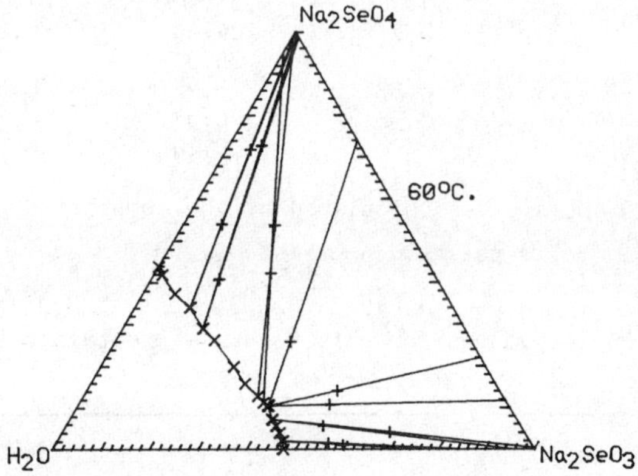

[a] Molalities calculated by the compiler.

[b] Solid phases: A - Na_2SeO_4, B - $Na_2SeO_4.10H_2O$, C - Na_2SeO_3, D - $Na_2SeO_3.5H_2O$

COMPONENTS:

1. Sodium selenite; Na_2SeO_3; [10102-18-8]
2. Sodium carbonate; Na_2CO_3; [497-19-8]
3. Water; H_2O; [7732-18-5]

ORIGINAL MEASUREMENTS:

Chimboulev, M.T.; Vassilev, H.

Dokl. Bolg. Akad. Nauk 1973, *26*, 1509-12.

VARIABLES:

Composition
Temperature: 303 - 363 K

PREPARED BY:

Mary R. Masson

EXPERIMENTAL VALUES:

Composition of equilibrium solutions

Na_2SeO_3 mass %	Na_2CO_3 mass %	Na_2SeO_3[a] mol/kg	Na_2CO_3[a] mol/kg	Solid[b] phase
Temperature = 30°C				
48.52	-	5.450	0.	B
45.68	3.27	5.174	0.604	B
43.10	6.03	4.899	1.118	B + C
40.83	7.14	4.538	1.295	C
37.17	8.62	3.965	1.500	C
31.94	11.87	3.287	1.993	C
28.63	14.71	2.922	2.449	C
26.55	16.10	2.677	2.649	C + D
24.26	16.33	2.361	2.593	D
18.74	19.15	1.745	2.909	D
14.06	21.90	1.270	3.226	D
10.21	24.46	0.904	3.532	D
8.5[c]	25.5	0.745	3.645	D + E
6.82	26.23	0.589	3.696	E
3.58	26.97	0.298	3.664	E
-	28.14	0.	3.695	E

[a] Molalities calculated by the compiler.

[b] Solid phases: A - Na_2SeO_3, B - $Na_2SeO_3.5H_2O$, C - $Na_2CO_3.H_2O$, D - $Na_2CO_3.7H_2O$, E - $Na_2CO_3.10H_2O$

[c] The values for this point were obtained by extrapolation.

(continued on next page)

AUXILIARY INFORMATION

METHOD APPARATUS/PROCEDURE:

Mixtures of twice recrystallized $Na_2SeO_3.5H_2O$ and $Na_2CO_3.10H_2O$ were dissolved isothermally in distilled water by continuous stirring in a glass vessel fitted with a hydraulic lock until equilibrium was reached (24 - 36 hr). Two weighed samples of the liquid phase were analysed for selenite gravimetrically (as zinc uranyl acetate).

The "bottom phase" (solid plus contaminating liquid) was also analysed for each point: the results are tabulated in the source. The compositions of the phases were determined by Schreinemakers' remainder method. The identities were confirmed by X-ray powder diffraction.

SOURCE AND PURITY OF MATERIALS:

Not stated.

ESTIMATED ERROR:

ESTIMATED ERROR:

Not stated, and no estimate possible from data available.

REFERENCES:

COMPONENTS:	ORIGINAL MEASUREMENTS:
1. Sodium selenite; Na_2SeO_3; [10102-18-8]	Chimboulev, M.T.; Vassilev, H.
2. Sodium carbonate; Na_2CO_3; [497-19-8]	*Dokl. Bolg. Akad. Nauk* 1973, *26*, 1509-12.
3. Water; H_2O; [7732-18-5]	

EXPERIMENTAL VALUES (continued):

Composition of equilibrium solutions

Na_2SeO_3 mass %	Na_2CO_3 mass %	Na_2SeO_3[a] mol/kg	Na_2CO_3[a] mol/kg	Solid[b] phase
Temperature = 60°C				
47.45	-	5.221	0.	A
44.80	2.74	4.938	0.493	A
41.93	5.80	4.638	1.047	A
40.19	7.43	4.437	1.338	A + C
38.02	8.51	4.112	1.502	C
34.79	10.12	3.652	1.733	C
29.10	13.07	2.910	2.132	C
23.17	16.69	2.228	2.618	C
15.85	21.94	1.473	3.327	C
8.32	25.18	0.723	3.572	C
-	31.42	0.	4.323	C
Temperature = 90°C				
46.10	-	4.946	0.	A
43.58	2.21	4.648	0.385	A
39.86	6.37	4.286	1.118	A
37.82	8.35	4.063	1.464	A + C
35.42	9.92	3.747	1.712	C
30.38	12.46	3.073	2.057	C
24.67	16.93	2.443	2.735	C
20.55	18.41	1.947	2.846	C
13.83	21.90	1.244	3.215	C
4.69	27.12	0.398	3.752	C
-	30.65	0.	4.170	C

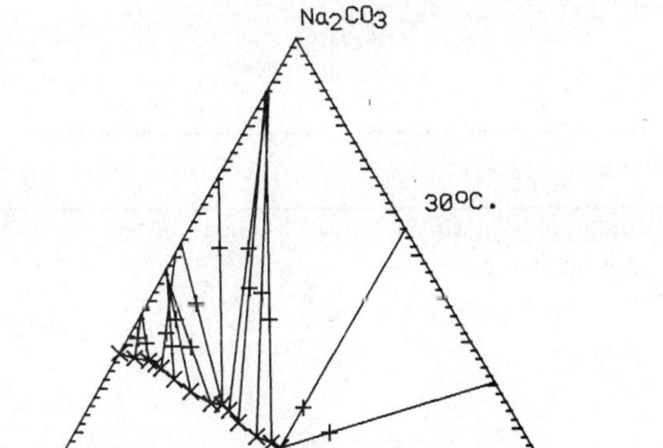

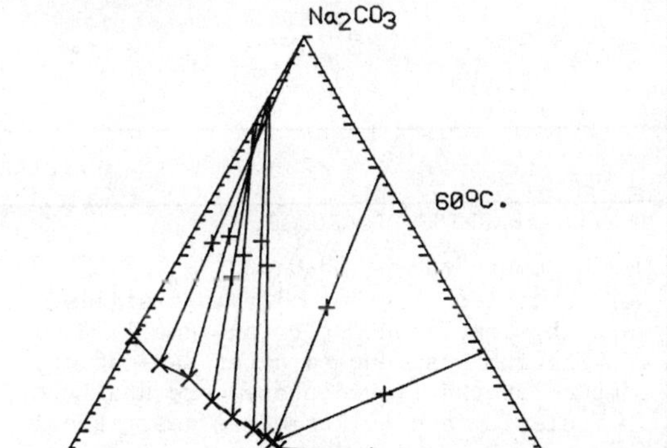

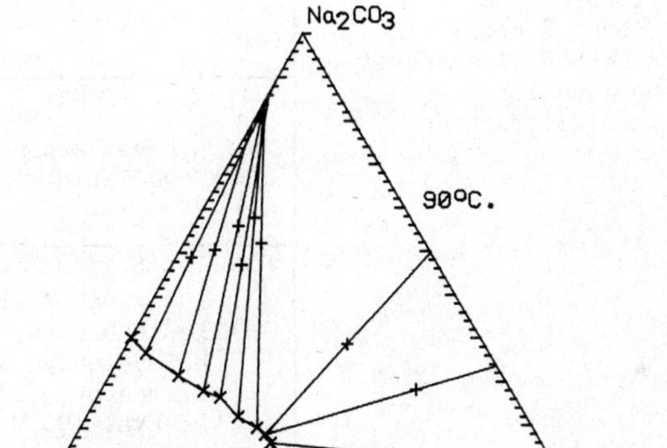

COMPONENTS:

1. Sodium selenite; Na_2SeO_3; [10102-18-8]
2. Sodium tellurite; Na_2TeO_3; [10102-20-2]
3. Water; H_2O; [7732-18-5]

ORIGINAL MEASUREMENTS:

Chimbulev, M.; Vasilev, Kh.

God. Vissh. Khim.-Tekhnol. Inst., Sofia 1977, 22, 247-254.

VARIABLES:

Concentrations of Na_2SeO_3 and Na_2TeO_3

Temperature: 303, 333 and 363 K

PREPARED BY:

Mary R. Masson

EXPERIMENTAL VALUES:

Composition of equilibrium solutions

Na_2SeO_3 mass %	Na_2TeO_3 mass %	Na_2SeO_3[a] mol/kg	Na_2TeO_3[a] mol/kg	Solid[b] phase
Temperature = 30°C				
48.52	-	5.450	0.	B
44.71	4.05	5.045	0.357	B
39.15	11.10	4.550	1.007	B
33.78	18.12	4.061	1.700	B
31.16	20.83	3.753	1.958	B + D
29.43	22.59	3.547	2.125	D
22.55	28.30	2.653	2.599	D
14.10	33.97	1.570	2.952	D
5.95	41.35	0.653	3.541	D
-	46.23	0.	3.880	D

a Molalities calculated by the compiler.

b Solid phases: A - Na_2SeO_3, B - $Na_2SeO_3.5H_2O$
C - Na_2TeO_3, D - $Na_2TeO_3.5H_2O$.

(continued on next page)

AUXILIARY INFORMATION

METHOD APPARATUS/PROCEDURE:

Mixtures of $Na_2SeO_3.5H_2O$ and $Na_2TeO_3.5H_2O$ were dissolved isothermally in distilled water by continuous stirring, until equilibrium was reached after 24 - 36 hr. Samples of the liquid phase were analysed for selenium and tellurium as described in (1).

The "bottom phase" (solid plus contaminatiing liquid) was also analysed for each point: the results are given in the source. The compositions of the solid phases were determined by Schreinemakers' remainder method. The identities were confirmed by X-ray powder diffraction.

SOURCE AND PURITY OF MATERIALS:

Not stated.

ESTIMATED ERROR:

Not stated, and no estimate possible from data available.

REFERENCES:

1. Knyazeva, R.N.; Florinskaya, A.A.; Lastukhinaya, Zh.Ya. *Metody opredeleniya i analiza redikh elementov*, Izd.-vo Akademii Nauk SSSR, Moskva, 1961, p.605.

COMPONENTS:	ORIGINAL MEASUREMENTS:
1. Sodium selenite; Na_2SeO_3; [10102-18-8]	Chimbulev, M.; Vasilev, Kh.
2. Sodium tellurite; Na_2TeO_3; [10102-20-2]	*God. Vissh. Khim.-Tekhnol. Inst., Sofia* 1977, *22*, 247-254.
3. Water; H_2O; [7732-18-5]	

EXPERIMENTAL VALUES (continued):

Na_2SeO_3 mass %	Na_2TeO_3 mass %	Na_2SeO_3[a] mol/kg	Na_2TeO_3[a] mol/kg	Solid[b] phase
Temperature = 60°C				
47.45	-	5.221	0.	A
41.38	7.73	4.702	0.686	A
33.61	16.68	3.910	1.514	A
27.15	23.30	3.168	2.122	A
23.26	30.52	2.910	2.980	A
20.05	33.28	2.484	3.218	A
17.95	36.20	2.264	3.563	A + D
15.80	38.54	2.001	3.809	D
9.97	43.31	1.234	4.184	D
4.64	47.10	0.556	4.405	D
-	51.68	0.	4.827	D
Temperature = 90°C				
46.10	-	4.946	0.	A
41.76	5.52	4.580	0.473	A
34.22	13.61	3.793	1.177	A
25.08	25.04	2.907	2.266	A
18.41	34.35	2.253	3.282	A
16.53	37.48	2.078	3.678	A + C
14.19	39.62	1.776	3.871	C
9.57	43.15	1.170	4.119	C
4.34	48.31	0.530	4.605	C
-	52.48	0.	4.984	C

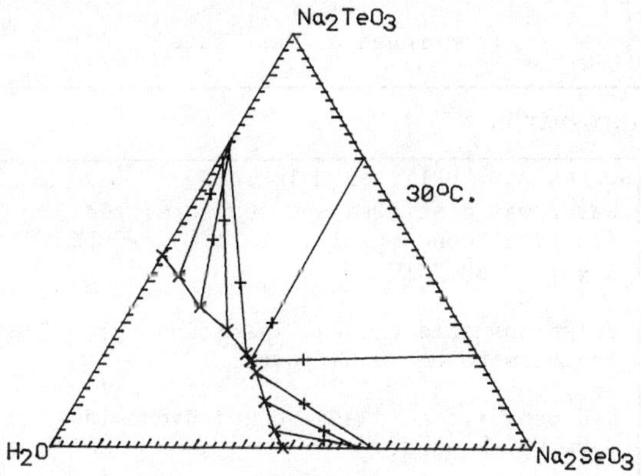

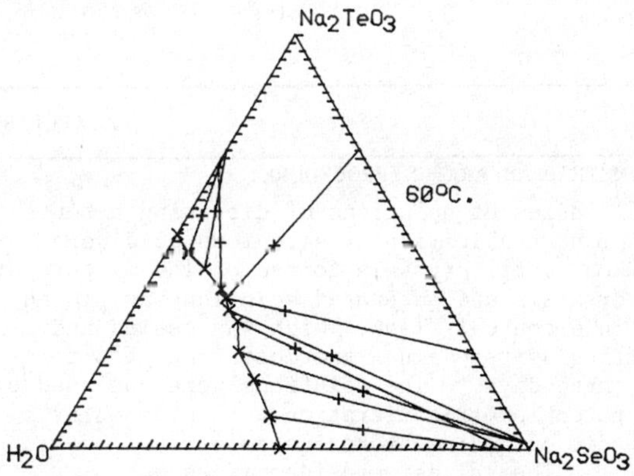

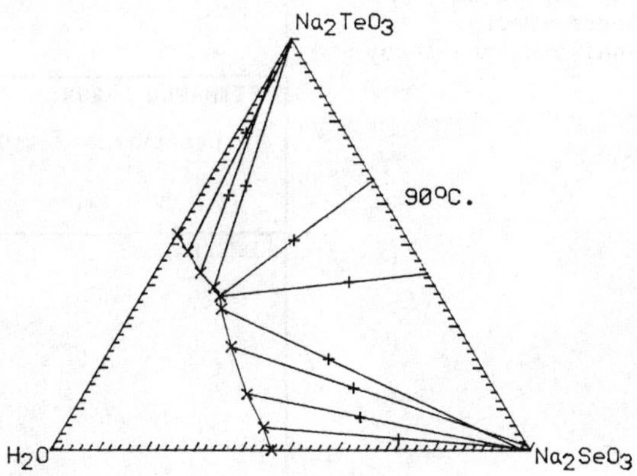

COMPONENTS:

1. Sodium selenite; Na_2SeO_3; [10102-18-8]
2. Selenious acid; H_2SeO_3; [7783-00-8]
3. Water; H_2O; [7732-18-5]

ORIGINAL MEASUREMENTS:

Sabbah, R.; Périnet, G.

J. Chim. Phys. 1965, *62*, 929-36.

VARIABLES:

Concentrations of the components
Temperature: 298 K

PREPARED BY:

Mary R. Masson

EXPERIMENTAL VALUES:

Composition of equilibrium solutions, mol/1000 g of solution, at 25°C

Na/Se	H_2SeO_3 total	H_2SeO_3 free	Na_2SeO_3	Solid phase[a]
0.0	6.23	6.23	0.0	A
0.047	6.37	6.22	0.15	A
0.072	6.39	6.16	0.23	A
0.080	6.43	6.17	0.26	A + B
0.089	6.38	6.10	0.28	B
0.111	6.29	5.94	0.35	B
0.131	6.17	5.76	0.41	B
0.152	6.00	5.54	0.46	B + C
0.156	5.89	5.42	0.47	C
0.256	5.08	4.42	0.66	C
0.398	4.29	3.43	0.86	C
0.502	3.98	2.98	1.00	C
0.642	3.76	2.55	1.21	C
0.848	4.10	2.36	1.74	C
0.906	4.50	2.46	2.04	C + D
0.930	4.54	2.43	2.11	D
1.08	4.30	2.13	2.17	D
1.22	4.18	1.62	2.56	D
1.28	4.16	1.50	2.66	D
1.35	4.24	1.40	2.84	D + E
1.41	4.01	1.19	2.82	E
1.44	3.91	1.10	2.81	E
1.65	3.35	0.59	2.76	F
1.94	2.86	0.09	2.77	F

[a] Solid phases:

A - H_2SeO_3

B - $Na_2SeO_3.7H_2SeO_3$

C - $Na_2SeO_3.3H_2SeO_3$

D - $Na_2SeO_3.H_2SeO_3$

E - Na_2SeO_3

F - $Na_2SeO_3.5H_2O$

(continued on next page)

AUXILIARY INFORMATION

METHOD APPARATUS/PROCEDURE:

A series of solutions of differing extents of neutralization of selenious acid were kept until crystals formed. The mixture of crystals and saturated solution was placed in a conical flask, which was sealed and then agitated in a thermostat at 25°C for three days. The solutions were analysed by potentiometric titration with hydrochloric acid or sodium hydroxide solution.

The solids were identified and characterized by thermogravimetry, differential thermal analysis, and X-ray diffraction.

SOURCE AND PURITY OF MATERIALS:

Water was distilled and demineralized. Its final conductivity at 25°C was about 2×10^{-5} $ohm^{-1}m^{-1}$.

Selenious acid (Fluka) was found by analysis to be 99.6% pure.

Hydrochloric acid and sodium hydroxide were 99.9% pure.

ESTIMATED ERROR:

Temperature: ± 0.05 K

REFERENCES:

COMPONENTS:	ORIGINAL MEASUREMENTS:
1. Sodium selenite; Na_2SeO_3; [10102-18-8]	Sabbah, R.; Périnet, G.
2. Selenious acid; H_2SeO_3; [7783-00-8]	*J. Chim. Phys.* 1965, *62*, 929-36.
3. Water; H_2O; [7732-18-5]	

EXPERIMENTAL VALUES (continued):

Compositions of solutions expressed in units of mass % and mol/kg

H_2SeO_3 [b] mass %	Na_2SeO_3 [b] mass %	H_2SeO_3 [c] mol/kg	Na_2SeO_3 [c] mol/kg
80.355	0.	31.712	0.
80.226	2.594	36.204	0.873
79.452	3.978	37.174	1.388
79.581	4.496	38.749	1.633
78.678	4.842	37.015	1.699
76.614	6.053	34.270	2.019
74.292	7.091	30.939	2.202
71.455	7.955	26.906	2.234
69.907	8.128	24.676	2.140
57.009	11.414	13.998	2.090
44.240	14.873	8.389	2.103
38.436	17.294	6.731	2.259
32.890	20.926	5.521	2.620
30.439	30.092	5.979	4.409
31.729	35.280	7.457	6.183
31.342	36.490	7.554	6.559
27.473	37.528	6.086	6.200
20.895	44.273	4.651	7.349
19.347	46.002	4.329	7.677
18.057	49.115	4.265	8.651
15.349	48.769	3.316	7.859
14.188	48.596	2.956	7.551
7.610	47.731	1.321	6.180
1.161	47.904	0.177	5.438

[b] Mass % values calculated by the compiler.

[c] Molalities calculated by the compiler.

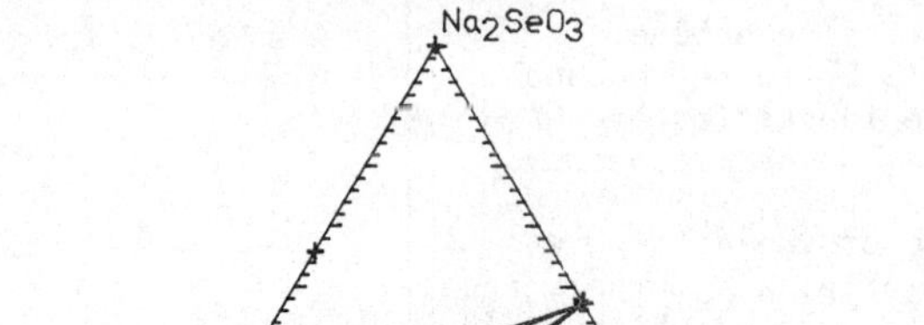

COMPONENTS:

1. Selenium dioxide; SeO_2; [7446-08-4]
2. Sodium hydroxide; NaOH; [1310-73-2]
3. Water; H_2O; [7732-18-5]

ORIGINAL MEASUREMENTS:

Rustamov, P.G.; Mardakhaev, B.N.

Azerb. Khim. Zh. 1963, 131-40.

VARIABLES:

Concentrations of the components
Temperature: 298 K

PREPARED BY:

Mary R. Masson

EXPERIMENTAL VALUES:

Composition of equilibrium solutions at 25°C

Na_2O mass %	SeO_2 mass %	H_2O mass %	Na_2O[a] mol/kg	SeO_2[a] mol/kg	Solid[b] phase
-	74.04	25.96	0.	25.704	A
1.61	70.04	28.35	0.916	22.265	A
1.61	70.02	28.37	0.916	22.243	A
1.86	68.70	29.44	1.019	21.031	A
3.11	65.70	31.19	1.609	18.984	A
3.72	63.80	32.48	1.848	17.703	A
4.19	64.10	31.71	2.132	18.218	A
3.95	62.10	33.95	1.877	16.485	A + B
3.44	62.74	33.82	1.641	16.719	A + B
5.56	46.50	47.94	1.871	8.742	B
6.51	42.30	51.19	2.052	7.447	B
7.86	36.10	56.04	2.263	5.806	B
10.15	41.16	48.69	3.363	7.618	B
12.87	44.10	43.03	4.826	9.236	B
4.35	52.70	42.95[c]	1.634	11.058	B + C
12.03	40.12	47.85	4.056	7.556	C
15.97	47.90	36.13	7.132	11.948	C
17.64	28.10	54.26[c]	5.245	4.667	C

(continued on next page)

AUXILIARY INFORMATION

METHOD APPARATUS/PROCEDURE:

An isothermal dissolution technique. Equilibrium took 6 - 12 hr to be reached.

Solid phases were identified by Schreinemakers' method and by microscopy.

Compiler's note: some of the data were illegible even in the original journal, and there were some obvious errors, which could be corrected (points marked "c"). Some other sets of data do not add up to 100%, but there is no way to tell which figure is in error (points marked "d").

SOURCE AND PURITY OF MATERIALS:

ESTIMATED ERROR:

Temperature: ±0.02 K
Analyses: no estimates possible.

REFERENCES:

COMPONENTS:	ORIGINAL MEASUREMENTS:
1. Selenium dioxide; SeO_2; [7446-08-4]	Rustamov, P.G.; Mardakhaev, B.N.
2. Sodium hydroxide; NaOH; [1310-73-2]	*Azerb. Khim. Zh.* 1963, 131-40.
3. Water; H_2O; [7732-18-5]	

EXPERIMENTAL VALUES (continued):

Na_2O mass %	SeO_2 mass %	H_2O mass %	Na_2O[a] mol/kg	SeO_2[a] mol/kg	Solid[b] phase
19.49	26.28	54.23	5.799	4.367	C
19.48	22.38	58.14	5.406	3.469	C
21.05	13.15	65.15[d]	5.161	1.801	C
19.44	9.21	71.35	4.396	1.163	?
20.17	8.60	71.23	4.569	1.088	?
21.79	8.95	69.26	5.076	1.165	?
23.17	8.72	68.11[c]	5.489	1.154	?
23.73	6.99	69.28	5.526	0.909	?
23.18	6.04	70.78	5.284	0.769	D
25.20	3.30	72.50[d]	5.686	0.416	D
27.52	2.03	70.45	6.303	0.260	D
29.84	0.82	69.34	6.943	0.107	D
30.54	0.47	68.99	7.142	0.061	D
34.39	0.47	65.14	8.518	0.065	D
38.03	0.34	61.63	9.956	0.050	D
40.59	0.59	58.82	11.134	0.090	D + E
40.35	0.60	59.05[c]	11.025	0.092	D + E
41.05	-	59.95	11.235	0.	E

Compiler's note: the computer plot for this system shows little resemblance to the diagram given in the paper, even when drawn in units of mol %, as in the paper. The diagram included here is drawn in the more usual mass % units, which show the system more clearly. The conclusions made by the authors as to the solid phases present seem rather dubious.

a Molalities calculated by the compiler.

b Solid phases: A - H_2SeO_3; B - $Na_2Se_4O_9.5H_2O$; C - $Na_2Se_2O_5.5H_2O$;
D - $Na_2SeO_3.5H_2O$; E - $NaOH.H_2O$

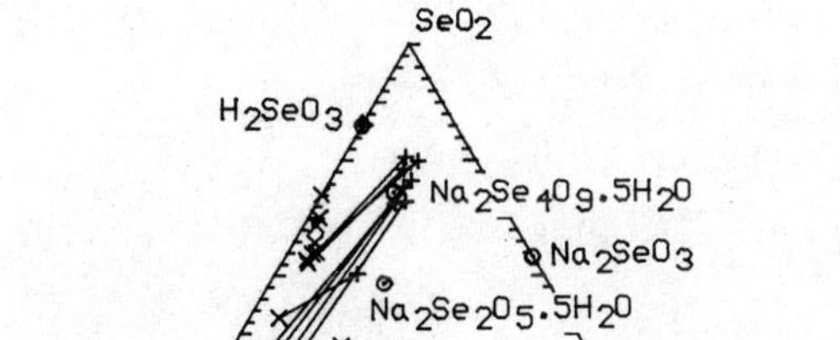

COMPONENTS:	EVALUATOR:
1. Potassium selenite; K_2SeO_3; [10431-47-7] 2. Water; H_2O; [7732-18-5]	Mary. R. Masson, Dept. of Chemistry, University of Aberdeen, Meston Walk, Old Aberdeen, AB9 2UE, Scotland, UK. June 1984.

CRITICAL EVALUATION:

The binary system potassium selenite - water was studied by Janickis (1,2), and data are also available from studies of ternary systems (3,4).

There are three regression equations, corresponding to the equilibria with ice, $K_2SeO_3.4H_2O$ [41292-06-2] and K_2SeO_3. For equation (3), two points had to be eliminated, but otherwise the data were in reasonable agreement.

The equations are (1) for 230 - 273 K, solid phase ice:

$$(T - 273.15) = -0.067 - 0.227y - 0.000154y^3 \qquad s = 0.44 \text{ (9 pts)}$$

or

$$y = 0.471 - 3.71(T - 293.2) - 0.105(T - 273.2)^2 - 0.00116(T - 273.2)^3 \qquad s = 1.01 \text{ (9 pts)}$$

(2) for 253 - 300 K, solid phase $K_2SeO_3.4H_2O$

$$y = 62.8 + 0.182(T - 273.2) + 0.00198(T - 273.2)^2 \qquad s = 0.111 \text{ (12 pts)}$$

and (3) for 273 - 353 K, solid phase K_2SeO_3:

$$y = 68.42 + 0.00007494(T - 273.2)^2 \qquad s = 0.037 \text{ (10 pts)}$$

where $y = 100w$ is the solubility expressed in mass % of K_2SeO_3, T is the temperature in K, and s is the standard deviation of the dependent variable about the regression line.

TENTATIVE SOLUBILITIES

The following tentative solubilities for K_2SeO_3 in water were calculated from equations (2) and (3).

COMPONENTS:	EVALUATOR:
1. Potassium selenite; K_2SeO_3; [10431-47-7] 2. Water; H_2O; [7732-18-5]	Mary. R. Masson, Dept. of Chemistry, University of Aberdeen, Meston Walk, Old Aberdeen, AB9 2UE, Scotland, UK. June 1984.

CRITICAL EVALUATION: (continued)

T/K	Solubility mass %	molality mol/kg
	$K_2SeO_3.4H_2O$	
253	60.0	7.31
263	61.2	7.69
273	62.8	8.23
283	64.8	8.97
293	67.2	9.99
300	69.2	10.95
	K_2SeO_3	
273	68.42	10.56
283	68.43	10.57
293	68.45	10.58
303	68.49	10.59
313	68.54	10.62
323	68.61	10.65
333	68.69	10.69
343	68.79	10.74
353	68.90	10.80

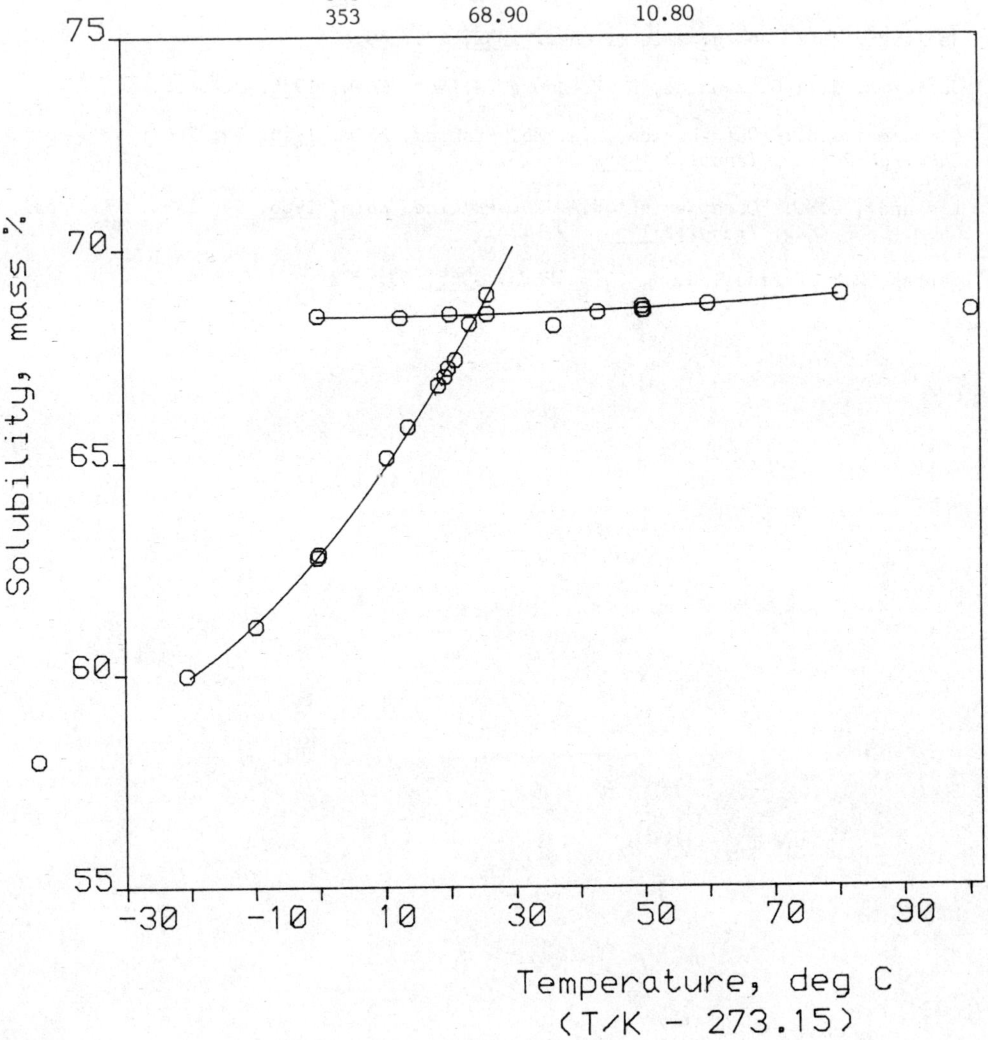

COMPONENTS:	EVALUATOR:
1. Potassium selenite; K_2SeO_3; [10431-47-7] 2. Water; H_2O; [7732-18-5]	Mary. R. Masson, Dept. of Chemistry, University of Aberdeen, Meston Walk, Old Aberdeen, AB9 2UE, Scotland, UK. June 1984.

CRITICAL EVALUATION: (continued)

TERNARY SYSTEMS

The ternary systems involving potassium sulfite (3) and ethanol (4) are straightforward and the data appear to be reliable. In the system involving selenious acid (5) three double salts are formed.

$K_2SeO_3.3H_2SeO_3$	[15457-71-3]	$(KH_3(SeO_3)_2)$
$K_2SeO_3.H_2SeO_3$	[7782-70-9]	$(KHSeO_3)$
$5K_2SeO_3.3H_2SeO_3$	[15457-72-4]	$(K_5H_3(SeO_3)_4)$

These salts are not identical to the ones formed in the corresponding sodium system.

REFERENCES

1. Janitzki, J. *Z. Anorg. Allgem. Chem.* 1932, *205*, 49.
2. Janickis, J.; Gutmanaite, H. *Z. Anorg. Allgem. Chem.* 1936, *227*, 1.
3. Klebanov, G.S.; Ostapkevich, N.A. *Zh. Priklad. Khim.* 1966, *39*, 2467; **J. Appl. Chem. USSR (Eng. Transl.)* 1966, *39*, 2315.
4. Klebanov, G.S.; Ostapkevich, N.A. *Zh. Priklad. Khim.* 1966, *39*, 1435; **J. Appl. Chem. USSR (Eng. Transl.)* 1966, *39*, 1342.
5. Sabbah, R.; Périnet, G. *J. Chim. Phys.* 1966, *63*, 332.

COMPONENTS:

1. Potassium selenite; K_2SeO_3; [10431-47-7]
2. Water; H_2O; [7732-18-5]

ORIGINAL MEASUREMENTS:

Janitzki, J.

Z. Anorg. Allgem. Chem. 1932, *205*, 49-75.

VARIABLES:

Temperature: 253 - 374 K

PREPARED BY:

Mary R. Masson

EXPERIMENTAL VALUES:

t/°C	K_2SeO_3 mass %	K_2SeO_3[a] mol/kg	Solid[b] phase
-20.5	59.98	7.305	B
- 9.7	61.14	7.669	B
- 0.2	62.76	8.214	B
+10.5	65.11	9.096	B
13.8	65.83	9.390	B
18.5	66.80	9.807	B
19.5	67.00	9.896	B
21.1	67.40	10.077	B
23.3	68.25	10.478	B
26.0	68.93	10.814	B
0.0	68.45	10.575	A
12.7	68.40	10.551	A
20.3	68.48	10.590	A
26.0	68.48	10.590	A
36.2	68.20	10.454	A
43.1	68.52	10.609	A
50.2	68.56	10.629	A
60.1	68.70	10.698	A
80.5	68.92	10.809	A
100.6	68.53	10.614	A

[a] Molalities calculated by the compiler.

[b] Solid phases: A - K_2SeO_3, B - $K_2SeO_3.4H_2O$

AUXILIARY INFORMATION

METHOD APPARATUS/PROCEDURE:

For each temperature, a saturated solution was prepared by stirring the salt in water inside a stoppered 4-cm diameter test-tube. Small samples of solution were removed at intervals for analysis, in order to test for attainment of equilibrium. The time required varied between 3 and 23 hr. The solutions were analysed for SeO_2 by the method of Norris and Fay (1).

The solid phases were identified by analysis.

SOURCE AND PURITY OF MATERIALS:

ESTIMATED ERROR:

Temperature: -20 - 0°C ±0.2°C, 0 - 60°C ±0.1°C, 60 - 110°C ±0.3°C
Analyses: no estimate possible.

REFERENCES:

1. Norris, J.F.; Fay, H. *Amer. Chem. J.* 1896, *18*, 703; 1900, *23*, 119.

COMPONENTS:

1. Potassium selenite; K_2SeO_3; [10431-47-7]
2. Water; H_2O; [7732-18-5]

ORIGINAL MEASUREMENTS:

Janickis, J.; Gutmanaite, H.

Z. Anorg. Allgem. Chem. 1936, *227*, 1-16.

VARIABLES:

Temperature: 230 - 273 K
Composition

PREPARED BY:

Mary R. Masson

EXPERIMENTAL VALUES:

Compositions of equilibrium solutions

t/°C	K_2SeO_3 mol/dm^3	K_2SeO_3 mass %	K_2SeO_3[a] mol/kg	Solid phase
- 0.262	0.05	1.019	0.0502	ice
- 0.500	0.1	2.021	0.1006	"
- 0.970	0.2	3.987	0.2024	"
- 2.375	0.5	9.54	0.514	"
- 5.03	1	17.87	1.060	"
-13.0	2	32.01	2.294	"
-23.6	3.194	45.13	4.01	"
-43.5	satd.	58.0	6.74	ice + K_2SeO_3

[a] Molalities calculated by the compiler.

AUXILIARY INFORMATION

METHOD APPARATUS/PROCEDURE:

Freezing points of prepared solutions were measured by use of a Beckman-type apparatus (1). Determinations were repeated until the desired reproducibility was attained. Each reported value is the mean of at least three determinations.

SOURCE AND PURITY OF MATERIALS:

Potasium selenite was prepared by neutralization of selenious acid with potassium hydroxide.

ESTIMATED ERROR:

Temperature reproducibility 0.5%

REFERENCES:

1. Ostwald, W.; Luther, R. *Hand- und Hilfsbuch zur Ausfuhrung physikochemischer Messungen*, 5th Ed., Akademische Verlag., Leipzig, 1931.

COMPONENTS:

1. Potassium selente; K_2SeO_3; [10431-47-7]
2. Potassium sulfite; K_2SO_3; [10117-38-1]
3. Water; H_2O; [7732-18-5]

ORIGINAL MEASUREMENTS:

Klebanov, G.S.; Ostapkevich, N.A.

Zh. Prikl. Khim. 1966, *39*, 2467-2470; **J. Appl. Chem. USSR (Eng. Transl)* 1966, *39*, 2315-8.

VARIABLES:

Two temperatures: 273 and 333 K
Potassium selenite and potassium sulfite concentrations.

PREPARED BY:

Mary R. Masson

EXPERIMENTAL VALUES:

Composition of equilibrium solutions

K_2SeO_3 mass %	K_2SO_3 mass %	K_2SeO_3[a] mol/kg	K_2SO_3[a] mol/kg	Solid[b] phase
Temperature = 0°C				
62.83	-	8.239	0.	B
56.02	3.97	6.825	0.627	B
50.64	8.00	5.968	1.222	B
49.51	10.50	6.035	1.659	B + C
45.10	12.03	5.128	1.773	C
35.21	20.25	3.853	2.873	C
20.30	32.50	2.096	4.351	C
12.86	38.51	1.289	5.004	C
-	51.40	0.	6.683	C
Temperature = 50°C				
68.56	-	10.629	0.	A
59.50	5.50	8.286	0.993	A
52.80	10.90	7.090	1.897	A
51.30	13.70	7.144	2.473	A + C
46.00	15.20	5.779	2.475	C
32.50	24.11	3.651	3.511	C
23.31	32.74	2.585	4.707	C
17.10	38.00	1.856	5.348	C
12.50	42.01	1.339	5.835	C
-	52.00	0.	6.845	C

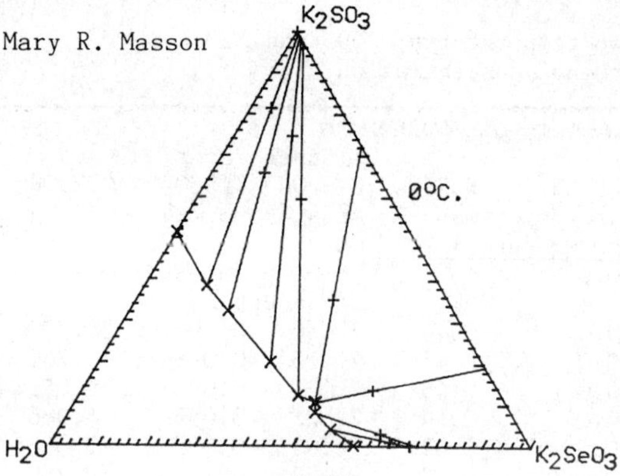

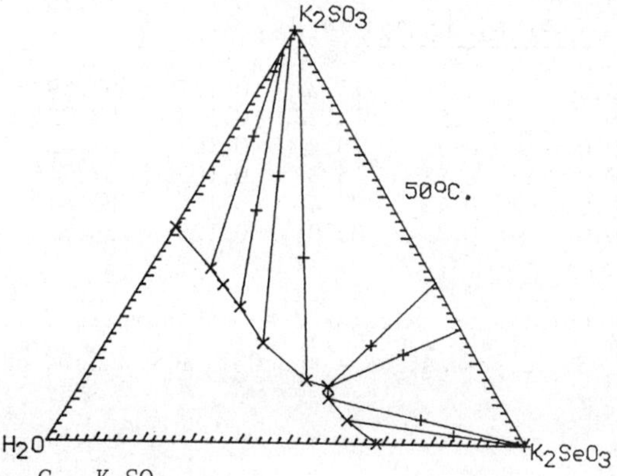

[a] Molalities calculated by the compiler.

[b] Solid phases: A - K_2SeO_3, B - $K_2SeO_3.4H_2O$, C - K_2SO_3

AUXILIARY INFORMATION

METHOD APPARATUS/PROCEDURE:

Mixtures of K_2SeO_3 and K_2SO_3 were dissolved isothermally in water contained in glass vessels with stirrers fitted with hydraulic seals. Trace amounts of p-phenylenediamine were added to the solutions to prevent aerial oxidation of the sulfite. 0°C was maintained by melting ice, and 50°C with the aid of a contact thermometer and electromagnetic relay. Equilibrium was reached after 2 days at 0°C and 30-35 hr at 50°C.

The solutions were analysed for selenite and sulfite as follows. For sulfite, sodium bicarbonate and excess of 0.1N iodine solution were added, then the solution was acidified with acetic acid, and the excess of iodine was titrated with thiosulfate. For selenite, sulfite was bound with formaldehyde, the solution was acidified with HCl, and KI was added. The iodine liberated was titrated with thiosulfate.

The solid residues were also analysed. The compositions of the solid phases were determined by Schreinemakers' remainder method.

SOURCE AND PURITY OF MATERIALS:

Potassium selenite was of reagent-grade quality, and potassium sulfite was prepared from "pure" grade potassium sulfite by salting out with alcohol (99.8% salt content).

ESTIMATED ERROR:

Temperature: ±0.1 K
Analyses: no estimate possible.

REFERENCES:

COMPONENTS:

1. Potassium selenite; K_2SeO_3; [10431-47-7]
2. Ethanol; C_2H_5OH; [64-17-5]
3. Water; H_2O; [7732-18-5]

ORIGINAL MEASUREMENTS:

Klebanov, G.S.; Ostapkevich, N.A.

Zh. Priklad. Khim. 1966, *39*, 1435-7; **J. Appl. Chem. USSR (Eng. Transl.)* 1966, *39*, 1342-4.

VARIABLES:

Two temperatures: 293 and 323 K
Ethanol concentration

PREPARED BY:

Mary R. Masson

EXPERIMENTAL VALUES:

Initial	Upper layer				Lower layer				Solid[b]
C_2H_5OH mass %	K_2SeO_3 mass %	C_2H_5OH mass %	K_2SeO_3[a] mol/kg	C_2H_5OH[a] mol/kg	K_2SeO_3 mass %	C_2H_5OH mass %	K_2SeO_3[a] mol/kg	C_2H_5OH[a] mol/kg	phase
Temperature = 20°C									
0.0	-	-	-	-	67.2	-	9.986	-	B
2.0	not enough sample				64.37	1.85	9.288	1.189	B
10.0	0.44	77.16	0.096	74.770	63.79	1.33	8.914	0.828	B
30.0	0.45	77.15	0.098	74.760	63.63	1.32	8.849	0.817	B
50.0	0.46	77.21	0.096	75.053	63.53	1.31	8.807	0.809	B
70.0	0.44	77.21	0.096	74.986	63.45	1.30	8.774	0.801	B
80.0	0.42	78.05	0.095	78.688	-	-	-	-	A
90.0	0.39	86.23	0.142	139.889	-	-	-	-	A
Temperature = 50°C									
0.0	-	-	-	-	68.65	-	10.674	-	A
10.0	0.53	75.32	0.107	67.698	65.01	1.65	9.504	1.074	A
30.0	0.53	75.33	0.107	67.735	64.91	1.64	9.459	1.064	A
50.0	0.53	75.41	0.107	68.032	65.00	1.63	9.494	1.060	A
70.0	0.52	75.34	0.105	67.744	65.02	1.64	9.506	1.068	A
80.0	0.45	78.48	0.104	80.849	-	-	-	-	A
90.0	0.35	85.81	0.123	134.581	-	-	-	-	A

[a] Molalities calculated by the compiler.

[b] Solid phases: A - K_2SeO_3, B - $K_2SeO_3 \cdot 4H_2O$

AUXILIARY INFORMATION

METHOD APPARATUS/PROCEDURE:

Aqueous ethanolic solutions were saturated with potassium selenite at 20 and 50°C in glass vessels (100-150 ml), fitted with hydraulic seals at 20°C and with reflux condensers at 50°C. Equilibrium was reached after 16-28 hr.

Selenite in the solutions and moist solids was determined iodometrically, and ethanol iodometrically after distillation from the samples.

SOURCE AND PURITY OF MATERIALS:

Reagent-grade potassium selenite was used. Ethanol and water were distilled twice.

ESTIMATED ERROR:

Temperature: ±0.1 K
Analyses: no estimate possible.

REFERENCES:

COMPONENTS:	ORIGINAL MEASUREMENTS:
1. Potassium selenite; K_2SeO_3; [10431-47-7] 2. Selenious acid; H_2SeO_3; [7783-00-8] 3. Water; H_2O; [7732-18-5]	Sabbah, R.; Périnet, G. *J. Chim. Phys.* 1966, *63*, 332-6.

VARIABLES:	PREPARED BY:
Temperature: 298 K Concentrations of the components	Mary R. Masson

EXPERIMENTAL VALUES:

Composition of equilibrium solutions, mol/1000 g of solution, at 25°C.

Na/Se	$[H_2SeO_3]$ total	$[H_2SeO_3]$ free	$[K_2SeO_3]$	Solid[a] phase
0.0	6.23	6.23	0.0	A
0.093$_6$	6.60	6.29	0.309	A
0.147	6.86	6.35	0.505	A
0.174	6.99	6.38	0.607	A
0.220	7.32	6.51	0.808	A
0.254	7.03	6.14	0.890	B
0.274	6.32	5.46	0.864	B
0.378	5.45	4.42	1.03	B
0.502	4.75	3.56	1.19	B
0.696	4.57	2.98	1.59	B
0.728	4.56	2.90	1.66	B
0.868	5.42	3.07	2.35	B + C
0.968	5.12	2.64	2.48	C
1.04	4.94	2.38	2.56	C
1.10	4.75	2.13	2.62	C
1.20	4.62	1.84	2.78	C
1.28	4.52	1.62	2.90	C + D
1.33	4.36	1.45	2.91	D
1.43	4.19	1.20	2.99	D
1.53	4.03	0.941	3.09	D
1.60	4.08	0.815	3.26	D
1.64	3.98	0.710	3.27	E
1.74	3.83	0.489	3.34	E
1.86	3.61	0.248	3.36	E
2.00	3.40	0.0	3.40	E

[a]Solid phases:

A – H_2SeO_3

B – $K_2SeO_3.3H_2SeO_3$

C – $K_2SeO_3.H_2SeO_3$

D – $5K_2SeO_3.3H_2SeO_3$

E – K_2SeO_3

(continued on next page)

AUXILIARY INFORMATION

METHOD APPARATUS/PROCEDURE:

A series of solutions of differing extents of neutralization of selenious acid were kept until crystals formed. The mixture of crystals and saturated solution was placed in a conical flask, which was sealed and then agitated in a thermostat at 25°C for up to a week. The solutions were analysed by potentiometric titration with hydrochloric acid or sodium hydroxide solution. Crystals could not be obtained from solutions with Na/Se between 0.22 and 0.25.

The solids were identified and characterized by thermogravimetry, differential thermal analysis, and X-ray diffraction; but some difficulties were encountered, owing to the deliquescence and hygroscopicity of some of the solid phases.

SOURCE AND PURITY OF MATERIALS:

Water was distilled and demineralized. Its final conductivity at 25°C was about 2×10^{-5} $ohm^{-1}m^{-1}$.

Selenious acid (Fluka) was found by analysis to be 99.6% pure.

ESTIMATED ERROR:

Temperature: ±0.05 K

REFERENCES:

COMPONENTS:	ORIGINAL MEASUREMENTS:
1. Potassium selenite; K_2SeO_3; [10431-47-7]	Sabbah, R.; Périnet, G.
2. Selenious acid; H_2SeO_3; [7783-00-8]	*J. Chim. Phys.* 1966, *63*, 332-6.
3. Water; H_2O; [7732-18-5]	

EXPERIMENTAL VALUES (continued):

Composition of solutions expressed in units of mass % and mol/kg

H_2SeO_3 [b] mass %	K_2SeO_3 [b] mass %	H_2SeO_3 [c] mol/kg	K_2SeO_3 [c] mol/kg
80.355	0.	31.712	0.
81.128	6.339	50.191	2.466
81.902	10.361	82.072	6.527
82.289	12.453	121.349	11.545
83.966	16.577	-	-
79.194	18.259	241.064	34.943
70.423	17.726	46.072	7.290
57.009	21.131	20.220	4.712
45.917	24.414	11.999	4.011
38.436	32.620	10.296	5.493
37.404	34.057	10.161	5.817
39.597	48.213	25.183	19.277
34.051	50.880	17.519	16.457
30.697	52.521	14.182	15.255
27.473	53.752	11.345	13.954
23.732	57.034	9.567	14.454
20.895	59.496	8.626	14.789
18.702	59.702	6.714	13.475
15.478	61.343	5.177	12.899
12.137	63.394	3.846	12.628
10.512	66.882	3.605	14.421
9.158	67.087	2.989	13.765
6.307	68.523	1.943	13.270
3.199	68.934	0.890	12.057
0.	69.754	0.	11.241

[b] Mass % values calculated by the compiler.

[c] Molalities calculated by the compiler.

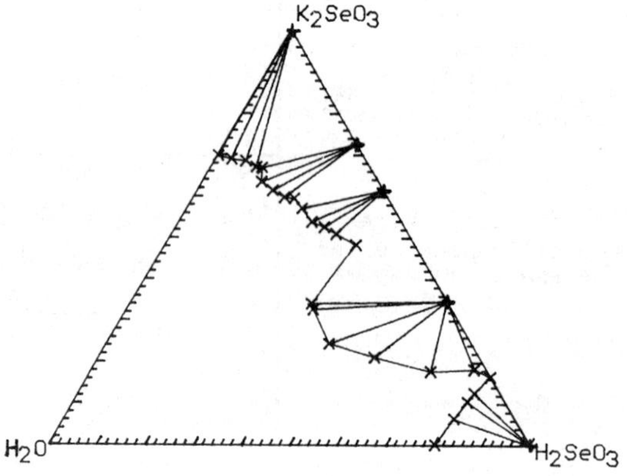

COMPONENTS:	EVALUATOR:
Alkali-metal selenites	Mary R. Masson, Dept. of Chemistry, University of Aberdeen, Meston Walk, Old Aberdeen, AB9 2UE, Scotland, UK. June 1984.

CRITICAL EVALUATION:

The data for sodium selenite and potassium selenite are discussed separately. For the following systems, only the binary systems of the compound with water have been studied.

Lithium selenite; Li_2SeO_3; [15593-51-8]		(1)
Sodium pyroselenite; $Na_2Se_2O_5$; [24458-98-8] (disodium diselenite)		(2,3)
Sodium trihydrogen diselenite; $NaH_3(SeO_3)_2$; [14013-56-0]		(2,3)
Potassium pyroselenite; $K_2Se_2O_5$; [12529-99-6] (dipotassium diselenite)		(2,3)
Potassium trihydrogen diselenite; $KH_3(SeO_3)_2$; [15457-71-3]		(2,3)
Ammonium selenite; $(NH_4)_2SeO_3$; [7783-19-9]		(3,4)
Ammonium pyroselenite; $(NH_4)_2Se_2O_5$; [13597-78-9] (diammonium diselenite)		(3,4)
Ammonium trihydrogen diselenite; $NH_4H_3(SeO_3)_2$; [25425-97-2]		(3,4)

The data (which are reported on the compilation pages) all appear to be reasonably reliable, but since there is only one study for each system, the data can just be regarded as tentative.

REFERENCES

1. Rosenheim, A.; Krause, L. *Z. Anorg. Allgem. Chem.* 1921, *118*, 177.
2. Janitzki, J. *Z. Anorg. Allgem. Chem.* 1932, *205*, 49.
3. Janickis, J. *Z. Anorg. Allgem. Chem.* 1934, *218*, 89.
4. Janickis, J.; Gutmanaite, H. *Z. Anorg. Allgem. Chem.* 1936, *227*, 1.

COMPONENTS:

1. Lithium selenite; Li_2SeO_3; [15593-51-8]
2. Water; H_2O; [7732-18-5]

ORIGINAL MEASUREMENTS:

Rosenheim, A.; Krause, L.

Z. Anorg. Allgem. Chem. 1921, *118*, 177-191.

VARIABLES:

Temperature: 273 - 373 K

PREPARED BY:

Mary. R. Masson

EXPERIMENTAL VALUES:

t/°C	Li_2SeO_3 mass %	Li_2SeO_3 g/100 ml water	Li_2SeO_3[a] mol/kg
0	19.99	24.99	1.774
25	16.76	20.17	1.430
47.5	14.53	16.99	1.207
60	12.75	14.62	1.038
100	9.05	9.94	0.707

[a] Molalities calculated by the compiler.

The solid phase was $4Li_2SeO_3.3H_2O$

AUXILIARY INFORMATION

METHOD APPARATUS/PROCEDURE:

A simple saturation procedure. Selenium was determined gravimetrically and iodometrically, and lithium alkalimetrically.

SOURCE AND PURITY OF MATERIALS:

ESTIMATED ERROR:

No estimates possible.

REFERENCES:

COMPONENTS:	ORIGINAL MEASUREMENTS:
1. Sodium pyroselenite; $Na_2Se_2O_5$; [24458-98-8] 2. Water; H_2O; [7732-18-5]	Janitzki, J. *Z. Anorg. Allgem. Chem.* <u>1932</u>, *205*, 49-75.
VARIABLES:	**PREPARED BY:**
Temperature: 264 - 369 K	Mary R. Masson

EXPERIMENTAL VALUES:

t/°C	$Na_2S_2O_5$ mass %	$Na_2S_2O_5$[a] mol/kg	Solid[b] phase
9.3	31.66	1.632	C
0.0	38.45	2.200	C
+ 8.1	43.97	2.764	C
14.6	49.19	3.410	C
20.0	54.02	4.138	C
25.4	60.32	5.355	C
27.8	63.52	6.133	C
27.0	62.52	5.876	B
28.8	62.69	5.918	B
28.8	62.88	5.967	B
31.5	63.86*	6.224	B
32.0	63.14	6.034	B
32.0	63.10	6.023	B
34.9	63.61	6.157	B
37.3	64.63*	6.436	B
37.3	63.98	6.257	B
39.7	64.16	6.306	B
40.1	64.27	6.336	B
45.2	65.15	6.585	B
50.0	65.98	6.831	B
59.9	67.40	7.282	B
79.2	72.26	9.175	B
89.0	74.36	10.215	B
91.8	75.32	10.750	B
92.8	75.67	10.955	B
93.8	75.95	11.124	A
96.0	76.05	11.185	A

(continued on next page)

AUXILIARY INFORMATION

METHOD APPARATUS/PROCEDURE:

For each temperature, a saturated solution was prepared by stirring the salt in water inside a stoppered 4-cm diameter test-tube. Small samples of solution were removed at intervals for analysis, in order to establish whether equilibrium had been attained. The time required varied between 2½ and 145 hr. The solutions were analysed for SeO_2 by the methods of Norris and Fay (1).

The solid phases were identified by analysis.

SOURCE AND PURITY OF MATERIALS:

ESTIMATED ERROR:

Temperature: -20 - 0°C ±0.2°C, 0 - 60°C ±0.1°C, 60 - 110°C ±0.3°C
Analyses: no estimate possible.

REFERENCES:

1. Norris, J.F.; Fay, H. *Amer. Chem. J.* <u>1896</u>, *18*, 703; <u>1900</u>, *23*, 119.

COMPONENTS:	ORIGINAL MEASUREMENTS:
1. Sodium pyroselenite; $Na_2Se_2O_5$; [24458-98-8] 2. Water; H_2O; [7732-18-5]	Janitzki, J. *Z. Anorg. Allgem. Chem.* 1932, *205*, 49-75.

EXPERIMENTAL VALUES (continued):

t/°C	$Na_2S_2O_5$ mass %	$Na_2S_2O_5$[a] mol/kg	Solid[b] phase
98.4	76.25	11.309	A
101.4	76.61	11.537	A
104.8	76.98	11.779	A
109.5	77.57	12.181	A

[a] Molalities calculated by the compiler.

[b] Solid phases: A - $Na_2Se_2O_5$, B - $NaHSeO_3$, C - $NaHSeO_3.3H_2O$

COMPONENTS:

1. Sodium pyroselenite; $Na_2Se_2O_5$; [24458-98-8]

2. Water; H_2O; [7732-18-5]

ORIGINAL MEASUREMENTS:

Janickis, J.; Gutmanaite, H.

Z. Anorg. Allgem. Chem. 1936, 225, 1-16.

VARIABLES:

Temperature: 264 - 273 K
Composition

PREPARED BY:

Mary R. Masson

EXPERIMENTAL VALUES:

Compositions of equilibrium solutions

t/°C	$Na_2Se_2O_5$ mol/dm^3	$Na_2Se_2O_5$ mass %	$Na_2Se_2O_5$[a] mol/kg	$NaHSeO_3$[a] mol/kg	Solid phase
-0.157	0.02	0.5665	0.0201	0.0401	ice
-0.360	0.05	1.406	0.0502	0.1005	"
-0.697	0.1	2.787	0.1010	0.2023	"
-1.34	0.2	5.44	0.2025	0.4066	"
-3.15	0.5	12.77	0.5154	1.041	"
-6.11	1	23.24	1.066	2.173	"
-9.3	satd.	31.66	1.631	3.361	ice + $NaHSeO_3.3H_2O$

[a] Molalities calculated by the compiler.

AUXILIARY INFORMATION

METHOD APPARATUS/PROCEDURE:

Freezing points of prepared solutions were measured by use of a Beckman-type apparatus (1). Determinations were repeated until the desired reproducibility was attained. Each reported value is the mean of at least three determinations.

SOURCE AND PURITY OF MATERIALS:

Sodium hydrogen selenite was prepared from selenious acid and sodium hydroxide.

ESTIMATED ERROR:

Temperature reproducibility 0.5%

REFERENCES:

1. Ostwald, W.; Luther, R. *Hand- und Hilfsbuch zur Ausfuhrung physikochemischer Messungen*, 5th Ed., Akademische Verlag., Leipzig, 1931.

COMPONENTS:

1. Sodium trihydrogen diselenite; $NaH_3(SeO_3)_2$; [14013-56-0]

2. Water; H_2O; [7732-18-5]

ORIGINAL MEASUREMENTS:

Janitzki, J.

Z. Anorg. Allgem. Chem. 1932, *205*, 49-75.

VARIABLES:

Temperature: 266 - 361 K

PREPARED BY:

Mary R. Masson

EXPERIMENTAL VALUES:

t/°C	$NaH_3(SeO_3)_2$ mass %	$NaH_3(SeO_3)_2$[a] mol/kg
- 6.9	37.12	2.109
+ 0.7	41.50	2.534
+13.5	48.83	3.409
+22.8	53.42	4.097
+32.0	58.17	4.968
+51.3	67.42	7.392
+69.6	75.88	11.238
+79.2	80.93	15.160
+88.0	83.95	18.685

[a] Molalities calculated by the compiler.

AUXILIARY INFORMATION

METHOD APPARATUS/PROCEDURE:

For each temperature, a saturated solution was prepared by stirring the salt in water inside a stoppered 4-cm diameter test-tube. Small samples of solution were removed at intervals, in order to test for attainment of equilibrium. The time required varied between 2 and 3 hr. The solutions were analysed for SeO_2 by the method of Norris and Fay (1).

SOURCE AND PURITY OF MATERIALS:

ESTIMATED ERROR:

Temperature: -20 - 0°C ±0.2°C, 0 - 60°C ±0.1°C, 60 - 110°C ±0.3°C

REFERENCES:

1. Norris, J.F.; Fay, H. *Amer. Chem. J.* 1896, *18*, 703; 1900, *23*, 119.

COMPONENTS:

1. Sodium trihydrogen diselenite; $NaH_3(SeO_3)_2$; [14013-56-0]
2. Water; H_2O; [7732-18-5]

ORIGINAL MEASUREMENTS:

Janickis, J.; Gutmanaite, H.

Z. Anorg. Allgem. Chem. 1936, *227*, 1-16.

VARIABLES:

Temperature: 264 - 273 K
Composition

PREPARED BY:

Mary R. Masson

EXPERIMENTAL VALUES:

Composition of equilibrium solutions

t/°C	$NaH_3(SeO_3)_2$ mol/dm^3	$NaH_3(SeO_3)_2$ mass %	$NaH_3(SeO_3)_2$[a] mol/kg	Solid phase
-0.132	0.02	0.560	0.0201	ice
-0.282	0.05	1.388	0.0503	"
-0.558	0.1	2.751	0.1010	"
-1.035	0.2	5.40	0.2037	"
-2.375	0.5	12.76	0.522	"
-4.38	1	23.43	1.093	"
-7.5	satd.	36.5	2.053	ice + $NaH_3(SeO_3)_2$
-8.4	2 (supersatd.)	46.0	2.415	?

[a] Molalities calculated by the compiler.

AUXILIARY INFORMATION

METHOD APPARATUS/PROCEDURE:

Freezing points of prepared solutions were measured by use of a Beckman-type apparatus (1). Determinations were repeated until the desired reproducibility was attained. Each reported value is the mean of at least three determinations.

SOURCE AND PURITY OF MATERIALS:

Sodium trihydrogen diselenite was prepared from selenious acid and sodium hydroxide.

ESTIMATED ERROR:

Temperature reproducibility 0.5%

REFERENCES:

1. Ostwald, W.; Luther, R. *Hand- und Hilfsbuch zur Ausfuhrung physikochemischer Messungen*, 5th Ed., Akademische Verlag., Leipzig, 1931.

COMPONENTS:

1. Potassium pyroselenite; $K_2Se_2O_5$; [12529-99-6]

2. Water; H_2O; [7732-18-5]

ORIGINAL MEASUREMENTS:

Janitzki, J.

Z. Anorg. Allgem. Chem. 1932, *205*, 49-75.

VARIABLES:

Temperature: 252 - 376 K

PREPARED BY:

Mary R. Masson

EXPERIMENTAL VALUES:

t/°C	$K_2Se_2O_5$ mass %	$K_2Se_2O_5$[a] mol/kg	Solid[b] phase
-20.6	73.52	8.783	B
-10.5	74.29	9.141	B
+ 0.2	75.87	9.946	B
+12.8	77.21	10.717	B
18.9	78.18	11.340	B
20.6	78.50	11.550	B
23.0	78.70	11.688	B
25.2	79.31	12.126	B
27.9	79.63	12.366	B
30.6	80.04	12.685	B
20.8	79.01	11.907	A
24.0	79.18	12.030	A
24.8	79.15	12.009	A
27.2	79.21	12.052	A
29.4	79.23	12.067	A
31.7	79.41	12.200	A
39.8	79.99	12.646	A
50.4	80.39	12.968	A
59.8	81.69	14.113	A
65.4	81.55	13.982	A
69.6	82.42	14.831	A
72.6	82.18	14.588	A
90.8	83.72	16.268	A
102.8	84.47	17.206	A

[a] Molalities calculated by the compiler.

[b] Solid phases: A - $K_2Se_2O_5$, B - $KHSeO_3$

AUXILIARY INFORMATION

METHOD APPARATUS/PROCEDURE:

For each temperature, a saturated solution was prepared by stirring the salt in water inside a stoppered 4-cm diameter test-tube. Small samples of solution were removed at intervals for analysis, in order to test for attainment of equilibrium. The time required varied between 3.33 and 114 hr. The solutions were analysed for SeO_2 by the method of Norris and Fay (1).

The solid phases were identified by analysis.

SOURCE AND PURITY OF MATERIALS:

ESTIMATED ERROR:

Temperature: -20 - 0°C ±0.2°C, 0 - 60°C ±0.1°C, 60 - 110°C ±0.3°C.

REFERENCES:

1. Norris, J.F.; Fay, H. *Amer. Chem. J.* 1896, *18*, 703; 1900, *23*, 119.

COMPONENTS:

1. Potassium pyroselenite; $K_2Se_2O_5$; [12529-99-6]

2. Water; H_2O; [7732-18-5]

ORIGINAL MEASUREMENTS:

Janickis, J.; Gutmanaite, H.

Z. Anorg. Allgem. Chem. 1936, 225, 1-16.

VARIABLES:

Temperature: 250 - 273 K
Composition

PREPARED BY:

Mary R. Masson

EXPERIMENTAL VALUES:

Composition of equilibrium solutions

t/°C	$K_2Se_2O_5$ mol/dm^3	$K_2Se_2O_5$ mass %	$K_2Se_2O_5$[a] mol/kg	$KHSeO_3$[a] mol/kg	Solid phase
- 0.166	0.02	0.630	0.0201	0.04015	ice
- 0.365	0.05	1.565	0.0503	0.1007	"
- 0.694	0.1	3.09	0.1009	0.2022	"
- 1.342	0.2	6.04	0.2033	0.4081	"
- 3.17	0.5	14.12	0.5202	1.050	"
- 6.285	1	25.55	1.086	2.215	"
-14.3	2	43.13	2.399	5.015	"
-22.7	2.67	52.55	3.504	7.48	"

[a] Molalities calculated by the compiler.

AUXILIARY INFORMATION

METHOD APPARATUS/PROCEDURE:

Freezing points of prepared solutions were measured by use of a Beckman-type apparatus (1). Determinations were repeated until the desired reproducibility was attained. Each reported value is the mean of at least three determinations.

SOURCE AND PURITY OF MATERIALS:

Potassium hydrogen selenite was prepared from selenious acid and potassium hydroxide.

ESTIMATED ERROR:

Temperature reproducibility 0.5%

REFERENCES:

1. Ostwald, W.; Luther, R. *Hand- und Hilfsbuch zur Ausfuhrung physikochemischer Messungen*, 5th Ed., Akademische Verlag., Leipzig, 1931.

COMPONENTS:

1. Potassium trihydrogen diselenite; $KH_3(SeO_3)_2$; [15457-71-3]
2. Water; H_2O; [7732-18-5]

ORIGINAL MEASUREMENTS:

Janitzki, J.

Z. Anorg. Allgem. Chem. 1932, *205*, 49-75.

VARIABLES:

Temperature: 266 - 333 K

PREPARED BY:

Mary R. Masson

EXPERIMENTAL VALUES:

t/°C	$KH_3(SeO_3)_2$ mass %	$KH_3(SeO_3)_2$[a] mol/kg
- 6.9	46.52	2.938
0.0	53.57	3.897
+11.9	63.20	5.801
20.3	68.65	7.397
31.0	75.71	10.529
40.2	80.30	13.769
50.8	85.55	19.999
59.4	89.65	29.259

[a] Molalities calculated by the compiler.

AUXILIARY INFORMATION

METHOD APPARATUS/PROCEDURE:

For each temperature, a saturated solution was prepared by stirring the salt in water inside a stoppered 4-cm diameter test-tube. Small samples of solution were removed at intervals, in order to test for attainment of equilibrium. The time required varied between 2 and 26 hr. The solutions were analysed for SeO_2 by the method of Norris and Fay (1).

SOURCE AND PURITY OF MATERIALS:

ESTIMATED ERROR:

Temperature: -20 - 0°C ±0.3°C, 0 - 60°C ±0.1°C, 60 - 110°C ±0.3°C.

REFERENCES:

1. Norris, J.F.; Fay, H. *Amer. Chem. J.* 1896, *18*, 703; 1900, *23*, 119.

COMPONENTS:

1. Potassium trihydrogen diselenite; $KH_3(SeO_3)_2$; [15457-71-3]
2. Water; H_2O; [7732-18-5]

ORIGINAL MEASUREMENTS:

Janickis, J.; Gutmanaite, H.

Z. Anorg. Allgem. Chem. 1936, *227*, 1-16.

VARIABLES:

Temperature: 265 - 273 K
Composition

PREPARED BY:

Mary R. Masson

EXPERIMENTAL VALUES:

Composition of equilibrium solutions

t/°C	$KH_3(SeO_3)_2$ mol/dm^3	$KH_3(SeO_3)_2$ mass %	$KH_3(SeO_3)_2$[a] mol/kg	Solid phase
-0.128	0.02	0.590	0.0201	ice
-0.288	0.05	1.467	0.0503	"
-0.544	0.1	2.905	0.1011	"
-1.027	0.2	5.69	0.2039	"
-2.302	0.5	13.43	0.524	"
-4.10	1	24.62	1.103	"
-7.42	2	42.3	2.478	"
-8.0	satd.	46.0	2.88	ice + $KH_3(SeO_3)_2$

[a] Molalities calculated by the compiler.

AUXILIARY INFORMATION

METHOD APPARATUS/PROCEDURE:

Freezing points of prepared solution were measured by use of a Beckman-type apparatus (1). Determinations were repeated until the desired reproducibility was attained. Each reported value is the mean of at least three determinations.

SOURCE AND PURITY OF MATERIALS:

Potassium trihydrogen diselenite was prepared from selenious acid and potassium hydroxide.

ESTIMATED ERROR:

Temperature reproducibility 0.5%

REFERENCES:

1. Ostwald, W.; Luther, R. *Hand- und Hilfsbuch zur Ausfuhrung physikochemischer Messungen*, 5th Ed., Akademische Verlag., Leipzig, 1931.

COMPONENTS:

1. Ammonium selenite; $(NH_4)_2SeO_3$; [7783-19-9]
2. Water; H_2O; [7732-18-5]

ORIGINAL MEASUREMENTS:

Janickis, J.

Z. Anorg. Allgem. Chem. 1934, *218*, 89-103.

VARIABLES:

Temperature: 253- 343 K

PREPARED BY:

Mary R. Masson

EXPERIMENTAL VALUES:

t/°C	$(NH_4)_2SeO_3$ mass %	$(NH_4)_2SeO_3$[a] mol/kg
-20.0	45.12	5.043
- 8.5	47.18	5.479
+ 1.0	49.21	5.943
14.0	51.99	6.642
25.0	54.70	7.406
32.0	56.00	7.806
35.2	57.13	8.174
43.0	59.90	9.162
50.0	62.31	10.140
70.0	69.08	13.703

[a] Molalities calculated by the compiler.

Solid phase: $(NH_4)_2SeO_3.H_2O$

AUXILIARY INFORMATION

METHOD APPARATUS/PROCEDURE:

For each temperature, a saturated solution was prepared by stirring the salt in water inside a stoppered 4-cm diameter test-tube. Small samples of solution were removed at intervals for analysis, in order to test for attainment of equilibrium. The time required varied between 1 and 15 hr. The solutions were analysed for SeO_2 by the method of Norris and Fay (1).

SOURCE AND PURITY OF MATERIALS:

ESTIMATED ERROR:

Temperature: -20 - 0°C ±0.2°C, 0 - 60°C ±0.1°C, 60 - 110°C ±0.3°C.

REFERENCES:

1. Norris, J.F.; Fay, H. *Amer. Chem. J.* 1896, *18*, 703; 1900, *23*, 119.

COMPONENTS:

1. Ammonium selenite; $(NH_4)_2SeO_3$; [7783-19-9]

2. Water; H_2O; [7732-18-5]

ORIGINAL MEASUREMENTS:

Janickis, J.; Gutmanaite, H.

Z. Anorg. Allgem. Chem. 1936, *227*, 1-16.

VARIABLES:

Temperature: 251 - 273 K
Composition

PREPARED BY:

Mary R. Masson

EXPERIMENTAL VALUES:

Composition of equilibrium solutions

t/°C	$(NH_4)_2SeO_3$ mol/dm^3	$(NH_4)_2SeO_3$ mass %	$(NH_4)_2SeO_3$[a] mol/kg	Solid phase
-0.105	0.02	0.326	0.0201	ice
-0.260	0.05	0.812	0.0502	"
-0.470	0.1	1.62	0.1007	"
-0.875	0.2	3.20	0.2028	"
-2.06	0.5	7.75	0.516	"
-4.08	1	14.80	1.066	"
-8.81	2	27.24	2.296	"
-15.95	3.173	39.88	4.07	"
-21.9	satd.	44.8	4.98	ice + $(NH_4)_2SeO_3.H_2O$

[a] Molalities calculated by the compiler.

AUXILIARY INFORMATION

METHOD APPARATUS/PROCEDURE:

Freezing points of prepared solutions were measured by use of a Beckman-type apparatus (1). Determinations were repeated until the desired reproducibility was attained. Each reported value is the mean of at least three determinations.

SOURCE AND PURITY OF MATERIALS:

Ammonium selenite was prepared by neutralization of selenious acid with ammonia solution.

ESTIMATED ERROR:

Temperature reproducibility 0.5%

REFERENCES:

1. Ostwald, W.; Luther, R. *Hand- und Hilfsbuch zur Ausfuhrung physikochemischer Messungen*, 5th Ed., Akademische Verlag., Leipzig, 1931.

COMPONENTS:

1. Ammonium pyroselenite; $(NH_4)_2Se_2O_5$; [13597-78-9]
2. Water; H_2O; [7732-18-5]

ORIGINAL MEASUREMENTS:

Janickis, J.

Z. Anorg. Allgem. Chem. 1934, *218*, 89-103.

VARIABLES:

Temperature: 258 - 343 K

PREPARED BY:

Mary R. Masson

EXPERIMENTAL VALUES:

t/°C	$(NH_4)_2Se_2O_5$ mass %	$(NH_4)_2Se_2O_5$[a] mol/kg	Solid phase
-15.0	49.62	3.595	B
-10.0	52.86	4.092	B
0.0	56.84	4.806	B
+15.0	66.65	7.294	B
20.0	69.50	8.316	B
25.0	73.24	9.989	B
30.0	79.74	14.364	B
32.0	82.29	16.958	B
32.0	86.23	22.855	A
33.2	86.35	23.088	A
34.0	86.43	23.245	A
45.1	87.23	24.930	A
57.2	88.78	28.878	A
70.1	90.56	35.012	A

[a] Molalities calculated by the compiler.

[b] Solid phases: A - $(NH_4)_2Se_2O_5$, B - $(NH_4)_2Se_2O_5.3H_2O$

AUXILIARY INFORMATION

METHOD APPARATUS/PROCEDURE:

For each temperature, a saturated solution was prepared by stirring the salt in water inside a stoppered 4-cm diameter test-tube. Small samples of solution were removed at intervals for analysis, in order to test for attainment of equilibrium. The time required varied between 2½ and 24 hr. The solutions were analysed for SeO_2 by the method of Norris and Fay (1).

The solid phases were identified by analysis.

SOURCE AND PURITY OF MATERIALS:

ESTIMATED ERROR:

Temperature: -20 - 0°C ±0.2°C, 0 - 60°C ±0.1°C, 60 - 110°C ±0.3°C.

REFERENCES:

1. Norris, J.F.; Fay, H. *Amer. Chem. J.* 1896, *18*, 703; 1900, *23*, 119.

COMPONENTS:

1. Ammonium pyroselenite; $(NH_4)_2Se_2O_5$; [13597-78-9]
2. Water; H_2O; [7732-18-5]

ORIGINAL MEASUREMENTS:

Janickis, J.; Gutmanaite, H.

Z. Anorg. Allgem. Chem. 1936, *225*, 1-16.

VARIABLES:

Temperature: 256 - 273 K
Composition

PREPARED BY:

Mary R. Masson

EXPERIMENTAL VALUES:

Composition of equilibrium solutions

t/°C	$(NH_4)_2Se_2O_5$ mol/dm^3	$(NH_4)_2Se_2O_5$ mass %	$(NH_4)_2Se_2O_5$[a] mol/kg	NH_4HSeO_3[a] mol/kg	Solid[b] phase
- 0.148	0.02	0.547	0.0201	0.0415	ice
- 0.373	0.05	1.36	0.0503	0.1007	"
- 0.697	0.1	2.69	0.1009	0.2021	"
- 1.365	0.2	5.29	0.204	0.410	"
- 3.18	0.5	12.57	0.525	1.059	"
- 6.23	1.004	23.35	1.112	2.269	"
-13.0	2.008	40.7	2.504	5.244	"
-16.9	satd.	49.0	3.507	7.49	ice + $(NH_4)_2Se_2O_5.3H_2O$

[a] Molalities calculated by the compiler.

AUXILIARY INFORMATION

METHOD APPARATUS/PROCEDURE:

Freezing points of prepared solutions were measured by use of a Beckman-type apparatus (1). Determinations were repeated until the desired reproducibility was attained. Each reported value is the mean of at least three determinations.

SOURCE AND PURITY OF MATERIALS:

Ammonium pyroselenite was prepared from selenious acid and ammonia solution.

ESTIMATED ERROR:

Temperature reproducibility 0.5%

REFERENCES:

1. Ostwald, W.; Luther, R. *Hand- und Hilfsbuch zur Ausfuhrung physikochemischer Messungen*, 5th Ed., Akademische Verlag., Leipzig, 1931.

COMPONENTS:

1. Ammonium trihydrogen diselenite; $NH_4H_3(SeO_3)_2$; [25425-97-2]
2. Water; H_2O; [7732-18-5]

ORIGINAL MEASUREMENTS:

Janickis. J.

Z. Anorg. Allgem. Chem. 1934, *218*, 89-103.

VARIABLES:

Temperature: 258 - 303 K

PREPARED BY:

Mary R. Masson

EXPERIMENTAL VALUES:

t/°C	$NH_4H_3(SeO_3)_2$ mass %	$NH_4H_3(SeO_3)_2$[a] mol/kg
-14.8	60.08	5.473
-10.3	64.58	6.631
- 5.8	68.70	7.982
+ 0.1	73.61	10.144
8.8	79.30	13.932
18.0	85.11	20.787
30.0	91.62	39.760

[a] Molalities calculated by the compiler.

AUXILIARY INFORMATION

METHOD APPARATUS/PROCEDURE:

For each temperature, a saturated solution was prepared by stirring the salt in water inside a stoppered 4-cm diameter test-tube. Small samples of solution were removed at intervals for analysis, in order to test for attainment of equilibrium. The time required varied between 2 and 14 hr. The solutions were analysed for SeO_2 by the method of Norris and Fay (1).

SOURCE AND PURITY OF MATERIALS:

ESTIMATED ERROR:

Temperature: -20 - 0°C ±0.2°C, 0 - 60°C ±0.1°C.

REFERENCES:

1. Norris, J.F.; Fay, H. *Amer. Chem. J.* 1896, *18*, 703; 1900, *23*, 119.

COMPONENTS:

1. Ammonium trihydrogen diselenite; $NH_4H_3(SeO_3)_2$; [25425-97-2]
2. Water; H_2O; [7732-18-5]

ORIGINAL MEASUREMENTS:

Janickis, J.; Gutmanaite, H.

Z. Anorg. Allgem. Chem. 1936, *227*, 1-16.

VARIABLES:

Temperature: 258 - 273 K
Composition

PREPARED BY:

Mary R. Masson

EXPERIMENTAL VALUES:

Composition of equilibrium solutions

t/°C	$NH_4H_3(SeO_3)_2$ mol/dm^3	$NH_4H_3(SeO_3)_2$ mass %	$NH_4H_3(SeO_3)_2$[a] mol/kg	Solid phase
-0.147	0.02	0.549	0.0201	ice
-0.335	0.05	1.365	0.0503	"
-0.595	0.1	2.71	0.1011	"
-1.055	0.2	5.32	0.2043	"
-2.435	0.5	12.94	0.540	"
-4.385	1	24.73	1.112	"
-8.6	2	40.89	2.516	"
-12.85	3	54.60	4.37	"
-14.8	satd.	60.08	5.46	ice + $NH_4H_3(SeO_3)_2$

[a] Molalities calculated by the compiler.

AUXILIARY INFORMATION

METHOD APPARATUS/PROCEDURE:

Freezing points of prepared solutions were measured by use of a Beckman-type apparatus (1). Determinations were repeated until the desired reproducibility was attained. Each reported value is the mean of at least three determinations.

SOURCE AND PURITY OF MATERIALS:

Ammonium trihydrogen diselenite was prepared from selenious acid and ammonia solution.

ESTIMATED ERROR:

Temperature reproducibility 0.5%

REFERENCES:

1. Ostwald, W.; Luther, R. *Hand- und Hilfsbuch zur Ausfuhrung physikochemischer Messungen*, 5th Ed., Akademische Verlag., Leipzig, 1931.

COMPONENTS:	EVALUATOR:
Sparingly soluble selenites	Mary R. Masson, Dept. of Chemistry, University of Aberdeen, Meston Walk, Old Aberdeen, AB9 2UE, Scotland, UK. July 1984.

CRITICAL EVALUATION:

Much of the data for the solubilities of sparingly soluble selenites come from three papers (1 - 3) by two groups of workers. None of these data are totally satisfactory because the experimental procedures are deficient in certain respects.

Ripan and Vericeanu (1) used the sensitive but unselective conductometric method to determine how much of the sparingly soluble salts had dissolved in water. Also, they neglected to make any correction for the hydrolysis of the selenite ion, or of any hydrolysable metal ion. Few replicates were done.

Chukhlantsev (2) and Chukhlantsev and Tomashevsky (3) determined the conditional solubilities in acid solutions, in order to increase the concentrations to be determined. This procedure has the disadvantage that the subsequent calculations to find the solubility in pure water require values for the acid dissociation constants of selenious acid for the relevant temperature and in the same medium, and only estimates of these are available. Again, few replicate determinations were done. In this case, the chemistry was reasonably selective.

The other work that has been done does not appear to be of much superior quality (4 - 18). Thus, even where 3 or 4 determinations have been made, it is impossible to make any reasoned choice between them. None of the values can be regarded as RECOMMENDED.

The solubilities are usually reported in terms of concentration constants, generally as K_{s0} values. No attempts have been made to calculate the corresponding thermodynamic constants, since the media are not adequately defined to allow calculation of activity coefficients.

Silver selenite has been studied by several authors (2, 11-15). The work of Chukhlantsev (2) has been discussed already. Lin and Pan (11) calculated pK_{s0} = 14.74 from work on the silver selenite electrode, but they made several errors in their calculation, and attempted recalculations gave nonsensical figures, so this work is rejected (not compiled). Selivanova's work (12) seems careful, but hydrolysis of selenite is neglected, and it is not certain whether equilibrium was really reached. Mehra and Gubeli (13,14) did extensive work, but they made little use of the data collected. The calculations of the compiler have shown that there is a lack of consistency between the results obtained by the different experimental techniques employed. The work of Chao and Cheng (15) gives a value that ought to be reasonably reliable, but it is not certain whether silver-ion activities or concentrations were used to calculate the solubility product. Also, the constant refers to a freshly precipitated solid.

COMPONENTS:	EVALUATOR:
Sparingly soluble selenites	Mary R. Masson, Dept. of Chemistry, University of Aberdeen, Meston Walk, Old Aberdeen, AB9 2UE, Scotland, UK. July 1984.

CRITICAL EVALUATION: (continued)

TERNARY SYSTEMS

The ternary systems barium selenite - selenium dioxide - water (5) and nickel selenite - selenium dioxide - water (9) have been studied. In the first case the solid phase $BaSe_2O_5$ [83534-22-9] was observed, and in the second case, $Ni(HSeO_3)_2.2H_2O$ [83753-29-1].

GENERAL COMMENTS

My feeling about all the data for the sparingly soluble selenites is that they do little more than give some general guidance as to the solubility behaviour that is to be expected for the various metal selenites. There is certainly scope for much more experimental investigation of these systems.

NOTE

The Chemical Abstracts reports of some Russian papers (19 - 23) suggest that these give solubility product data, but there is no original experimental solubility work reported in these papers. They report calculations of 'theoretical' solubility products from other thermodynamic data, and comparison of these with literature values; and also correlations of other thermodynamic parameters.

REPORTED VALUES

Ion	K_{s0}	pK_{s0}	Other constants	Ref.
Be^{2+}	1.1×10^{-8} mol^2dm^{-6}	7.96 a		1
Mg^{2+}	1.9×10^{-6} mol^2dm^{-6}	5.72 a		1
	1.29×10^{-5} mol^2dm^{-6}	4.89 b		2
Ca^{2+}	1.8×10^{-6} mol^2dm^{-6}	5.74 a		1
	2.96×10^{-6} mol^2dm^{-6}	5.53 b		2
Sr^{2+}	7.9×10^{-6} mol^2dm^{-6}	5.10 a		1
	1.82×10^{-6} mol^2dm^{-6}	5.74 b		2
	8.4×10^{-7} mol^2dm^{-6}	6.07 b		2 *
	4.23×10^{-6} mol^2dm^{-6}	5.37 c		4
Ba^{2+}	6.2×10^{-6} mol^2dm^{-6}	6.21 a		1
	4.07×10^{-7} mol^2dm^{-6}	6.39 c		4
	4.45×10^{-7} mol^2dm^{-6}	6.35 d		5
Mn^{2+}	1.2×10^{-7} mol^2dm^{-6}	6.9 b		3

COMPONENTS:	EVALUATOR:
Sparingly soluble selenites	Mary R. Masson, Dept. of Chemistry, University of Aberdeen, Meston Walk, Old Aberdeen, AB9 2UE, Scotland, UK. July 1984.

CRITICAL EVALUATION: (continued)

Ion	K_{s0}		pK_{s0}	Other constants	Ref.
Fe^{3+}	2.0×10^{-31} mol^{5}dm^{-15}		30.7 b		2
	3.7×10^{-36} mol^{5}dm^{-15}		35.4 e		7
Co^{2+}	1.2×10^{-7} mol^{2}dm^{-6}		6.92 a		1
	1.6×10^{-7} mol^{2}dm^{-6}		6.8 b		3
	1.05×10^{-8} mol^{2}dm^{-6}		7.98		8
	1.17×10^{-6} mol^{2}dm^{-6}	(I = 0.3)	5.93 e	$K_{instab} = 5.3 \times 10^{-4}$	8
	1.14×10^{-7} mol^{2}dm^{-6}	(I = 0.01)	6.94 e	$K_{instab} = 6.25 \times 10^{-6}$	8
Ni^{2+}	5.1×10^{-6} mol^{2}dm^{-6}		5.29 a		1
	1.0×10^{-5} mol^{2}dm^{-6}		5.0 b		3
Cu^{2+}	3.2×10^{-8} mol^{2}dm^{-6}		7.49 a		1
	2.09×10^{-8} mol^{2}dm^{-6}		7.68 b		2
	1.02×10^{-8} mol^{2}dm^{-6}		7.99 e		10
Ag^{+}	9.7×10^{-16} mol^{3}dm^{-9}		15.01 b		2
	2.85×10^{-16} mol^{3}dm^{-9}		15.55 c		12
	1.43×10^{-16} mol^{3}dm^{-9}		15.84 c	recalc. value	
	2.63×10^{-16} mol^{3}dm^{-9}		15.58 c	(alk. data) I = 1	13,14
	2.34×10^{-16} mol^{3}dm^{-9}		15.63 c	(all data) I = 1	
				$\beta_1 = 2.63 \times 10^2$ c	
				$\beta_2 = 5.75 \times 10^3$ c	
	3.55×10^{-16} mol^{3}dm^{-9}		15.45 b	I = 0.1	15
Zn^{2+}	1.9×10^{-8} mol^{2}dm^{-6}		7.71 a		1
	2.58×10^{-7} mol^{2}dm^{-6}		6.59 b		2
Cd^{2+}	6.0×10^{-9} mol^{2}dm^{-6}		8.22 a		1
	1.29×10^{-9} mol^{2}dm^{-6}		8.89 b		2
	4.0×10^{-10} mol^{2}dm^{-6}		9.40 e		16
Hg_2^{2+}	2.3×10^{-15} mol^{2}dm^{-6}		14.64 b		3
	8.7×10^{-14} mol^{2}dm^{-6}	???	13.1 e		16

COMPONENTS:	EVALUATOR:
Sparingly soluble selenites	Mary R. Masson, Dept. of Chemistry, University of Aberdeen, Meston Walk, Old Aberdeen, AB9 2UE, Scotland, UK. July 1984.

CRITICAL EVALUATION: (continued)

Ion	K_{s0}	pK_{s0}	Other constants	Ref.
Hg^{2+}	5.76×10^{-15} mol^2dm^{-6}	14.24 e		16
	–	c	$K_{s2} = 4.42 \times 10^{-2}$ (pK_{s2} - 1.35)	17
	2.16×10^{-15} mol^2dm^{-6}	14.67 c	K_f for $Hg(SeO_3)_2^{2-}$ $= 10^{12.48}$ $mol^{-2}dm^6$ $K_{s2} = 10^{-1.36}$	18
Pb^{2+}	3.4×10^{-12} mol^2dm^{-6}	11.5 b		3
	3.0×10^{-7} mol^2dm^{-6}	6.52 d		5

Temperatures: a - 291.2 K, b - 293.2 K, c - 298.2 K, d - 283.2 K,
e - not stated, presumably ambient (293.2 - 298.2 K, usually)

REFERENCES

1. Ripan, R,; Vericeanu, G. *Studia Univ. Babes-Bolyai, Ser. Chim.* 1968, *13*, 31.
2. Chukhlantsev, V.G. *Zh. Neorg. Khim.* 1956, *1*, 2300; **Russ. J. Inorg. Chem. (Eng. Transl.)* 1956, *1*, 132.
3. Chukhlantsev, V.G.; Tomashevsky, G.P. *Zh. Anal. Khim.* 1957, *12*, 296; **J. Anal. Chem. USSR* 1957, *12*, 303.
4. Leschinskaya, Z.I.; Averbukh, M.A.; Selivanova, N.M. *Zh. Fiz. Khim.* 1965, *39*, 2036; **Russ. J. Phys. Chem. (Eng. Transl.)* 1965, *39*, 1082.
5. Dolique, R. *Bull. Soc. Chim. France* 1943, *10*, M50.
6. Neal, J.L., Jr.; McCrosky, C.R. *J. Am. Chem. Soc.* 1938, *60*, 911.
7. Pinaev, G.F.; Volkova, V.P. *Obshch. Prikl. Khim.* 1970, 33.
8. Pyatnitskii, I.V.; Durdyev, M. *Ukr. Khim. Zh.* 1966, *32*, 77.
9. Ebert, M.; Mička, Z.; Peková, I. *Collect. Czech. Chem. Commun.* 1982, *47*, 2069.
10. Geilmann, W.; Wrigge, W. *Z. Anorg. Allgem. Chem* 1931, *197*, 353.
11. Lin, J.-L.; Pan. K. *J. Chinese Chem. Soc. (Formosa)* 1961, *8*, 14.

COMPONENTS:	EVALUATOR:
Sparingly soluble selenites	Mary R. Masson, Dept. of Chemistry, University of Aberdeen, Meston Walk, Old Aberdeen, AB9 2UE, Scotland, UK. July 1984.

CRITICAL EVALUATION: (continued)

12. Selivanova, N.M.; Leschinskaya, Z.L.; Klushina, T.V. *Zh. Fiz. Khim.* 1962, *36*, 1349; **Russ. J. Phys. Chem. (Eng. Transl.)* 1962, *36*, 719.

13. Mehra, M.C.; Gubeli, A.O. *Radiochem. Radioanal. Lett.* 1969, 2, 1968.

14. Mehra, M.C. *Dissertation*, *Laval University*, Quebec P.Q., 1968.

15. Chao, E.E.; Cheng, K.L. *Anal. Chem.* 1976, *48*, 267.

16. Redman, M.J.; Harvey, W.W. *J. Less-Common Met.* 1967, *12*, 395-404.

17. Rosenheim, A.; Pritze, M. *Z. Anorg. Chem* 1909, *63*, 275.

18. Toropova, V.F. *Zh. Neorg. Khim.* 1957, 2, 515; **Russ. J. Inorg. Chem. (Eng. Transl.)* 1957, 2, 68.

19. Erdenbaeva, M.I.; Mamonova, G.F. *Ivz. Akad. Nauk Kaz. SSR, Ser. Khim.* 1964, *14*, 41.

20. Erdenbaeva, M.I. *Vestn. Akad. Nauk Kaz. SSR* 1975, (2), 51.

21. Erdenbaeva, M.I. *Viniti Deposited Document* No. 3215-77.

22. Bakeeva, S.S.; Buketov, E.A.; Pashinkin, A.S. *Tr. Khim.-Met. Inst., Akad. Nauk Kaz. SSR* 1967, 4, 3.

23. Buketov, E.A. *Vestn. Akad. Nauk Kaz. SSR* 1965, *21*, 30.

COMPONENTS:	ORIGINAL MEASUREMENTS:
1. Beryllium selenite; $BeSeO_3$; [38333-62-9] 2. Water; H_2O; [7732-18-5]	Ripan, R.; Vericeanu, G. *Studia Univ. Babes-Bolyai, Ser. Chim.* 1968, *13*, 31-37.
VARIABLES:	PREPARED BY:
One temperature: 291 K	Mary R. Masson

EXPERIMENTAL VALUES:

All concentrations are expressed in units of mol dm^{-3}

Concentration	K_{s0} mol^2dm^6	Mean K_{s0}	pK_{s0}
8.501×10^{-5}	7.2×10^{-9}	$1.1 \pm 0.5 \times 10^{-8}$ mol^2dm^{-6}	7.96
8.451×10^{-5}	7.1×10^{-9}		
9.150×10^{-5}	8.4×10^{-9}		
1.141×10^{-4}	1.3×10^{-8}		
1.186×10^{-4}	1.4×10^{-8}		
1.189×10^{-4}	1.4×10^{-8}		

The concentration c in the saturated solution was calculated from the measured conductivity κ from the equation

$$c = \frac{1000\kappa}{\Lambda°}$$

Compiler's note

Neither in the determination of the ionic conductivity of the selenite ion nor in the evaluation of the solubility product was hydrolysis of the selenite ion taken into account. This would give rise to errors, since, for example, in a 0.001*M* solution, $[SeO_3^{2-}]$ = 0.000955*M*, $[HSeO_3^-]$ = 0.000045*M* and $[OH^-]$ = 0.000045*M*, and hydroxide and hydrogen selenite have different ionic conductivities from selenite. If the ionic conductivity of hydrogen selenite were known, the experimental results could have been interpreted correctly (cf. ref. 2), but this value does not seem to be available.

However, because the calibration and sample solutions had concentrations of about the same order of magnitude, the errors would cancel to some extent, but the K_{s0} value cannot be regarded as reliable.

AUXILIARY INFORMATION

METHOD APPARATUS/PROCEDURE:

The ionic conductivity of the selenite ion in water was determined by measuring the mean molar conductivities of a series of aqueous solutions of lithium selenite with concentrations below 2×10^{-3} mol dm^{-3}. The mean molar conductivity at infinite dilution was found by extrapolation to be $172 \pm 2.9\ \Omega^{-1}cm^2mol^{-1}$ at 18°C. At that temperature, the ionic conductivity of the lithium ion is 33.4 $\Omega^{-1}cm^2mol^{-1}$, so the ionic conductivity of the selenite ion is 105.2 $\Omega^{-1}cm^2mol^{-1}$.

The well washed beryllium selenite was mixed with 150 ml of water and agitated intermittently for 1 hr. From this, 10-20 ml was taken in the conductivity cell and the conductivity measured. The sampling and measurement were repeated until a constant value for the conductivity was obtained.

The value of $\Lambda° = (\lambda_+ + \lambda_-)$, the molar conductivity, was calculated with λ_- = 105.2 and λ_+ = 74.8 $\Omega^{-1}cm^2mol^{-1}$.

SOURCE AND PURITY OF MATERIALS:

Lithium selenite was prepared from sublimed selenium dioxide by neutralization with lithium hydroxide. An approximately 0.05*M* solution was standardized gravimetrically, and this was used to prepare the various dilutions.

The composition of the selenite was checked by analysis.

ESTIMATED ERROR:

Temperature: ±0.5 K

Error in K_{s0} (2*s*) = 0.5×10^{-8} (compiler)

REFERENCES:

1. Landolt-Bornstein *Physicalisch-Chemische Tabellen II* 1923, p. 1105.
2. Monk, C.B. *J. Chem. Soc.* 1949, 429.

COMPONENTS:

1. Magnesium selenite; $MgSeO_3$; [15593-61-0]
2. Water; H_2O; [7732-18-5]

ORIGINAL MEASUREMENTS:

Ripan, R.; Vericeanu, G

Studia Univ. Babes-Bolyai, Ser. Chim. 1968, *13*, 31-37.

VARIABLES:

One temperature: 291 K

PREPARED BY:

Mary R. Masson

EXPERIMENTAL VALUES:

All concentrations are expressed in units of mol dm^{-3}

Concentration	K_{s0} mol^2dm^6	Mean K_{s0}	pK_{s0}
1.281×10^{-3}	1.64×10^{-6}		
1.288×10^{-3}	1.66×10^{-6}		
1.323×10^{-3}	1.75×10^{-6}		
1.449×10^{-3}	2.10×10^{-6}	$1.9 \pm 0.5 \times 10^{-6}$ mol^2dm^{-6}	5.72
1.435×10^{-3}	2.13×10^{-6}		
1.497×10^{-3}	2.24×10^{-6}		

The concentration c in the saturated solution was calculated from the measured conductivity κ from the equation

$$c = \frac{1000\kappa}{\Lambda°}$$

Compiler's note

Neither in the determination of the ionic conductivity of the selenite ion nor in the evaluation of the solubility product was hydrolysis of the selenite ion taken into account. This would give rise to errors, since, for example, in a 0.001*M* solution, $[SeO_3^{2-}] = 0.000955M$, $[HSeO_3^-] = 0.000045M$ and $[OH^-] = 0.000045M$, and hydroxide and hydrogen selenite have different ionic conductivities from selenite. If the ionic conductivity of hydrogen selenite were known, the experimental results could have been interpreted correctly (cf. ref. 2), but this value does not seem to be available.

However, because the calibration and sample solutions had concentrations of about the same order of magnitude, the errors would cancel to some extent, but the K_{s0} value cannot be regarded as reliable.

AUXILIARY INFORMATION

METHOD APPARATUS/PROCEDURE:

The ionic conductivity of the selenite ion in water was determined by measuring the mean molar conductivities of a series of aqueous solutions of lithium selenite with concentrations below 2×10^{-3} mol dm^{-3}. The mean molar conductivity at infinite dilution was found by extrapolation to be $172 \pm 2.9\ \Omega^{-1}cm^2mol^{-1}$ at 18°C. At that temperature, the ionic conductivity of the lithium ion is $33.4\ \Omega^{-1}cm^2mol^{-1}$, so the ionic conductivity of the selenite ion is $105.2\ \Omega^{-1}cm^2mol^{-1}$.

The well washed magnesium selenite was mixed with 150 ml of water and agitated intermittently for 1 hr. From this, 10-20 ml was taken in the conductivity cell and the conductivity measured. The sampling and measurement were repeated until a constant value for the conductivity was obtained.

The value of $\Lambda° = (\lambda_+ + \lambda_-)$, the molar conductivity, was calculated with $\lambda_- = 105.2$ and $\lambda_+ = 90\ \Omega^{-1}cm^2mol^{-1}$.

SOURCE AND PURITY OF MATERIALS:

Lithium selenite was prepared from sublimed selenium dioxide by neutralization with lithium hydroxide. An approximately 0.05*M* solution was standardized gravimetrically, and this was used to prepare the various dilutions.

The composition of the selenite was checked by analysis.

ESTIMATED ERROR:

Temperature: ±0.5 K
Error in K_{s0} (2*s*) = 0.5×10^{-6} (compiler)

REFERENCES:

1. Landolt-Bornstein *Physicalisch-Chemische Tabellen II* 1923, p. 1105.
2. Monk, C.B. *J. Chem. Soc.* 1949, 429.

COMPONENTS:	ORIGINAL MEASUREMENTS:
1. Magnesium selenite; $MgSeO_3$; [15593-61-0] 2a. Hydrochloric acid; HCl; [7647-01-0] 2b. Nitric acid; HNO_3; [7697-37-2] 3. Water; H_2O; [7732-18-5]	Chukhlantsev, V.G. *Zh. Neorg. Khim.* 1956, *1*, 2300-5; **Russ. J. Inorg. Chem. (Eng. Transl.)* 1956, *1*, 132-8.
VARIABLES:	PREPARED BY:
HCl and HNO_3 concentrations One temperature: 293 K	Mary R. Masson

EXPERIMENTAL VALUES:

All concentrations are expressed in units of mol dm^{-3}

Soln.	Initial pH	Final pH	$[Mg^{2+}]$	pMg	log $\alpha_{L(H)}$	$p[SeO_3^{2-}]$	pK_{s0}
HCl	3.12	8.60	3.59×10^{-3}	2.45	0.12	2.57	5.02
	2.79	8.11	5.56×10^{-3}	2.25	0.25	2.50	4.75
	2.01	6.41	2.08×10^{-2}	1.68	1.60	3.28	4.96
HNO_3	2.88	8.20	5.28×10^{-3}	2.28	0.24	2.52	4.80
	1.98	5.98	3.48×10^{-2}	1.46	2.00	3.46	4.92
	2.42	7.26	9.14×10^{-3}	2.04	0.80	2.84	4.88

The average value is $K_{s0} = 1.29 \times 10^{-5}$ $mol^2 dm^{-6}$.
($pK_{s0} = 4.89$)

Notes.

$[Se_{tot}] = [Mg^{2+}]$ and $[SeO_3^{2-}] = [Se_{tot}]/\alpha_{L(H)}$
where $\alpha_{L(H)} = (1 + [H^+]/K_2 + [H^+]^2/K_1K_2)$
and the acid dissociation constants have the values $K_1 = 4 \times 10^{-3}$ and $K_2 = 1.0 \times 10^{-8}$ (ref. 1).

AUXILIARY INFORMATION

METHOD APPARATUS/PROCEDURE:

Solutions of hydrochloric and nitric acids were saturated with magnesium selenite by stirring in a thermostat at 20°C for 8 hr. The remaining solid phase was removed by centrifugation, then the pH was measured ("Moskip" pH meter, to 0.01 pH unit) and the magnesium concentration was measured (method not stated).

SOURCE AND PURITY OF MATERIALS:

C.P.-grade reagents were used.
Crystalline magnesium selenite was precipitated when 0.5*N* magnesium chloride was mixed with stoichiometric amounts of sodium selenite. The precipitate was washed with water and dried at 40°C. Magnesium was determined gravimetrically as the pyrophosphate, and selenium gravimetrically as the element.

ESTIMATED ERROR:

The spread in the results is 0.27 of a log unit.
Temperature: probably ±0.05 K

REFERENCES:

1. Rumpf, P. *Compt. Rendu* 1933, *197*, 686.

COMPONENTS:	ORIGINAL MEASUREMENTS:
1. Calcium selenite; $CaSeO_3$; [13780-18-2] 2. Water; H_2O; [7732-18-5]	Ripan, R.; Vericeanu, G. *Studia Univ. Babes-Bolyai, Ser. Chim.* 1968, *13*, 31-37.
VARIABLES:	PREPARED BY:
One temperature: 291 K	Mary R. Masson

EXPERIMENTAL VALUES:

All concentrations are expressed in units of mol dm^{-3}

Concentration	K_{s0} mol dm^{-6}	Mean K_{s0}	pK_{s0}
1.540×10^{-3}	2.37×10^{-6}	$1.8 \pm 1.1 \times 10^{-6}$ $mol^2 dm^{-6}$	5.74
1.557×10^{-3}	2.42×10^{-6}		
1.581×10^{-3}	2.50×10^{-6}		
1.095×10^{-3}	1.20×10^{-6}		
1.127×10^{-3}	1.26×10^{-6}		
1.052×10^{-3}	1.10×10^{-6}		

The concentration c in the saturated solution was calculated from the measured conductivity κ from the equation

$$c = \frac{1000\kappa}{\Lambda^\circ}$$

Compiler's note

Neither in the determination of the ionic conductivity of the selenite ion nor in the evaluation of the solubility product was hydrolysis of the selenite ion taken into account. This would give rise to errors, since, for example, in a 0.001*M* solution, $[SeO_3^{2-}] = 0.000955M$, $[HSeO_3^-] = 0.000045M$ and $[OH^-] = 0.000045M$, and hydroxide and hydrogen selenite have different ionic conductivities from selenite. If the ionic conductivity of hydrogen selenite were known, the experimental results could have been interpreted correctly (cf. ref. 2), but this value does not seem to be available.

However, because the calibration and sample solutions had concentrations of about the same order of magnitude, the errors would cancel to some extent, but the K_{s0} value cannot be regarded as reliable.

AUXILIARY INFORMATION

METHOD APPARATUS/PROCEDURE:

The ionic conductivity of the selenite ion in water was determined by measuring the mean molar conductivities of a series of aqueous solutions of lithium selenite with concentrations below 2×10^{-3} mol dm^{-3}. The mean molar conductivity at infinite dilution was found by extrapolation to be $172 \pm 2.9\ \Omega^{-1} cm^2 mol^{-1}$ at 18°C. At that temperature, the ionic conductivity of the lithium ion is $33.4\ \Omega^{-1} cm^2 mol^{-1}$, so the ionic conductivity of the selenite ion is $105.2\ \Omega^{-1} cm^2 mol^{-1}$.

The well washed calcium selenite was mixed with 150 ml of water and agitated intermittently for 1 hr. From this, 10-20 ml was taken in the conductivity cell and the conductivity measured. The sampling and measurement were repeated until a constant value for the conductivity was obtained.

The value of $\Lambda^\circ = (\lambda_+ + \lambda_-)$, the molar conductivity, was calculated with $\lambda_- = 105.2$ and $\lambda_+ = 102\ \Omega^{-1} cm^2 mol^{-1}$.

SOURCE AND PURITY OF MATERIALS:

Lithium selenite was prepared from sublimed selenium dioxide by neutralization with lithium hydroxide. An approximately 0.05*M* solution was standardized gravimetrically, and this was used to prepare the various dilutions.

The composition of the selenite was checked by analysis.

ESTIMATED ERROR:

Temperature: ± 0.5 K
Error in K_{s0} (2*s*) = 1.1×10^{-6} (compiler)

REFERENCES:

1. Landolt-Bornstein *Physicalisch-Chemische Tabellen II* 1923, p. 1105.
2. Monk, C.B. *J. Chem. Soc.* 1949, 429.

COMPONENTS:	ORIGINAL MEASUREMENTS:
1. Calcium selenite; $CaSeO_3$; [13780-18-2] 2a. Hydrochloric acid; HCl; [7647-01-0] 2b. Nitric acid; HNO_3; [7697-37-2] 3. Water; H_2O; [7732-18-5]	Chukhlantsev, V.G. *Zh. Neorg. Khim.* 1956, *1*, 2300-5; **Russ. J. Inorg. Chem. (Eng. Transl.)* 1956, *1*, 132-8.
VARIABLES:	PREPARED BY:
HCl and HNO_3 concentrations One temperature: 293 K	Mary R. Masson

EXPERIMENTAL VALUES:

All concentrations are expressed in units of mol dm^{-3}

Soln.	Initial pH	Final pH	$[Ca^{2+}]$	pCa	$\log \alpha_{L(H)}$	$p[SeO_3^{2-}]$	pK_{s0}
HCl	3.12	8.39	1.72×10^{-3}	2.77	0.17	2.94	5.71
	2.79	7.96	3.1×10^{-3}	2.51	0.30	2.81	5.32
	2.01	5.89	1.92×10^{-2}	1.72	2.10	3.82	5.54
HNO_3	2.88	7.87	2.54×10^{-3}	2.60	0.38	2.98	5.58
	1.98	5.56	3.14×10^{-2}	1.50	2.40	3.90	5.40
	2.42	6.64	7.09×10^{-2}	2.15	1.35	3.50	5.65

The average value for K_{s0} is 2.96×10^{-6} mol^2dm^{-6}.

(pK_{s0} = 5.53)

Notes.

$[Se_{tot}] = [Ca^{2+}]$ and $[SeO_3^{2-}] = [Se_{tot}]/\alpha_{L(H)}$

where $\alpha_{L(H)} = (1 + [H^+]/K_2 + [H^+]^2/K_2K_1)$

and the acid dissociation constants have the values $K_1 = 4 \times 10^{-3}$ and $K_2 = 1.0 \times 10^{-8}$ (ref. 1).

AUXILIARY INFORMATION

METHOD APPARATUS/PROCEDURE:

Solutions of hydrochloric and nitric acids were saturated with calcium selenite by stirring in a thermostat at 20°C for 8 hr. The remaining solid phase was removed by centrifugation, then the pH was measured ("Moskip" pH meter, to 0.01 pH unit) and the calcium concentration was measured (method not stated).

SOURCE AND PURITY OF MATERIALS:

C.P.-grade reagents were used. Calcium selenite was precipitated when 0.1*N* calcium chloride was mixed with a stoichiometric amount of sodium selenite, at 50 - 60°C. The precipitate was washed with water and dried at 40°C. Calcium was determined by precipitating the oxalate, then titrating with permanganate, and selenium gravimetrically as the element, after precipitation with hydrazine.

ESTIMATED ERROR:

The spread in the results is 0.31 of a log unit.

Temperature: probably ±0.05 K

REFERENCES:

1. Rumpf, P. *Compt. Rendu* 1933, *197*, 686.

COMPONENTS:	ORIGINAL MEASUREMENTS:
1. Strontium selenite; $SrSeO_3$; [14590-38-6] 2. Water; H_2O; [7732-18-5]	Ripan, R.; Vericeanu, G. *Studia Univ. Babes-Bolyai, Ser. Chim.* 1968, *13*, 31-37.
VARIABLES:	PREPARED BY:
One temperature: 291 K	Mary R. Masson

EXPERIMENTAL VALUES:

All concentrations are expressed in units of mol dm^{-3}

Concentration	K_{s0} mol^2 dm^{-6}	Mean K_{s0}	pK_{s0}
2.832×10^{-3}	8.0×10^{-6}	$7.9 \pm 0.8 \times 10^{-6}$ mol^2 dm^{-6}	5.10
2.915×10^{-3}	8.5×10^{-6}		
2.893×10^{-3}	8.4×10^{-6}		
2.770×10^{-3}	7.7×10^{-6}		
2.746×10^{-3}	7.5×10^{-6}		
2.743×10^{-3}	7.5×10^{-6}		

The concentration *c* in the saturated solution was calculated from the measured conductivity κ from the equation

$$c = \frac{1000\kappa}{\Lambda^\circ}$$

Compiler's note

Neither in the determination of the ionic conductivity of the selenite ion nor in the evaluation of the solubility product was hydrolysis of the selenite ion taken into account. This would give rise to errors, since, for example, in a 0.001*M* solution, $[SeO_3^{2-}] = 0.000955M$, $[HSeO_3^-] = 0.000045M$ and $[OH^-] = 0.000045M$, and hydroxide and hydrogen selenite have different ionic conductivities from selenite. If the ionic conductivity of hydrogen selenite were known, the experimental results could have been interpreted correctly (cf. ref. 2), but this value does not seem to be available.

However, because the calibration and sample solutions had concentrations of about the same order of magnitude, the errors would cancel to some extent, but the K_{s0} value cannot be regarded as reliable.

AUXILIARY INFORMATION

METHOD APPARATUS/PROCEDURE:

The ionic conductivity of the selenite ion in water was determined by measuring the mean molar conductivities of a series of aqueous solutions of lithium selenite with concentrations below 2×10^{-3} mol dm^{-3}. The mean molar conductivity at infinite dilution was found by extrapolation to be $172 \pm 2.9\ \Omega^{-1} cm^2 mol^{-1}$ at 18°C. At that temperature, the ionic conductivity of the lithium ion is $33.4\ \Omega^{-1} cm^2 mol^{-1}$, so the ionic conductivity of the selenite ion is $105.2\ \Omega^{-1} cm^2 mol^{-1}$.

The well washed strontium selenite was mixed with 150 ml of water and agitated intermittently for 1 hr. From this, 10-20 ml was taken in the conductivity cell and the conductivity measured. The sampling and measurement were repeated until a constant value for the conductivity was obtained.

The value of $\Lambda^\circ = (\lambda_+ + \lambda_-)$, the molar conductivity, was calculated with $\lambda_- = 105.2$ and $\lambda_+ = 102\ \Omega^{-1} cm^2 mol^{-1}$.

SOURCE AND PURITY OF MATERIALS:

Lithium selenite was prepared from sublimed selenium dioxide by neutralization with lithium hydroxide. An approximately 0.05*M* solution was standardized gravimetrically, and this was used to prepare the various dilutions.

The composition of the selenite was checked by analysis.

ESTIMATED ERROR:

Temperature: ± 0.5 K

Error in K_{s0} (2*s*) = 0.8×10^{-6} (compiler)

REFERENCES:

1. Landolt-Bornstein *Physicalisch-Chemische Tabellen II* 1923, p. 1105.
2. Monk, C.B. *J. Chem. Soc.* 1949, 429.

COMPONENTS:

1. Strontium selenite; $SrSeO_3$; [14590-38-6]
2a. Hydrochloric acid; HCl; [7647-01-0]
2b. Nitric acid; HNO_3; [7697-37-2]
3. Water; H_2O; [7732-18-5]

ORIGINAL MEASUREMENTS:

Chukhlantsev, V.G.

Zh. Neorg. Khim. 1956, *1*, 2300-5; **Russ. J. Inorg. Chem. (Eng. Transl.)* 1956, *1*, 132-8.

VARIABLES:

HCl and HNO_3 concentrations

One temperature: 293 K

PREPARED BY:

Mary R. Masson

EXPERIMENTAL VALUES:

All concentrations are expressed in units of mol dm^{-3}

Soln.	Initial pH	Final pH	$[Sr^{2+}]$	pSr	$\log \alpha_{L(H)}$	$p[SeO_3^{2-}]$	pK_{s0}
HCl	3.12	8.49	1.18×10^{-3}	2.93	0.15	3.08	6.01
	2.79	7.76	2.61×10^{-3}	2.59	0.40	2.99	5.58
	2.01	5.87	1.6×10^{-2}	1.80	2.15	3.85	5.65
HNO_3	1.98	5.69	1.84×10^{-2}	1.74	2.30	4.03	5.77
	2.88	7.97	1.89×10^{-3}	2.72	0.30	3.02	5.74
	2.45	6.79	5.89×10^{-3}	2.23	1.21	3.44	5.67

The average value is $K_{s0} = 1.82 \times 10^{-6}$ mol^2dm^{-6}.

($pK_{s0} = 5.74$)

Notes.

$[Se_{tot}] = [Sr^{2+}]$ and $[SeO_3^{2-}] = [Se_{tot}]/\alpha_{L(H)}$

where $\alpha_{L(H)} = (1 + [H^+]/K_2 + [H^+]^2/K_1K_2)$

and the acid dissociation constants have the values $K_1 = 4 \times 10^{-3}$ and $K_2 = 1.0 \times 10^{-8}$ (ref. 1).

AUXILIARY INFORMATION

METHOD APPARATUS/PROCEDURE:

Solutions of hydrochloric and nitric acids were saturated with strontium selenite by stirring in a thermostat at 20°C for 8 hr. The remaining solid phase was removed by centrifugation, then the pH was measured ("Moskip" pH meter, to 0.01 pH unit) and the strontium concentration was measured (method not stated).

SOURCE AND PURITY OF MATERIALS:

C.P.-grade reagents were used. Strontium chloride (0.1*N*) was mixed with 0.1*N* sodium selenite to precipitate crystalline strontium selenite, which was washed with water and dried at 40°C. Strontium was determined by precipitating the oxalate, then titrating with permanganate, and selenium was determined gravimetrically.

ESTIMATED ERROR:

The spread in the results is 0.43 of a log unit.

Temperature: probably ±0.05 K

REFERENCES:

1. Rumpf, P. *Compt. Rendu* 1933, *197*, 686.

COMPONENTS:

1. Strontium selenite; $SrSeO_3$; [14590-38-6]
2. Water; H_2O; [7732-18-5]

ORIGINAL MEASUREMENTS:

Chukhlantsev, V.G.

Zh. Neorg. Khim. 1956, *1*, 2300-5; **Russ. J. Inorg. Chem. (Eng. Transl.)* 1956, *1*, 132-8.

VARIABLES:

One temperature: 293 K

PREPARED BY:

Mary R. Masson

EXPERIMENTAL VALUES:

All concentrations are expressed in units of mol dm^{-3}

Number	A	V	$[Sr^{2+}]$	$[Se_{tot}]$	K_{s0}, $mol^2\ dm^{-6}$
1	289	1	9.57×10^{-4}	9.57×10^{-4}	9.1×10^{-7}
2	270	1	8.93×10^{-4}	8.93×10^{-4}	8.0×10^{-7}
3	277	1	9.17×10^{-4}	9.17×10^{-4}	8.4×10^{-7}

The average value is $K_{s0} = 8.5 \times 10^{-7}$ $mol^2\ dm^{-6}$. ($pK_{s0} = 6.07$).

The author has neglected to allow for hydrolysis of the selenite ion in this part of the experiment.

COMMENTS AND/OR ADDITIONAL DATA

The compiler has recalculated the results to take account of the hydrolysis of the selenite ions: the computer program HALTAFALL (2) was used. The values for the dissociation constants were from ref. (3).

Number	$[Sr^{2+}]$	$[Se_{tot}]$	$[SeO_3^{2-}]$	K_{s0}, $mol^2\ dm^{-6}$	Mean K_{s0}, $mol^2\ dm^{-6}$
1	9.57×10^{-4}	9.57×10^{-4}	9.27×10^{-4}	8.9×10^{-7}	
2	8.93×10^{-4}	8.93×10^{-4}	8.64×10^{-4}	7.7×10^{-7}	
3	9.17×10^{-4}	9.17×10^{-4}	8.87×10^{-4}	8.1×10^{-7}	8.2×10^{-7}

AUXILIARY INFORMATION

METHOD APPARATUS/PROCEDURE:

The radioactive isotope ^{89}Sr was used. Radioactivity was measured on a "B" counter. The results represent the average values of four determinations (two duplicates in two positions), and the values were corrected for background radiation. The experimental procedure is described in (1).
Calculations were carried out using the formula

$$L_i = \frac{A}{I_{sp} \cdot VM}$$

where A is the radioactivity of the sample in cpm, I_{sp} is the specific radioactivity of a standard solution in cpm/mg Sr, V is the volume in ml of the saturated radioactive solution, M is the molecular weight of strontium, and L_i is the strontium concentration.

[Compiler: presumably, the procedure was to prepare a sample of strontium selenite containing ^{89}Sr, and then to make a saturated aqueous solution of this solid, and determine its radioactivity as described above.]

SOURCE AND PURITY OF MATERIALS:

ESTIMATED ERROR:

The spread in the results is 0.06 of a log unit.
Temperature: probably ± 0.05 K

REFERENCES:

1. Spitsyn, V.I.; Kodochogov, P.N.; *et al.*, *Methods Involving Use of Radioactive Tracers*, p. 234, Izd. AN SSSR, Moskva, 1955.
2. Ingri, N.; Kakolowicz, W.; Sillén, L.G.; Warnqvist, B. *Talanta* 1967, *14*, 1261.
3. Rumpf. P. *Compt. Rendu* 1933, *197*, 686.

COMPONENTS:	ORIGINAL MEASUREMENTS:
1. Strontium selenite; $SrSeO_3$; [14590-38-6] 2. Water; H_2O; [7732-18-5]	Leschinskaya, Z.I.; Averbukh, M.A.; Selivanova, N.M. *Zh. Fiz. Khim.* 1965, *39*, 2036-8; *Russ. J. Phys. Chem.* 1965, *39*, 1082-3.
VARIABLES:	PREPARED BY:
One temperature: 298 K	Mary R. Masson

EXPERIMENTAL VALUES:

All concentrations are expressed in units of mol dm^{-3} at 25°C.

Time of mixing (days) at 25°C	Concentration of Sr	K_{s0}, (authors) $mol^2\ dm^{-6}$	Concentration of SeO_3^{2-}	K_{s0} (compiler) $mol^2\ dm^{-6}$
3	1.4×10^{-3}		1.35×10^{-3}	
5	1.8×10^{-3}		1.74×10^{-3}	
7	2.07×10^{-3}		2.01×10^{-3}	
10	2.09×10^{-3}		2.03×10^{-3}	
12	2.11×10^{-3}		2.05×10^{-3}	
Mean	2.09×10^{-3}	4.37×10^{-6}	2.03×10^{-3}	4.23×10^{-6}
		$pK_{s0} = 5.36$		$pK_{s0} = 5.37$

In their calculations, the authors omitted to allow for hydrolysis of the selenite ion. The compiler recalculated the results to take account of this, making use of the program HALTAFALL (1). The dissociation constant values were those of Hagisawa (2), namely $K_{a1} = 10^{-2.62}$ mol dm^{-3} and $K_{a2} = 10^{-8.32}$ mol dm^{-3}. For this system, the neglect of hydrolysis did not have a great influence on the results.

AUXILIARY INFORMATION

METHOD APPARATUS/PROCEDURE:

Solid strontium selenite was equilibrated with water for 10 -12 days, in a thermostat at 25°C. The saturated solution was filtered through a glass filter, then the concn. of strontium was determined by flame spectrophotometry, with an experimental error of 1.0 - 1.5% of the quantity measured.

SOURCE AND PURITY OF MATERIALS:

Strontium selenite was made by the reactions of aqueous solutions of strontium chloride and sodium selenite in stoichiometric proportions at 25°C.

ESTIMATED ERROR:

Error in temperature probably ±0.1 K (cf. ref. 3).
Error in K_{s0} (2*s*) = $\pm 0.20 \times 10^{-6}$ (based on results for days 7 - 12).

REFERENCES:

1. Ingri, N.; Kakolowicz, W.; Sillén, L.G.; Warnqvist, B. *Talanta* 1967, *14*, 1261.
2. Hagisawa, H. *Bull. Inst. Phys. Chem. Res., Tokyo* 1939, *18*, 648.
3. Selivanova, N.M.; Leschinskaya, Z.I.; Klushina, T.V. *Zh. Fiz. Khim.* 1962, *36*, 1349.

COMPONENTS:

1. Barium selenite; $BaSeO_3$; [13718-59-7]
2. Water; H_2O; [7732-18-5]

ORIGINAL MEASUREMENTS:

Ripan, R.; Vericeanu, G.

Studia Univ. Babes-Bolyai, Ser. Chim. 1968, *13*, 31-37.

VARIABLES:

One temperature: 291 K

PREPARED BY:

Mary R. Masson

EXPERIMENTAL VALUES:

All concentrations are expressed in units of mol dm^{-3}.

Concentration	K_{s0} $mol^2\ dm^{-6}$	Mean K_{s0}	pK_{s0}
7.661×10^{-4}	5.3×10^{-7}	$6.2 \pm 1.2 \times 10^{-6}$	6.21
7.711×10^{-4}	6.0×10^{-7}		
7.612×10^{-4}	5.8×10^{-7}	$mol^2\ dm^{-6}$	
8.086×10^{-4}	6.5×10^{-7}		
8.230×10^{-4}	6.8×10^{-7}		
8.190×10^{-4}	6.7×10^{-7}		

The concentration c in the saturated solution was calculated from the measured conductivity κ from the equation

$$c = \frac{1000\kappa}{\Lambda^\circ}$$

Compiler's note

Neither in the determination of the ionic conductivity of the selenite ion nor in the evaluation of the solubility product was hydrolysis of the selenite ion taken into account. This would give rise to errors, since, for example, in a 0.001*M* solution, $[SeO_3^{2-}] = 0.000955M$, $[HSeO_3^-] = 0.000045M$ and $[OH^-] = 0.000045M$, and hydroxide and hydrogen selenite have different ionic conductivities from selenite. If the ionic conductivity of hydrogen selenite were known, the experimental results could have been interpreted correctly (cf. ref. 2), but this value does not seem to be available.

However, because the calibration and sample solutions had concentrations of about the same order of magnitude, the errors would cancel to some extent, but the K_{s0} value cannot be regarded as reliable.

AUXILIARY INFORMATION

METHOD APPARATUS/PROCEDURE:

The ionic conductivity of the selenite ion in water was determined by measuring the mean molar conductivities of a series of aqueous solutions of lithium selenite with concentrations below 2×10^{-3} mol dm^{-3}. The mean molar conductivity at infinite dilution was found by extrapolation to be $172 \pm 2.9\ \Omega^{-1} cm^2 mol^{-1}$ at 18°C. At that temperature, the ionic conductivity of the lithium ion is $33.4\ \Omega^{-1} cm^2 mol^{-1}$, so the ionic conductivity of the selenite ion is $105.2\ \Omega^{-1} cm^2 mol^{-1}$.

The well washed barium selenite was mixed with 150 ml of water and agitated intermittently for 1 hr. From this, 10-20 ml was taken in the conductivity cell and the conductivity measured. The sampling and measurement were repeated until a constant value for the conductivity was obtained.

The value of $\Lambda^\circ = (\lambda_+ + \lambda_-)$, the molar conductivity, was calculated with $\lambda_- = 105.2$ and $\lambda_+ = 110\ \Omega^{-1} cm^2 mol^{-1}$.

SOURCE AND PURITY OF MATERIALS:

Lithium selenite was prepared from sublimed selenium dioxide by neutralization with lithium hydroxide. An approximately 0.05*M* solution was standardized gravimetrically, and this was used to prepare the various dilutions.

The composition of the selenite was checked by analysis.

ESTIMATED ERROR:

Temperature: ± 0.5 K

Error in K_{s0} $(2s) = 1.2 \times 10^{-6}$ (compiler)

REFERENCES:

1. Landolt-Bornstein *Physicalisch-Chemische Tabellen II* 1923, p. 1105.
2. Monk, C.B. *J. Chem. Soc.* 1949, 429.

COMPONENTS:

1. Barium selenite; $BaSeO_3$; [13718-59-7]
2. Water; H_2O; [7732-18-5]

ORIGINAL MEASUREMENTS:

Leschinskaya, Z.I.; Averbukh, M.A.; Selivanova, N.M.

Zh. Fiz. Khim. 1965, *39*, 2036-8; *Russ. J. Phys. Chem.* 1965, *39*, 1082-3.

VARIABLES:

One temperature: 298 K

PREPARED BY:

Mary R. Masson

EXPERIMENTAL VALUES:

All concentrations are expressed in units of mol dm^{-3} at 25°C.

Time of mixing (days) at 25°C	Concentration of Ba^{2+}	K_{s0}, (authors) $mol^2\ dm^{-6}$	Concentration of SeO_3^{2-}	K_{s0} (compiler) $mol^2\ dm^{-6}$
3				
5	5.82×10^{-4}		5.48×10^{-4}	
9	6.06×10^{-4}		5.71×10^{-4}	
12	6.25×10^{-4}		5.90×10^{-4}	
15	6.53×10^{-4}		6.17×10^{-4}	
21	6.59×10^{-4}		6.23×10^{-4}	
Mean	6.56×10^{-4}	4.30×10^{-7}	6.20×10^{-4}	4.07×10^{-7}
		$pK_{s0} = 6.37$		$pK_{s0} = 6.39$

In their calculations, the authors omitted to allow for hydrolysis of the selenite ion. The compiler recalculated the results to take account of this, making use of the program HALTAFALL (1). The dissociation constant values were those of Hagisawa (2), namely $K_{a1} = 10^{-2.62}$ mol dm^{-3} and $K_{a2} = 10^{-8.32}$ mol dm^{-3}. For this system, the neglect of hydrolysis did not have a great influence on the results.

AUXILIARY INFORMATION

METHOD APPARATUS/PROCEDURE:

Solid barium selenite was equilibrated with water for 18 -20 days, in a thermostat at 25°C. The saturated solution was filtered through a glass filter, then the concn. of barium was determined by flame spectrophotometry, with an experimental error of 1.0 - 1.5% of the quantity measured.

SOURCE AND PURITY OF MATERIALS:

Barium selenite was prepared by the reaction of stoichiometric amounts of barium chloride and sodium selenite at 25°C.

ESTIMATED ERROR:

Error in temperature probably ±0.1 K (cf. ref. 3).
Error in K_{s0} (2*s*) = $\pm 0.09 \times 10^{-7}$ (based on results for days 15 - 21).

REFERENCES:

1. Ingri, N.; Kakolowicz, W.; Sillén, L.G.; Warnqvist, B. *Talanta* 1967, *14*, 1261.
2. Hagisawa, H. *Bull. Inst. Phys. Chem. Res., Tokyo* 1939, *18*, 648.
3. Selivanova, N.M.; Leschinskaya, Z.I.; Klushina, T.V. *Zh. Fiz. Khim.* 1962, *36*, 1349.

COMPONENTS:

1. Barium selenite; $BaSeO_3$; [13718-59-7]
2. Water; H_2O; [7732-18-5]

ORIGINAL MEASUREMENTS:

Dolique, R.

Bull. Soc. Chim. France 1943, *10*, M50.

VARIABLES:

One temperature: 283 K

PREPARED BY:

Mary R. Masson

EXPERIMENTAL VALUES:

The author found the solubility of barium selenite in water to be 9.53 mg/100 ml after 9 days, 13.75 mg/100 ml after 26 days, and 17.63 mg after 39 days.

The concentration after 39 days can be expressed as 6.67×10^{-4} mol dm^{-3}; this would give a value of 4.45×10^{-7} mol^2 dm^{-6} for K_{s0} (pK_{s0} = 6.35), if hydrolysis is neglected. The temperature of the determination was 10°C.

AUXILIARY INFORMATION

METHOD APPARATUS/PROCEDURE:

The barium selenite was agitated in water for the time given above. The concentration of selenium in solution was then determined by a method developed by Dolique, Perahia and Roca.

SOURCE AND PURITY OF MATERIALS:

Barium selenite was prepared by the traditional method of "double decomposition", thoroughly washed, but not dried.

ESTIMATED ERROR:

Temperature: ±1 K
Solubility: no estimate available.

REFERENCES:

COMPONENTS:	ORIGINAL MEASUREMENTS:
1. Barium selenite; $BaSeO_3$; [13718-59-7] 2. Selenium dioxide; SeO_2; [7446-08-4] 3. Water; H_2O; [7732-18-5]	Neal, J.L., Jr.; McCrosky, C.R. *J. Am. Chem. Soc.* 1938, *60*, 911-4.
VARIABLES:	PREPARED BY:
Concentrations of the components Three temperatures	Mary R. Masson

EXPERIMENTAL VALUES: Composition of equilibrium solutions

SeO_2 mass %	$BaSeO_3$ mass %	SeO_2[a] mol/kg	$BaSeO_3$[a] mol/kg	Solid[b] phase
Temperature = 0°C				
58.9	0.00	12.915	0.	A
58.8	1.24	13.261	0.117	A
58.7	2.51	13.638	0.245	A
58.7	3.51	13.999	0.351	A + P
58.6	3.54	13.949	0.354	A + P
58.6	3.53	13.946	0.353	A + P
58.7	3.51	13.999	0.351	A + P
57.8	3.42	13.432	0.334	P
52.8	2.77	10.710	0.236	P
49.7	2.44	9.359	0.193	P
44.3	1.99	7.433	0.140	P
35.6	1.44	5.096	0.087	P
30.6	1.18	4.042	0.065	P
26.6	1.10	3.316	0.058	P
17.9	0.89	1.986	0.041	P
8.80	0.69	0.876	0.029	P
4.57	0.58	0.434	0.023	P
2.04	0.58	0.189	0.023	P
1.11	0.65	0.102	0.025	P
0.28	0.64	0.025	0.024	P
0.27	0.64	0.025	0.024	P + N
0.17	0.39	0.015	0.015	N
0.00	0.005	0.	0.000	N

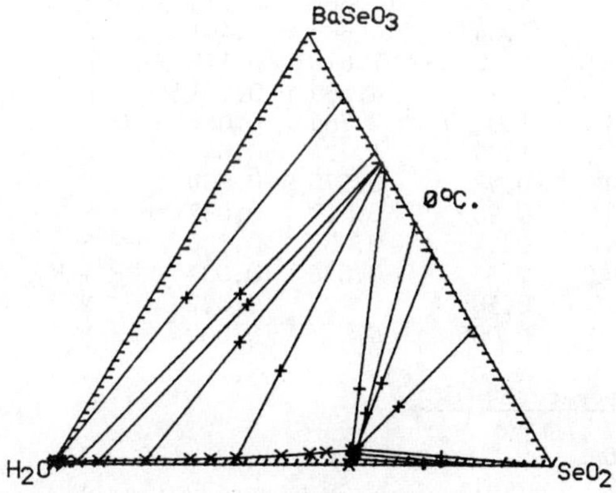

(continued on next page)

AUXILIARY INFORMATION

METHOD APPARATUS/PROCEDURE:

Weighed amounts of solids, slightly more than necessary to saturate the proposed amount of solution, were placed in a sample bottle and a measured volume of water was added. The sealed bottles, kept in a thermostat, were shaken by hand at frequent intervals. Samples were taken at intervals of from 3 - 14 days, depending on the rate of attainment of saturation, which was slowest for the most dilute solutions. Samples were removed by pipettes, through filters, transferred to tared 100-ml standard flasks, then weighed and diluted to the mark with water. In portions removed by pipette, barium was determined by precipitation as the sulfate, and total selenious acid by the method of Coleman and McCrosky (1). The remaining solid phase was examined by microscope or by X-ray diffraction to determine its nature.

SOURCE AND PURITY OF MATERIALS:

Barium selenite: barium carbonate was dissolved in nitric acid, and the salt was recrystallized several times. The purified nitrate was treated with a slight excess of selenious acid, then the selenite was precipitated by addition of ammonia, washed free of ammonia, and dried.
Selenious acid: selenium dioxide was prepared by oxidation of black selenium and purified by sublimation. This was dissolved in the minimum of hot water. Crystals separated on cooling and inoculation.

Barium pyroselenite: the normal selenite was digested with an excess of selenious acid at 50°C for several weeks to ensure complete conversion.

ESTIMATED ERROR:

Temperature was controlled to ±0.1 K
The maximum deviations from the mean compositions found were 0.1% in 3.5% for barium selenite and 0.25% in 75% for H_2SeO_3.

REFERENCES:

1. Coleman, W.C.; McCrosky, C.R. *Ind. Eng. Chem., Anal. Ed.* 1937, *9*, 431.

COMPONENTS:	ORIGINAL MEASUREMENTS:
1. Barium selenite; $BaSeO_3$; [13718-59-7]	Neal, J.L., Jr.; McCrosky, C.R.
2. Selenium dioxide; SeO_2; [7446-08-4]	*J. Am. Chem. Soc.* 1938, *60*, 911-4.
3. Water; H_2O; [7732-18-5]	

EXPERIMENTAL VALUES (continued):

Composition of equilibrium solutions

Temperature = 25°C

SeO_2 mass %	$BaSeO_3$ mass %	SeO_2[a] mol/kg	$BaSeO_3$[a] mol/kg	Solid[b] phase
68.8	0.00	19.873	0.	A
68.3	1.54	20.409	0.193	A
67.7	3.13	20.916	0.406	A
67.3	4.61	21.592	0.621	A
67.2	5.03	21.809	0.685	A + P
67.1	5.12	21.768	0.697	A + P
67.2	5.04	21.816	0.687	A + P
63.6	4.36	17.889	0.515	P
59.9	3.84	14.888	0.401	P
44.7	2.42	7.618	0.173	P
33.2	1.74	4.599	0.101	P
23.3	1.33	2.786	0.067	P
13.5	1.09	1.424	0.048	P
5.94	0.97	0.575	0.039	P
1.27	0.96	0.117	0.037	P
0.41	0.95	0.037	0.036	P
0.42	0.97	0.038	0.037	P + N
0.25	0.58	0.023	0.022	N
0.00	0.005	0.	0.000	N

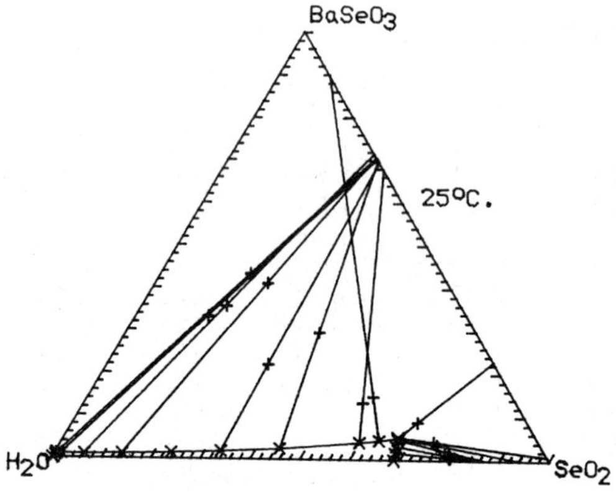

Temperature = 50°C

SeO_2 mass %	$BaSeO_3$ mass %	SeO_2[a] mol/kg	$BaSeO_3$[a] mol/kg	Solid[b] phase
76.6	0.00	29.502	0.	A
76.2	1.51	30.809	0.256	A
75.4	2.75	31.100	0.476	A
74.8	3.66	31.296	0.643	A
73.8	5.80	32.603	1.076	A
73.5	6.67	33.404	1.273	A + P
73.4	6.59	33.058	1.246	A + P
71.8	6.42	29.710	1.115	P
68.3	5.32	23.333	0.763	P
57.8	3.90	13.601	0.385	P
44.7	2.81	7.675	0.203	P
37.1	2.38	5.525	0.149	P
24.7	1.80	3.029	0.093	P
13.7	1.45	1.455	0.065	P
4.87	1.28	0.468	0.052	P
1.34	1.16	0.124	0.045	P
0.51	1.18	0.047	0.045	P
0.52	1.18	0.048	0.045	P + N
0.29	0.67	0.026	0.026	N
0.00	0.005	0.	0.000	N

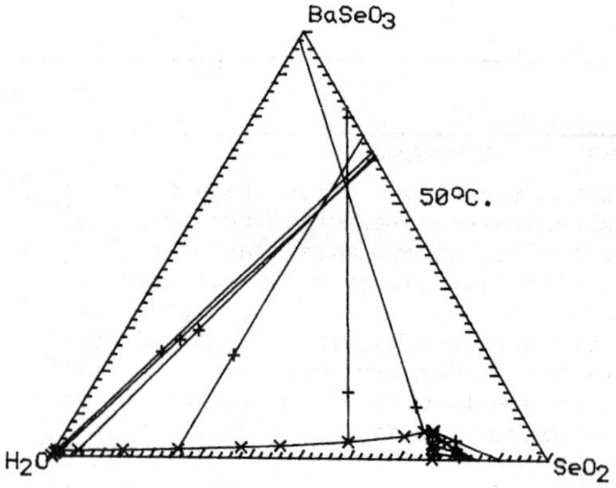

[a] Molalities calculated by the compiler.

[b] Solid phases: A - H_2SeO_3, P - $BaSe_2O_5$, N - $BaSeO_3$

COMPONENTS:

1. Manganese(II) selenite; $MnSeO_3$; [15702-34-8]
2a. Sulfuric acid; H_2SO_4; [7664-93-9]
2b. Hydrochloric acid; HCl; [7647-01-0]
3. Water; H_2O; [7732-18-5]

ORIGINAL MEASUREMENTS:

Chukhlantsev, V.G.; Tomashevsky, G.P.

Zh. Anal. Khim. 1957, *12*, 296-301;
**J. Anal. Chem. USSR* 1957, *12*, 303-9.

VARIABLES:

Sulfuric and hydrochloric acid concentrations
One temperature: 293 K

PREPARED BY:

Mary R. Masson

EXPERIMENTAL VALUES:

All concentrations are expressed in units of mol dm^{-3}.

Soln.	Initial pH	Final pH	$[Mn^{2+}]$	pMn	log $\alpha_{L(H)}$	$p[SeO_3^{2-}]$	pK_{s0}
H_2SO_4	2.69	6.07	3.1×10^{-3}	2.51	1.94	4.45	6.96
	2.06	4.35	2.8×10^{-2}	1.55	3.63	5.18	6.73
HCl	2.79	6.24	2.2×10^{-3}	2.66	1.77	4.43	7.09
	2.41	5.37	6.8×10^{-3}	2.17	2.60	4.77	6.94
	2.05	4.42	2.1×10^{-2}	1.68	3.58	5.26	6.94

The average value is $K_{s0} = 1.2 \times 10^{-7}$ mol^2dm^{-6}.
($pK_{s0} = 6.9$)

Notes.

$[Se_{tot}] = [Mn^{2+}]$ and $[SeO_3^{2-}] = [Se_{tot}]/\alpha_{L(H)}$
where $\alpha_{L(H)} = (1 + [H^+]/K_2 + [H^+]^2/K_1K_2)$
and the acid dissociation constants have the values $K_1 = 4 \times 10^{-3}$ and $K_2 = 1.0 \times 10^{-8}$ (ref. 1).

AUXILIARY INFORMATION

METHOD APPARATUS/PROCEDURE:

Solutions of sulfuric and hydrochloric acid were saturated with manganese(II) selenite by shaking in a thermostat at 20 ±0.05°C for 8 hr. The remaining solid phase was removed by centrifugation, then the pH was measured ("Moskip" pH meter, to 0.01 pH unit) and the manganese concentration was determined spectrophotometrically, after oxidation to Mn(VII) with ammonium persulfate.

SOURCE AND PURITY OF MATERIALS:

C.P.-grade reagents were used. Manganese selenite was prepared by mixing stoichiometric amounts of 0.2*N* manganese(II) sulfate and 0.2*N* sodium selenite, and heating the solution to 50 - 70°C. After prolonged standing, the precipitate was washed with water and dried at 40°C. Manganese was determined gravimetrically as the sulfate, and selenium as the element.

ESTIMATED ERROR:

$\pm 0.4 \times 10^{-7}$. (The spread in the results is 0.36 of a log unit.)
Temperature: ±0.05 K

REFERENCES:

1. Rumpf, P. *Compt. Rendu* 1933, *197*, 686.

COMPONENTS:

1. Iron(III) selenite; $Fe_2(SeO_3)_3$; [15857-44-0]

2a. Nitric acid; HNO_3; [7697-37-2]

2b. Sulfuric acid; H_2SO_4; [7664-93-9]

3. Water; H_2O; [7732-18-5]

ORIGINAL MEASUREMENTS:

Chukhlantsev, V.G.; Tomashevsky, G.P.

Zh. Anal. Khim. 1957, *12*, 296-301;
**J. Anal. Chem. USSR* 1957, *12*, 303-9.

VARIABLES:

Nitric and sulfuric acid concentrations
One temperature: 293 K

PREPARED BY:

Mary R. Masson

EXPERIMENTAL VALUES:

All concentrations are expressed in units of mol dm^{-3}.

Soln.	Initial pH	Final pH	$[Fe^{3+}]$	pFe	$\log \alpha_{L(H)}$	$p[SeO_3^{2-}]$	pK_{s0}
HNO_3	2.41	2.68	1.1×10^{-3}	2.96	5.50	8.28	30.76
	2.05	2.12	3.4×10^{-3}	2.47	6.33	8.61	30.77
H_2SO_4	2.74	2.97	8.2×10^{-4}	3.09	5.13	8.04	30.30
	2.13	2.22	2.6×10^{-3}	2.58	6.20	8.61	30.99
	2.08	2.17	2.9×10^{-3}	2.54	6.27	8.63	30.97

The average value is $K_{s0} = 2.0 \times 10^{-31}$ $mol^5 dm^{15}$.
($pK_{s0} = 30.7$)

Notes.

$[Se_{tot}] = 1.5 \times [Fe^{3+}]$ and $[SeO_3^{2-}] = [Se_{tot}]/\alpha_{L(H)}$
where $\alpha_{L(H)} = (1 + [H^+]/K_2 + [H^+]^2/K_1K_2)$
and the acid dissociation constants have the values $K_1 = 4 \times 10^{-3}$ and
$K_2 = 1.0 \times 10^{-8}$ (ref. 1).

AUXILIARY INFORMATION

METHOD APPARATUS/PROCEDURE:

Solutions of nitric and sulfuric acid were saturated with iron(III) selenite by shaking in a thermostat at 20 $\pm$0.05°C for 8 hr. The remaining solid phase was removed by centrifugation, then the pH was measured ("Moskip" pH meter, to 0.01 pH unit) and the iron(III) concentration was determined spectrophotometrically.

SOURCE AND PURITY OF MATERIALS:

C.P.-grade reagents were used. Stoichiometric amounts of 0.2*N* iron(III) sulfate and 0.1*N* sodium selenite were mixed: the pH of the solution was 5 - 6. After 24 hr, the precipitate was separated by centrifugation and decanting, and dried at 40°C. Iron was determined titrimetrically after reduction with zinc amalgam, and selenium gravimetrically as the element.

ESTIMATED ERROR:

$\pm 1.7 \times 10^{-31}$. (The spread in the results is 0.69 of a log unit.)
Temperature: $\pm$0.05 K

REFERENCES:

1. Rumpf, P. *Compt. Rendu* 1933, *197*, 686.

COMPONENTS:

1. Iron(III) selenite; $Fe_2(SeO_3)_3$; [15857-44-0]

2. Water; H_2O; [7732-18-5]

ORIGINAL MEASUREMENTS:

Pinaev, G.F.; Volkova, V.P.

Obshch. Prikl. Khim. 1970, 33-9.

VARIABLES:

One temperature (room)
Concentrations of the components

PREPARED BY:

Mary R. Masson

EXPERIMENTAL VALUES:

Concs. in mother liq., g/l. Fe	Se	$[Fe^{3+}]$ x 10^4 mol dm^{-3}	$[Se_{tot}]$ x 10^3 mol dm^{-3}	Final pH	$\alpha_{L(H)}$ x 10^{-7}	$[SeO_3^{2-}]$ x 10^{11} mol dm^{-3}	K_{s0} x 10^{36} mol^5 dm^{-15}
0.011	0.3	1.97	3.80	1.31	5.10	7.44	0.016
0.047	0.527	8.42	6.67	1.36	4.10	16.3	3.06
0.025	0.807	4.48	10.2	1.37	3.92	26.0	3.54
0.004	1.412	0.716	9.07	1.42	3.15	28.8	0.122
0.211	0.205	37.8	2.60	1.34	4.47	5.81	2.80
0.337	0.139	60.4	1.76	1.37	3.92	4.49	3.30
0.432	0.104	77.4	1.32	1.44	2.89	4.57	5.73

Mean K_{s0} = 3.7 x 10^{-36} mol^5 dm^{-15} (omitting first and fourth results)
(pK_{s0} = 35.43)

The concentrations in g/l. and the final pH values were given in the original paper. The other columns of data were calculated by the compiler.

Note: $[SeO_3^{2-}] = [Se_{tot}]/\alpha_{L(H)}$
where $\alpha_{L(H)} = (1 + [H^+]/K_2 + [H^+]^2/K_1K_2)$, with $K_1 = 10^{-2.35}$ and $K_2 = 10^{-7.94}$

AUXILIARY INFORMATION

METHOD APPARATUS/PROCEDURE:

This paper describes preparative work on iron(III) selenite, but some analytical data on the mother liquor in one experiment allowed solubility information to be extracted. Various volumes of ferric chloride (at pH 2.1) and selenious acid (pH 2.0) were mixed, and the pH was measured after the precipitate had formed. Both solutions had a constant ionic strength of 3.0 mol dm^{-3}. This was done at room temperature.

SOURCE AND PURITY OF MATERIALS:

ESTIMATED ERROR:

Error in K_{s0} (2*s*) = 2.5 x 10^{-36}
(0.4 log unit)

REFERENCES:

COMPONENTS:

1. Cobalt selenite; $CoSeO_3$; [10026-23-0]
2. Water; H_2O; [7732-18-5]

ORIGINAL MEASUREMENTS:

Ripan, R.; Vericeanu, G.

Studia Univ. Babes-Bolyai, Ser. Chim. 1968, *13*, 31-37.

VARIABLES:

One temperature: 291 K

PREPARED BY:

Mary R. Masson

EXPERIMENTAL VALUES:

All concentrations are expressed in units of mol dm^{-3}.

Concentration	K_{s0} / $mol^2\ dm^{-6}$	Mean K_{s0}	pK_{s0}
3.160×10^{-4}	0.999×10^{-7}	$1.2 \pm 0.2 \times 10^{-7}$ $mol^2\ dm^{-6}$	6.92
3.310×10^{-4}	1.095×10^{-7}		
3.425×10^{-4}	1.173×10^{-7}		
3.511×10^{-4}	1.233×10^{-7}		
3.564×10^{-4}	1.270×10^{-7}		
3.528×10^{-4}	1.245×10^{-7}		

The concentration c in the saturated solution was calculated from the measured conductivity κ from the equation

$$c = \frac{1000\kappa}{\Lambda^\circ}$$

Compiler's note

Neither in the determination of the ionic conductivity of the selenite ion nor in the evaluation of the solubility product was hydrolysis of the selenite ion taken into account. This would give rise to errors, since, for example, in a 0.001*M* solution, $[SeO_3^{2-}]$ = 0.000955*M*, $[HSeO_3^-]$ = 0.000045*M* and $[OH^-]$ = 0.000045*M*, and hydroxide and hydrogen selenite have different ionic conductivities from selenite. If the ionic conductivity of hydrogen selenite were known, the experimental results could have been interpreted correctly (cf. ref. 2), but this value does not seem to be available.

However, because the calibration and sample solutions had concentrations of about the same order of magnitude, the errors would cancel to some extent, but the K_{s0} value cannot be regarded as reliable.

AUXILIARY INFORMATION

METHOD APPARATUS/PROCEDURE:

The ionic conductivity of the selenite ion in water was determined by measuring the mean molar conductivities of a series of aqueous solutions of lithium selenite with concentrations below 2×10^{-3} mol dm^{-3}. The mean molar conductivity at infinite dilution was found by extrapolation to be $172 \pm 2.9\ \Omega^{-1}cm^2mol^{-1}$ at 18°C. At that temperature, the ionic conductivity of the lithium ion is $33.4\ \Omega^{-1}cm^2mol^{-1}$, so the ionic conductivity of the selenite ion is $105.2\ \Omega^{-1}cm^2mol^{-1}$.

The well washed cobalt selenite was mixed with 150 ml of water and agitated intermittently for 1 hr. From this, 10-20 ml was taken in the conductivity cell and the conductivity measured. The sampling and measurement were repeated until a constant value for the conductivity was obtained.

The value of $\Lambda^\circ = (\lambda_+ + \lambda_-)$, the molar conductivity, was calculated with $\lambda_- = 105.2$ and $\lambda_+ = 90\ \Omega^{-1}cm^2mol^{-1}$.

SOURCE AND PURITY OF MATERIALS:

Lithium selenite was prepared from sublimed selenium dioxide by neutralization with lithium hydroxide. An approximately 0.05*M* solution was standardized gravimetrically, and this was used to prepare the various dilutions.

The composition of the selenite was checked by analysis.

ESTIMATED ERROR:

Temperature: ± 0.5 K
Error in K_{s0} $(2s) = 0.2 \times 10^{-7}$ (compiler)

REFERENCES:

1. Landolt-Bornstein *Physicalisch-Chemische Tabellen II* 1923, p. 1105.
2. Monk, C.B. *J. Chem. Soc.* 1949, 429.

COMPONENTS:

1. Cobalt(II) selenite; $CoSeO_3$; [10026-23-0]

2a. Nitric acid; HNO_3; [7697-37-2]

2b. Sulfuric acid; H_2SO_4; [7664-93-9]

3. Water; H_2O; [7732-18-5]

ORIGINAL MEASUREMENTS:

Chukhlantsev, V.G.; Tomashevsky, G.P.

Zh. Anal. Khim. 1957, *12*, 296-301;
**J. Anal. Chem. USSR* 1957, *12*, 303-9.

VARIABLES:

Nitric and sulfuric acid concentrations
One temperature: 293 K

PREPARED BY:

Mary R. Masson

EXPERIMENTAL VALUES:

All concentrations are expressed in units of mol dm^{-3}.

Soln.	Initial pH	Final pH	$[Co^{2+}]$	pCo	$\log \alpha_{L(H)}$	$p[SeO_3^{2-}]$	pK_{s0}
HNO_3	2.41	5.69	7.8×10^{-3}	2.11	2.30	4.41	6.52
	2.06	4.20	3.5×10^{-2}	1.46	3.80	5.26	6.72
H_2SO_4	2.74	6.39	2.0×10^{-3}	2.70	1.60	4.30	7.00
	2.13	4.30	2.1×10^{-2}	1.68	3.70	5.38	7.06
	2.08	4.25	2.7×10^{-2}	1.57	3.75	5.32	6.89

The average value is $K_{s0} = 1.6 \times 10^{-7}$ $mol^2 dm^{-6}$.
($pK_{s0} = 6.8$)

Notes.

$[Se_{tot}] = [Co^{2+}]$ and $[SeO_3^{2-}] = [Se_{tot}]/\alpha_{L(H)}$
where $\alpha_{L(H)} = (1 + [H^+]/K_2 + [H^+]^2/K_1K_2)$
and the acid dissociation constants have the values $K_1 = 4 \times 10^{-3}$ and $K_2 = 1.0 \times 10^{-8}$ (ref. 1).

AUXILIARY INFORMATION

METHOD APPARATUS/PROCEDURE:

Solutions of nitric and sulfuric acid were saturated with cobalt selenite by shaking in a thermostat at 20±0.05°C for 8 hr. The remaining solid phase was removed by centrifugation, then the pH was measured ("Moskip" pH meter, to 0.01 pH unit) and the cobalt concentration was determined by gravimetry of the sulfate.

SOURCE AND PURITY OF MATERIALS:

C.P.-grade reagents were used. Cobalt selenite was prepared by adding 0.1*N* sodium selenite (a 3% excess) to 0.1*N* cobalt nitrate at 50 - 60°C. After prolonged standing, the precipitate was separated by centrifugation and decanting, then it was dried at 40°C. Cobalt was determined gravimetrically as the sulfate, and selenium as the element.

ESTIMATED ERROR:

$\pm 0.8 \times 10^{-7}$. (The spread in the results is 0.64 of a log unit.)
Temperature: ±0.0 K

REFERENCES:

1. Rumpf, P. *Compt. Rendu* 1933, *197*, 686.

COMPONENTS:

1. Cobalt selenite; $CoSeO_3$; [10026-23-0]
2. Water; H_2O; [7732-18-5]

ORIGINAL MEASUREMENTS:

Pyatnitskii, I.V.; Durdyev, M.

Ukr. Khim. Zh. 1966, *32*, 77-81. *(English translation, pp. 57-61.)

VARIABLES:

One temperature (room?)
pH varied

PREPARED BY:

Mary R. Masson

EXPERIMENTAL VALUES:

$[CoSeO_3]$ g/l.	pH	$[Co^{2+}]$ x 10^4 mol dm^{-3}	$\alpha_{L(H)}$	$[SeO_3^{2-}]$ x 10^5 mol dm^{-3}	K_{s0} x 10^8 mol^2 dm^{-6}
0.037	7.55	1.99	3.45	5.77	1.15
0.031	7.65	1.67	2.95	5.66	0.945
0.034	7.60	1.83	3.19	5.73	1.05

$$\text{Mean } K_{s0} = 1.05 \times 10^{-8} \text{ mol}^2 \text{ dm}^{-6}$$

$$pK_{s0} = 7.98$$

Columns 1 and 2 give data from the original paper; the other data were calculated by the compiler, since there appeared to be errors in the original calculations.

AUXILIARY INFORMATION

METHOD APPARATUS/PROCEDURE:

Small quantities of cobalt selenite were placed in a flask, water was added, and the flask was shaken until equilibrium was reached (2 hr). Cobalt in the filtrate was determined photometrically with nitroso-R-salt, and the pH of the solution was measured (LP-58 pH meter).

The solubility measurements appear to have been made at room temperature.

SOURCE AND PURITY OF MATERIALS:

Cobalt selenite dihydrate was prepared by adding a slight excess of sodium or ammonium selenite to a solution of cobalt sulfate. The precipitate was heated at 60°C for 1 hr, then filtered off after 3-4 hr. It was dried at 105-110°C.

ESTIMATED ERROR:

The spread in the K_{s0} results is 0.2×10^{-8}.

REFERENCES:

COMPONENTS:

1. Cobalt selenite; $CoSeO_3$; [10026-23-0]
2. Ammonium selenite; $(NH_4)_2SeO_3$; [7783-19-9]
3. Potassium nitrate; KNO_3; [7757-79-1]
4. Water; H_2O; [7732-18-5]

ORIGINAL MEASUREMENTS:

Pyatnitskii, I.V.; Durdyev, M.

Ukr. Khim. Zh. 1966, *32*, 77-81. (English translation, pp. 57-61.)

VARIABLES:

One temperature (room?)
pH and selenite concentration varied

PREPARED BY:

Mary R. Masson

EXPERIMENTAL VALUES:

All concentrations are expressed in units of mol dm^{-3}.

Solubility data for ionic strength 0.3 mol dm^{-3}

$[CoSeO_3]$ x 10^4	$[SeO_3^{2-}]$ x 10^2	K_{s0} x 10^6 mol^2 dm^{-6}
1.71	0.56	0.91
1.51	1.15	1.38
1.37	2.30	1.58
1.49	4.60	1.37
1.83	9.20	0.99
2.81	18.4	0.79

$[SeO_3^{2-}]$ for minimum solubility = 0.023

$K_{instab} = ([SeO_3^{2-}]_{min})^2$

$= 5.3 \times 10^{-4}$

Mean $K_{s0} = 1.17 \times 10^{-6}$ ($pK_{s0} = 5.93$)

Solubility data for ionic strength 0.01 mol dm^{-3}

pH	$[CoSeO_3]$ x 10^4	$[SeO_3^{2-}]$ x 10^3	K_{s0} x 10^7 mol^2 dm^{-6}
8.25	1.19	1.0	1.02
8.4	1.05	1.95	1.23
8.45	1.05	2.9	1.29
8.55	1.19	3.9	1.35
8.55	1.15	4.8	1.18
8.55	1.24	5.7	1.14
8.55	1.36	6.7	1.12
8.55	1.42	7.7	1.04
8.55	1.52	8.6	1.02
8.55	1.73	9.6	1.05

$[SeO_3^{2-}]$ for minimum solubility = 2.5×10^{-3}

$K_{instab} = 6.25 \times 10^{-6}$

Mean $K_{s0} = 1.14 \times 10^{-7}$

($pK_{s0} = 6.94$)

AUXILIARY INFORMATION

METHOD APPARATUS/PROCEDURE:

Solids and solutions were equilibrated in flasks with shaking for 2 hr. Cobalt in the filtrate was determined with nitroso-R-salt photometrically, and the pH was measured (LP-58 pH meter). The measurements appear to have been made at room temperature.

The solubility product values were calculated from the formula:

$$K_{s0} = \frac{[Co^{2+}].K_{instab}.[SeO_3^{2-}]}{K_{instab} + [SeO_3^{2-}]^2}$$

Compiler's note: these two sets of results do not appear to be consistent with each other or with the water solubility data. Minor corrections to the calculations made no significant difference, so the authors' original figures are reproduced. The differences appear rather large to be caused simply by the ionic-strength effect.

SOURCE AND PURITY OF MATERIALS:

Cobalt selenite dihydrate was prepared by adding a slight excess of sodium or ammonium selenite to a solution of cobalt sulfate. The precipitate was heated for 1 hr at 60°C, then filtered off after 3-4 hr. It was dried at 105-110°C.

ESTIMATED ERROR:

At $I = 0.3$, $s = 0.3 \times 10^{-6}$ for K_{s0}

At $I = 0.01$, $s = 0.12 \times 10^{-7}$ for K_{s0}

REFERENCES:

COMPONENTS:

1. Nickel selenite; $NiSeO_3$; [10101-96-9]
2. Water; H_2O; [7732-18-5]

ORIGINAL MEASUREMENTS:

Ripan, R.; Vericeanu, G.

Studia Univ. Babes-Bolyai, Ser. Chim. 1968, *13*, 31-37.

VARIABLES:

One temperature: 291 K

PREPARED BY:

Mary R. Masson

EXPERIMENTAL VALUES:

All concentrations are expressed in units of mol dm^{-3}.

Concentration	K_{s0} / $mol^2\ dm^{-6}$	Mean K_{s0}	pK_{s0}
2.040×10^{-3}	4.16×10^{-6}	$5.1 \pm 1.5 \times 10^{-6}$ $mol^2\ dm^{-6}$	5.29
2.132×10^{-3}	4.55×10^{-6}		
2.133×10^{-3}	4.56×10^{-6}		
2.450×10^{-3}	6.0×10^{-6}		
2.365×10^{-3}	5.6×10^{-6}		
2.385×10^{-3}	5.6×10^{-6}		

The concentration c in the saturated solution was calculated from the measured conductivity κ from the equation

$$c = \frac{1000\kappa}{\Lambda^\circ}$$

Compiler's note

Neither in the determination of the ionic conductivity of the selenite ion nor in the evaluation of the solubility product was hydrolysis of the selenite ion taken into account. This would give rise to errors, since, for example, in a 0.001*M* solution, $[SeO_3^{2-}] = 0.000955M$, $[HSeO_3^-] = 0.000045M$ and $[OH^-] = 0.000045M$, and hydroxide and hydrogen selenite have different ionic conductivities from selenite. If the ionic conductivity of hydrogen selenite were known, the experimental results could have been interpreted correctly (cf. ref. 2), but this value does not seem to be available.

However, because the calibration and sample solutions had concentrations of about the same order of magnitude, the errors would cancel to some extent, but the K_{s0} value cannot be regarded as reliable.

AUXILIARY INFORMATION

METHOD APPARATUS/PROCEDURE:

The ionic conductivity of the selenite ion in water was determined by measuring the mean molar conductivities of a series of aqueous solutions of lithium selenite with concentrations below 2×10^{-3} mol dm^{-3}. The mean molar conductivity at infinite dilution was found by extrapolation to be $172 \pm 2.9\ \Omega^{-1} cm^2 mol^{-1}$ at 18°C. At that temperature, the ionic conductivity of the lithium ion is $33.4\ \Omega^{-1} cm^2 mol^{-1}$, so the ionic conductivity of the selenite ion is $105.2\ \Omega^{-1} cm^2 mol^{-1}$.

The well washed nickel selenite was mixed with 150 ml of water and agitated intermittently for 1 hr. From this, 10-20 ml was taken in the conductivity cell and the conductivity measured. The sampling and measurement were repeated until a constant value for the conductivity was obtained.

The value of $\Lambda^\circ = (\lambda_+ + \lambda_-)$, the molar conductivity, was calculated with $\lambda_- = 105.2$ and $\lambda_+ = 90\ \Omega^{-1} cm^2 mol^{-1}$.

SOURCE AND PURITY OF MATERIALS:

Lithium selenite was prepared from sublimed selenium dioxide by neutralization with lithium hydroxide. An approximately 0.05*M* solution was standardized gravimetrically, and this was used to prepare the various dilutions.

The composition of the selenite was checked by analysis.

ESTIMATED ERROR:

Temperature: ± 0.5 K

Error in K_{s0} (2*s*) = 1.5×10^{-6} (compiler)

REFERENCES:

1. Landolt-Bornstein *Physicalisch-Chemische Tabellen II* 1923, p. 1105.
2. Monk, C.B. *J. Chem. Soc.* 1949, 429.

COMPONENTS:

1. Nickel selenite; $NiSeO_3$; [10101-96-9]
2a. Nitric acid; HNO_3; [7697-37-2]
2b. Sulfuric acid; H_2SO_4; [7664-93-9]
3. Water; H_2O; [7732-18-5]

ORIGINAL MEASUREMENTS:

Chukhlantsev, V.G.; Tomashevsky, G.P.

Zh. Anal. Khim. 1957, *12*, 296-301;
**J. Anal. Chem. USSR* 1957, *12*, 303-9.

VARIABLES:

Nitric and sulfuric acid concentrations
One temperature: 293 K

PREPARED BY:

Mary R. Masson

EXPERIMENTAL VALUES:

All concentrations are expressed in units of mol dm^{-3}.

Soln.	Initial pH	Final pH	$[Ni^{2+}]$	pNi	log $\alpha_{L(H)}$	$p[SeO_3^{2-}]$	pK_{s0}
HNO_3	2.41	7.12	8.8×10^{-3}	2.06	0.91	2.97	5.03
	2.05	6.20	2.6×10^{-2}	1.59	1.80	3.39	4.98
H_3SO_4	2.74	7.68	5.1×10^{-3}	2.29	0.50	2.79	5.08
	2.13	6.38	2.1×10^{-2}	1.67	1.63	3.30	4.97
	2.08	6.26	2.5×10^{-2}	1.60	1.72	3.32	4.92

The average value is $K_{s0} = 1.0 \times 10^{-5}$ $mol^2 dm^{-6}$.
($pK_{s0} = 5.0$)

Notes.

$[Se_{tot}] = [Ni^{2+}]$ and $[SeO_3^{2-}] = [Se_{tot}]/\alpha_{L(H)}$
where $\alpha_{L(H)} = (1 + [H^+]/K_2 + [H^+]^2/K_1K_2)$
and the acid dissociation constants have the values $K_1 = 4 \times 10^{-3}$ and $K_2 = 1.0 \times 10^{-8}$ (ref. 1).

AUXILIARY INFORMATION

METHOD APPARATUS/PROCEDURE:

Solutions of nitric and sulfuric acids were saturated with nickel selenite by shaking in a thermostat at 20±0.05°C for 8 hr. The remaining solid phase was removed by centrifugation, then the pH was measured ("Moskip" pH meter, to 0.01 pH unit) and the nickel concentration was determined gravimetrically with dimethylglyoxime.

SOURCE AND PURITY OF MATERIALS:

C.P.-grade reagents were used. Nickel selenite was prepared by mixing a 0.1*N* solution of nickel sulfate with a 5% excess of 0.1*N* sodium selenite. After 24 hr, the precipitate was separated by centrifugation and decanting, then it was dried at 40°C. Nickel was determined gravimetrically as the dimethylglyoximate, and selenium as the element after precipitation with hydrazine.

ESTIMATED ERROR:

$\pm 0.1 \times 10^{-5}$. (The spread in the results is 0.16 of a log unit.)
Temperature: ±0.05 K

REFERENCES:

1. Rumpf, P. *Compt. Rendu* 1933, *197*, 686.

COMPONENTS:	ORIGINAL MEASUREMENTS:
1. Nickel selenite; $NiSeO_3$; [10101-96-9] 2. Selenium dioxide; SeO_2; [7446-08-4] 3. Water; H_2O; [7732-18-5]	Ebert, M.; Mička, Z.; Peková, I. *Collect. Czech. Chem. Commun.* 1982, 47, 2069-76.
VARIABLES:	**PREPARED BY:**
One temperature: 298 K	Mary R. Masson

EXPERIMENTAL VALUES:

The solution at point P has the composition 7.1% $NiSeO_3$, 17.9% SeO_2 and 75.0% H_2O. The solution at point E has the composition 8.0% $NiSeO_3$, 64.8% SeO_2 and 27.2% H_2O. The other data were presented only in graphical form.

The concentration at the two points expressed as molalities[a] are:

for P, $NiSeO_3$ = 0.510 mol/kg, SeO_2 = 2.151 mol/kg;

for E, $NiSeO_3$ = 1.584 mol/kg, SeO_2 = 21.470 mol/kg.

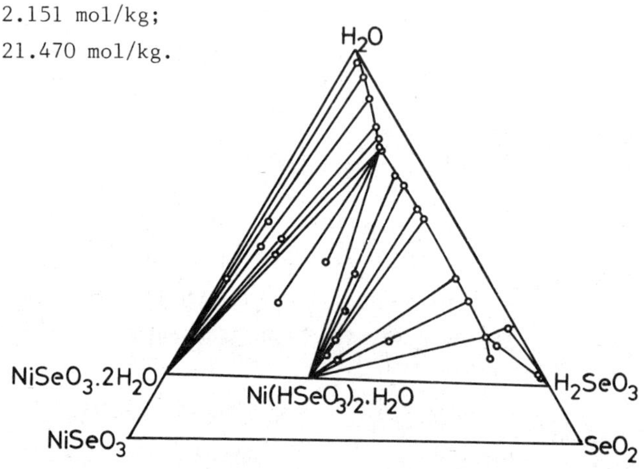

[a] Molalities calculated by the compiler.

Reprinted by permission

AUXILIARY INFORMATION

METHOD APPARATUS/PROCEDURE:	SOURCE AND PURITY OF MATERIALS:
Schreinemakers' wet residue method was used to study the solubility in this system. It required 5 - 12 weeks for equilibrium to be established. Selenium(IV) was determined iodometrically (1), and nickel by titration with EDTA, with murexide as indicator (2).	$NiSeO_3.2H_2O$ was prepared by dropwise addition of 0.06*M* sodium selenite solution to boiling 0.05*M* nickel chloride solution. Reagent-grade materials (Lachema, Brno) were used.
	ESTIMATED ERROR: No estimates possible.
	REFERENCES: 1. Ganitskii, M.Zh.; Zelinokrajte, V.I. *Zh. Neorg. Khim.* 1957, 2, 134. 2. Přibil, R. *Applied Complexometry*, Pergamon, Oxford, 1982.

COMPONENTS:

1. Copper(II) selenite; $CuSeO_3$; [10214-40-1]
2. Water; H_2O; [7732-18-5]

ORIGINAL MEASUREMENTS:

Ripan, R.; Vericeanu, G.

Studia Univ. Babes-Bolyai, Ser. Chim. 1968, *13*, 31-37.

VARIABLES:

One temperature: 291 K

PREPARED BY:

Mary R. Masson

EXPERIMENTAL VALUES:

All concentrations are expressed in units of mol dm^{-3}.

Concentration	K_{s0} / $mol^2\ dm^{-6}$	Mean K_{s0}	pK_{s0}
1.747×10^{-4}	2.9×10^{-8}	$3.2 \pm 0.4 \times 10^{-8}$	7.49
1.741×10^{-4}	3.1×10^{-8}		
1.855×10^{-4}	3.4×10^{-8}	$mol^2\ dm^{-6}$	
1.848×10^{-4}	3.4×10^{-8}		
1.820×10^{-4}	3.3×10^{-8}		
1.880×10^{-4}	3.4×10^{-8}		

The concentration c in the saturated solution was calculated from the measured conductivity κ from the equation

$$c = \frac{1000\kappa}{\Lambda°}$$

Compiler's note

Neither in the determination of the ionic conductivity of the selenite ion nor in the evaluation of the solubility product was hydrolysis of the selenite ion taken into account. This would give rise to errors, since, for example, in a 0.001*M* solution, $[SeO_3^{2-}]$ = 0.000955*M*, $[HSeO_3^-]$ = 0.000045*M* and $[OH^-]$ = 0.000045*M*, and hydroxide and hydrogen selenite have different ionic conductivities from selenite. If the ionic conductivity of hydrogen selenite were known, the experimental results could have been interpreted correctly (cf. ref. 2), but this value does not seem to be available.

However, because the calibration and sample solutions had concentrations of about the same order of magnitude, the errors would cancel to some extent, but the K_{s0} value cannot be regarded as reliable.

AUXILIARY INFORMATION

METHOD APPARATUS/PROCEDURE:

The ionic conductivity of the selenite ion in water was determined by measuring the mean molar conductivities of a series of aqueous solutions of lithium selenite with concentrations below 2×10^{-3} mol dm^{-3}. The mean molar conductivity at infinite dilution was found by extrapolation to be $172 \pm 2.9\ \Omega^{-1} cm^2 mol^{-1}$ at 18°C. At that temperature, the ionic conductivity of the lithium ion is $33.4\ \Omega^{-1} cm^2 mol^{-1}$, so the ionic conductivity of the selenite ion is $105.2\ \Omega^{-1} cm^2 mol^{-1}$.

The well washed copper selenite was mixed with 150 ml of water and agitated intermittently for 1 hr. From this, 10-20 ml was taken in the conductivity cell and the conductivity measured. The sampling and measurement were repeated until a constant value for the conductivity was obtained.

The value of $\Lambda° = (\lambda_+ + \lambda_-)$, the molar conductivity, was calculated with $\lambda_- = 105.2$ and $\lambda_+ = 90.6\ \Omega^{-1} cm^2 mol^{-1}$.

SOURCE AND PURITY OF MATERIALS:

Lithium selenite was prepared from sublimed selenium dioxide by neutralization with lithium hydroxide. An approximately 0.05*M* solution was standardized gravimetrically, and this was used to prepare the various dilutions.

The composition of the selenite was checked by analysis.

ESTIMATED ERROR:

Temperature: ±0.5 K

Error in K_{s0} (2*s*) = 0.4×10^{-8} (compiler)

REFERENCES:

1. Landolt-Bornstein *Physicalisch-Chemische Tabellen II* 1923, p. 1105.
2. Monk, C.B. *J. Chem. Soc.* 1949, 429.

COMPONENTS:

1. Copper selenite; $CuSeO_3$; [10214-40-1]

2a. Nitric acid; HNO_3; [7697-37-2]

2b. Sulfuric acid; H_2SO_4; [7664-93-9]

3. Water; H_2O; [7732-18-5]

ORIGINAL MEASUREMENTS:

Chukhlantsev, V.G.

Zh. Neorg. Khim. 1956, *1*, 2300-5; **Russ. J. Inorg. Chem.* 1956, *1*, 132-8.

VARIABLES:

HNO_3 and H_2SO_4 concentrations

One temperature: 293 K

PREPARED BY:

Mary R. Masson

EXPERIMENTAL VALUES:

All concentrations are expressed in units of mol dm^{-3}.

Soln.	Initial pH	Final pH	$[Cu^{2+}]$	pCu	$\log \alpha_{L(H)}$	$p[SeO_3^{2-}]$	pK_{s0}
HNO_3	2.97	6.39	5.38×10^{-4}	3.27	1.60	4.87	8.14
	2.27	5.63	2.12×10^{-3}	2.67	2.37	5.04	7.71
	2.00	5.40	3.41×10^{-3}	2.47	2.61	5.08	7.55
H_2SO_4	2.53	5.86	1.11×10^{-3}	2.96	2.12	5.08	8.04
	2.12	5.58	3.54×10^{-3}	2.45	2.40	4.85	7.30
	2.03	5.45	4.03×10^{-3}	2.40	2.55	4.95	7.35

The average value is $K_{s0} = 2.09 \times 10^{-8}$ mol^2dm^{-6}.

($pK_{s0} = 7.68$)

Notes.

$[Se_{tot}] = [Cu^{2+}]$ and $[SeO_3^{2-}] = [Se_{tot}]/\alpha_{L(H)}$

where $\alpha_{L(H)} = (1 + [H^+]/K_2 + [H^+]^2/K_1K_2)$

and the acid dissociation constants have the values $K_1 = 4 \times 10^{-3}$ and $K_2 = 1.0 \times 10^{-8}$ (ref. 1).

AUXILIARY INFORMATION

METHOD APPARATUS/PROCEDURE:

Solutions of nitric and sulfuric acids were saturated with copper selenite by stirring in a thermostat at 20°C for 8 hr. The remaining solid phase was removed by centrifugation, then the pH was measured ("Moskip" pH meter, to 0.01 pH unit) and the copper concentration was determined (method not stated).

SOURCE AND PURITY OF MATERIALS:

C.P.-grade reagents were used. Copper selenite was prepared by mixing a 0.1*N* solution of copper sulfate with a 5% excess of 0.1*N* sodium selenite solution, in the cold. After 24 hours standing, the precipitate was separated by centrifugation and washed with water. Copper was determined as the element after precipitation on a platinum net, and selenium was determined gravimetrically as the element.

ESTIMATED ERROR:

The spread in the results is 0.84 of a log unit.

Temperature: probably ±0.05 K

REFERENCES:

1. Rumpf, P. *Compt. Rendu* 1933, *197*, 686.

COMPONENTS:

1. Copper(II) selenite; $CuSeO_3$; [10214-40-1]
2a. Ammonia solution; [1336-21-6]
2b. Hydrochloric acid; [7647-01-0]
2c. Sulfuric acid; [7697-37-2]
2d. Acetic acid;
3. Water; H_2O; [7732-18-5]

ORIGINAL MEASUREMENTS:

Geilman, W.; Wrigge, W.

Z. Anorg. Allgem. Chem. 1931, *197*, 353-63.

VARIABLES:

One temperature, presumably ambient.

PREPARED BY:

Mary R. Masson

EXPERIMENTAL VALUES:

The concentrations of Cu and Se are expressed in units of mg/250 ml.

Normality* of Solvent	Ammoia solution		Hydrochloric acid		Sulfuric acid		Acetic acid	
	[Cu]	[Se]	[Cu]	[Se]	[Cu]	[Se]	[Cu]	[Se]
0.0 (water)	1.6	2.0	1.6	2.0	1.6	2.0	1.6	2.0
0.01	2.2	13.8	38.9	48.9	55.8	68.7	3.2	3.9
0.02	6.6	26.1	92.6	115.0	109.0	134.7	13.4	16.6
0.05	45.5	116.5	233.2	288.0	244.8	301.4	19.0	28.8
0.10	181.1	285.1	560.6	698.4	560.6	698.4	25.1	31.3
0.20	560.6	698.4	"	"	"	"	32.1	40.0
0.50	"	"	"	"	"	"	46.2	58.4
1.00	"	"	"	"	"	"	60.6	75.7
2.00	"	"	"	"	"	"	85.7	106.5

a For ammonia solution, hydrochloric acid and acetic acid, the normality is equal to the concentration expressed in mol dm^{-3}. For sulfuric acid, the normality must be divided by 2 to obtain the concentration in mol dm^{-3}.

A value for the solubility product can be obtained from the data for aqueous solutions. 1.6 mg/250 ml or copper = 0.0064 g/l. = 0.0001007*M* and 2.0 mg/250 ml of Se = 0.008 g/l. = 0.0001013*M*.
Thus, the solubility product is

$$K_{s0} = 1.02 \times 10^{-8} \text{ mol}^2 \text{ dm}^{-6} \quad (pK_{s0} = 7.99).$$

AUXILIARY INFORMATION

METHOD APPARATUS/PROCEDURE:

The copper selenite dihydrate (2-g samples) was digested with 250 ml of each solvent, for 24 hr. Selenium in the filtrate was determined as the element after reduction with sulfur dioxide in concentrated hydrochloric acid medium. Dissolved copper was determined in the filtrate from the selenium determination by precipitation with hydrogen sulfide.

Note: data are also given from experiments on the relative rates of dissolution of copper selenite and copper sulfide, in the presence and absence of air or hydrogen peroxide.

SOURCE AND PURITY OF MATERIALS:

Not stated.

ESTIMATED ERROR:

No estimates possible.

REFERENCES:

COMPONENTS:

1. Silver selenite; Ag_2SeO_3; [7784-05-6]

2a. Nitric acid; HNO_3; [7697-37-2]

2b. Sulfuric acid; H_2SO_4; [7664-93-9]

3. Water; H_2O; [7732-18-5]

ORIGINAL MEASUREMENTS:

Chukhlantsev, V.G.

Zh. Neorg. Khim. 1956, *1*, 2300; **Russ. J. Inorg. Chem.* 1956, *1*, 132.

VARIABLES:

NHO_3 and H_2SO_4 concentrations

One temperature: 293 K

PREPARED BY:

Mary R. Masson

EXPERIMENTAL VALUES:

Concentrations are expressed in units of mol dm^{-3}.

Soln.	Initial pH	Final pH	$[Ag^+]$	pAg	log $\alpha_{L(H)}$	$p[SeO_3^{2-}]$	pK_{s0}
HNO_3	2.24	2.43	8.52×10^{-4}	3.07	5.83	9.20	15.34
	2.00	2.27	1.59×10^{-3}	2.80	6.10	9.20	14.80
H_2SO_4	2.35	2.61	6.12×10^{-4}	3.21	5.60	9.11	15.53
	2.06	2.31	1.51×10^{-3}	2.82	6.05	9.17	14.81

The average value is $K_{s0} = 9.7 \times 10^{-16}$ mol^3dm^{-9}.
($pK_{s0} = 15.01$)

Notes.

$[Se_{tot}] = [Ag^{2+}]$ and $[SeO_3^{2-}] = [Se_{tot}]/\alpha_{L(H)}$
where $\alpha_{L(H)} = (1 + [H^+]/K_2 + [H^+]^2/K_1K_2)$ (refs. 1 and 2) and the acid dissociation constants have the values $K_1 = 4 \times 10^{-3}$ mol dm^{-3} and $K_2 = 1.0 \times 10^{-8}$ mol dm^{-3} (ref. 1).

AUXILIARY INFORMATION

METHOD APPARATUS/PROCEDURE:

Solutions of nitric and sulfuric acids were saturated with silver selenite by stirring in a thermostat at 20°C for 8 hr. The remaining solid phase was removed by centrifugation, then the pH was measured ("Moskip" pH meter, to 0.01 pH unit) and the silver concentration was determined (method not stated).

SOURCE AND PURITY OF MATERIALS:

C.P.-grade reagents were used. Silver selenite was prepared by mixing stoichiometric amounts of 0.1*N* solutions of selenious acid and silver nitrate in the dark, then the precipitate was washed with water and dried at 40°C. Silver was determined by the Volhard method and selenium gravimetrically after precipitation with hydrazine.

ESTIMATED ERROR:

The spread in the results is 0.73 of a log unit.

REFERENCES:

1. Rumpf, P. *Compt. Rendu* 1933, *197*, 686.

COMPONENTS:

1. Silver selenite; Ag_2SeO_3; [7784-05-6]
2. Water; H_2O; [7732-18-5]

ORIGINAL MEASUREMENTS:

Selivanova, N.M.; Leshchinskaya, Z.L.; Klushina, T.V.

Zhur. Fiz. Khim. 1962, *36*, 1349; **Russ. J. Phys. Chem.* 1962, *36*, 719.

VARIABLES:

One temperature: 298 K

PREPARED BY:

Mary R. Masson

EXPERIMENTAL VALUES:

All concentrations are expressed in units of mol dm^{-3}.

Time of mixing (days) at 25°C	Concentration of Ag^+	Solubility Product	
6	8.26×10^{-6}		
7	8.28×10^{-6}		Hydrolysis of selenite neglected.
7	8.30×10^{-6}		
8	8.30×10^{-6}		
10	8.34×10^{-6}		Has equilibrium been reached?
Mean	8.31×10^{-6}	2.85×10^{-16} $mol^3\ dm^{-9}$	

The compiler has recalculated the results to take account of the hydrolysis of the selenite ions; the computer program HALTAFALL (2) was used, and the values for the dissociation constants were those of Hagisawa (3) - $pK_1 = 2.62$, $pK_2 = 8.32$.

Concentration of Ag^+	Total conc. of selenite	Conc. of SeO_3^{2-}	Solubility Product
8.26×10^{-6}	4.13×10^{-6}	2.058×10^{-6}	
8.28×10^{-6}	4.14×10^{-6}	2.065×10^{-6}	
8.30×10^{-6}	4.15×10^{-6}	2.072×10^{-6}	
8.30×10^{-6}	4.15×10^{-6}	2.072×10^{-6}	
8.34×10^{-6}	4.17×10^{-6}	2.085×10^{-6}	
8.31×10^{-6}	4.155×10^{-6}	2.075×10^{-6}	1.433×10^{-16} $mol^3\ dm^{-9}$ ($pK_{s0} = 15.84$)

It should be noted that the calculation of $[SeO_3^{2-}]$ assumes that the water used for dissolution of the silver selenite was extremely pure and had a pH not significantly differing from 7. If the pH was significantly different from 7 (e.g. because of the presence of dissolved carbon dioxide) hydrolysis of the selenite ion would have occurred to a greater or lesser extent.

AUXILIARY INFORMATION

METHOD APPARATUS/PROCEDURE:

Solid silver selenite was equilibrated with water for 7 - 10 days, in a thermostat at 25°C ± 0.1°C. The concentration of silver in the solution was determined turbidimetrically as follows. To 2 - 10 ml of the solution were added 10 drops of a 1% gelatin solution and 2 ml of HCl (1:1), and the solution was diluted to 100 ml. After 15 min, the absorbance was measured (blue filter) (ref. 1).

SOURCE AND PURITY OF MATERIALS:

Silver selenite was prepared by mixing 0.1*N* solutions of silver nitrate and sodium selenite in stoichiometric proportions, washing several times with warm water, and drying at 40°C.

ESTIMATED ERROR:

$s = 0.03 \times 10^{-16}$ $mol^3 dm^{-9}$ (estimated from the range of results on days 7- 10).
$s = 0.02 \times 10^{-16}$ (for recalculated results)

REFERENCES:

1. Selivanova, N.M.; Zubova, G.A.; Finkel'shtein, E.I. *Zh. Fiz. Khim.* 1959, *33*, 2365; *Russ. J. Phys. Chem.* 1959, *33*, 430.
2. Ingri, N.; Kakolowicz, W.; Sillén, L.G.; Warnqvist, B. *Talanta* 1967, *14*, 1261.
3. Hagisawa, H. *Bull. Inst. Phys. Chem. Res. Tokyo*, 1939, *18*, 648.

COMPONENTS:

1. Silver selenite; Ag_2SeO_3; [7784-05-6]
2. Sodium perchlorate; $NaClO_4$; [7601-89-0]
3. Selenous acid; H_2SeO_3; [7783-00-8]
4. Water; H_2O; [7732-18-5]

(Experiment A)

ORIGINAL MEASUREMENTS:

1. Mehra, M.C.; Gubeli, A.O. *Radiochem. Radioanal. Lett.* 1969, 2, 61.
2. Mehra, M.C. *Dissertation* Laval University, Quebec P.Q., 1968.

VARIABLES:

One temperature: 298 K

pH was varied by addition of $HClO_4$ or NaOH

Ionic strength was kept at 1 mol dm^{-3}

PREPARED BY:

Mary R. Masson

EXPERIMENTAL VALUES:

All concentrations are expressed in terms of -log of mol dm^{-3}.

pH	pAg	pH	pAg	pH	pAg	pH	pAg	pH	pAg
1.20	3.22	1.60	3.51	2.57	4.30	6.30	6.10	12.55	7.14
1.25	3.24	1.67	3.56	2.75	4.47	6.85	6.47	12.75	7.11
1.28	3.26	1.70	3.64	2.78	4.47	7.65	6.77	12.77	7.09
1.30	3.31	1.75	3.70	3.05	4.64	7.65	6.77	12.80	7.16
1.35	3.34	1.85	3.73	3.45	4.84	8.20	6.94	13.00	7.08
1.37	3.34	1.95	3.81	3.70	4.98	8.80	7.03	13.00	7.06
1.45	3.37	2.00	3.91	4.00	5.22	9.17	7.08	13.00	7.12
1.47	3.43	2.25	4.07	4.30	5.22	10.30	7.16	13.05	7.11
1.50	3.43	2.35	4.18	5.12	5.77	11.50	7.25	13.07	7.04
1.55	3.48	2.40	4.27	5.82	6.13	11.65	7.20		
1.57	3.49	2.48	4.32	6.05	5.86	11.95	7.20		

From the last 13 results, when pH > pK_2 for H_2SeO_3, the author calculated that pK_{s0} = 2pAg = 2pSeO_3 = 15.58, $s = \pm 0.12$. (This assumed that pSeO_3 remained constant at the initial value of 1.32.)

The acid dissociation constants of selenous acid were also evaluated from this set of data: the values found were $pK_1 = 2.26$ and $pK_2 = 8.12$.

(continued on next page)

AUXILIARY INFORMATION

METHOD APPARATUS/PROCEDURE:

Precipitations were done in 100-ml standard flasks kept in a water bath at 25°C. The ligand was always added last. The initial concentration of silver ions was $4.79 \times 10^{-3} M$ and of selenous acid $4.79 \times 10^{-2} M$. The pH was adjusted to the required value by addition of sodium hydroxide or perchloric acid. Sealed flasks were equilibrated for six days. pH and pAg were measured potentiometrically in the aqueous phase after removal of the solid phase by filtration through a frit, under nitrogen. The potentiometric determination of silver was done with a silver metal electrode and a calomel electrode filled with 1*M* sodium perchlorate solution. $E°$ for this cell was determined by measuring its potential with solutions of known [Ag^+] in 1*M* sodium perchlorate.

SOURCE AND PURITY OF MATERIALS:

The water used was demineralized and deoxygenated, and stored under nitrogen. The selenous acid used was of reagent grade.

ESTIMATED ERROR:

s = 0.2 log units for all the data. (compiler)

REFERENCES:

COMPONENTS:	ORIGINAL MEASUREMENTS:
1. Silver selenite; Ag_2SeO_3; [7784-05-6] 2. Sodium perchlorate; $NaClO_4$; [7601-89-0] 3. Selenous acid; H_2SeO_3; [7783-00-8] 4. Water; H_2O; [7732-18-5] (Experiment A)	1. Mehra, M.C.; Gubeli, A.O. *Radiochem. Radioanal. Lett.* 1969, 2, 61. 2. Mehra, M.C. *Dissertation* Laval University, Quebec P.Q., 1968. (continued)

COMMENTS AND/OR ADDITIONAL DATA

The compiler used the values for K_1 and K_2 determined by the authors to allow a value for K_{s0} based on the complete set of data to be evaluated. That is, a value for $[SeO_3^{2-}]$ was calculated for each point as follows.

Amount of Ag precipitated = initial Ag - free Ag

Total Se in solution = initial Se - precip. Se = initial Se - (ppt Ag/2)

$[SeO_3^{2-}]$ = (Total Se in solution)/$(1 + [H^+]/K_2 + [H^+]^2/K_1K_2)$.

The $K_{s0} = [Ag^+]^2[SeO_3^{2-}]$.

A value of 2.36×10^{-16} $mol^3\ dm^{-9}$ was found for K_{s0} (pK_{s0} = 15.63). The standard deviation estimate, *s* expressed in logarithmic terms is 0.2.

COMPONENTS:

1. Silver selenite; Ag_2SeO_3; [7784-05-6]
2. Sodium perchlorate; $NaClO_4$; [7601-89-0]
3. Selenous acid; H_2SeO_3; [7783-00-8]
4. Water; H_2O; [7732-18-5]

(Experiment B)

ORIGINAL MEASUREMENTS:

1. Mehra, M.C.; Gubeli, A.O. *Radiochem. Radioanal. Lett.* 1969, 2, 61.
2. Mehra, M.C. *Dissertation* Laval University, Quebec P.Q., 1968.

VARIABLES:

One temperature: 298 K
Total selenous acid concentration was varied, and also pH, by addition of $HClO_4$ or NaOH. The ionic strength was kept at 1 mol dm^{-3}.

(SUMMARY)

PREPARED BY:

Mary R. Masson

EXPERIMENTAL VALUES:

Concentrations are expressed in terms of -log of mol dm^{-3}.

pH	pAg_{tot} $[H_2SeO_3]_{tot}$ = 0.00646 mol dm^{-3}	pAg_{tot} $[H_2SeO_3]_{tot}$ = 0.1 mol dm^{-3}	pAg_{tot} $[H_2SeO_3]_{tot}$ = 0.1995 mol dm^{-3}
1.0	2.26	-	-
1.5	2.76	3.16	-
2.0	3.18	3.66	-
2.5	3.57	4.10	4.42
3.0	3.93	4.42	4.61
3.5	4.18	4.68	4.80
4.0	4.42	4.90	(5.00)
4.5	(4.64)	(5.07)	(5.19)
5.0	(4.87)	(5.24)	(5.38)
5.5	(5.10)	(5.41)	(5.57)
6.0	5.32	5.58	5.75
6.5	5.49	5.65	5.76
7.0	5.60	5.65	5.77
7.5	5.65	5.59	5.48
8.0	5.65	5.43	5.22
8.5	5.65	5.28	5.05
9.0 - 12.0	5.65	5.20	5.00

The compiler summarized the results by plotting all the authors' experimental points, drawing smooth curves through them, then abstracting values at regular intervals of pH. Values in brackets are in regions of pH where there were no experimental points, but where interpolation seems to be justified.

The initial concentration of silver ions was 3.63×10^{-3} mol dm^{-3} for the first two sets of results, and 3.98×10^{-4} mol^{-3} for the third. (continued on next page)

AUXILIARY INFORMATION

METHOD APPARATUS/PROCEDURE:

Precipitations were done in 100-ml standard flasks kept in a water bath at 25°C. The ligand was always added last. The pH was adjusted by means of $HClO_4$ and NaOH only. Sealed flasks were equilibrated for six days. A radioactive silver solution was used to enable the total concentration of silver in solution to be determined by scintillation spectrometry. Radioactivity of three 5-ml samples of filtered aqueous phase was measured by means of a Philips single γ-spectrometer, with a well-type NaI/Tl crystal. Count rates were observed under the photopeak. pH was determined potentiometrically in the aqueous phase.

SOURCE AND PURITY OF MATERIALS:

The silver isotope used was ^{110}Ag, with a half-life of 249 days, E_β = 0.54 MeV, and E_γ = 0.66, 0.88 MeV. The radiotracer was mixed with inactive solution in such a proportion that the count rate at 1000-fold dilution would be measurable. A minimum count rate of 20 cpm/ml at a calculated dilution to $10^{-8}M$ was always maintained. In the final samples, a minimum of 1000 counts was taken. The calibration graph was redrawn for every experiment.

ESTIMATED ERROR:

REFERENCES:

1. Ingri, N.; Kakolowicz, W.; Sillén, L.G.; Warnqvist, B. *Talanta* 1967, *14*, 1261.

COMPONENTS:	ORIGINAL MEASUREMENTS:
1. Silver selenite; Ag_2SeO_3; [7784-05-6] 2. Sodium perchlorate; $NaClO_4$; [7601-89-0] 3. Selenous acid; H_2SeO_3; [7783-00-8] 4. Water; H_2O; [7732-18-5] (Experiment B)	1. Mehra, M.C.; Gubeli, A.O. *Radiochem. Radioanal. Lett.* 1969, 2, 61. 2. Mehra, M.C. *Dissertation* Laval University, Quebec P.Q., 1968. (continued)

COMMENTS AND/OR ADDITIONAL DATA

From this set of data, the authors conclude that the concentration of silver in solution starts increasing at around pH 8, and soon reaches a value that is independent of pH. The behaviour at pH values below 8 conforms to that expected if no complexes are formed between silver ions and H_2SeO_3 or $HSeO_3^-$. However, the enhanced solubility at higher pH values suggests that complexes must be formed between silver ions and SeO_3^{2-}.

The compiler did calculations with this set (B) of data similar to those done with set (A), but with an additional correction for complex formation. The values found for pK_{s0} were 14.72 (s = 0.5), 14.86 (s = 0.4) and 15.06 (s =0.35) for the three selenite concentrations.

The compiler has used the values calculated by the authors for pK_{s0}, β_1 and β_2 to do a back-calculation of the solubility expressed as $[Ag]_{tot}$. The calculated values for pAg corresponding to listed experimental ones are tabulated below.

pH	$[H_2SeO_3]_{tot}$ = 0.00646 mol dm^{-3}	$[H_2SeO_3]_{tot}$ = 0.1 mol dm^{-3}	$[H_2SeO_3]_{tot}$ = 0.1995 mol dm^{-3}
1.0	2.49	3.09	3.40
1.5	2.93	3.56	3.72
2.0	3.35	4.00	4.15
2.5	3.72	4.38	4.53
3.0	4.03	4.69	4.84
3.5	4.30	4.97	5.12
4.0	4.56	5.22	5.37
4.5	4.81	5.47	5.62
5.0	5.06	5.72	5.86
5.5	5.31	5.95	6.08
6.0	5.56	6.15	6.23
6.5	5.80	6.26	6.25
7.0	6.01	6.22	6.10
7.5	6.18	6.02	5.79
8.0	6.26	5.75	5.44
8.5	6.26	5.54	5.19
9.0	6.26	5.43	5.06
10.0	6.25	5.38	5.00
11.0 - 12.0	6.25	5.37	4.99

The computer program HALTAFALL (1) was used for the calculations.

COMPONENTS:

1. Silver selenite; Ag_2SeO_3; [7784-05-6]
2. Sodium perchlorate; $NaClO_4$; [7601-89-0]
3. Selenous acid; H_2SeO_3; [7783-00-8]
4. Water; H_2O; [7732-18-5]

(Experiment C)

ORIGINAL MEASUREMENTS:

1. Mehra, M.C.; Gubeli, A.O. *Radiochem. Radioanal. Lett.* 1969, 2, 61.
2. Mehra, M.C. *Dissertation* Laval University, Quebec P.Q., 1968.

VARIABLES:

One temperature: 298 K
The total selenite concentration was varied.
The pH and ionic strength were kept constant.

PREPARED BY:

Mary R. Masson

EXPERIMENTAL VALUES:

Concentrations are expressed in units of mol dm^{-3}.

$[SeO_3^{2-}]_{tot}$	pAg_{tot}	$[SeO_3^{2-}]_{tot}$	pAg_{tot}
0.001	6.00	0.100	5.36
0.003	6.22	0.200	4.97
0.005	6.15	0.300	4.75
0.008	6.15	0.400	4.55
0.010	6.17	0.500	4.42
0.020	5.97	0.600	4.32
0.030	5.92	0.700	4.25
0.050	5.72	0.800	4.19
0.070	5.62	0.900	4.18
0.090	5.42		

From this set of data, the authors evaluated the stability constants for the complexes $Ag(SeO_3)^-$ and $Ag(SeO_3)_2^{3-}$. The values found were log β_1 = 2.42 and log β_2 = 3.76. The value for pK_{s0} of 15.58, as calculated previously by these authors, was used in the calculation.

Note: $\beta_1 = \dfrac{[Ag(SeO_3)^-]}{[Ag^+][SeO_3^{2-}]}$ $\beta_2 = \dfrac{[Ag(SeO_3)_2^{3-}]}{[Ag^+][SeO_3^{2-}]^2}$

AUXILIARY INFORMATION

METHOD APPARATUS/PROCEDURE:

Precipitations were done in 100-ml standard flasks kept in a water bath at 25°C. The ligand was always added last. The pH was adjusted to 9.40-9.75 by addition of $HClO_4$ and NaOH only. Sealed flasks were equilibrated for six days. A radioactive silver solution was used to enable the total concentration of silver in solution to be determined by scintillation spectrometry. The radioactivity of three 5-ml samples of filtered aqueous phase was measured by means of a Philips single γ-spectrometer, with a well-type NaI/Tl crystal. Count rates were observed under the photopeak. The calibration graph was redrawn for every experiment.

SOURCE AND PURITY OF MATERIALS:

The silver isotope used was ^{110}Ag, with a half-life of 249 days, E_β = 0.54 MeV, and E_γ = 0.66, 0.88 MeV. It was obtained from Atomic Energy of Canada Ltd., Chalk River, Ontario, Canada. The radiotracer was mixed with inactive solution in such a proportion that the count rate at 1000-fold dilution would be measurable. A minimum count rate of 20 cpm/ml at a calculated dilution to $10^{-8}M$ was always maintained. In the final samples, a minimum of 1000 counts was taken.

ESTIMATED ERROR:

s = 0.12 for log β_1 and s = 0.05 for log β_2. (compiler)

REFERENCES:

COMPONENTS:	ORIGINAL MEASUREMENTS:
1. Silver selenite; Ag_2SeO_3; [7784-05-6] 2. Sodium perchlorate; $NaClO_4$; [7601-89-0] 3. Selenous acid; H_2SeO_3; [7783-00-8] 4. Water; H_2O; [7732-18-5] (Experiment C)	1. Mehra, M.C.; Gubeli, A.O. *Radiochem. Radioanal. Lett.* 1969, *2*, 61. 2. Mehra, M.C. *Dissertation* Laval University, Quebec P.Q., 1968.

COMMENTS AND/OR ADDITIONAL INFORMATION

The authors deduced that the two complexes formed are $AgSeO_3^-$ and $Ag(SeO_3)_2^{3-}$ from the behaviour of the solubility, expressed in terms of $[Ag]_{tot}$ as a function of the concentration of selenite. A plot of log $[Ag]_{tot}$ *vs.* $[SeO_3^{2-}]$ has two distinct linear regions with slopes of -0.5 and -1.5. A value for β_2 was calculated from the data corresponding to a slope of -1.5, by assuming that essentially all the silver in solution would be present as $Ag(SeO_3)_2^{3-}$; i.e. $[Ag(SeO_3)_2^{3-}] = [Ag]_{tot}$. A value for β_1 was calculated then from the data corresponding to a slope of -0.5, by setting $[Ag]_{tot} = [AgSeO_3^-] + [Ag(SeO_3)_2^-]$, and utilizing the value already calculated for β_2 to calculate $[Ag(SeO_3)_2^{3-}]$. For both calculations, $[Ag^+]$ was calculated from $K_{s0} = [Ag^+]^2 \times [SeO_3^{2-}]$.

Compiler's comments: the compiler felt that the extensive sets of data from experiments B and C should be further utilized to confirm and improve the value derived for K_{s0} from the first series of experiments (A). However, this was not found to be possible; in fact, the sets of data are not consistent, and even the three subsets of data in the second series of experiments are not consistent with one another. The conclusion has to be that one or other, or both, of the experimental procedures must be faulty. In view of the lack of consistency even among the data of the second series of experiments, it seems that the radiochemical technique is the one that is more likely to be faulty. Also, the potentiometric technique used in the first series would have been expected to be more reliable in any case. It is therefore suggested tentatively that the results from series one be regarded as reliable, but that the values for log β_1 and log β_2 be regarded with suspicion.

COMPONENTS:	ORIGINAL MEASUREMENTS:
1. Silver selenite; Ag_2SeO_3; [7784-05-6] 2. Water; H_2O; [7732-18-5]	Chao, E.E.; Cheng, K.L. *Anal. Chem.* 1976, *48*, 267.
VARIABLES:	PREPARED BY:
One temperature: 293 K	Mary R. Masson

EXPERIMENTAL VALUES:

The ionic strength was constant at 0.1 mol dm^{-3} (medium not stated)

$pK_{s0} = 15.45 \pm 0.15$ $K_{s0} = 3.55 \times 10^{-16}$ $mol^3\ dm^{-9}$

Concentrations are given in units of mol dm^{-3}.

Compiler's note

The values used for the acid dissociation constants of selenious acid are not given, but if the determination was done at pH 11.0, as it was for silver arsenite (1), the values would have an almost negligible influence on the value obtained for the solubility product. Therefore, this value is probably a reasonably good estimate of the concentration solubility product.

The value would refer to a freshly precipitated solid, and may therefore be expected to differ from values found by equilibrium of solutions with aged solids.

AUXILIARY INFORMATION

METHOD APPARATUS/PROCEDURE:

The solubility product was determined from data obtained by potentiometric titration of a selenite solution with a silver nitrate solution. Silver ion activities were measured by means of an Orion silver sulfide electrode (94-16) and an Orion double junction reference electrode (90-02). Emf readings were taken with a Corning model 10 pH meter with expanded scale.
Method of calculation is given in ref. (1). This involved determining, from the *E* value, pAg at the point of incipient precipitation of silver selenite.

SOURCE AND PURITY OF MATERIALS:

Reagent-grade chemicals were used.

ESTIMATED ERROR:

Range in $pK_{s0} = \pm 0.15$

REFERENCES:

1. Chao, E.E. *Ph.D. Dissertation* University of Missouri, Kansas City, Mo. 1975.

COMPONENTS:

1. Zinc selenite; $ZnSeO_3$; [13597-46-1]
2a. Nitric acid; HNO_3; [7697-37-2]
2b. Sulfuric acid; H_2SO_4; [7664-93-9]
3. Water; H_2O; [7732-18-5]

ORIGINAL MEASUREMENTS:

Chukhlantsev, V.G.

Zh. Neorg. Khim. 1956, *1*, 2300-5; **Russ. J. Inorg. Chem. (Eng. Transl.)* 1956, *1*, 132-8.

VARIABLES:

HNO_3 and H_2SO_4 concentrations

One temperature: 293 K

PREPARED BY:

Mary R. Masson

EXPERIMENTAL VALUES:

All concentrations are expressed in units of mol dm^{-3}.

Soln.	Initial pH	Final pH	$[Zn^{2+}]$	pZn	log $\alpha_{L(H)}$	$p[SeO_3^{2-}]$	pK_{s0}
HNO_3	2.97	7.28	1.82×10^{-3}	2.74	0.80	3.54	6.28
	2.27	5.84	7.27×10^{-3}	2.14	2.15	4.29	6.43
	2.00	5.09	1.15×10^{-2}	1.94	2.90	4.84	6.78
H_2SO_4	2.53	6.52	2.93×10^{-3}	2.53	1.47	4.00	6.53
	2.11	5.41	9.06×10^{-3}	2.04	2.60	4.64	6.68
	2.03	5.18	1.08×10^{-2}	1.97	2.80	4.77	6.84

The average value is $K_{s0} = 2.58 \times 10^{-7}$ $mol^2 dm^{-6}$.
($pK_{s0} = 6.59$)

Notes.

$[Se_{tot}] = [Zn^{2+}]$ and $[SeO_3^{2-}] = [Se_{tot}]/\alpha_{L(H)}$

where $\alpha_{L(H)} = (1 + [H^+]/K_2 + [H^+]^2/K_1K_2)$

and the acid dissociation constants have the values $K_1 = 4 \times 10^{-3}$ and $K_2 = 1.0 \times 10^{-8}$ (ref. 1).

AUXILIARY INFORMATION

METHOD APPARATUS/PROCEDURE:

Solutions of nitric and sulfuric acids were saturated with zinc selenite by stirring in a thermostat at 20°C for 8 hr. The remaining solid phase was removed by centrifugation, then the pH was measured ("Moskip" pH meter, to 0.01 pH unit) and the zinc concentration was determined (method not stated).

SOURCE AND PURITY OF MATERIALS:

C.P.-grade reagents were used. Zinc selenite was precipitated at 50 - 60°C by mixing a 0.1*N* solution of sodium selenite with the stoichiometric amount of zinc sulfate. The precipitate crystallized on prolonged keeping in the mother liquor. It was washed with water and dried at 40°C. Zinc was determined gravimetrically after precipitation with hydroxyquinoline, and selenium gravimetrically as the element.

ESTIMATED ERROR:

The spread in the results is 0.56 of a log unit.

Temperature: probably ±0.05 K

REFERENCES:

1. Rumpf, P. *Compt. Rendu* 1933, *197*, 686.

COMPONENTS:

1. Zinc selenite; $ZnSeO_3$; [13597-46-1]
2. Water; H_2O; [7732-18-5]

ORIGINAL MEASUREMENTS:

Ripan, R.; Vericeanu, G.

Studia Univ. Babes-Bolyai, Ser. Chim. 1968, *13*, 31-37.

VARIABLES:

One temperature: 291 K

PREPARED BY:

Mary R. Masson

EXPERIMENTAL VALUES:

All concentrations are expressed in units of mol dm^{-3}.

Concentration	K_{s0} $mol^2\ dm^{-6}$	Mean K_{s0}	pK_{s0}
1.457×10^{-4}	2.1×10^{-8}	$1.9 \pm 0.3 \times 10^{-8}$	7.71
1.480×10^{-4}	2.2×10^{-8}		
1.430×10^{-4}	2.0×10^{-8}	$mol^2\ dm^{-6}$	
1.335×10^{-4}	1.8×10^{-8}		
1.331×10^{-4}	1.7×10^{-8}		
1.331×10^{-4}	1.75×10^{-8}		

The concentration c in the saturated solution was calculated from the measured conductivity κ from the equation

$$c = \frac{1000\kappa}{\Lambda^\circ}$$

Compiler's note

Neither in the determination of the ionic conductivity of the selenite ion nor in the evaluation of the solubility product was hydrolysis of the selenite ion taken into account. This would give rise to errors, since, for example, in a 0.001*M* solution, $[SeO_3^{2-}]$ = 0.000955*M*, $[HSeO_3^-]$ = 0.000045*M* and $[OH^-]$ = 0.000045*M*, and hydroxide and hydrogen selenite have different ionic conductivities from selenite. If the ionic conductivity of hydrogen selenite were known, the experimental results could have been interpreted correctly (cf. ref. 2), but this value does not seem to be available.

However, because the calibration and sample solutions had concentrations of about the same order of magnitude, the errors would cancel to some extent, but the K_{s0} value cannot be regarded as reliable.

AUXILIARY INFORMATION

METHOD APPARATUS/PROCEDURE:

The ionic conductivity of the selenite ion in water was determined by measuring the mean molar conductivities of a series of aqueous solutions of lithium selenite with concentrations below 2×10^{-3} mol dm^{-3}. The mean molar conductivity at infinite dilution was found by extrapolation to be $172 \pm 2.9\ \Omega^{-1}cm^2mol^{-1}$ at 18°C. At that temperature, the ionic conductivity of the lithium ion is $33.4\ \Omega^{-1}cm^2mol^{-1}$, so the ionic conductivity of the selenite ion is $105.2\ \Omega^{-1}cm^2mol^{-1}$.

The well washed zinc selenite was mixed with 150 ml of water and agitated intermittently for 1 hr. From this, 10-20 ml was taken in the conductivity cell and the conductivity measured. The sampling and measurement were repeated until a constant value for the conductivity was obtained.

The value of $\Lambda^\circ = (\lambda_+ + \lambda_-)$, the molar conductivity, was calculated with $\lambda_- = 105.2$ and $\lambda_+ = 90\ \Omega^{-1}cm^2mol^{-1}$.

SOURCE AND PURITY OF MATERIALS:

Lithium selenite was prepared from sublimed selenium dioxide by neutralization with lithium hydroxide. An approximately 0.05*M* solution was standardized gravimetrically, and this was used to prepare the various dilutions.

The composition of the selenite was checked by analysis.

ESTIMATED ERROR:

Temperature: ±0.5 K

Error in K_{s0} (2*s*) = 1.5×10^{-6} (compiler)

REFERENCES:

1. Landolt-Bornstein *Physicalisch-Chemische Tabellen II* 1923, p. 1105.
2. Monk, C.B. *J. Chem. Soc.* 1949, 429.

COMPONENTS:	ORIGINAL MEASUREMENTS:
1. Cadmium selenite; $CdSeO_3$; [13814-59-0] 2. Water; H_2O; [7732-18-5]	Ripan, R.; Vericeanu, G. *Studia Univ. Babes-Bolyai, Ser. Chim.* 1968, *13*, 31-37.
VARIABLES:	PREPARED BY:
One temperature: 291 K	Mary R. Masson

EXPERIMENTAL VALUES:

All concentrations are expressed in units of mol dm^{-3}.

Concentration	K_{s0} / $mol^2\ dm^{-6}$	Mean K_{s0}	pK_{s0}
8.152×10^{-5}	6.6×10^{-9}	$6.0 \pm 1.1 \times 10^{-9}$	8.22
8.075×10^{-5}	6.5×10^{-9}		
7.107×10^{-5}	6.7×10^{-9}	$mol^2\ dm^{-6}$	
7.306×10^{-5}	5.3×10^{-9}		
7.356×10^{-5}	5.4×10^{-9}		
7.887×10^{-5}	5.5×10^{-9}		

The concentration c in the saturated solution was calculated from the measured conductivity κ from the equation

$$c = \frac{1000\kappa}{\Lambda^\circ}$$

Compiler's note

Neither in the determination of the ionic conductivity of the selenite ion nor in the evaluation of the solubility product was hydrolysis of the selenite ion taken into account. This would give rise to errors, since, for example, in a 0.001*M* solution, $[SeO_3^{2-}] = 0.000955M$, $[HSeO_3^-] = 0.000045M$ and $[OH^-] = 0.000045M$, and hydroxide and hydrogen selenite have different ionic conductivities from selenite. If the ionic conductivity of hydrogen selenite were known, the experimental results could have been interpreted correctly (cf. ref. 2), but this value does not seem to be available.

However, because the calibration and sample solutions had concentrations of about the same order of magnitude, the errors would cancel to some extent, but the K_{s0} value cannot be regarded as reliable.

AUXILIARY INFORMATION

METHOD APPARATUS/PROCEDURE:

The ionic conductivity of the selenite ion in water was determined by measuring the mean molar conductivities of a series of aqueous solutions of lithium selenite with concentrations below 2×10^{-3} mol dm^{-3}. The mean molar conductivity at infinite dilution was found by extrapolation to be $172 \pm 2.9\ \Omega^{-1} cm^2 mol^{-1}$ at 18°C. At that temperature, the ionic conductivity of the lithium ion is $33.4\ \Omega^{-1} cm^2 mol^{-1}$, so the ionic conductivity of the selenite ion is $105.2\ \Omega^{-1} cm^2 mol^{-1}$.

The well washed cadmium selenite was mixed with 150 ml of water and agitated intermittently for 1 hr. From this, 10-20 ml was taken in the conductivity cell and the conductivity measured. The sampling and measurement were repeated until a constant value for the conductivity was obtained.

The value of $\Lambda^\circ = (\lambda_+ + \lambda_-)$, the molar conductivity, was calculated with $\lambda_- = 105.2$ and $\lambda_+ = 90.2\ \Omega^{-1} cm^2 mol^{-1}$.

SOURCE AND PURITY OF MATERIALS:

Lithium selenite was prepared from sublimed selenium dioxide by neutralization with lithium hydroxide. An approximately 0.05*M* solution was standardized gravimetrically, and this was used to prepare the various dilutions.

The composition of the selenite was checked by analysis.

ESTIMATED ERROR:

Temperature: ±0.5 K

Error in K_{s0} (2*s*) = 1.1×10^{-9} (compiler)

REFERENCES:

1. Landolt-Bornstein *Physikalisch-Chemische Tabellen II* 1923, p. 1105.
2. Monk, C.B. *J. Chem. Soc.* 1949, 429.

COMPONENTS:

1. Cadmium selenite; $CdSeO_3$; [13814-59-0]

2a. Nitric acid; HNO_3; [7697-37-2]

2b. Sulfuric acid; H_2SO_4; [7664-93-9]

3. Water; H_2O; [7732-18-5]

ORIGINAL MEASUREMENTS:

Chukhlantsev, V.G.

Zh. Neorg. Khim. 1956, *1*, 2300-5; **Russ. J. Inorg. Chem. (Eng. Transl.)* 1956, *1*, 132-8.

VARIABLES:

HNO_3 and H_2SO_4 concentrations

One temperature: 293 K

PREPARED BY:

Mary R. Masson

EXPERIMENTAL VALUES:

All concentrations are expressed in units of mol dm^{-3}.

Soln.	Initial pH	Final pH	$[Cd^{2+}]$	pCd	log $\alpha_{L(H)}$	$p[SeO_3^{2-}]$	pK_{s0}
HNO_3	2.97	4.86	1.87×10^{-3}	2.73	3.13	5.86	8.59
	2.27	3.28	7.85×10^{-3}	2.11	4.77	6.88	8.99
	2.00	2.86	1.52×10^{-2}	1.82	5.24	7.06	8.88
H_2SO_4	2.53	4.23	3.03×10^{-3}	2.52	3.75	6.27	8.79
	2.12	3.02	9.25×10^{-3}	2.03	5.10	7.13	9.14
	2.03	2.91	1.22×10^{-2}	1.92	5.17	7.07	8.99

The average value is $K_{s0} = 1.29 \times 10^{-9}$ $mol^2 dm^{-6}$.

($pK_{s0} = 8.89$)

Notes.

$[Se_{tot}] = [Cd^{2+}]$ and $[SeO_3^{2-}] = [Se_{tot}]/\alpha_{L(H)}$

where $\alpha_{L(H)} = (1 + [H^+]/K_2 + [H^+]^2/K_1K_2)$

and the acid dissociation constants have the values $K_1 = 4 \times 10^{-3}$ and $K_2 = 1.0 \times 10^{-8}$ (ref. 1).

AUXILIARY INFORMATION

METHOD APPARATUS/PROCEDURE:

Solutions of nitric and sulfuric acids were saturated with cadmium selenite by stirring in a thermostat at 20°C for 8 hr. The remaining solid phase was removed by centrifugation, then the pH was measured ("Moskip" pH meter, to 0.01 pH unit) and the cadmium concentration was determined (method not stated).

SOURCE AND PURITY OF MATERIALS:

C.P.-grade reagents were used. Cadmium selenite was prepared by mixing 0.1*N* sodium selenite solution with a stoichiometric amount of 3% cadmium chloride solution. The precipitate was removed by centrifugation and dried at 40°C. Cadmium was determined gravimetrically as the sulfate, and selenium gravimetrically as the element.

ESTIMATED ERROR:

The spread in the results is 0.55 of a log unit.

Temperature: probably ±0.05 K

REFERENCES:

1. Rumpf, P. *Compt. Rendu* 1933, *197*, 686.

COMPONENTS:	ORIGINAL MEASUREMENTS:
1. Cadmium selenite; $CdSeO_3$; [13814-59-0] 2. Water; H_2O; [7732-18-5]	Redman, M.J.; Harvey, W.W. *J. Less-Common Met.* <u>1967</u>, *12*, 395-404.
VARIABLES:	PREPARED BY:
One temperature, probably 293 or 298 K	Mary R. Masson

EXPERIMENTAL VALUES:

A solution in contact with a precipitate of $CdSeO_3$ was found to contain 2.35×10^{-5} mol dm^{-3} cadmium(II) and 1.70×10^{-3} mol dm^{-3} total selenite at pH 6.0.

A value for the solubility product can be calculated as follows (compiler).

At pH 6.0, log $\alpha_{L(H)} = 2.00$ when $K_1 = 4 \times 10^{-3}$, $K_2 = 1 \times 10^{-8}$ (ref. 1).

Then, since $[SeO_3^{2-}] = [Se_{tot}]/\alpha_{L(H)} = 1.7 \times 10^{-5}$, and $[Cd^{2+}] = 2.35 \times 10^{-5}$ mol dm^{-3},

$K_{s0} = 4.0 \times 10^{-10}$ mol^2 dm^{-6}, and $pK_{s0} = 9.40$.

The temperature of the determination is not stated.

AUXILIARY INFORMATION

METHOD APPARATUS/PROCEDURE:	SOURCE AND PURITY OF MATERIALS:
The filtrate after precipitation of cadmium selenite was analysed for cadmium with dithizone, and for selenite with 3,3'-diaminobenzidine.	Not stated. ESTIMATED ERROR: Errors of $\pm 1\%$ in determination of Cd^{2+} and selenite, and of ± 0.1 pH unit would cause an error of $\pm 1.0 \times 10^{-10}$ in K_{s0}. (± 0.12 in pK_{s0}). REFERENCES: 1. Rumpf, P. *Compt. Rendu* <u>1933</u>, *197*, 686.

COMPONENTS:

1. Mercury(I) selenite; Hg_2SeO_3; [15855-76-2]
2a. Sulfuric acid; H_2SO_4; [7664-93-9]
2b. Nitric acid; HNO_3; [7697-37-2]
3. Water; H_2O; [7732-18-5]

ORIGINAL MEASUREMENTS:

Chukhlantsev, V.G.; Tomashevsky, G.P.

Zh. Anal. Khim. 1957, *12*, 296-301;
**J. Anal. Chem. USSR* 1957, *12*, 303-9.

VARIABLES:

Sulfuric and nitric acid concentrations
One temperature: 293 K

PREPARED BY:

Mary R. Masson

EXPERIMENTAL VALUES:

All concentrations are expressed in units of mol dm^{-3}.

Soln.	Initial pH	Final pH	$[Hg^+]$	pHg	$\log \alpha_{L(H)}$	$p[SeO_3^{2-}]$	pK_{s0}
H_2SO_4	2.69	2.72	5.8×10^{-5}	4.24	5.49	10.05	18.53
	2.06	2.20	1.3×10^{-4}	3.89	6.20	10.39	18.17
HNO_3	2.95	3.05	2.6×10^{-5}	4.58	5.03	9.92	19.08
	2.40	2.59	7.2×10^{-5}	4.14	5.66	10.10	18.38
	2.08	2.22	1.1×10^{-4}	3.96	6.18	10.44	18.36

The average value is $K_{s0} = 3.8 \times 10^{-19}$ $mol^3 dm^{-9}$.

$(pK_{s0} = 18.42)$

Notes.

$[Se_{tot}] = [Hg^+]$ and $[SeO_3^{2-}] = [Se_{tot}]/\alpha_{L(H)}$

where $\alpha_{L(H)} = (1 + [H^+]/K_2 + [H^+]^2/K_1K_2)$

and the acid dissociation constants have the values $K_1 = 4 \times 10^{-3}$ and $K_2 = 1.0 \times 10^{-8}$ (ref. 1).

(continued on next page)

AUXILIARY INFORMATION

METHOD APPARATUS/PROCEDURE:

Solutions of nitric and sulfuric acids were saturated with mercury(I) selenite by shaking in a thermostat at 20±0.05°C for 8 hr. The remaining solid phase was removed by centrifugation, then the pH was measured ("Moskip" pH meter, to 0.01 pH unit) and the mercury(I) concentration was determined by titration with 0.01*N* ammonium thiocyanate.

SOURCE AND PURITY OF MATERIALS:

C.P.-grade reagents wereused. Mercury(I) selenite was prepared by mixing 0.1*N* solutions of mercury(I) nitrate and sodium selenite in the cold. The precipitate was washed with water and dried at 20°C. Mercury was determined by titration with ammonium thiocyanate, and selenium was determined gravimetrically as the element after precipitation with hydrazine.

ESTIMATED ERROR:

$\pm 2.2 \times 10^{-19}$. (The spread in the results is 0.91 of a log unit.)
Temperature: ±0.05 K

REFERENCES:

1. Rumpf, P. *Compt. Rendu* 1933, *197*, 686.

COMPONENTS:
1. Mercury(I) selenite; Hg_2SeO_3; [15855-76-2]

2a. Sulfuric acid; H_2SO_4; [7664-93-9]

2b. Nitric acid; HNO_3; [7697-37-2]

3. Water; H_2O; [7732-18-5]

ORIGINAL MEASUREMENTS:

Chukhlantsev, V.G.; Tomashevsky, G.P.

Zh. Anal. Khim. 1957, *12*, 296-301;
**J. Anal. Chem. USSR* 1957, *12*, 303-9.

COMMENTS AND/OR ADDITIONAL DATA:

The authors were in error in treating the mercury(I)) ion as Hg^+, instead of as Hg_2^{2+}, and in expressing K_{s0} as $[Hg^+]^2.[SeO_3^{2-}]$ instead of as $[Hg_2^{2+}].[SeO_3^{2-}]$. The compiler hs recalculated the results in the correct form.

Soln.	Initial pH	Final pH	$[Hg_2^{2+}]$	pHg_2^{2+}	log $\alpha_{L(H)}$	$p[SeO_3^{2-}]$	pK_{s0}
H_2SO_4	2.69	2.72	2.9×10^{-5}	4.54	5.49	10.05	14.59
	2.06	2.20	6.5×10^{-5}	4.19	6.20	10.39	14.58
HNO_3	2.95	3.05	1.3×10^{-5}	4.89	5.03	9.92	14.81
	2.40	2.59	3.6×10^{-5}	4.44	5.66	10.10	14.54
	2.08	2.22	5.5×10^{-5}	4.26	6.18	10.44	14.70

The average value is $K_{s0} = 2.3 \times 10^{-15}$ mol^2 dm^{-6}.

$pK_{s0} = 14.64$.

COMPONENTS:

1. Mercury(I) selenite; Hg_2SeO_3; [15855--76-2]

2. Water; H_2O; [7732-18-5]

ORIGINAL MEASUREMENTS:

Redman, M.J.; Harvey, W.W.

J. Less-Common Met. 1967, *12*, 395-404.

VARIABLES:

One temperature, probably 293 or 298 K

PREPARED BY:

Mary R. Masson

EXPERIMENTAL VALUES:

A solution in contact with a precipitate of Hg_2SeO_3 was found to contain 4.97×10^{-5} mol dm^{-3} mercury(I) and 7×10^{-3} mol dm^{-3} total selenite at pH 3.4, and 4.57×10^{-4} mol dm^{-3} mercury(I) and 1.3×10^{-5} mol dm^{-3} total selenite at pH 2.2. Values for the solubility product can be calculated as follows (compiler).

pH	log $\alpha_{L(H)}$	[Se_{tot}]	[SeO_3^{2-}]	[Hg(I)]	[Hg^{2+}]	K_{s0} mol^2 dm^{-6}	pK_{s0}
3.4	4.64	7×10^{-3}	1.6×10^{-7}	4.97×10^{-5}	2.48×10^{-5}	4.0×10^{-12}	11.40
2.2	6.21	$<1.3 \times 10^{-3}$	2.98×10^{-10}	4.57×10^{-4}	2.27×10^{-4}	$<1.8 \times 10^{-15}$	>14.74

If it is assumed that the wrong pH values have been assigned to the two solutions, the values 12.97 and >13.17 can be calculated for pK_{s0} and 1.07×10^{-13} and $<6.76 \times 10^{-14}$ mol^2 dm^{-6} for K_{s0}.

The temperature of the determination is not stated.

*Calculated with $K_1 = 4 \times 10^{-4}$ and $K_2 = 1 \times 10^{-8}$ (ref. 1).

AUXILIARY INFORMATION

METHOD APPARATUS/PROCEDURE:

The results were obtained from the analysis of the filtrates obtained in two experiments. In one, the precipitation was done with excess of selenite, and in the other equivalent amounts were used. Mercury(I) was determined with dithizone after oxidation to mercury(II) with conc. nitric acid; and selenite was determined with 3,3'-diaminobenzidine.

SOURCE AND PURITY OF MATERIALS:

Not stated.

ESTIMATED ERROR:

REFERENCES:

1. Rumpf, P. *Compt. Rendu* 1933, *197*, 686.

COMPONENTS:

1. Mercury(II) selenite; $HgSeO_3$; [14459-36-0]
2. Water; H_2O; [7732-18-5]

ORIGINAL MEASUREMENTS:

Redman, M.J.; Harvey, W.W.

J. Less-Common Met. 1967, *12*, 395-404.

VARIABLES:

One temperature, probably 293 or 298 K

PREPARED BY:

Mary R. Masson

EXPERIMENTAL VALUES:

A solution in contact with a precipitate of $HgSeO_3$ was found to contain 7.83×10^{-4} mol dm^{-3} Hg^{2+} and 1.20×10^{-5} mol dm^{-3} total selenite at pH 2.2.

The solubility product can be calculated from this as follows (compiler).

At pH 2.2, log $\alpha_{L(H)}$ = 6.21 when the acid dissociation constants have the values $K_1 = 4 \times 10^{-3}$ and $K_2 = 1.0 \times 10^{-8}$ (ref. 1). Then, since $[SeO_3^{2-}] = [Se_{tot}]/\alpha_{L(H)} = 7.36 \times 10^{-12}$, and $[Hg^{2+}] = 7.83 \times 10^{-4}$ mol dm^{-3}, $K_{s0} = 5.76 \times 10^{-15}$ mol^2 dm^{-6}.

(pK_{s0} = 14.24.)

The temperature of the determination is not stated.

AUXILIARY INFORMATION

METHOD APPARATUS/PROCEDURE:

A solution of 0.02*M* $Hg(NO_3)_2$/0.01*M* HNO_3 was mixed with a 0.05*M* solution of K_2SeO_3. The precipitate was filtered off and the filtrate was analysed for mercury(II) and total selenite. Mercury(II) was determined with dithizone, and selenite by means of the reaction with 3,3'-diaminobenzidine.

SOURCE AND PURITY OF MATERIALS:

Not stated.

ESTIMATED ERROR:

Errors of ± 1% in the determination of Hg^{2+} and selenite, and an error of ±0.1 unit in the determination of pH would cause an error of approx. $\pm 2.5 \times 10^{-15}$ in K_{s0} (±0.2 in p*K*)

REFERENCES:

1. Rumpf, P. *Compt. Rendu* 1933, *197*, 686.

COMPONENTS:

1. Mercury(II) selenite; $HgSeO_3$; [14459-36-0]
2. Sodium selenite; Na_2SeO_3; [10102-18-8]
3. Water; H_2O; [7732-18-5]

ORIGINAL MEASUREMENTS:

Rosenheim, A.; Pritze, M.

Z. Anorg. Chem. 1909, *63*, 275-81.

VARIABLES:

Concentration of sodium selenite
One temperature: 298 K

PREPARED BY:

Mary R. Masson

EXPERIMENTAL VALUES:

All concentrations are expressed in units of mol dm^{-3}.

$[Na_2SeO_3]$ (total)	Solution wt., g	HgS found, g	$[Hg(SeO_3)_2^{2-}]$ x 10^2	"K" x 10^2	$[SeO_3^{2-}]^a$ (free)	$K_{s2}{}^a$ x 10^2
2.0	5.7554 3.4477	0.2762 0.0640	8.35	4.18	1.92	4.36
1.0	4.6239 9.3640	0.0462 0.0911	4.25	4.25	0.96	4.44
0.5	7.5800 9.6050	0.0381 0.0471	2.14	4.28	0.48	4.47
0.25	7.5130 7.2401	0.0288 0.0264	1.61	(6.44)	0.23	(6.88)
0.125	6.8250 9.4330	0.0151 0.0214	0.97	(7.76)	0.115	(8.41)
0.0625	9.7856 7.4048	0.0121 0.0098	0.55	(8.80)	0.057	(9.65)

[a] Calculated by compiler.

The mean value of K_{s2} is 4.42 x 10^{-2}

pK_{s2} = 1.35

Note: $K_{s2} = [Hg(SeO_3)_2^{2-}]/[SeO_3^{2-}]$, "$K$" = $[Hg(SeO_3)_2^{2-}]$/(total selenite)

AUXILIARY INFORMATION

METHOD APPARATUS/PROCEDURE:

Solutions of sodium selenite of various concentrations were saturated with mercuric selenite at 25°C. Samples of the saturated solutions were diluted, raised to boiling, then 1 - 2 g of potassium cyanide was added, and the boiling continued for 5 min. Hydrogen sulfide was passed into the solutions until all the mercury had been precipitated, then the mercuric sulfide precipitates were collected on Gooch crucibles and weighed.

SOURCE AND PURITY OF MATERIALS:

Mercuric selenite was prepared by mixing solutions of sodium selenite and mercuric nitrate.

ESTIMATED ERROR:

The differences between the duplicate determinations range from 2.3 to 7.2%.

REFERENCES:

COMPONENTS:

1. Mercury(II) selenite; $HgSeO_3$; [14459-36-0]
2. Nitric acid; HNO_3; [7697-37-2]
3. Sodium nitrate; $NaNO_3$; [7631-99-4]
4. Water; H_2O; [7732-18-5]

ORIGINAL MEASUREMENTS:

Toropova, V.F.

Zh. Neorg. Khim. 1957, 2, 515-22; *Russ. J. Inorg. Chem. (Eng. Transl.)* 1957, 2, 63-76.

VARIABLES:

HNO_3 concentration

One temperature: 298 K

PREPARED BY:

Mary R. Masson

EXPERIMENTAL VALUES:

All concentrations are expressed in units of mol dm^{-3}.

pH	$[Hg^{2+}]$	$[SeO_3^{2-}]^a$	K_{s0}^a, $mol^2 dm^{-6}$	$[SeO_3^{2-}]^b$	K_{s0}^b, $mol^2 dm^{-6}$
1.00	9×10^{-4}	1.6×10^{-11}	1.4×10^{-14}	2.31×10^{-12}	2.08×10^{-15}
1.26	5×10^{-4}	2.7×10^{-11}	1.6×10^{-14}	4.16×10^{-12}	2.08×10^{-15}
1.30	4.7×10^{-4}	3.1×10^{-11}	1.4×10^{-14}	4.70×10^{-12}	2.21×10^{-15}

[a] Results quoted by author; constants for H_2SeO_3, $K_1 = 3.5 \times 10^{-3}$, $K_2 = 5 \times 10^{-8}$

[b] Results recalculated by compiler; constants for H_2SeO_3 - $K_1 = 2.63 \times 10^{-3}$, $K_2 = 1.0 \times 10^{-8}$ (Ref. 1).

The average value for K_{s0}, as calculated by author, is 1.45×10^{-14} mol^2 dm^{-6}

The average value for K_{s0}, as recalculated by compiler, is 2.16×10^{-15} mol^2 dm^{-6}

(pK_{s0} = 13.84 and 14.67).

Note $[Se_{tot}] = [Hg^{2+}]$ and $[SeO_3^{2-}] = [Se_{tot}]/\alpha_{L(H)}$

where $\alpha_{L(H)} = (1 + [H^+]/K_2 + [H^+]^2/K_1K_2)$

AUXILIARY INFORMATION

METHOD APPARATUS/PROCEDURE:

A 1*M* solution of sodium nitrate, acidified with nitric or sulfuric acid, was saturated with mercury selenite by shaking at 25 $\pm$ 0.05°C until equilibrium was established (after 6 hr). The solution was analysed for mercury polarographically, after separation of selenium as the element.

SOURCE AND PURITY OF MATERIALS:

A small excess of sodium selenite was reacted with a solution of mercuric nitrate. The precipitate was washed with distilled water, and dried at 105°C. Selenium was determined polarographically as the selenosulfate ion after precipitation as the element, and mercury gravimetrically the sulfide after separation of the selenium.

ESTIMATED ERROR:

The spread in the results for pK_{s0} is 0.04 of a log unit.

Temperature: $\pm$0.05 K

REFERENCES:

1. Sabbah, R.; Carpeni, G. *J. Chim. Phys.* 1956, 63, 1549.

COMPONENTS:

1. Mercury(II) selenite; $HgSeO_3$; [14459-36-0]
2. Sodium selenite; Na_2SeO_3; [10102-18-8]
3. Sodium nitrate; $NaNO_3$; [7631-14-5]
4. Water; H_2O; [7732-18-5]

ORIGINAL MEASUREMENTS:

Toropova, V.F.

Zh. Neorg. Khim. 1957, 2, 515-22; *Russ. J. Inorg. Chem. (Eng. Transl.)* 1957, 2, 63-76.

VARIABLES:

Sodium selenite concentration

PREPARED BY:

Mary R. Masson

EXPERIMENTAL VALUES:

All concentrations are expressed in units of mol dm^{-3}.
The results presented here were read from the author's Fig. 2 by the compiler. However, the -log *C* axis is labelled as follows (at equal spacings): 1.8, 2.0, 2.2, 2.3, 2.4, 2.6, 2.8. Thus it is not certain whether the data presented here are correct.

-log *A*	-log *C*
0.71	2.0
0.975	2.205
1.20	2.305
1.295	2.345
1.62	2.515
2.0	2.815

The author calculated that the log of the formation constant of $Hg(SeO_3)_2^{2-}$ is 12.48, whence, since pK_{s0} = 13.84, pK_{s2} = 1.36. However, the slope of the graph is not, as stated by the author, close to 1, especially when the data are replotted on a correctly labelled graph. It is not at all clear just how the data should best be interpreted.

$-\log A = -\log [SeO_3^{2-}]$

$-\log C = -\log [Hg(SeO_3)_2^{2-}]$

AUXILIARY INFORMATION

METHOD APPARATUS/PROCEDURE:

Solutions of ionic strength equal to unity, containing varying concentrations of sodium selenite and sodium nitrate were saturated with mercuric selenite, by shaking at 25 ± 0.05°C until equilibrium was established. The complex ion concentration, C_M, was taken as equal to the total concentration of mercury ions in the saturated solution (determined polarographically). The equilibrium selenite concentration was calculated from the total concentration.

SOURCE AND PURITY OF MATERIALS:

A small excess of sodium selenite was reacted with a solution of mercuric nitrate. The precipitate was washed with distilled water, and dried at 105°C. Selenium was determined polarographically as the selenosulfate ion after precipitation as the element, and mercury gravimetrically as the sulfide after separation of the selenium.

ESTIMATED ERROR:

No estimates possible.

REFERENCES:

COMPONENTS:	ORIGINAL MEASUREMENTS:
1. Lead(II) selenite; $PbSeO_3$; [7488-51-9] 2. Hydrochloric acid; HCl; [7647-01-0] 3. Nitric acid; HNO_3; [7697-37-2] 4. Water; H_2O; [7732-18-5]	Chukhlantsev, V.G.; Tomashevsky, G.P. *Zh. Anal. Khim.* 1957, *12*, 296-301; **J. Anal. Chem. USSR* 1957, *12*, 303-9.
VARIABLES: Hydrochloric and nitric acid concentrations One temperature: 293 K	PREPARED BY: Mary R. Masson

EXPERIMENTAL VALUES:

All concentrations are expressed in units of mol dm^{-3}.

Soln.	Initial pH	Final pH	$[Pb^{2+}]$	pPb	$\log \alpha_{L(H)}$	$p[SeO_3^{2-}]$	pK_{s0}
HCl	2.05	2.21	2.9×10^{-3}	2.54	6.20	8.74	11.28
	2.79	3.12	4.2×10^{-4}	3.38	4.95	8.33	11.71
HNO_3	2.95	3.38	3.1×10^{-4}	3.47	4.66	8.13	11.60
	2.40	2.85	8.3×10^{-4}	3.08	5.25	8.33	11.41
	2.08	2.25	2.3×10^{-3}	2.64	6.15	8.79	11.43

The average value is $K_{s0} = 3.4 \times 10^{-12}$ $mol^2 dm^{-6}$.

($pK_{s0} = 11.5$)

Notes.

$[Se_{tot}] = [Pb^{2+}]$ and $[SeO_3^{2-}] = [Se_{tot}]/\alpha_{L(H)}$

where $\alpha_{L(H)} = (1 + [H^+]/K_2 + [H^+]^2/K_1K_2)$

and the acid dissociation constants have the values $K_1 = 4 \times 10^{-3}$ and $K_2 = 1.0 \times 10^{-8}$ (ref. 1).

AUXILIARY INFORMATION

METHOD APPARATUS/PROCEDURE:

Solutions of hydrochloric and nitric acid were saturated with lead(II) selenite by shaking in a thermostat at 20±0.05°C for 8 hr. The remaining solid phase was removed by centrifugation, then the pH was measured ("Moskip" pH meter, to 0.01 pH unit) and the lead concentration was determined by gravimetry of the sulfate.

SOURCE AND PURITY OF MATERIALS:

C.P.-grade reagents were used. Lead selenite was prepared by mixing a 0.2*N* solution of lead acetate acidified with acetic acid with a 0.1*N* solution of sodium selenite in stoichiometric proportions at 50 - 60°C. The precipitate was washed with water and dried at 40°C. Lead was determined gravimetrically as the sulfate, and selenium as the element.

ESTIMATED ERROR:

$\pm 1.3 \times 10^{-12}$. (The spread in the results is 0.43 of a log unit.)
Temperature: ±0.05 K

REFERENCES:

1. Rumpf, P. *Compt. Rendu* 1933, *197*, 686.

COMPONENTS:

1. Lead selenite; $PbSeO_3$; [7488-51-9]
2. Water; H_2O; [7732-18-5]

ORIGINAL MEASUREMENTS:

Dolique, R.

Bull. Soc.Chim. France 1943, *10*, 50.

VARIABLES:

One temperature: 283 K

PREPARED BY:

Mary R. Masson

EXPERIMENTAL VALUES:

The author found the solubility of lead selenite in water at 10°C to be about 18.3 mg in 100 ml; after 7 days agitation in 10°C, the amount dissolved had reached only 11 mg in 100 ml.

18.3 mg/100 ml can be expressed as 5.48×10^{-4} mol dm^{-3}; this would give a value of 3.0×10^{-7} mol^2 dm^{-6} for K_{s0} ($pK_{s0} = 6.52$) if hydrolysis is neglected.

AUXILIARY INFORMATION

METHOD APPARATUS/PROCEDURE:

The lead selenite was agitated in water, then the concentration of selenium in solution was determined by a method developed by Dolique, Perahia and Roca.

SOURCE AND PURITY OF MATERIALS:

Lead selenite was prepared by traditional method of "double decomposition", thoroughly washed, but not dried.

ESTIMATED ERROR:

Temperature: ±1 K
Solubility: no estimate possible.

REFERENCES:

COMPONENTS:	EVALUATOR:
Tellurites	Mary R. Masson, Dept. of Chemistry, University of Aberdeen, Meston Walk, Old Aberdeen, AB9 2UE, Scotland, UK. July 1984.

CRITICAL EVALUATION:

LITHIUM TELLURITE

The binary system lithium tellurite - water has been studied only once (1). The regression equation for the data is

$$y = 23.5 - 0.505(T - 273.2) + 0.00542(T - 273.2)^2 - 0.0000234(T - 273.2)^3$$

$s = 0.08$ (6 points)

where $y = 100w$ is the concentration of lithium tellurite in mass %, T is the temperature in K, and s is the standard deviation of the dependent variable about the regression line.

TENTATIVE (SMOOTHED) VALUES

T/K	Solubility mass %	molality mol/kg
303.2	12.60	0.761
313.2	10.47	0.617
323.2	8.88	0.514
333.2	7.66	0.438
343.2	6.68	0.378
353.2	5.81	0.326

SODIUM TELLURITE

There has been no study of the binary system sodium tellurite - water, but some data are available from ternary systems (2 - 5). It was not possible to fit a satisfctory regression equation to the available data, because of the lack of good agreement and the small number of points available. The solid phase was $Na_2TeO_3.5H_2O$ [22451-06-5], except at 363.2 K, where it was the anhydrous salt.

TENTATIVE VALUES

T/K	Solubility mass %	molality mol/kg	Ref.
298.2	45.03	3.697	3,4 (mean)
303.2	46.23	3.880	5
333.2	51.68	4.827	5
343.2	55.54	5.638	2
363.2	52.46	4.980	5

COMPONENTS:	EVALUATOR:
Tellurites	Mary R. Masson, Dept. of Chemistry, University of Aberdeen, Meston Walk, Old Aberdeen, AB9 2UE, Scotland, UK. July 1984.

CRITICAL EVALUATION: (continued)

TERNARY SYSTEMS

The ternary systems studied were sodium tellurite - sodium hydroxide - water (2,3), sodium tellurite - sodium carbonate - water (4,5), sodium tellurite - sulfuric acid - water (6), sodium tellurite - perchloric acid - water (7), sodium selenite - sodium tellurite - water (8) and sodium telllurite - ethanol - water (9). No comparisons were possible.

OTHER TELLURITES

The solubility of cesium tellurite was found to be 67.65% (molality 1.550 mol/kg) at 291.1 K (10).

The solubilities of some sparingly soluble tellurites are reported to be as follows:

Ion	K_{s0}	pK_{s0}	T/K	Ref.
Ba^{2+}	8.24×10^{-8} to 2.17×10^{-5} mol^2dm^{-6}	4.66 - 7.08	298.2	11
Co^{2+}	3.1×10^{-7} mol^2dm^{-6}	6.51	298.2?	12
Ni^{2+}	2.34×10^{-10} mol^2dm^{-6} (HCl)	9.63	298.2?	12
	3.84×10^{-10} mol^2dm^{-6} (H_2SO_4)	9.42	298.2?	12
Cu^{2+}	1.11×10^{10} mol^2dm^{-6} (H_2SO_4)	9.95	298.2?	12
	1.6×10^{-11} mol^2dm^{-6} (HCl)	10.80	298.2?	12
Ag^{+}	3.7×10^{-3} mol^3dm^{-9}	2.43??	298.2	13
	1.41×10^{-18} mol^3dm^{-9}	17.85	298.2	14
	1.17×10^{-18} mol^3dm^{-9} (recalc.)	17.93	298.2	
	8.71×10^{-19} mol^3dm^{-9}	18.06	293.2	15
Pb^{2+}	4.07×10^{-11} to 5.93×10^{-8} mol^2dm^{-6}	7.23 - 10.99	298.2	11

GENERAL COMMENTS

It is evident that there is very little information available on the solubility of tellurites, and unfortunately much of what is available appears to be not very reliable. Of the silver tellurite values, the values reported in (14) and (15) are in reasonable agreement, so it seems that the values in (13) should be rejected.

COMPONENTS:	EVALUATOR:
Tellurites	Mary R. Masson, Dept. of Chemistry, University of Aberdeen, Meston Walk, Old Aberdeen, AB9 2UE, Scotland, UK. July 1984.

CRITICAL EVALUATION: (continued)

REFERENCES

1. Breusov, O.N.; Revzina, T.V.; Druz, N.A. *Zh. Neorg. Khim.* 1965, *10*, 1990; **Russ. J. Inorg. Chem. (Eng. Transl.)* 1965, *10*, 1084.

2. Lavut, E.A.; Vorob'eva, O.I.; Shulgina, I.M. *Zh. Neorg. Khim.* 1961, *6*, 2758; **Russ. J. Inorg. Chem. (Eng. Transl.)* 1961, *6*, 1394.

3. Lavut, E.A.; Vorob'eva, O.I. *Zh. Neorg. Khim.* 1960, *5*, 1813; **Russ. J. Inorg. Chem. (Eng. Transl.)* 1960, *5*, 880.

4. Kunev, D.K.; Vassilev, H. *C.R. Acad. Bulg. Sci.* 1968, *21*, 233.

5. Chimbulev, M.; Vasilev, Kh.; Kunev, D. *Khim. Ind. (Sofia)* 1973, *45*, 71.

6. Babayan, G.G.; Kapantsyan, E.E.; Arutyunyan, M.G.; Akopyan, Z.A. *Arm. Khim. Zh.* 1973, *26*, 467.

7. Masson, M.R. *J. Inorg. Nucl. Chem.* 1976, *38*, 545.

8. Chimbulev, M.; Vasilev, Kh. *God. Vissh. Khim.-Tekhnol. Inst., Sofia* 1977, 22, 247.

9. Vorob'eva, O.I.; Lavut, E.A. *Zh. Neorg. Khim.* 1957, 2, 1154; **Russ. J. Inorg. Chem. (Eng. Transl.)* 1957, 2, 261.

10. Lavut, E.A. *Vestn. Mosk. Univ. Ser. II, Khim.* 1966, *21*, 91.

11. Ganelina, E.Sh.; Merzon, V.V.; Biryukov, V.P. *Izv. Vyssh. Ucheb. Zaved. Khim. Khim. Tekhnol.* 1969, *12*, 1465.

12. Ganelina, E.Sh. *Zh. Priklad. Khim.* 1967, *40*, 1019; **J. Appl. Chem. USSR (Eng. Transl.)* 1967, *40*, 983.

13. Ganelina, E.Sh.; Pozhidaeva, T.N. *Zh. Priklad. Khim.* 1965, *38*, 2210; **J. Appl. Chem. (Eng. Transl.)* 1965, *38*, 2168.

14. Mehra, M.C.; Kahn, S.M. *Can. J. Chem.* 1972, *50*, 1788.

15. Chao, E.E.; Cheng, K.L. *Anal. Chem.* 1976, *48*, 267.

COMPONENTS:

1. Lithium tellurite; Li_2TeO_3; [14929-69-2]

2. Water; H_2O; [7732-18-5]

ORIGINAL MEASUREMENTS:

Breusov, O.N.; Revzina, T.V.; Druz, N.A.

Zh. Neorg. Khim. 1965, *10*, 1990-2;
**Russ. J. Inorg. Chem.* 1965, *10*, 1084-5.

VARIABLES:

Temperature: 303 - 353 K

PREPARED BY:

Mary R. Masson

EXPERIMENTAL VALUES:

t/°C	Li_2O mass %	TeO_2 mass %	Li_2TeO_3[a] mass %	Li_2O : TeO_2 molar ratio	Li_2TeO_3[b] mol/kg
30	1.99	10.69			
30	1.99	10.70	12.65 ± 0.05	0.994:1	0.764
40	1.65	8.79			
40	1.64	8.70			
40	1.65	8.79	10.44 ± 0.01	1.001:1	0.615
50	1.41	7.58			
50	1.39	7.59			
50	1.43	7.57	8.96 ± 0.04	0.994:1	0.519
60	1.22	6.50			
60	1.22	6.48			
60	1.22	6.52	7.72 ± 0.01	1.006:1	0.442
70	1.05	5.63			
70	1.00	5.62			
70	1.10	5.64	6.66 ± 0.02	0.996:1	0.377
80	0.92	4.97			
80	0.915	4.97			
80	0.92	4.97	5.86 ± 004	0.987:1	0.329

[a] Mean value calculated from Li_2O and TeO_2 concentrations.

[b] Molalities calculated by the compiler.

AUXILIARY INFORMATION

METHOD APPARATUS/PROCEDURE:

Isothermal dissolution: equilibrium was established within 12 hr. Solutions were presumably analysed in the same way as the lithium tellurite.

The lithium tellurite had been purified by dissolving it in water, filtering, and evaporating in an atmosphere free of carbon dioxide.

SOURCE AND PURITY OF MATERIALS:

Lithium tellurite was prepared by treating a solution of analytical grade lithium hydroxide with an excess of freshly precipitated tellurium dioxide, filtering off the residue, and evaporating the solution almost to dryness. TeO_2 was determined by oxidation with excess of standard dichromate, and back-titration with iron(II). Lithium oxide was determined by titration with HCl to Methyl Orange.

ESTIMATED ERROR:

Temperature: ±0.1 K
Analyses: see table of solubilities

REFERENCES:

COMPONENTS:	ORIGINAL MEASUREMENTS:
1. Sodium tellurite; Na_2TeO_3; [10102-20-2] 2. Sodium hydroxide; NaOH; [1310-73-2] 3. Water; H_2O; [7732-18-5]	Lavut, E.A.; Vorob'eva, O.I. *Zh. Neorg. Khim.* 1960, 5, 1813-8; **Russ. J. Inorg. Chem.* 1960, 5, 880-2.
VARIABLES:	PREPARED BY:
Concentrations of the components One temperature: 298 K	Mary R. Masson

EXPERIMENTAL VALUES:

Solubility in the $Na_2O - TeO_2 - H_2O$ system at 25°C

Na_2TeO_3 mass %	NaOH mass %	Na_2O mass %	TeO_2 mass %	Na_2O[a] mol/kg	TeO_2[a] mol/kg	Solid[b] phase
-	52.97	41.05	0.	11.235	0.	A
0.35	51.69	40.17	0.25	10.878	0.026	A
0.35	50.66	39.36	0.25	10.516	0.026	A + B
0.69	43.62	34.00	0.50	8.375	0.048	B
0.79	41.79	32.60	0.57	7.870	0.053	B
0.92	39.39	30.79	0.65	7.246	0.059	B
2.71	35.37	28.17	1.95	6.504	0.175	B
6.32	32.27	26.78	4.55	6.292	0.415	C
8.16	26.50	22.82	5.88	5.164	0.517	C
23.30	12.48	16.19	16.78	3.897	1.569	C
37.59	3.65	13.35	27.07	3.615	2.847	C
44.95	-	12.56	32.39	3.681	3.687	C
45.17	-	12.57	32.60	3.699	3.725	C

(continued on next page)

AUXILIARY INFORMATION

METHOD APPARATUS/PROCEDURE:	SOURCE AND PURITY OF MATERIALS:
The solubility of sodium tellurite in concentrated sodium hydroxide solutions was measured in Plexiglas vessels: equilibrium was reached in 2 - 3 days. The solutions and the solid phases were analysed for Te by the chromate method, and for sodium by alkalimetry or by precipitation as sodium zinc uranyl acetate. The solid phases were identified by Schreinemakers' method (2). To study the region where polytellurites may be expected, sodium tellurite solutions were mixed with tellurium dioxide or a previously synthesized polytellurite.	Tellurium dioxide and sodium tellurite were freshly prepared (1). ESTIMATED ERROR: No estimates possible. REFERENCES: 1. Vorob'eva, O.I.; Lavut, E.A. *Zh. Neorg. Khim.* 1958, 3, 2006.

COMPONENTS:	ORIGINAL MEASUREMENTS:
1. Sodium tellurite; Na_2TeO_3; [10102-20-2] 2. Sodium hydroxide; NaOH; [1310-73-2] 3. Water; H_2O; [7732-18-5]	Lavut, E.A.; Vorob'eva, O.I. *Zh. Neorg. Khim.* 1960, 5, 1813-8; **Russ. J. Inorg. Chem.* 1960, 5, 880-2.

EXPERIMENTAL VALUES (continued):

Na_2O mass %	TeO_2 mass %	Na_2O[a] mol/kg	TeO_2[a] mol/kg	Solid[b] phase
12.79	34.60	3.922	4.121	C
12.90	36.30	4.097	4.477	C + D
12.78	35.75	4.006	4.352	D
12.62	34.91	3.881	4.169	D
11.93	34.53	3.595	4.041	E
11.67	33.23	3.417	3.779	E
10.75	31.33	2.995	3.389	E
10.00	27.67	2.589	2.782	F
9.46	24.70	2.318	2.351	F
7.45	20.34	1.665	1.765	F
5.82	16.89	1.215	1.369	F
5.37	14.80	1.085	1.162	F
4.46	13.35	0.876	1.018	F
3.62	10.52	0.680	0.768	F
1.47	4.48	0.252	0.298	F
1.09	3.36	0.184	0.220	F
1.05	2.75	0.176	0.179	F
0.78	2.43	0.130	0.157	G
0.65	2.03	0.108	0.131	G

[a] Molalities calculated by the compiler.

[b] Solid phases: A - $NaOH.H_2O$, B - Na_2TeO_3, C - $Na_2TeO_3.5H_2O$, D - $Na_2TeO_2O_5.3H_2O$, E - $Na_2Te_3O_7.5H_2O$, F - $Na_2Te_4O_9.5H_2O$, G - TeO_2

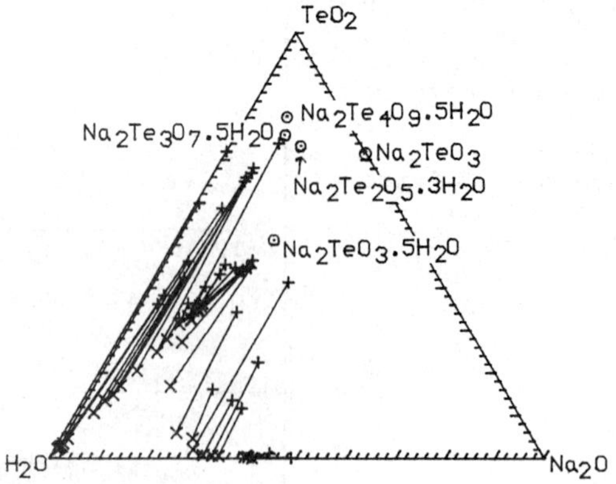

COMPONENTS:

1. Sodium tellurite; Na_2TeO_3; [10102-20-2]
2. Sodium hydroxide; NaOH; [1310-73-2]
3. Water; H_2O; [7732-18-5]

ORIGINAL MEASUREMENTS:

Lavut, E.A.; Vorob'eva, O.I.; Shul'gina, I.M.

Zh. Neorg. Khim. 1961, *6*, 2758-61;
**Russ. J. Inorg. Chem.* 1961, *6*, 1394-6.

VARIABLES:

Concentrations of the components
One temperature: 343 K

PREPARED BY:

Mary R. Masson

EXPERIMENTAL VALUES:

Solubility in the Na_2O - TeO_2 - H_2O system at 70°C

Na_2TeO_3 mass %	NaOH mass %	Na_2O mass %	TeO_2 mass %	Na_2O[a] mol/kg	TeO_2[a] mol/kg	Solid[b] phase
0.79	44.68	34.80	0.57	8.687	0.055	B
11.58	25.44	22.96	8.34	5.392	0.761	B
24.92	16.47	19.73	17.95	4.772	1.776	B
38.13	9.72	18.19	27.46	5.400	3.166	B
43.12	6.81	17.35	31.06	5.426	3.772	C
46.13	4.94	16.73	33.23	5.394	4.161	C
50.58	2.66	16.21	36.44	5.523	4.822	C
52.90	1.08	15.64	38.11	5.456	5.163	C
53.70	0.30	15.25	38.76	5.350	5.281	C
55.53	-	15.54	39.99	5.638	5.634	C

(continued on next page)

AUXILIARY INFORMATION

METHOD APPARATUS/PROCEDURE:

Isothermal dissolution and isothermal crystallization methods were used to determine the solubilities. Glass or Perspex vessels were employed; equilibrium was usually reached after 8 - 10 hr. The solutions and the solid phases, after separation, were analysed for tellurium and sodium (1), and the compositions of the solid phases were established by Schreinemakers' method.

SOURCE AND PURITY OF MATERIALS:

Tellurium dioxide, sodium tellurite and sodium tetratellurite were synthesized.

ESTIMATED ERROR:

No estimates possible.

REFERENCES:

1. Lavut, E.A.; Vorob'eva, O.I. *Zh. Neorg. Khim.* 1960, *3*, 1813.

COMPONENTS:	ORIGINAL MEASUREMENTS:
1. Sodium tellurite; Na_2TeO_3; [10102-20-2]	Lavut, E.A.; Vorob'eva, O.I.; Shul'gina, I.M.
2. Sodium hydroxide; NaOH; [1310-73-2]	*Zh. Neorg. Khim.* 1961, *6*, 2758-61;
3. Water; H_2O; [7732-18-5]	**Russ. J. Inorg. Chem.* 1961, *6*, 1394-6.

EXPERIMENTAL VALUES (continued):

Na_2O mass %	TeO_2 mass %	Na_2O[a] mol/kg	TeO_2[a] mol/kg	Mole ratio $Na_2O:TeO_2$	Solid[b] phase
15.54	39.99	5.638	5.634	1:1.0	C
15.22	40.75	5.577	5.799	1:1.04	C + D
15.22	40.33	5.524	5.685	1:1.02	C + D
14.05	39.26	4.855	5.269	1:1.08	D
13.92	37.73	4.645	4.889	1:1.05	D
13.32	37.14	4.338	4.697	1:1.08	D
11.93	35.02	3.628	4.136	1:1.13	D
10.93	34.64	3.240	3.988	1:1.23	F
10.90	32.52	3.108	3.601	1:1.16	F
7.27	22.74	1.676	2.036	1:1.22	F
5.71	18.15	1.210	1.494	1:1.23	F
2.28	8.06	0.410	0.563	1:1.36	F
1.73	6.14	0.303	0.418	1:1.37	F
1.05	3.62	0.178	0.238	1:1.34	F
0.70	2.73	0.117	0.177	1:1.51	G

[a] Molalities calculated by the compiler.

[b] Solid phases: B - Na_2TeO_3, C - $Na_2TeO_3.5H_2O$, D - $Na_2Te_2O_5.3H_2O$, F- $Na_2Te_3O_7.5H_2O$, G - TeO_2

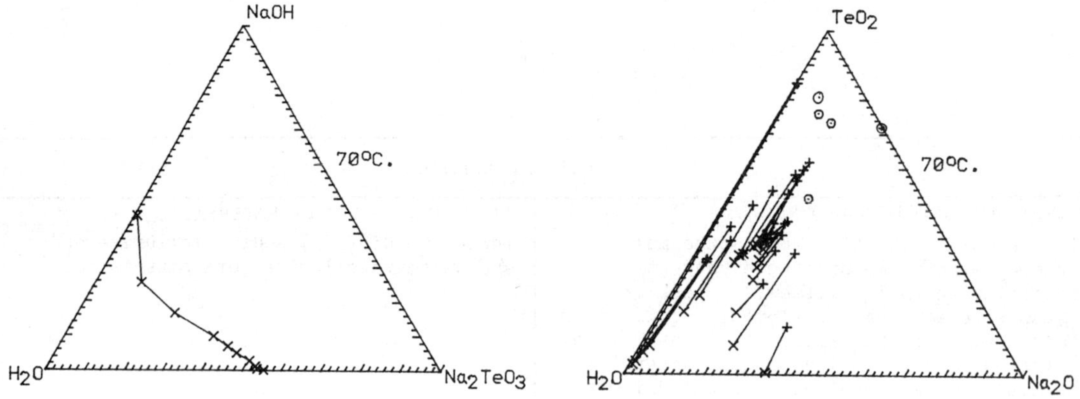

⊙ - see page 408

COMPONENTS:	ORIGINAL MEASUREMENTS:
1. Sodium tellurite; Na_2TeO_3; [10102-20-2] 2. Sodium carbonate; Na_2CO_3; [497-19-8] 3. Water; H_2O; [7732-18-5]	Kunev, D.K.; Vassilev, H. *C.R. Acad. Bulg. Sci.* 1968, *21*, 233-5.

VARIABLES:	PREPARED BY:
Concentrations of the components One temperature: 298 K	Mary R. Masson

EXPERIMENTAL VALUES:

Composition of the solution phase at 25°C

Na_2TeO_3 mass %	Na_2CO_3 mass %	Na_2TeO_3 mol/kg	Na_2CO_3[a] mol/kg	Solid[b] phase
44.97	-	3.688	0.0	A
40.97	1.15	3.195	0.187	A
38.12	4.01	2.973	0.654	B
36.08	6.14	2.818	1.003	B
33.12	8.29	2.551	1.335	B
30.80	11.03	2.390	1.789	B
28.71	13.76	2.252	2.257	B
27.02	14.23	1.076	2.285	C
25.92	14.76	1.972	2.348	C
24.84	15.92	1.892	2.535	C

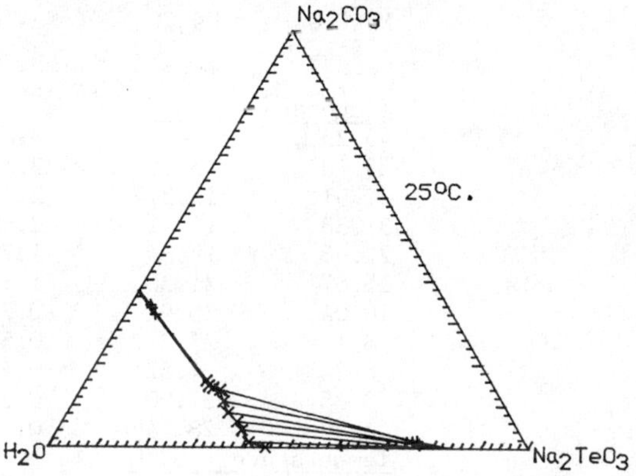

[a] Molalities calculated by the compiler.

[b] Solid phases: A - $Na_2TeO_3.5H_2O$, B - $Na_2TeO_3.3H_2O$, C - $Na_2CO_3.10H_2O$

AUXILIARY INFORMATION

METHOD APPARATUS/PROCEDURE:

Isothermal dissolution and crystallization were used to determine the solubilities in the system. Equilibrium was established in 18 - 20 hr. Glass covered vessels were used, and they were kept in a thermostat at 25 ± 0.5°C while equilibrium was established. The liquid and solid phases were analysed for tellurium and sodium by titration with dichromate and acid, respectively. The solid phases were identified by the method of Schreinemakers.

SOURCE AND PURITY OF MATERIALS:

Sodium tellurite was prepared by sintering equivalent amounts of sodium carbonate and tellurium dioxide at 550° in an atmosphere of CO_2. The TeO_2 used in the synthesis was repeatedly purified by dissolution in sodium hydroxide, neutralization to pH 9 - 10, where impurities precipitated as hydroxides, then neutralization of the filtrate to pH 5.5 to precipitate TeO_2.

ESTIMATED ERROR:

Temperature: ±0.5 K

REFERENCES:

COMPONENTS:

1. Sodium tellurite; Na_2TeO_3; [10102-20-2]
2. Sodium carbonate; Na_2CO_3; [497-19-8]
3. Water; H_2O; [7732-18-5]

ORIGINAL MEASUREMENTS:

Chimbulev, M.; Vasilev, Kh.; Kunev, D.

Khim. Ind. (Sofia) 1973, *45*, 71-3.

VARIABLES:

Concentrations of the components
Three temperatures: 303, 333 and 363 K

PREPARED BY:

Mary R. Masson

EXPERIMENTAL VALUES: Composition of the solution phase

Na_2TeO_3 mass %	Na_2CO_3 mass %	Na_2TeO_3[a] mol/kg	Na_2CO_3[a] mol/kg	Solid[b] phase
Temperature = 30°C				
46.23	-	3.800	0.	A
39.10	5.68	3.196	0.970	A
32.91	12.37	2.714	2.133	A
31.33	14.52	2.611	2.530	A + D
22.98	17.84	1.752	2.844	D
15.67	21.30	1.122	3.188	D
10.04	25.91	0.707	3.817	D
8.1	27.2	0.565	3.966	D + C
6.24	27.52	0.425	3.920	C
3.35	27.90	0.220	3.829	C
-	28.14	0.	3.695	C
Temperature = 60°C				
51.68		4.827	0.	A
48.53	2.03	4.430	0.387	A
46.07	4.26	4.186	0.809	A
44.15	6.73	4.056	1.293	A + E
41.61	8.69	3.778	1.650	E
33.02	10.46	2.637	1.746	E
27.14	14.32	2.092	2.308	E
21.10	17.65	1.555	2.719	E
14.21	20.97	0.989	3.052	E
7.96	25.58	0.541	3.631	E
-	31.42	0.	4.323	E

(continued on next page)

AUXILIARY INFORMATION

METHOD APPARATUS/PROCEDURE:

The method of isothermal dissolution was used. Equilibrium required 8 - 12 hr. The solid phases were identified by the method of Streinemekers, and by X-ray methods.

SOURCE AND PURITY OF MATERIALS:

ESTIMATED ERROR:

Temperature: ±0.1 K

REFERENCES:

COMPONENTS:	ORIGINAL MEASUREMENTS:
1. Sodium tellurite; Na_2TeO_3; [10102-20-2] 2. Sodium carbonate; Na_2CO_3; [497-19-8] 3. Water; H_2O; [7732-18-5]	Chimbulev, M.; Vasilev, Kh.; Kunev, D. *Khim. Ind. (Sofia)* 1973, *45*, 71-3.

EXPERIMENTAL VALUES (continued):

Na_2TeO_3 mass %	Na_2CO_3 mass %	Na_2TeO_3[a] mol/kg	Na_2CO_3[a] mol/kg	Solid[b] phase
Temperature = 90°C				
52.46		4.980	0.	F
50.79	1.71	4.826	0.340	F
48.65	3.57	4.595	0.705	F
47.53	5.44	4.561	1.091	F + E
44.92	6.28	4.154	1.214	E
39.07	8.84	3.385	1.601	E
31.86	11.51	2.539	1.918	E
23.34	16.94	1.764	2.676	E
15.81	19.87	1.109	2.915	E
6.59	26.28	0.443	3.694	E
-	30.15	0.	4.072	E

[a] Molalities calculated by the compiler.

[b] Solid phases: A - $Na_2TeO_3.5H_2O$, B - $Na_2TeO_3.3H_2O$, C - $Na_2CO_3.10H_2O$, D- $Na_2CO_3.5H_2O$, E - $Na_2CO_3.H_2O$, F - Na_2TeO_3

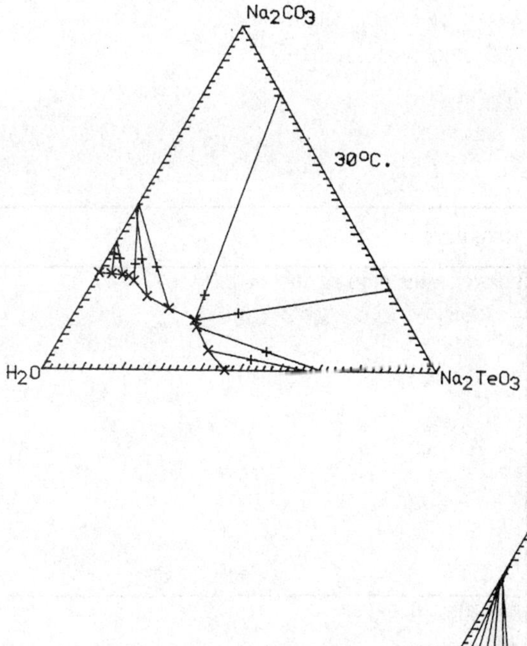

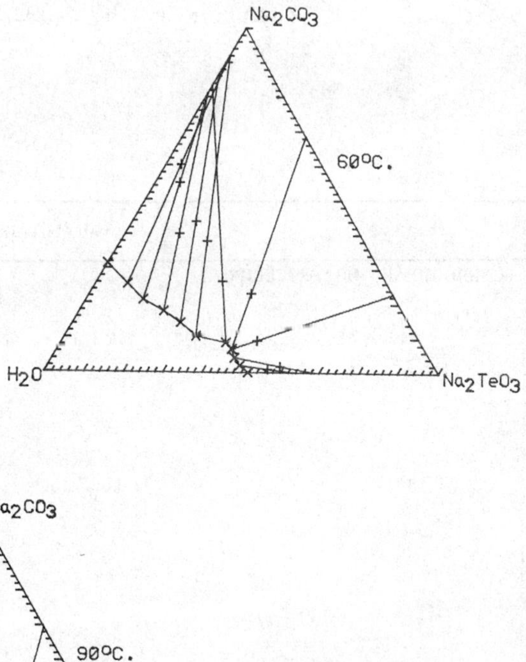

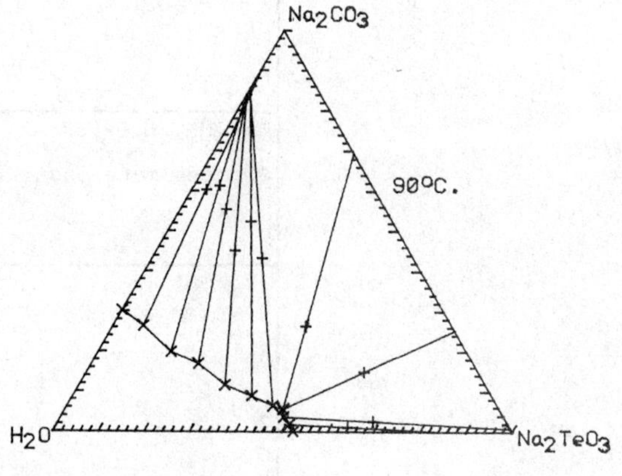

COMPONENTS:	ORIGINAL MEASUREMENTS:
1. Sodium tellurite; Na_2TeO_3; [10102-20-2] 2. Sulfuric acid; H_2SO_4; [7664-93-9] 3. Water; H_2O; [7732-18-5]	Babayan, G.G.; Kapantsyan, E.E.; Arutyunyan, M.G.; Akopyan, Z.A. *Arm. Khim. Zh.* 1973, *26*, 467-72.
VARIABLES:	**PREPARED BY:**
One temperature: 298 K Concentrations of the components	Mary R. Masson

EXPERIMENTAL VALUES:

The results were presented only in the form of a diagram.

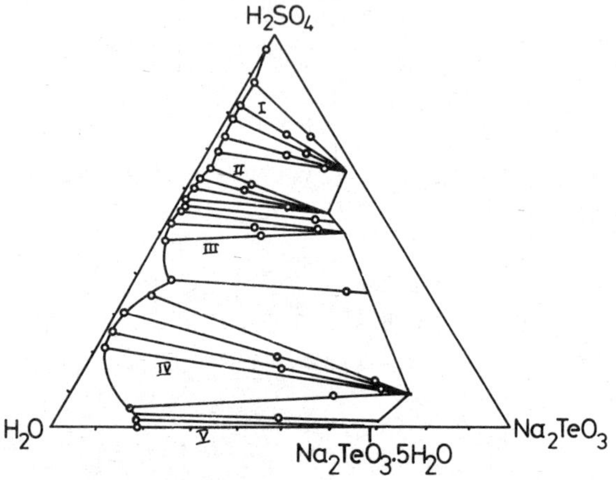

Solid phases: I - $4H_2SO_4.Na_2TeO_3.H_2O$; II - $4H_2SO_4.Na_2TeO_3.4H_2O$;
III - $3H_2SO_4.Na_2TeO_3.3H_2O$; IV - $H_2SO_4.3Na_2TeO_3.10H_2O$;
V - $Na_2TeO_3.5H_2O$

AUXILIARY INFORMATION

METHOD APPARATUS/PROCEDURE:

SOURCE AND PURITY OF MATERIALS:

ESTIMATED ERROR:

No estimate possible.

REFERENCES:

COMPONENTS:	ORIGINAL MEASUREMENTS:
1. Sodium tellurite; Na_2TeO_3; [10102-20-2] 2. Perchloric acid; $HClO_4$; [7601-90-3] 3. Water; H_2O; [7732-18-5]	Masson, M.R. *J. Inorg. Nucl. Chem.* 1976, *38*, 545-8. Masson, M.R. *unpublished data*
VARIABLES:	PREPARED BY:
Temperature: 293 - 300 K Composition	Mary R. Masson

EXPERIMENTAL VALUES:

Concentrations are expressed in terms of mol dm^{-3}

Minimum solubility	$pK^H_{H_3L}$	$pK^H_{H_2L}$	Temp.	Equil. time	Solid phase
1.2×10^{-5}	2.7	6.3	20	1 week	TeO_2
1.2×10^{-5}	2.9	6.3	25	1 week	TeO_2
1.15×10^{-5}	3.0	6.1	30	1 week	TeO_2
5.0×10^{-4}	2.8	6.25	20	5 min	"H_2TeO_3"
3.5×10^{-4}	2.7	6.2	30	30 min	"H_2TeO_3"

The solubility at any particular pH can be found from the equation (1)

$$\log (S/S_o - 1) = pH - pK$$

where S is the solubility to be found, S_o is the minimum solubility, and pK is $pK^H_{H_3L}$ for pH-values between 2 and 4, and $pK^H_{H_2L}$ for pH-values between 5 and 7.5. The solubility between pH 4 and 5 is approximately equal to the minimum value.

(continued on next page)

AUXILIARY INFORMATION

METHOD APPARATUS/PROCEDURE:

Solutions of sodium tellurite were adjusted to a range of pH-values by addition of perchloric acid. After the necessary equilibration time, aliquots were removed, filtered, then analysed for tellurite by titration with potassium permanganate (2) or silver nitrate (3).

$$K^H_{H_3L} = \frac{[H_2TeO_3][H^+]}{[H_3TeO_3^+]}$$

$$K^H_{H_2L} = \frac{[HTeO_3^-][H^+]}{[H_2TeO_3]}$$

SOURCE AND PURITY OF MATERIALS:

Sodium tellurite was prepared from Koch-Light electronic grade tellurium dioxide (99.998% pure). Other reagents were AnalaR grade.

ESTIMATED ERROR:

Temperature: ± 0.1 K
pK-values: ± 0.2 - 0.3
Analyses: $\pm 10\%$ at 10^{-5} mol dm^{-3} level, $\pm$ 1% at higher levels.

REFERENCES:

1. Krebs, H.A.; Speakman, J.C. *J. Chem. Soc.* 1945, 593.
2. Issa, I.M.; Awad, S.A. *Analyst* 1953, *78*, 487.
3. Masson, M.R. *Mikrochim. Acta* 1976 I, 399.

COMPONENTS:

1. Sodium tellurite; Na_2TeO_3; [10102-20-2]
2. Perchloric acid; $HClO_4$; [7601-90-3]
3. Water; H_2O; [7732-18-5]

ORIGINAL MEASUREMENTS:

Masson, M.R.

J. Inorg. Nucl. Chem. 1976, *38*, 545-8.

Masson, M.R. *unpublished data*

EXPERIMENTAL VALUES (continued):

Solubility of "TeO_2"

pH	concentration x 10^5
Temperature = 20°C	
8.25	446
8.25	235
8.14	137
7.60	30.8
7.27	10.8
7.11	8.4
6.80	5.45
6.27	1.68
4.19	1.40
4.17	1.31
3.84	1.21
3.44	1.48
2.77	2.05
2.43	3.0
Temperature = 25°C	
8.46	451
8.45	227
8.29	152
7.62	31.4
7.30	12.2
7.02	8.1
6.73	5.85
6.68	4.2
6.18	1.9
4.21	1.6
4.13	1.1
3.87	1.2
3.46	1.55
2.81	2.9
2.43	3.55
Temperature = 30°C	
8.44	479
8.28	264.5
8.16	163.5
7.32	35
7.22	12.7
6.84	7.1
6.48	5.9
6.29	2.3
6.02	1.9
4.15	2.3
4.00	1.1
3.83	1.9
3.34	2.5
2.76	2.6
2.34	6.1

Solubility of "H_2TeO_3"

pH	concentration x 10^4
Temperature = 20°C	
8.12	187
7.83	93.5
7.63	74.8
7.42	46.7
7.13	26.7
6.88	18.7
6.30	9.35
5.83	6.23
4.95	4.67
4.29	3.74
3.41	2.41
2.98	9.00
2.57	12.2
2.49	12.84
1.685	108
Temperature = 30°C	
7.42	47.9
7.28	36.4
7.04	26.45
6.77	16.35
6.16	6.35
5.61	4.35
4.88	3.75
4.50	3.55
4.38	3.80
3.85	3.9
3.80	3.75
3.57	3.9
3.44	3.95
3.21	4.7
2.61	6.3
2.33	10.6

Note: all concentrations are given in mol dm^{-3}.

COMPONENTS:

1. Sodium tellurite; Na_2TeO_3; [10102-20-2]
2. Ethanol; C_2H_5OH; [64-17-5]
3. Water; H_2O; [7732-18-5]

ORIGINAL MEASUREMENTS:

Vorob'eva, O.I.; Lavut, E.A.

Zh. Neorg. Khim. 1957, 2, 1154-1157;
**Russ. J. Inorg. Chem. (Eng. Transl.)* 1957, 2, 261.

VARIABLES:

Concentrations of the components
One temperature: 298 K

PREPARED BY:

Mary R. Masson

EXPERIMENTAL VALUES:

Initial	Ethanol layer				Aqueous layer				Solid[b]
C_2H_5OH	C_2H_5OH	Na_2TeO_3	C_2H_5OH[a]	Na_2TeO_3[a]	C_2H_5OH	Na_2TeO_3	C_2H_5OH[a]	Na_2TeO_3[a]	phase
% v/v	mass %	mass %	mol/kg	mol/kg	mass %	mass %	mol/kg	mol/kg	
100	95.70	-	483.087	0.	No separate layer				A
96	91.52		234.262	0.	"				B
91.4	88.58	0.10	169.852	0.040	"				B
90.0	85.50	0.10	128.880	0.031	"				B
85.3	77.09	0.20	73.682	0.040	"				B
80.0	72.75	0.20	58.378	0.033	"				B
76.8	70.64	0.50	53.130	0.078	"				B
72.0	62.43	1.00	37.055	0.123	"				B
64.0	61.63	1.30	56.087	0.158	3.24	40.80	1.257	3.290	B
57.6	61.00	1.30	35.121	0.156	3.28	40.64	1.270	3.270	B
48.0	60.57	2.07	35.191	0.250	3.26	40.87	1.267	3.301	B
48.0	61.07	2.05	35.943	0.251	3.24	40.78	1.256	3.287	B
	43.45	6.48	18.836	0.584	7.18	36.35	2.760	2.905	none
	29.11	15.46	11.399	1.259	15.15	27.27	5.711	2.137	none
	28.12	16.82	11.086	1.379	17.45	24.72	6.550	1.929	none
8.7	-	-	-	-	2.98	41.74	1.170	3.407	B
0.0	-	-	-	-	-	44.97	0.	3.688	B

[a] Molalities calculated by the compiler.

[b] Solid phases: A - $Na_2TeO_3.xH_2O$, B - $Na_2TeO_3.5H_2O$

(continued on next page)

AUXILIARY INFORMATION

METHOD APPARATUS/PROCEDURE:

Reaction mixtures were placed in sealed glass ampoules, which were equilibrated for at least 30 days, with shaking. Weighed samples were dissolved in water, then the ethanol was distilled off and determined iodometrically. Tellurite was determined by the periodate method (1) or chromatographically (2). Sodium was determined gravimetrically as the zinc uranyl acetate, after prior precipitation of tellurium dioxide. The compositions of the solid residues were determined by Schreinemaker's method.

SOURCE AND PURITY OF MATERIALS:

Sodium tellurite was prepared by dissolving tellurium dioxide in 20% aqueous sodium hydroxide in stoichiometric proportions. A fivefold amount of ethanol was added to precipitate sodium tellurite pentahydrate. An excess amount of sodium hydroxide was found not to disturb the reaction.

ESTIMATED ERROR:

No estimates possible.

REFERENCES:

1. Syrokomskii, V.S.; Knyazeva, R.N. *Zavod. Lab.* 1950, *16*, 1041.
2. Schrenk, W.T. and Browning, B.L. *J. Am. Chem. Soc.* 1926, *48*, 139.

COMPONENTS:	ORIGINAL MEASUREMENTS:
1. Sodium tellurite; Na_2TeO_3; [10102-20-2] 2. Ethanol; C_2H_5OH; [64-17-5] 3. Water; H_2O; [7732-18-5]	Vorob'eva, O.I.; Lavut, E.A. *Zh. Neorg. Khim.* 1957, 2, 1154-1157; **Russ. J. Inorg. Chem. (Eng. Transl.)* 1957, 2, 261.

EXPERIMENTAL VALUES (continued):

Ethanol layer

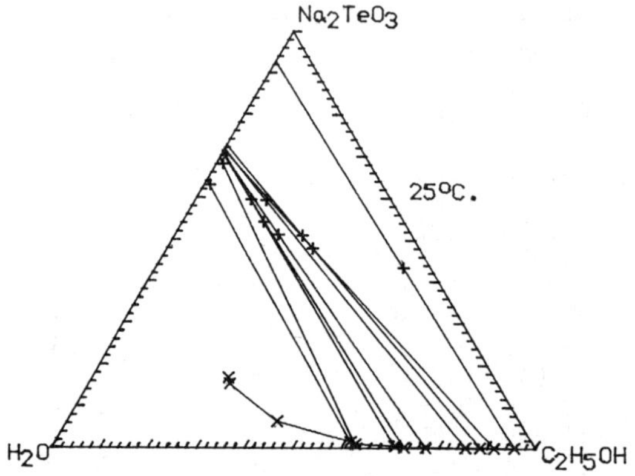

Aqueous layer

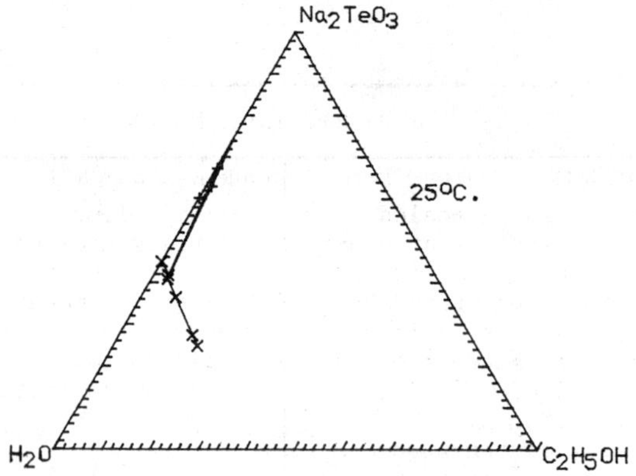

COMPONENTS:

1. Cesium tellurite; Cs_2TeO_3; [15899-92-0]

2. Water; H_2O; [7732-18-5]

ORIGINAL MEASUREMENTS:

Lavut, E.A.

Vestn. Mosk. Univ. Ser. II, Khim. 1966, *21*, 91-3. (English translation pp. 225-6).

VARIABLES:

One temperature: 291 K

PREPARED BY:

Mary R. Masson

EXPERIMENTAL VALUES:

Solubility of Cs_2TeO_3 in water at 18°C

Wt. taken, g	0.2931	0.3397
Te, %	19.63	19.49
Cs, %	41.63	41.48
Cs/Te	2.04	2.04

The solubility of cesium tellurite, calculated for the anhydrous salt, is 67.65% at 18°C. (Molality[a] = 1.550 mol/kg).

Prolonged treatment of cesium tellurite pentahydrate with absolute ethanol resulted in decomposition of the salt. Prolonged treatment with water results in some hydrolysis of the tellurite ion. Cesium tetratellurite pentahydrate was found to be insoluble in water, and to decompose in boiling water to CsOH and TeO_2.

[a] Molality calculated by the compiler.

AUXILIARY INFORMATION

METHOD APPARATUS/PROCEDURE:

Not stated.

SOURCE AND PURITY OF MATERIALS:

A weighed specimen of freshly precipitated tellurium dioxide was dissolved with heating in an approx. 20% aqueous solution of cesium hydroxide. The solution was concentrated by evaporation, and the residue was treated with absolute ethanol. The alcohol solution was separated from the residue by filtration, and the residue was dried over H_2SO_4 and KOH. The product was shown to be cesium tellurite pentahydrate.

ESTIMATED ERROR:

No estimates possible.

REFERENCES:

COMPONENTS:

1. Barium tellurite; $BaTeO_3$; [14899-38-8]
2a. Nitric acid; HNO_3; [7697-37-2]
2b. Hydrochloric acid; HCl; [7647-01-0]
3a. Sodium nitrate; $NaNO_3$; [7631-99-4]
3b. Sodium chloride; NaCl; [7647-14-5]
4. Water; H_2O; [7732-18-5]

ORIGINAL MEASUREMENTS:

Ganelina, E.Sh.; Merzon, V.V.; Biryukov, V.P.

Izv. Vyssh. Ucheb. Zaved. Khim. Khim. Tekhnol. 1969, *12*, 1465-7.

VARIABLES:

One temperature: 298 K
pH varied.

PREPARED BY:

Mary R. Masson

EXPERIMENTAL VALUES:

pH	$[Ba^{2+}] \times 10^3$ / mol dm^{-3}	$\alpha_{L(H)}$	$K_{s0} \times 10^4$ / mol dm^{-6}	$\alpha^*_{L(H)}$	K^*_{s0} / mol^2 dm^{-6}	
8.90	11.0	1.0070	1.2	5.578	2.7×10^{-5}	nitrate medium
8.10	4.42	1.0341	0.18	30.116	6.49×10^{-7}	
8.75	8.35	1.0101	0.69	7.470	9.33×10^{-6}	
8.55	6.28	1.0163	0.39	11.268	3.5×10^{-6}	
7.59	2.83	2.475	0.032	97.21	8.24×10^{-8}	chloride medium
7.68	2.64	2.252	0.031	79.76	8.85×10^{-8}	

The starred (*) values were recalculated by the compiler, since the author had used erroneous values for the dissociation constants of tellurous acid (from (1)). The compiler used values from (2).

Note: $[Te_{tot}] = [Ba^{2+}]$ and $[TeO_3^{2-}] = [Te_{tot}]/\alpha_{L(H)}$

The inconsistencies in the results may be caused by interference by atmospheric carbon dioxide, which can cause precipitation of barium carbonate in solutions as acidic as pH 6.1 (log K_{s0} for barium carbonate is -9.4) (3).

AUXILIARY INFORMATION

METHOD APPARATUS/PROCEDURE:

Barium tellurite was stirred with nitric and hyrochloric acid solutions of various concentrations until equilibrium was established. The pH was determined by means of an LPU-01 instrument and a glass electrode. The barium concentration was determined by complexometric titration in ammonia buffer, with Eriochrome Black T as indicator.

SOURCE AND PURITY OF MATERIALS:

Barium tellurite was prepared by reaction of sodium tellurite with barium nitrate.

ESTIMATED ERROR:

The spread in K_{s0} values is very large; a value for *s* would not be meaningful.

REFERENCES:

1. Blanc, E. *J. Chem. Phys.* 1920, *18*, 40.
2. Masson, M.R. *J. Inorg. Nucl. Chem.* 1976, *38*, 545-8.
3. Kragten, J. *Atlas of Metal-Ligand Equilibria in Aqueous Solution* Horwood, Chichester, 1977.

COMPONENTS:

1. Cobalt tellurite; $CoTeO_3$; [15851-44-2]
2. Sulfuric acid; H_2SO_4; [7664-93-9]
3. Water; H_2O; [7732-18-5]

ORIGINAL MEASUREMENTS:

Ganelina, E.Sh.

Zh. Priklad. Khim. 1967, *40*, 1019-24;
**J. Appl. Chem. USSR (Eng. Transl.)* 1967, *40*, 983-7.

VARIABLES:

One temperature, probably 298 K
pH varied.

PREPARED BY:

Mary R. Masson

EXPERIMENTAL VALUES:

pH	$[Co^{2+}] \times 10^3$ / mol dm^{-3}	Author $\alpha_{L(H)}$	Author $K_{s0} \times 10^5$ / mol^2 dm^{-6}	Compiler $\alpha_{L(H)}$	Compiler $K_{s0} \times 10^7$ / mol^2 dm^{-6}
6.4	25.0	22.7	2.8	2138	2.92
6.75	22.6	10.7	4.8	785	6.51
6.35	18.8	25.6	1.3	2490	1.42
7.0	8.0	6.9	0.92	408	1.57
			Mean = 1.95×10^{-5}		Mean = 3.1×10^{-7} pK_{s0} = 6.51

The results calculated by the author by using acid dissociation constants said to be from (1) are given, along with values calculated by the compiler using constants from (2), which should be more reliable.

Note: $[Te_{tot}] = [Co^{2+}]$ and $[TeO_3^{2-}] = [Te_{tot}]/\alpha_{L(H)}$

The author does not give the temperature at which the investigations were done. The work on barium and lead tellurites was done at 25°C, and this work was probably done at this temperature also.

AUXILIARY INFORMATION

METHOD APPARATUS/PROCEDURE:

Cobalt tellurite was stirred with sulfuric acid solutions of various concentrations until equilibrium was established. The solution pH was measured by means of an LPU-01 instrument with a glass electrode. Cobalt in the filtrate was determined colorimetrically as the nitroso-R salt complex.

SOURCE AND PURITY OF MATERIALS:

Cobalt tellurite was prepared by the exchange reaction between sodium tellurite and a cobalt salt. The precipitate was dried over H_2SO_4 and analysed for cobalt, tellurium, and water of crystallization.

ESTIMATED ERROR:

Error in K_{s0} (2*s*) = 4.8×10^{-7}

REFERENCES:

1. Blanc, E. *J. Chem. Phys.* 1920, *18*, 40.
2. Masson, M.R. *J. Inorg. Nucl. Chem.* 1976 *38*, 545-8.

COMPONENTS:

1. Nickel tellurite; $NiTeO_3$; [15851-51-2]
2. Hydrochloric acid; HCl; [7647-01-0]
3. Sulfuric acid; H_2SO_4; [7664-93-9]
4. Water; H_2O; [7732-18-5]

ORIGINAL MEASUREMENTS:

Ganelina, E.Sh.

Zh. Priklad. Khim. 1967, *40*, 1019-24;
**J. Appl. Chem. USSR (Eng. Transl.)* 1967, *40*, 983-7.

VARIABLES:

One temperature, probably 298 K
pH varied.

PREPARED BY:

Mary R. Masson

EXPERIMENTAL VALUES:

		Author		Compiler		
pH	$[Ni^{2+}] \times 10^3$ mol dm^{-3}	$\alpha_{L(H)}$	$K_{s0} \times 10^8$ $mol^2\ dm^{-6}$	$\alpha_{L(H)}$	$K_{s0} \times 10^{10}$ $mol^2\ dm^{-6}$	
5.2	7.1	2.66×10^3	1.9	1.97×10^5	2.56	
5.8	1.1	71.0	1.7	1.67×10^4	0.724	hydrochloric
6.1	0.9	34.1	2.4	5.64×10^3	1.44	acid
7.3	0.3	4.42	2.0	1.94×10^2	4.64	
		Mean = 2.0×10^{-8}		Mean = 2.34×10^{-10}		pK_{s0} = 9.63
6.1	1.4	36.8	5.3	5.64×10^3	3.48	
6.6	0.7	12.1	4.1	1.19×10^3	4.12	sulfuric
7.0	0.4	5.52	2.9	4.08×10^2	3.92	acid
		Mean = 4.1×10^{-8}		Mean = 3.84×10^{-10}		pK_{s0} = 9.42

The results calculated by the author using acid dissociation constants said to be from (1) are given, along with values calculated by the compiler using constants from (2), which should be more reliable.

Note: $[Te_{tot}] = [Ni^{2+}]$ and $[TeO_3^{2-}] = [Te_{tot}]/\alpha_{L(H)}$

The author does not state the temperature at which the investigations were done. The work on barium and lead tellurites was done at 25°C, and this work was probably done at this temperature.

AUXILIARY INFORMATION

METHOD APPARATUS/PROCEDURE:

Nickel tellurite was stirred with solutions of hydrochloric or sulfuric acid of various concentrations until equilibrium was established. The solution pH was measured by means of an LPU-01 instrument with a glass electrode. Nickel in the filtrate was determined gravimetrically as the dimethylglyoximate.

SOURCE AND PURITY OF MATERIALS:

Nickel tellurite was prepared by the exchange reaction between sodium tellurite and a nickel salt. The precipitate was dried over H_2SO_4 and analysed for nickel, tellurium and water of crystallization.

ESTIMATED ERROR:

Error in K_{s0} (2*s*)
$= 1.9 \times 10^{-10}$ (hydrochloric acid)
$= 3.8 \times 10^{-11}$ (sulfuric acid)

REFERENCES:

1. Blanc, E. *J. Chem. Phys.* 1920, *18*, 40.
2. Masson, M.R. *J. Inorg. Nucl. Chem.* 1976 *38*, 545-8.

COMPONENTS:

1. Copper tellurite; $CuTeO_3$; [13812-58-3]
2. Hydrochloric acid; HCl; [7647-01-0]
3. Sulfuric acid; H_2SO_4; [7664-93-9]
4. Water; H_2O; [7732-18-5]

ORIGINAL MEASUREMENTS:

Ganelina, E.Sh.

Zh. Priklad. Khim. 1967, *40*, 1019-24;
**J. Appl. Chem. USSR (Eng. Transl.)* 1967, *40*, 983-7.

VARIABLES:

One temperature, probably 298 K
pH varied.

PREPARED BY:

Mary R. Masson

EXPERIMENTAL VALUES:

pH	$[Cu^{2+}] \times 10^2$ / mol dm^{-3}	Author $\alpha_{L(H)}$	Author $K_{s0} \times 10^7$ / mol^2 dm^{-6}	Compiler $\alpha_{L(H)}$	Compiler $K_{s0} \times 10^{11}$ / mol^2 dm^{-6}	
3.78	7.2	3.24×10^5	0.16	1.21×10^8	4.29	Sulfuric acid
4.04	5.4	1.27×10^5	0.23	3.66×10^7	7.96	
4.44	3.5	3.5×10^4	0.35	5.89×10^6	20.8	
4.61	2.6	1.78×10^4	0.38	2.72×10^6	24.9	
4.26	1.7	1.81×10^4	0.16	1.34×10^7	2.16	
4.54	1.56	1.43×10^4	0.17	3.74×10^6	6.51	
		Mean =	0.24×10^{-7}	Mean =	11.1×10^{-11}	pK_{s0} = 9.95
4.25	1.19	2.21×10^3	0.64	1.40×10^7	1.01	Hydrochloric acid
4.45	1.73	4.34×10^3	0.69	5.62×10^6	5.32	
3.85	2.91	1.97×10^4	0.43	8.76×10^7	0.967	
4.42	0.84	3.92×10^3	0.18	6.45×10^6	1.09	
4.04	1.52	1.05×10^4	0.22	3.66×10^7	0.631	
4.50	0.62	3.49×10^3	0.11	4.48×10^6	0.783	
		Mean =	0.38×10^{-7}	Mean =	1.6×10^{-11}	pK_{s0} = 10.80

The results calculated by the author by using acid dissociation constants said to be from (1) are given along with values calculated by the compiler using constants from (2), which should be more reliable.

The author does not state the temperature at which the investigations were done. The work on barium and lead tellurites was done at 25°C, and this work was probably done at the same temperature.

AUXILIARY INFORMATION

METHOD APPARATUS/PROCEDURE:

Copper tellurite was stirred with solutions of hydrochloric or sulfuric acid of various concentrations until equilibrium was established. The solution pH was measured by means of an LPU-01 instrument with a glass electrode. Copper in the filtrate was determined gravimetrically as CuCNS.

SOURCE AND PURITY OF MATERIALS:

Copper tellurite was prepared by the exchange reaction between sodium tellurite and a copper salt. The precipitate was dried over H_2SO_4 and analysed for copper, tellurium and water of crystallization.

ESTIMATED ERROR:

The error in K_{s0} is very large, possibly because of the low solubility of tellurous acid at low pH.

REFERENCES:

1. Blanc, E. *J. Chem. Phys.* 1920, *18*, 40.
2. Masson, M.R. *J. Inorg. Nucl. Chem.* 1976 *38*, 545-8.

COMPONENTS:

1. Silver tellurite; Ag_2TeO_3; [15122-56-2]
2. Water; H_2O; [7732-18-5]

ORIGINAL MEASUREMENTS:

Ganelina, E.Sh.; Pozhidaeva, T.N.

Zh. Priklad. Khim. 1965, *38*, 2210-6;
**J. Appl. Chem. USSR (Eng. Transl.)* 1965, *38*, 2168-73.

VARIABLES:

One temperature: 298 K
pH varied.

PREPARED BY:

Mary R. Masson

EXPERIMENTAL VALUES:

The summary table given by the authors reports the following values:

Ag_2TeO_3	$K_{s0} = 3.7 \times 10^{-3}$ mol^3 dm^{-9}
$Ag_2TeO_3.H_2TeO_3$	$K_{s0} = 1.12 \times 10^{-8}$
$Ag_2TeO_3.5AgOH$	$K_{s0} = 2.3 \times 10^{-6}$

Compiler's comments

This paper is very confusing, and it is very difficult to see what the authors did in their calculations. They do not seem to realise anything about the solubility and acid-base behaviour of tellurous acid, since they report values for solubility products measured at around pH 3, where tellurite exists mainly in the form of H_2TeO_3 (or TeO_2), which has very low solubility in water. Attempts at recalculation were unsuccessful.

The compiler does not think that these authors have given conclusive evidence for the existence of $Ag_2TeO_3.H_2TeO_3$ or $Ag_2TeO_3.5AgOH$.

AUXILIARY INFORMATION

METHOD APPARATUS/PROCEDURE:

Samples were equilibrated with water or water plus any of KNO_3, H_2TeO_3, $AgNO_3$, Na_2TeO_3, for 3 - 4 months.
Equilibrium pH values were measured with an LP-5 potentiometer fitted with a glass electrode.

SOURCE AND PURITY OF MATERIALS:

Silver tellurite was prepared either by adding sodium tellurite to silver nitrate, or by adding silver nitrate to sodium tellurite.

ESTIMATED ERROR:

Nothing stated.

REFERENCES:

COMPONENTS:

1. Silver tellurite; Ag_2TeO_3; [15122-56-2]
2. Sodium tellurite; Na_2TeO_3; [10102-20-0]
3. Sodium perchlorate; $NaClO_4$; [7601-89-0]
4. Water; H_2O; [7732-18-5]

ORIGINAL MEASUREMENTS:

Mehra, M.C.; Kahn, S.M.

Can. J. Chem. 1972, *50*, 1788-91.

VARIABLES:

One temperature: 298 K
Tellurite concentrations and pH varied.
Ionic strength constant at 1 mol dm^{-3}.

PREPARED BY:

Mary R. Masson

EXPERIMENTAL VALUES:

The authors calculated the following value for the solubility product of silver tellurite (concentrations expressed in mol dm^{-3}):

$$pK_{s0} = 17.85 \qquad (K_{s0} = 1.41 \times 10^{-18}\ mol^3\ dm^{-9})$$

Unfortunately, the authors assumed values of 2.52 and 7.7 for pK_1 and pK_2 of tellurous acid, and these values have been shown to be seriously in error (1,2). More sensible values for ionic strength 1 mol dm^{-3} are 5.89 and 8.91 (2). Insufficient data are given to allow proper recalculation. If a mean pH of determination is taken to be 9.6, since the range 9.4 - 9.8 was used for the data in Fig. 1, the value of the solubility product can be corrected to:

$$pK_{s0} = 17.93 \qquad (K_{s0} = 1.17 \times 10^{-18}\ mol^3\ dm^{-9})$$

The authors also give values for formation constants of postulated complexes $Ag(TeO_3)_2^{3-}$ and $[Ag_2TeO_3]_{aq}$, but the compiler believes that this work is totally invalidated by the assumption of the incorrect values for the acid dissociation constants for tellurous acid.

AUXILIARY INFORMATION

METHOD APPARATUS/PROCEDURE:

The system was equilibrated for 5 days. Experimental details and the method of analysis are given in (3). The radiometric solubility represents the total solubility, and the potentiometric solubility represents the ionic solubility of Ag_2TeO_3.

SOURCE AND PURITY OF MATERIALS:

All reagents were of analytical grade. Doubly distilled demineralized water was used throughout. Silver tellurite was produced *in situ* by mixing radioactive silver nitrate and sodium tellurite at the desired acidity and tellurite concentration. The radiotracer Ag-110 was obtained from AECL Chalk River, Ontario.

ESTIMATED ERROR:

Range in pK_{s0} = $\pm$0.11 (authors)

REFERENCES:

1. Nazarenko, V.A.; Shitareva, G.G.; Poluektova, E.N. *Russ. J. Inorg. Chem.* 1973, *18*, 609.
2. Masson, M.R. *J. Inorg. Nucl. Chem.* 1976, *38*, 545-548.
3. Mehra, M.C.; Gubeli, A.O. *Can. J. Chem.* 1970, *48*, 3491.

COMPONENTS:

1. Silver tellurite; Ag_2TeO_3; [15122-56-2]
2. Water; H_2O; [7732-18-5]

ORIGINAL MEASUREMENTS:

Chao, E.E.; Cheng, K.L.

Anal. Chem. 1976, *48*, 267-271.

VARIABLES:

One temperature: 293 K

PREPARED BY:

Mary R. Masson

EXPERIMENTAL VALUES:

Concentrations are expressed in units of mol dm^{-3}.

The ionic strength was constant at 0.1 mol dm^{-3} (medium not stated).

$pK_{s0} = 18.06 \pm 0.07$ $\quad\quad$ $K_{s0} = 8.71 \times 10^{-19}$ $mol^3\ dm^{-9}$

Compiler's note

The values used for the acid dissociation constants of tellurous acid are not stated, but if the determination was done at pH 11.0, as it was for silver arsenite (1), the values used would have only a very small influence on the value obtained for the solubility product. Therefore, this value is probably a reasonably reliable estimate of the concentration solubility product.

The value would refer to a freshly precipitated solid, and might therefore differ somewhat from values obtained from equilibration of solutions with aged solids.

AUXILIARY INFORMATION

METHOD APPARATUS/PROCEDURE:

The solubility product was determined from data obtained by potentiometric titration of a tellurite solution with a silver nitrate solution. Silver ion activities were measured by means of an Orion silver sulfide electrode (94-16) and an Orion double junction reference electrode (90-02). Emf readings were taken with a Corning model 10 pH meter with expanded scale.
Method of calculation is given in ref. (1).

This involved determining, from the *E* value, pAg at the point of incipient precipitation of silver tellurite.

SOURCE AND PURITY OF MATERIALS:

Reagent-grade chemicals were used.

ESTIMATED ERROR:

Range in $pK_{s0} = \pm 0.07$

REFERENCES:

1. Chao, E.E. *Ph.D. Dissertation* University of Missouri, Kasas City, Mo. 1975.

COMPONENTS:	ORIGINAL MEASUREMENTS:
1. Lead tellurite; $PbTeO_3$; [15851-47-5] 2. Nitric acid; HNO_3; [7697-37-2] 3. Sodium nitrate; $NaNO_3$; [7631-99-4] 4. Water; H_2O; [7732-18-5]	Ganelina, E.Sh.; Merzon, V.V.; Biryukov, V.P. *Izv. Vyssh. Ucheb. Zaved. Khim. Khim. Technol.* 1969, *12*, 1465-7.
VARIABLES:	PREPARED BY:
One temperature: 298 K pH varied.	Mary R. Masson

EXPERIMENTAL VALUES:

	Author			Compiler		
pH	$[Pb_{tot}] \times 10^3$ mol dm^{-3}	$\alpha_{L(H)}$	$K_{s0} \times 10^7$ mol^2 dm^{-6}	$\alpha_{L(H)}$	$\alpha_{Pb(OH)}$	K_{s0} mol^2 dm^{-6}
4.50	13.5	1.67×10^3	1.1	4.48×10^6	1.0	4.07×10^{-11}
5.55	3.5	1.54×10^2	0.62	4.49×10^4	1.005	2.72×10^{-10}
6.40	1.62	21.14	3.62	2.14×10^3	1.035	1.18×10^{-9}
7.47	0.50	2.849	0.87	1.29×10^2	1.412	1.37×10^{-9}
7.73	2.7	2.132	39.1	70.12	1.757	5.92×10^{-8}
7.78	2.62	2.004	36.5	62.46	1.852	5.93×10^{-8}

The results calculated by the author by using acid dissociation constants from (1) are given, along with values calculated by the compiler using constants from (2), which differ considerably. The values calculated by the compiler also take account of the hydrolysis of the lead ions (3). Only the last two values of K_{s0} are in agreement. The results obtained at lower pH values are likely to be unreliable for two reasons:
(1) tellurous acid has a very low solubility at pH values below 7, and
(2) soluble but inert polynuclear lead hydroxo complexes are formed in the pH region 4 - 9, and these could disturb analyses or equilibria (3).

Note: $[Te_{tot}] = [Pb_{tot}]$,
$[TeO_3^{2-}] = [Te_{tot}]/\alpha_{L(H)}$ and $[Pb^{2+}] = [Pb_{tot}]/\alpha_{Pb(OH)}$

AUXILIARY INFORMATION

METHOD APPARATUS/PROCEDURE:

Lead tellurite was stirred with nitric acid solutions of various concentrations until equilibrium was established. The pH was determined by means of an LPU-01 instrument and a glass electrode. The lead concentration was determined by complexometric titration in ammonia buffer, with Eriochrome Black T as indicator.

SOURCE AND PURITY OF MATERIALS:

Lead tellurite was prepared by reaction of sodium tellurite with lead nitrate.

ESTIMATED ERROR:

The spread in K_{s0} values is very large; a value for *s* would not be meaningful.

REFERENCES:

1. Blanc, E. *J. Chem. Phys.* 1920, *18*, 40.
2. Masson, M.R. *J. Inorg. Nucl. Chem.* 1976, *38*, 545-8.
3. Kragten, J. *Atlas of Metal-Ligand Equilibria in Aqueous Solution* Horwood, Chichester, 1977.

SYSTEM INDEX

Page numbers preceded by E refer to evaluation texts whereas those not preceded by E refer to compilation tables. All compounds are listed as in Chemical Abstracts. For example, sodium sulfite is listed as sulfurous acid, disodium salt.

REGISTRY NUMBER INDEX

Page numbers preceded by E refer to evaluation texts whereas those not preceded by E refer to compilation tables.

AUTHOR INDEX

Page numbers preceded by E refer to evaluation texts whereas those not preceded by E refer to compilation tables.